Mean square within

$$\text{MSW} = \frac{\text{SSW}}{\text{dfw}}$$

Mean square between

$$\text{MSB} = \frac{\text{SSB}}{\text{dfb}}$$

F ratio

$$F = \frac{\text{MSB}}{\text{MSW}}$$

CHAPTER 11

Chi square

$$\chi^2(\text{obtained}) = \sum \frac{(f_o - f_e)^2}{f_e}$$

CHAPTER 12

Phi

$$\phi = \sqrt{\frac{\chi^2}{N}}$$

Cramer's V

$$V = \sqrt{\frac{\chi^2}{(N)(\text{Minimum of } r - 1, c - 1)}}$$

Lambda

$$\lambda = \frac{E_1 - E_2}{E_1}$$

CHAPTER 13

Gamma

$$G = \frac{N_s - N_d}{N_s + N_d}$$

Spearman's rho

$$r_s = 1 - \frac{6\sum D^2}{N(N^2 - 1)}$$

CHAPTER 14

Least-squares regression line

$$Y = a + bX$$

Slope

$$b = \frac{\sum(X - \overline{X})(Y - \overline{Y})}{\sum(X - \overline{X})^2}$$

Y intercept

$$a = \overline{Y} - b\overline{X}$$

Pearson's r

$$r = \frac{\sum(X - \overline{X})(Y - \overline{Y})}{\sqrt{[\sum(X - \overline{X})^2][\sum(Y - \overline{Y})^2]}}$$

CHAPTER 16

Partial correlation coefficient

$$r_{yx.z} = \frac{r_{yx} - (r_{yz})(r_{xz})}{\sqrt{1 - r_{yz}^2}\sqrt{1 - r_{xz}^2}}$$

Least-squares multiple regression line

$$Y = a + b_1 X_1 + b_2 X_2$$

Partial slope for X_1

$$b_1 = \left(\frac{s_y}{s_1}\right)\left(\frac{r_{y1} - r_{y2}r_{12}}{1 - r_{12}^2}\right)$$

Partial slope for X_2

$$b_2 = \left(\frac{s_y}{s_2}\right)\left(\frac{r_{y2} - r_{y1}r_{12}}{1 - r_{12}^2}\right)$$

Y intercept

$$a = \overline{Y} - b_1\overline{X}_1 - b_2\overline{X}_2$$

Beta-weight for X_1

$$b_1^* = b_1\left(\frac{s_1}{s_y}\right)$$

Beta-weight for X_2

$$b_2^* = b_2\left(\frac{s_2}{s_y}\right)$$

Standardized least-squares regression line

$$Z_y = b_1^* Z_1 + b_2^* Z_2$$

Coefficient of multiple determination

$$R^2 = r_{y1}^2 + r_{y2.1}^2(1 - r_{y1}^2)$$

STATISTICS

A Tool for Social Research

WADSWORTH
CENGAGE Learning

Australia · Brazil · Japan · Korea · Mexico
Singapore · Spain · United Kingdom
United States

WADSWORTH
CENGAGE Learning

**Statistics: A Tool for Social Research,
Ninth Edition**
Joseph F. Healey

Acquiring Editor: Erin Mitchell

Developmental Editor: Kristin Makarewycz

Assistant Editor: John Chell

Editorial Assistant: Mallory Ortberg

Media Editor: Melanie Cregger

Marketing Manager: Andrew Keay

Marketing Assistant: Dimitri Hagnere

Marketing Communications Manager:
Tami Strang

Content Project Manager: Cheri Palmer

Design Director: Rob Hugel

Art Director: Caryl Gorska

Print Buyer: Karen Hunt

Rights Acquisitions Specialist:
Tom McDonough

Production Service: MPS Limited, a
Macmillan Company

Text Designer: Carole Lawson

Text Researcher: PreMedia Global

Copy Editor: David Heath

Illustrator: MPS Limited, a Macmillan
Company

Cover Designer: Tim Heraldo/RHDG Design

Cover Image: © Chris Jordan

Compositor: MPS Limited, a Macmillan
Company

For product information and technology assistance, contact us at
Cengage Learning Customer & Sales Support, 1-800-354-9706.
For permission to use material from this text or product,
submit all requests online at **www.cengage.com/permissions**.
Further permissions questions can be e-mailed to
permissionrequest@cengage.com.

Library of Congress Control Number: 2001012345

ISBN-13: 978-1-111-18636-4

ISBN-10: 1-111-18636-7

Wadsworth
20 Davis Drive
Belmont, CA 94002-3098
USA

Cengage Learning is a leading provider of customized learning solutions with office locations around the globe, including Singapore, the United Kingdom, Australia, Mexico, Brazil, and Japan. Locate your local office at **www.cengage.com/global**.

Cengage Learning products are represented in Canada by Nelson Education, Ltd.

To learn more about Wadsworth, visit **www.cengage.com/wadsworth**.

Purchase any of our products at your local college store or at our preferred online store **www.cengagebrain.com**.

Printed in the United States of America
1 2 3 4 5 6 7 14 13 12 11

Brief Contents

Detailed Contents

Preface

Statistics are part of the everyday language of sociology and other social sciences (including political science, social work, public administration, criminal justice, urban studies, and gerontology). These disciplines are research-based and routinely use statistics to express knowledge and to discuss theory and research. To join the conversations being conducted in these disciplines, you must be literate in the vocabulary of research, data analysis, and scientific thinking. Knowledge of statistics will enable you to understand the professional research literature, conduct quantitative research yourself, contribute to the growing body of social science knowledge, and reach your full potential as a social scientist.

Although essential, learning (and teaching) statistics can be a challenge. Students in social science statistics courses typically have a wide range of mathematical backgrounds and an equally diverse set of career goals. They are often puzzled about the relevance of statistics for them, and, not infrequently, there is some math anxiety to deal with. This text introduces statistical analysis for the social sciences while addressing these realities.

The text makes minimal assumptions about mathematical background (the ability to read a simple formula is sufficient preparation for virtually all of the material in the text), and a variety of special features help students analyze data successfully. The text has been written especially for sociology and social work programs but is sufficiently flexible to be used in any program with a social science base.

The text is written at an intermediate level and its goal is show the relevance and value of statistics for the social sciences. I emphasize interpretation and understanding statistics in the context of social science research, but I have not sacrificed comprehensive coverage or statistical correctness. Mathematical explanations are kept at an elementary level, as is appropriate in a first exposure to social statistics. For example, I do not treat formal probability theory per se in the text.[1] Rather, the background necessary for an understanding of inferential statistics is introduced, informally and intuitively, in Chapters 5 and 6 while considering the concepts of the normal curve and the sampling distribution.

The text does not claim that statistics are "fun" or that the material can be mastered without considerable effort. At the same time, students are not overwhelmed with abstract proofs, formula derivations, or mathematical theory, which can needlessly frustrate the learning experience at this level.

Goal of the Text

The goal of this text is to develop basic statistical literacy. The statistically literate person understands and appreciates the role of statistics in the research process, is competent to perform basic calculations, and can read and appreciate

---

[1] A presentation of probability is available at the Web site for this text for those who are interested.

the professional research literature in his or her field as well as any research reports he or she may encounter in everyday life. This goal has not changed since the first edition of this text. However, in recognition of the fact that "mere computation" has become less of a challenge in this high-tech age, this edition continues to increase the stress on interpretation and computer applications while deemphasizing computation. This will be apparent in several ways. For example, the feature called "Interpreting Statistics" has been continued and updated. These non-computational sections are included in about half of the chapters and present detailed examples of "what to say after the statistics have been calculated." They use real data and real research situations to illustrate the process of developing meaning and understanding, and they exemplify how statistics can be used to answer important questions. The issues addressed in these sections include workplace surveillance, the gender gap in income, and the correlates of street crime.

Also, in recognition of the fact that modern technology has rendered long, tedious hand calculation obsolete, the end-of-chapter problems feature smaller, easier-to-handle data sets, although more challenging problems are also included in the chapters and in the Cumulative Exercises at the end of each part. Computer applications using SPSS are presented at the end of virtually every chapter. The data set used in the computer exercises has been updated to the 2008 General Social Survey.

A further goal of this ninth edition is to provide more examples of everyday applications of statistics. A new boxed feature entitled "Statistics in Everyday Life" appears in each chapter and highlights the relevance of statistics in the real world.

This edition continues to focus on the development of basic statistical literacy, the three aspects of which provide a framework for discussing the additional features of this text.

1. An Appreciation of Statistics.   A statistically literate person understands the relevance of statistics for social research, can analyze and interpret the meaning of a statistical test, and can select an appropriate statistic for a given purpose and a given set of data. This textbook develops these qualities, within the constraints imposed by the introductory nature of the course, in the following ways.

- *The relevance of statistics.* Chapter 1 includes a discussion of the role of statistics in social research and stresses their usefulness as ways of analyzing and manipulating data and answering research questions. Each example problem is framed in the context of a research situation. A question is posed and then, with the aid of a statistic, answered. The relevance of statistics for answering questions is thus stressed throughout the text. This central theme of usefulness is further reinforced by a series of "Applying Statistics" boxes, each of which illustrates some specific way statistics can be used to answer questions.

   Almost all end-of-chapter problems are labeled by the social science discipline or discipline from which they are drawn: $\boxed{\text{SOC}}$ for sociology, $\boxed{\text{SW}}$ for social work, $\boxed{\text{PS}}$ for political science, $\boxed{\text{CJ}}$ for criminal justice, $\boxed{\text{PA}}$ for public administration, and $\boxed{\text{GER}}$ for gerontology. By identifying problems with specific disciplines, students can more easily see the relevance of statistics to their own academic interests. (Not incidentally, they will also see that the disciplines have a large subject matter in common.)

- *Interpreting statistics.* For most students, interpretation—saying what statistics mean—is a big challenge. The ability to interpret statistics can be developed only by exposure and experience. To provide exposure, I have been careful, in the example problems, to express the meaning of the statistic in terms of the original research question. To provide experience, the end-of-chapter problems almost always call for an interpretation of the statistic calculated. To provide examples, many of the Answers to Odd-Numbered Computational Problems in the back of the text are expressed in words as well as numbers. The "Interpreting Statistics" sections provide additional, detailed examples of how to express the meaning of statistics.

- *Using statistics: Ideas for research projects.* Appendix E offers ideas for independent data-analysis projects for students. The projects require students to use SPSS to analyze a data set. They can be assigned at intervals throughout the semester or at the end of the course. Each project provides an opportunity for students to practice and apply their statistical skills and, above all, to exercise their ability to understand and interpret the meaning of the statistics they produce.

2. Computational Competence. Students should emerge from their first course in statistics with the ability to perform elementary forms of data analysis—to execute a series of calculations and arrive at the correct answer. While computers and calculators have made computation less of an issue today, computation is inseparable from statistics, so I have included a number of features to help students cope with these mathematical challenges.

- *"One Step at a Time" boxes* for each statistic break down computation into individual steps for maximum clarity and ease.

- *Extensive problem sets* are provided at the end of each chapter. For the most part, these problems use fictitious data and are designed for ease of computation.

- *Cumulative exercises* are included at the end of each part to provide practice in choosing, computing, and analyzing statistics. These exercises present only data sets and research questions. Students must choose appropriate statistics as part of the exercise.

- *Solutions* to odd-numbered computational problems are provided so that students may check their answers.

- *SPSS for Windows* continues as a feature (as in previous editions), to give students access to the computational power of the computer. This is explained in more detail later.

3. The Ability to Read the Professional Social Science Literature. The statistically literate person can comprehend and critically appreciate research reports written by others. The development of this skill is a particular problem at the introductory level since (1) the vocabulary of professional researchers is so much more concise than the language of the textbook, and (2) the statistics featured in the literature are more advanced than those covered at the introductory level. To help bridge this gap, I have included, beginning in Chapter 1, a series of boxes labeled "Reading Statistics." In each such box, I briefly describe the reporting style typically used for the statistic in question and try to alert students about what to expect when they approach the professional literature. These inserts have been updated in this edition and include excerpts from the research literature that illustrate how statistics are actually applied and interpreted by social scientists.

Additional Features

A number of other features make the text more meaningful for students and more useful for instructors.

- *Readability and clarity.* The writing style is informal and accessible to students without ignoring the traditional vocabulary of statistics. Problems and examples have been written to maximize student interest and to focus on issues of concern and significance. For the more difficult material (such as hypothesis testing), students are first walked through an example problem before being confronted by formal terminology and concepts. Each chapter ends with a summary of major points and formulas and a glossary of important concepts. A list of frequently used formulas inside the front cover and a glossary of symbols inside the back cover can be used for quick reference.

- *Organization and coverage.* The text is divided into four parts, with most of the coverage devoted to univariate descriptive statistics, inferential statistics, and bivariate measures of association. The distinction between description and inference is introduced in the first chapter and maintained throughout the text. In selecting statistics for inclusion, I have tried to strike a balance between the essential concepts with which students must be familiar and the amount of material students can reasonably be expected to learn in their first (and perhaps only) statistics course, all the while bearing in mind that different instructors will naturally wish to stress different aspects of the subject. Thus, the text covers a full gamut of the usual statistics, with each chapter broken into subsections so that instructors may choose the particular statistics they wish to include. In this edition, the text has been shortened and streamlined by moving some infrequently used techniques and statistical procedures to the companion Web site.

- *Learning objectives.* Learning objectives are stated at the beginning of each chapter. These are intended to serve as "study guides" and to help students identify and focus on the most important material.

- *Review of mathematical skills.* A comprehensive review of all of the mathematical skills that will be used in this text is included as a Prologue. Students who are inexperienced or out of practice with mathematics can study this review early in the course and/or refer to it as needed. A self-test is included so that students can check their level of preparation for the course.

- *Statistical techniques and end-of-chapter problems are explicitly linked.* After a technique is introduced, students are directed to specific problems for practice and review. The "how-to-do-it" aspects of calculation are reinforced immediately and clearly.

- *End-of-chapter problems are organized progressively.* Simpler problems with small data sets are presented first. Often, explicit instructions or hints accompany the first several problems in a set. The problems gradually become more challenging and require more decision making by the student (e.g., choosing the most appropriate statistic for a certain situation). Thus, each problem set develops problem-solving abilities gradually and progressively.

- *Computer applications.* This text integrates SPSS, the leading social science statistics package, to help students take advantage of the power of the computer. Appendix F provides an introduction to SPSS. The demonstrations at the end of each chapter explain how to use the statistical package to produce the statistics presented in the chapter. Student exercises analyzing data with SPSS are also included. The student version of SPSS is available as a supplement to this text.

- *Realistic, up-to-date data.* The database for computer applications in the text is a shortened version of the 2008 General Social Survey. This database will give students the opportunity to practice their statistical skills on "real-life" data. The database is described in Appendix G.

Supplements

Several ancillaries accompany the text:

- *Student Companion Web Site.* The Web site for this text includes a more extensive treatment of probability, some less frequently used techniques, a table of random numbers, hypothesis tests for ordinal-level variables, "find-the-test" flowcharts, and a number of other features. To access these and additional course materials, please visit www.cengagebrain.com. At the CengageBrain.com home page, search for the ISBN of your title (from the back cover of your book) using the search box at the top of the page. This will take you to the product page where these resources can be found.

- *Online PowerPoint® Slides.* A revised series of PowerPoint slides allows instructors to present class lectures and presentations discussing chapter-by-chapter content.

- *Online Instructor's Manual/Testbank.* The Instructor's Manual includes chapter summaries, a test item file of multiple-choice questions, answers to even-numbered computational problems, and step-by-step solutions to selected problems. In addition, the Instructor's Manual includes cumulative exercises (with answers) that can be used for testing purposes. To access these instructor resources, please login to your account at http://login.cengage.com.

- *Study Guide.* The Study Guide, written by Professor Roy Barnes, contains additional examples to illuminate basic principles, review problems with detailed answers, SPSS projects, and multiple-choice questions and answers that complement but do not duplicate the test item file.

Changes to the Ninth Edition

The following are the most important changes in the ninth edition.

- The text has been completely redesigned to enhance readability and ease of use.
- Former Chapters 12 and 13 have been combined into a single chapter (Chapter 12, entitled "Association Between Variables Measured at the Nominal Level").
- In response to reviewers' suggestions to include more examples of everyday applications of statistics, a new feature entitled "Statistics in Everyday Life" has been added in each chapter. This feature provides examples of the relevance and usefulness of statistics in a wide variety of situations, ranging from telephone solicitations to the effect of the Internet on the crime of rape, demonstrating to students that statistics are not just abstract mathematical constructs but have practical value in the real world of government, education, business, media, politics, sports, and so forth.
- The end-of-chapter SPSS exercises are in a new format. Previous editions provided extensive examples of the use of SPSS and interpretation of SPSS output. Now, students are required to be much more actively involved in the exercises, stating research questions, selecting appropriate variables, and interpreting results.
- The sections of some chapters have been divided into subsections for ease of comprehension.
- The data set used in the text has been updated to the 2008 General Social Survey. Likewise, the data cited in the text and in end-of-chapter problems have been updated.
- There is more emphasis on interpretation throughout the text.
- The "Reading Statistics" inserts have been updated, along with the data sets used for examples and end-of-chapter problems.
- The steps in the "One Step at a Time" feature have been renumbered and reorganized to improve clarity and ease of use.
- The numbering of chapter sections has been deleted to streamline the layout.
- Titles have been added to the "Applying Statistics" feature (formerly called "Application" boxes) to call attention to the topic being highlighted.

CHAPTER-BY-CHAPTER CHANGES

The following is a list of changes at the chapter level.

Chapter 1, Introduction

- New "Statistics in Everyday Life" box on "Push Polls"
- New "Statistics in Everyday Life" box on "TVs"
- New "Statistics in Everyday Life" box on "Gay Marriage"
- Added new feature: You Are the Researcher ("Introduction")

Chapter 2, Basic Descriptive Statistics: Percentages, Ratios and Rates, Tables, Charts, and Graphs

- Updated "Applying Statistics 2.4 *Percentage Change*"
- Updated Table 2.4, "Projected Population Growth for Six Nations, 2000–2050" and accompanying explanation
- Added "Statistics in Everyday Life" box on "Rates"
- Updated Table 2.18, "Distribution of Income by Household, United States, 2008" and accompanying explanation
- Changed title of subsection to "Applying Frequency Distributions for Interval-Ratio Variables: An Example"
- Added "Statistics in Everyday Life" box on "Divorce"
- Updated Figure 2.7 and accompanying explanation
- Updated information in subsection "Interpreting Statistics: Changing Patterns of Workplace Surveillance"
- Updated Tables 2.21 and 2.22 and Figures 2.8 and 2.9
- Updated "Reading Statistics" feature
- Revised Problem 2.2
- Updated data in Problem 2.4
- Updated Figures 2.10-2.12
- Added new feature: You Are the Researcher ("Culture Wars")

Chapter 3, Measures of Central Tendency

- New "Statistics in Everyday Life" box on "Religions"
- Reformatted "One Step at a Time" feature for finding the median
- Reformatted subsections under "The Mean"
- New "Statistics in Everyday Life" feature on "Global McDonalds"
- Updated Problems 3.4, 3.6, 3.11, and 3.15
- Added new feature: You Are the Researcher ("Typical American")

Chapter 4, Measures of Dispersion

- Updated Table 4.1 on p. 90
- New "Statistics in Everyday Life" box on "Changing Religion"
- Changed and updated Table 4.3, "Percent of Population Aged 25 and Older with a College Degree, 2007"
- New "Statistics in Everyday Life" Insert on "Life Span"
- Updated "Applying Statistics 4.2 *Describing Dispersion*"
- Updated and changed "Reading Statistics 3" feature
- Updated and revised section on "Interpreting Statistics: The Changing Distribution of Income in the United States"
- Updated Figures 4.2 and 4.3

- Updated Problems 4.5, 4.8, 4.9, 4.10, and 4.11
- Added new feature: You Are the Researcher ("Following Up")

Chapter 5, The Normal Curve

- New "Statistics in Everyday Life" feature on "IQ Scores"
- Reformatted "One Step at a Time" feature for "Finding Areas Between Z Scores"
- New "Statistics in Everyday Life" box on "Games of Chance"
- Deleted SPSS exercise at end of chapter

Chapter 6, Introduction to Inferential Statistics: Sampling and the Sampling Distribution

- New "Statistics in Everyday Life" feature on "Sample Size"
- New "Statistics in Everyday Life" feature on "Phone Solicitors"
- Reformatted subsections of "The Sampling Distribution"
- New "Statistics in Everyday Life" box on "GSS Tracking Trends"
- Added new feature: You Are the Researcher ("Estimating Parameters")

Chapter 7, Estimation Procedures

- Reformatted subsections of "Estimation Selection Criteria"
- Reformatted subsections of "Interval Estimation Procedures"
- Reformatted "One Step at a Time" feature for confidence intervals for means
- New "Statistics in Everyday Life" box on "Happiest Days"
- Reformatted "One Step at a Time" feature for confidence intervals for proportions
- Updated data for "Applying Statistics 7.3 *Estimating Population Proportions*"
- Updated and revised section on "Interpreting Statistics: Predicting Election Outcomes and Judging the Performance of the President"
- Updated Tables 7.5 and 7.6 and Figure 7.6
- New "Statistics in Everyday Life" box on "Dewey Defeats Truman"
- Revised Problems 7.4, 7.6, 7.8, 7.10, and 7.18
- Updated Problem 7.16
- Added new feature: You Are the Researcher ("Estimating the Characteristics of the Typical American")

Chapter 8, Hypothesis Testing I: The One-Sample Case

- New "Statistics in Everyday Life" feature on "Approving Drugs—FDA"
- New "Statistics in Everyday Life" feature on "Gambling Analogy"
- Reformatted subsections in "One versus Two tailed tests" section
- New "Statistics in Everyday Life" box on "Alpha = 0.05"

- Reformatted subsections in "Student *t* Tests" section
- Reformatted "One Step at a Time" feature for *t* tests: separated Steps 4 and 5
- Reformatted subsections in "Testing Sample Proportions" section
- Reformatted "One Step at a Time" feature for sample proportions: separated Steps 4 and 5
- Revised Problems 8.2, 8.4, 8.6, and 8.10

Chapter 9, Hypothesis Testing II: The Two-Sample Case

- Reformatted "One Step at a Time" feature for testing large sample means: separated Steps 4 and 5
- New "Statistics in Everyday Life" box on "Reading Scores"
- Reformatted "One Step at a Time" feature for testing small sample means: separated Steps 4 and 5
- Reformatted "One Step at a Time" feature for testing large sample proportions: separated Steps 4 and 5
- New "Statistics in Everyday Life" box on "Political Polarization"
- Updated and revised section on "Interpreting Statistics: Are There Significant Differences in Income Between Men and Women?"
- New "Statistics in Everyday Life" box on "Gender Gap"
- Revised and updated "Reading Statistics 5: Hypothesis Testing"
- Updated data for Applying Statistics 9.3
- Revised Problems 9.4 and 9.14
- Added new feature: You Are the Researcher ("Gender Gaps and Support for Traditional Gender Roles")

Chapter 10, Hypothesis Testing III: The Analysis of Variance

- New "Statistics in Everyday Life" feature on "Dieting"
- Updated and revised section on "Interpreting Statistics: Sexual Activity by Marital Status"
- Updated Tables 10.5, 10.6, 10.7, and 10.8
- New "Statistics in Everyday Life" feature on "Lying"
- Revised and updated "Reading Statistics 6: Hypothesis Testing"
- Revised Problem 10.4
- Added new feature: You Are the Researcher ("Liberals and Conservatives")

Chapter 11, Hypothesis Testing IV: Chi Square

- New "Statistics in Everyday Life" feature on "Gender Gap in Politics"
- Revised and updated section on "Interpreting Statistics: Family Values and Social Class" using new variables
- Revised and updated "Applying Statistics 11.2 *Cross-National Differences in Socialization Values*"

- Rewrote "Reading Statistics 7: A Profile of Children Who Attack Parents"
- Added new feature: You Are the Researcher ("Understanding Political Beliefs")

Chapter 12, Association Between Variables Measured at the Nominal Level

- Rewrote the chapter to incorporate Chapter 13 from the 8th edition
- New "Statistics in Everyday Life" box on "Professional Football"
- Updated and rewrote "Reading Statistics 10" and incorporated it as a new section ("The Importance of Percentages and Some Errors of Interpretation")
- New "Statistics in Everyday Life" box on "Free Speech and Muslim Clergy"
- Rewrote "Reading Statistics 8: Smoking Study"
- Revised Problem 12.4
- Added new feature: You Are the Researcher ("Understanding Political Beliefs, Part II")

Chapter 13, Association Between Variables Measured at the Ordinal Level

- Rewrote and updated former Section 14.5 as "Interpreting Statistics with Bivariate Tables: What Is the Relationship Between Industrialization and Political and Civil Rights?"
- Added Table 13.9, "Political and Civil Rights by Income Level"
- New "Statistics in Everyday Life" feature on "Attitudes Toward the Federal Government"
- New "Statistics in Everyday Life" feature on "Quality of Life by Nation"
- Revised Problem 13.6
- Added new feature: You Are the Researcher ("Who Approves of Premarital Sex?")

Chapter 14, Association Between Variables Measured at the Interval-Ratio Level

- New "Statistics in Everyday Life" feature on "Household Division of Labor"
- New "Statistics in Everyday Life" feature on "Regression and Prediction"
- Reformatted and rewrote former Section 15.4, "Computing and Interpreting the Regression Coefficients (a and b)"
- New "Statistics in Everyday Life" feature on "Popularity"
- Revised and updated section on the Correlation Matrix
- Added Table 14.5
- Rewrote and updated "Applying Statistics 14.2 *Regression and Correlation*"
- Rewrote and updated former Section 15.10 as "Interpreting Statistics: The Correlates of Crime"
- Updated Table 14.6 and Figure 14.8

- Updated Problems 14.5, 14.6, and 14.8
- Added new feature: You Are the Researcher ("Who Watches TV? Who Succeeds in Life?")

Chapter 15, Elaborating Bivariate Tables

- New "Statistics in Everyday Life" box on "Spurious Relationships"
- New "Statistics in Everyday Life" box on "Interaction"
- Rewrote and updated former Section 16.7 as "Interpreting Statistics: Analyzing the Sources of Democracy"
- Added Tables 15.6 and 15.7
- Revised and updated Problem 15.8
- Added new feature: You Are the Researcher ("A Multivariate Analysis of Sexual Attitudes")

Chapter 16, Partial Correlation and Multiple Regression and Correlation

- Reformatted "Partial Correlation" section
- New "Statistics in Everyday Life" feature on "Education and Income"
- New "Statistics in Everyday Life" feature on "Rape and the Internet"
- Revised and updated "Applying Statistics 16.2"
- Rewrote and updated former Section 17.6 as "Interpreting Statistics: Another Look at the Correlates of Crime"
- Updated Tables 16.3, 16.4, and 16.5
- New "Statistics in Everyday Life" feature on "Baseball"
- Updated Problems 17.4 and 17.7
- Added new feature: You Are the Researcher ("A Multivariate Analysis of TV Viewing and Success")

The text has been thoroughly reviewed for clarity and readability. As with previous editions, my goal is to offer a comprehensive, flexible, and student-oriented book that will provide a challenging first exposure to social statistics.

Acknowledgments

This text has been in development, in one form or another, for over 25 years. An enormous number of people have made contributions, both great and small, to this project and, at the risk of inadvertently omitting someone, I am bound to at least attempt to acknowledge my many debts.

This edition reflects the thoughtful guidance of Erin Mitchell, my acquiring sponsoring editor at Wadsworth Cengage Learning, and Mary Falcon, my development editor, and I thank them both for their contributions. I also thank many others at Wadsworth—including Andrew Keay, Cheri Palmer, Caryl Gorska, Kristin Makarewycz, John Chell, Melanie Cregger, and Mallory Ortberg—for their work on the ninth edition.

Much of whatever integrity and quality this book has is a direct result of the very thorough (and often highly critical) reviews that have been conducted over the years. I am consistently impressed by the sensitivity of my colleagues to the needs of the students, and, for their assistance in preparing this edition, I would like to thank the following reviewers: Robert Carini, *University of Louisville*; Stephanie Dodge, *University of Hawaii at Hilo*; Scott Feld, *Purdue University*; Jeremy Forbis, *University of Dayton*; Xuanning Fu, *California State University at Fresno*; Brea L. Perry, *University of Kentucky*; Tammy L. Reedy-Strother, *University of Kentucky*; and Mark Riddle, *University of Northern Colorado*. Whatever failings are contained in the text are, of course, my responsibility and are probably the result of my occasional decisions not to follow the advice of my colleagues.

I would like to thank the instructors who made statistics understandable to me (Professors Satoshi Ito, Noelie Herzog, and Ed Erikson) and all of my colleagues at Christopher Newport University for their support and encouragement (especially Professors F. Samuel Bauer, Stephanie Byrd, Cheryl Chambers, Robert Durel, Marcus Griffin, Mai Lan Gustafsson, Kai Heiddemann, Ruth Kernodle, Michael Lewis, Marion Manton, Eileen O'Brien, Lea Pellet, Eduardo Perez, Virginia Purtle, Andrea Timmer, and Linda Waldron). I owe a special debt of gratitude to Professor Roy Barnes of the University of Michigan-Flint for his help with the "Interpreting Statistics" feature. Also, I thank all of my students for their patience and thoughtful feedback, and I am grateful to the Literary Executor of the late Sir Ronald A. Fisher, F.R.S., to Dr. Frank Yates, F.R.S., and to Longman Group Ltd., London, for permission to reprint Appendixes B, C, and D from their book *Statistical Tables for Biological, Agricultural and Medical Research* (6th edition, 1974).

Finally, I want to acknowledge the support of my family and rededicate this work to them. I have the extreme good fortune to be a member of an extended family that is remarkable in many ways and that continues to increase in size. Although I cannot list everyone, I would especially like to thank the older generation (my mother, Alice T. Healey, who, sadly, passed away as this edition was nearing completion), the next generation (my sons, Kevin and Christopher, my daughters-in-law, Jennifer and Jessica), the new members (my wife, Patricia Healey, and Christopher and Jennifer Schroen and Kate and Matt Cowell), and the youngest generation (Benjamin, Caroline, and Isabelle Healey and Abigail Cowell).

Prologue:
Basic Mathematics Review

You will probably be relieved to hear that this text, your first exposure to statistics for social science research, does not stress computation per se. While you will encounter many numbers to work with and numerous formulas to solve, the major emphasis will be on understanding the role of statistics in research and the logic by which we attempt to answer research questions empirically. You will also find that, in this text, the example problems and many of the homework problems have been intentionally simplified so that the computations will not unduly impede the task of understanding the statistics themselves.

On the other hand, you may regret to learn that there is, inevitably, some arithmetic that you simply cannot avoid if you want to master this material. It is likely that some of you haven't had any math in a long time, others have convinced themselves that they just cannot do math under any circumstances, and still others are just rusty and out of practice. All of you will find that mathematical operations that might seem complex and intimidating at first glance can be broken down into simple steps. If you have forgotten how to cope with some of these steps or are unfamiliar with these operations, this prologue is designed to ease you into the skills you will need to do all of the computations in this textbook. Also, you can use this section for review whenever you feel uncomfortable with the mathematics in the chapters to come.

Calculators and Computers

A calculator is a virtual necessity for this text. Even the simplest, least expensive model will save you time and effort and is definitely worth the investment. However, I recommend that you consider investing in a more sophisticated calculator with memories and preprogrammed functions, especially the statistical models that can compute means and standard deviations automatically. Calculators with these capabilities are available for around $20.00 to $30.00 and will almost certainly be worth the small effort it will take to learn to use them.

In the same vein, there are several computerized statistical packages (or **statpaks**) commonly available on college campuses that can further enhance your statistical and research capabilities. The most widely used of these is the Statistical Package for the Social Sciences (**SPSS**). This program comes in a student version that is available bundled with this text (for a small fee). Statistical packages like SPSS are many times more powerful than even the most sophisticated handheld calculators, and it will be well worth your time to learn how to use them because they will eventually save you time and effort.

SPSS is introduced in Appendix F of this text, and exercises at the end of almost every chapter will show you how to use the program to generate and interpret the statistics covered in the chapter.

There are many other programs that can help you generate accurate statistical results with a minimum of effort and time. Even spreadsheet programs such as Microsoft Excel, which is included in many versions of Microsoft Office, have some statistical capabilities. You should be aware that all of these programs (other than the simplest calculators) will require some effort to learn, but the rewards will be worth the effort.

In summary, you should find a way at the beginning of this course—with a calculator, a statpak, or both—to minimize the tedium and hassle of mere computing. This will permit you to devote maximum effort to the truly important goal of increasing your understanding of the meaning of statistics in particular and social research in general.

Variables and Symbols

Statistics are a set of techniques by which we can describe, analyze, and manipulate variables. A **variable** is a trait that can change value from case to case or from time to time. Examples of variables include height, weight, level of prejudice, and political party preference. The possible values or scores associated with a given variable might be numerous (for example, income) or relatively few (for example, gender). I will often use symbols, usually the letter X, to refer to variables in general or to a specific variable.

Sometimes we will need to refer to a specific value or set of values of a variable. This is usually done with the aid of subscripts. So the symbol X_1 (read "X-sub-one") would refer to the first score in a set of scores, X_2 ("X-sub-two") to the second score, and so forth. Also, we will use the subscript i to refer to all the scores in a set. Thus, the symbol X_i ("X-sub-eye") refers to all of the scores associated with a given variable (for example, the test grades of a particular class).

Operations

You are all familiar with the four basic mathematical operations of addition, subtraction, multiplication, and division and the standard symbols $(+, -, \times, \div)$ used to denote them. I should remind you that multiplication and division can be symbolized in a variety of ways. For example, the operation of multiplying some number a by some number b may be symbolized in (at least) six different ways:

$$a \times b$$
$$a \cdot b$$
$$a * b$$
$$ab$$
$$a(b)$$
$$(a)(b)$$

In this text, we will commonly use the "adjacent symbols" format (that is, ab), the conventional times sign ($\times$), or adjacent parentheses to indicate multiplication. On most calculators and computers, the asterisk ($*$) is the symbol for multiplication.

The operation of division can also be expressed in several different ways. In this text, we will use either of these two methods:

$$a/b \quad \text{or} \quad \frac{a}{b}$$

Several of the formulas with which we will be working require us to find the square of a number. To do this, multiply the number by itself. This operation is symbolized as X^2 (read "X squared"), which is the same thing as $(X)(X)$. If X has a value of 4, then

$$X^2 = (X)(X) = (4)(4) = 16$$

or we could say that "4 squared is 16."

The square root of a number is the value that, when multiplied by itself, results in the original number. So the square root of 16 is 4 because $(4)(4)$ is 16. The operation of finding the square root of a number is symbolized as

$$\sqrt{X}$$

A final operation with which you should be familiar is summation, or the addition of the scores associated with a particular variable. When a formula requires the addition of a series of scores, this operation is usually symbolized as ΣX_i. Σ is the uppercase Greek letter sigma and stands for "the summation of." So the combination of symbols ΣX_i means "the summation of all the scores" and directs us to add the value of all the scores for that variable. If four people had family sizes of 2, 4, 5, and 7, then the summation of these four scores for this variable could be symbolized as

$$\Sigma X_i = 2 + 4 + 5 + 7 = 18$$

The symbol Σ is an operator, just like the $+$ and $\times$ signs. It directs us to add all of the scores on the variable indicated by the X symbol.

There are two other common uses of the summation sign, and, unfortunately, the symbols denoting these uses are not, at first glance, sharply different from each other or from the symbol used earlier. A little practice and some careful attention to these various meanings should minimize the confusion. The first set of symbols is ΣX_i^2, which means "the sum of the squared scores." This quantity is found by *first* squaring each of the scores and *then* adding the squared scores together. A second common set of symbols will be $(\Sigma X_i)^2$, which means "the sum of the scores, squared." This quantity is found by *first* summing the scores and *then* squaring the total.

These distinctions might be confusing at first, so let's see if an example helps to clarify the situation. Suppose we had a set of three scores: 10, 12, and 13. So

$$X_i = 10, 12, 13$$

The sum of these scores would be indicated as

$$\Sigma X_i = 10 + 12 + 13 = 35$$

The sum of the squared scores would be

$$\Sigma X_i^2 = (10)^2 + (12)^2 + (13)^2 = 100 + 144 + 169 = 413$$

Take careful note of the order of operations here. First the scores are squared one at a time and then the squared scores are added. This is a completely different operation from squaring the sum of the scores:

$$(\Sigma X_i)^2 = (10 + 12 + 13)^2 = (35)^2 = 1225$$

To find this quantity, first the scores are summed and then the total of all the scores is squared. The squared sum of the scores (1225) is *not* the same as the sum of the squared scores (413). In summary, the operations associated with each set of symbols can be summarized as follows:

Symbols	Operations
ΣX_i	Add the scores
ΣX_i^2	First square the scores and then add the squared scores
$(\Sigma X_i)^2$	First add the scores and then square the total

Operations with Negative Numbers

A number can be either positive (if it is preceded by a + sign or by no sign at all) or negative (if it is preceded by a − sign). Positive numbers are greater than zero, and negative numbers are less than zero. It is very important to keep track of signs because they will affect the outcome of virtually every mathematical operation. This section briefly summarizes the relevant rules for dealing with negative numbers.

First, adding a negative number is the same as subtraction. For example,

$$3 + (-1) = 3 - 1 = 2$$

Second, subtraction changes the sign of a negative number:

$$3 - (-1) = 3 + 1 = 4$$

Note the importance of keeping track of signs here. If you neglected to change the sign of the negative number in the second expression, you would get the wrong answer.

For multiplication and division, you need to be aware of various combinations of negative and positive numbers. Ignoring the case of all positive numbers, this leaves several possible combinations. A negative number times a positive number results in a negative value:

$$(-3)(4) = -12$$

or

$$(3)(-4) = -12$$

A negative number multiplied by a negative number is always positive:

$$(-3)(-4) = 12$$

Division follows the same patterns. If there is a negative number in the calculations, the answer will be negative. If both numbers are negative, the answer will be positive. So

$$\frac{-4}{2} = -2$$

and

$$\frac{4}{-2} = -2$$

but

$$\frac{-4}{-2} = 2$$

Negative numbers do not have square roots, since multiplying a number by itself cannot result in a negative value. Squaring a negative number always results in a positive value (see the multiplication rules earlier).

Accuracy and Rounding Off

A possible source of confusion in computation involves accuracy and rounding off. People work at different levels of accuracy and precision and, for this reason alone, may arrive at different answers to problems. This is important because our answers can be at least slightly different if you work at one level of precision and I (or your instructor or your study partner) work at another. You may sometimes think you've gotten the wrong answer when all you've really done is round off at a different place in the calculations or in a different way.

There are two issues here: when to round off and how to round off. In this text, I have followed the convention of working in as much accuracy as my calculator or statistics package will allow and then rounding off to two places of accuracy (two places beyond, or to the right of, the decimal point) only at the very end. If a set of calculations is lengthy and requires the reporting of intermediate sums or subtotals, I will round the subtotals off to two places as I go.

In terms of how to round off, begin by looking at the digit immediately to the right of the last digit you want to retain. If you want to round off to 100ths (two places beyond the decimal point), look at the digit in the 1000ths place (three places beyond the decimal point). If that digit is 5 or more, round up. For example, 23.346 would round off to 23.35. If the digit to the right is less than 5, round down. So, 23.343 would become 23.34.

Let's look at some more examples of how to follow these rules of rounding. If you are calculating the mean value of a set of test scores and your calculator shows a final value of 83.459067 and you want to round off to two places, look at the digit three places beyond the decimal point. In this case the value is 9 (greater than 5), so we would round the second digit beyond the decimal point up and report the mean as 83.46. If the value had been 83.453067, we would have reported our final answer as 83.45.

Formulas, Complex Operations, and the Order of Operations

A mathematical formula is a set of directions, stated in general symbols, for calculating a particular statistic. To "solve a formula," you replace the symbols with the proper values and then perform a series of calculations. Even the most complex formula can be simplified by breaking the operations down into smaller steps. Working through these steps requires some knowledge of general procedure and the rules of precedence of mathematical operations. This is because the order in which you perform calculations may affect your final answer. Consider the following expression:

$$2 + 3(4)$$

If you add first, you will evaluate the expression as

$$5(4) = 20$$

but if you multiply first, the expression becomes

$$2 + 12 = 14$$

Obviously, it is crucial to complete the steps of a calculation in the correct order.

The basic rules of precedence are to find all squares and square roots first, then do all multiplication and division, and finally complete all addition and subtraction. So the expression

$$8 + 2 \times 2^2/2$$

would be evaluated as

$$8 + 2 \times \frac{4}{2} = 8 + \frac{8}{2} = 8 + 4 = 12$$

The rules of precedence may be overridden by parentheses. Solve all expressions within parentheses before applying the rules stated earlier. For most of the complex formulas in this text, the order of calculations will be controlled by the parentheses. Consider the following expression:

$$(8 + 2) - 4(3)^2/(8 - 6)$$

Resolving the parenthetical expressions first, we would have

$$(10) - 4 \times 9/(2) = 10 - 36/2 = 10 - 18 = -8$$

Without the parentheses, the same expression would be evaluated as

$$
\begin{aligned}
8 + 2 - 4 \times 3^2/8 - 6 &= 8 + 2 - 4 \times 9/8 - 6 \\
&= 8 + 2 - 36/8 - 6 \\
&= 8 + 2 - 4.5 - 6 \\
&= 10 - 10.5 \\
&= -0.5
\end{aligned}
$$

A final operation you will encounter in some formulas in this text involves denominators of fractions that themselves contain fractions. In this situation, solve the fraction in the denominator first and then complete the division. For example,

$$\frac{15 - 9}{6/2}$$

would become

$$\frac{15 - 9}{6/2} = \frac{6}{3} = 2$$

When you are confronted with complex expressions such as these, don't be intimidated. If you're patient with yourself and work through them step by step, beginning with the parenthetical expression, even the most imposing formulas can be managed.

Exercises

You can use the following problems as a "self-test" on the material presented in this review. If you can handle these problems, you're ready to do all of the arithmetic in this text. If you have difficulty with any of these problems, please review the appropriate section of this prologue. You might also want to use this section as an opportunity to become more familiar with your calculator. The answers are given immediately following these exercises, along with commentary and some reminders.

1. Complete each of the following:
   a. $17 \times 3 =$
   b. $17(3) =$
   c. $(17)(3) =$
   d. $17/3 =$
   e. $(42)^2 =$
   f. $\sqrt{113} =$

2. For the set of scores (X_i) of 50, 55, 60, 65, and 70, evaluate each of the following expressions:

$\Sigma X_i =$

$\Sigma X_i^2 =$

$(\Sigma X_i)^2 =$

3. Complete each of the following:
   a. $17 + (-3) + (4) + (-2) =$
   b. $15 - 3 - (-5) + 2 =$
   c. $(-27)(54) =$
   d. $(113)(-2) =$
   e. $(-14)(-100) =$
   f. $-34/-2 =$
   g. $322/-11 =$
   h. $\sqrt{-2} =$
   i. $(-17)^2 =$

4. Round off each of the following to two places beyond the decimal point:
   a. 17.17532
   b. 43.119
   c. 1076.77337
   d. 32.4641152301
   e. 32.4751152301

5. Evaluate each of the following:
   a. $(3 + 7)/10 =$
   b. $3 + 7/10 =$
   c. $\dfrac{(4 - 3) + (7 + 2)(3)}{(4 + 5)(10)} =$
   d. $\dfrac{22 + 44}{15/3} =$

Answers to Exercises

1. a. 51   b. 51   c. 51

(The obvious purpose of these first three problems is to remind you that there are several different ways of expressing multiplication.)

   d. 5.67 (Note the rounding off.)   e. 1764   f. 10.63

2. The first expression is "the sum of the scores," so this operation would be

$$\Sigma X_i = 50 + 55 + 60 + 65 + 70 = 300$$

The second expression is the "sum of the squared scores." So

$$\Sigma X_i^2 = (50)^2 + (55)^2 + (60)^2 + (65)^2 + (70)^2$$
$$\Sigma X_i^2 = 2500 + 3025 + 3600 + 4225 + 4900$$
$$\Sigma X_i^2 = 18,250$$

The third expression is "the sum of the scores, squared":

$$(\Sigma X_i)^2 = (50 + 55 + 60 + 65 + 70)^2$$
$$(\Sigma X_i)^2 = (300)^2$$
$$(\Sigma X_i)^2 = 90,000$$

Remember that ΣX_i^2 and $(\Sigma X_i)^2$ are two completely different expressions with very different values.

3. a. 16   **b.** 19 (Remember to change the sign of -5.)   **c.** -1458

   d. -226 **e.** 1400   **f.** 17   **g.** -29.27

   h. Your calculator probably gave you some sort of error message for this problem, since negative numbers do not have square roots.

   i. 289

4. a. 17.18   **b.** 43.12   **c.** 1076.77

   d. 32.46   **e.** 32.48

5. a. 1   **b.** 3.7 (Note again the importance of parentheses.)   **c.** 0.31

   d. 13.2

1 Introduction

LEARNING OBJECTIVES By the end of this chapter, you will be able to

1. Describe the limited but crucial role of statistics in social research.
2. Distinguish between three applications of statistics and identify situations in which each is appropriate.
3. Distinguish between discrete and continuous variables and cite examples of each.
4. Identify and describe three levels of measurement and cite examples of variables from each.

Students sometimes wonder about the value of studying statistics. What, after all, do numbers and statistics have to do with understanding people and society? The value of statistics will become clear as we move from chapter to chapter but, for now, we can demonstrate the importance of statistics by considering **research**. Scientists conduct research to answer questions, examine ideas, and test theories. Research is a disciplined inquiry that can take numerous forms. Statistics are relevant for **quantitative research** projects, or projects that collect information in the form of numbers or **data**. **Statistics**, then, are a set of mathematical techniques used by social scientists to organize and manipulate data for the purpose of answering questions and testing theories.

What is so important about learning how to manipulate data? On one hand, some of the most important and enlightening works in the social sciences do not utilize statistics at all. There is nothing magical about data and statistics. The mere presence of numbers guarantees nothing about the quality of a scientific inquiry. On the other hand, data can be the most trustworthy information available to the researcher and, consequently, deserve special attention. Data that have been carefully collected and thoughtfully analyzed can be the strongest, most objective foundations for building theory and enhancing our understanding of the social world.

Let us be very clear about one point: It is never enough merely to gather data (or, for that matter, any kind of information). Even the most carefully collected numerical information does not and cannot speak for itself. The researcher must be able to use statistics effectively to organize, evaluate, and analyze the data. Without a good understanding of the principles of statistical analysis, the researcher will be unable to make sense of the data. Without the appropriate application of statistics, data will be useless.

Statistics are an indispensable tool for the social sciences. They provide the scientist with some of the most useful techniques for evaluating ideas and testing theory. The next section describes the relationships between theory, research, and statistics in more detail.

The Role of Statistics in Scientific Inquiry

Figure 1.1 graphically represents the role of statistics in the research process. The diagram is based on the thinking of Walter Wallace and illustrates a useful conception of how the knowledge base of any scientific enterprise grows and develops. One point the diagram makes is that scientific theory and research continually shape each other. Statistics are one of the most important means by which research and theory interact. Let's take a closer look at the process.

A JOURNEY THROUGH THE SCIENTIFIC PROCESS

Since Figure 1.1 is circular, having neither a beginning nor an end, we could start our journey at any point. For the sake of convenience, let's begin at the top and follow the arrows around the circle. A **theory** is an explanation of the relationships among phenomena. People naturally (and endlessly) wonder about problems in society (like prejudice, poverty, child abuse, and serial murders), and they develop explanations ("low levels of education cause prejudice") in their attempts to understand. Unlike our everyday, informal explanations, scientific theory is subject to a rigorous testing process. Let's take racial prejudice as an example to illustrate how the research process works.

Theory. What causes racial prejudice? One possible answer to this question is provided by a theory of prejudice stated over 50 years ago by social psychologist Gordon Allport.[1] The theory states that prejudice will decrease in certain kinds of situations: those in which members of different groups have equal status and cooperate with each other to work toward mutual goals. The more equal and cooperative the contact, the more likely people will see each other as

FIGURE 1.1 The Wheel of Science

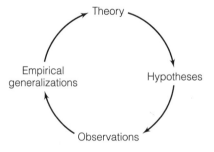

Source: Adapted from Walter Wallace, *The Logic of Science in Sociology* (Chicago: Aldine-Atherton, 1971).

[1]Allport, Gordon, 1954. *The Nature of Prejudice*. Reading, MA: Addison-Wesley. This theory is often called "the contact hypothesis." For recent attempts to test this theory, see: McLaren, Lauren. 2003. "Anti-Immigrant Prejudice in Europe: Contact, Threat Perception, and Preferences for the Exclusion of Migrants." *Social Forces*, 81:909–937; Pettigrew, Thomas. 1997. "Generalized Intergroup Contact Effects on Prejudice." *Personality and Social Psychology Bulletin*, 23:173–185, and Sigelman, Lee, and Susan Welch. 1993. "The Contact Hypothesis Revisited: Black-White Interaction and Positive Racial Attitudes." *Social Forces*, 71:781–795.

individuals and not as representatives of a particular group. For example, we might predict that members of a racially mixed athletic team who work together to achieve victory would experience a decline in prejudice. On the other hand, when different groups compete for jobs, housing, or other resources, prejudice will tend to increase. Allport's theory is not a complete explanation of prejudice, of course, but it will serve to illustrate social science research.

Before moving on, let's examine theory in a little more detail. Note that Allport's theory is stated in terms of causal relationships between two variables. A **variable** is any trait that can change values from case to case. Examples of variables would be gender, age, income, and political party affiliation. A theory may identify some variables as causes and others as effects or results. In the language of science, the causes are called **independent variables** and the effects or result variables are called **dependent variables**. In our theory, equal-status contact would be the independent variable (or the cause) and prejudice would be the dependent variable (the result or effect). In other words, the theory argues that prejudice *depends on* (or is caused by) the extent to which a person participates in equal-status, cooperative contacts with members of other groups.

Diagrams can be a useful way of representing the relationships between variables:

$$Equal\ Status\ Contact \rightarrow Prejudice$$
$$Independent\ variable \rightarrow Dependent\ Variable$$
$$X \rightarrow Y$$

The arrow represents the direction of the causal relationship, and "X" and "Y" are general symbols for the independent and dependent variables, respectively.

Hypotheses. So far, we have a theory of prejudice and independent and dependent variables. What we don't know yet is whether the theory is true or false. To find out, we need to compare our theory with the facts; we need to do some research. Our next steps would be to define our terms and ideas. One problem we often face in research is that theories are complex and abstract and we need to be very specific to conduct a valid test. Often, we do this by deriving a hypothesis from the theory: A **hypothesis** is a statement about the relationship between variables that is specific and exact.

For example, if we wished to test Allport's theory, we would have to say exactly what we mean by prejudice and we would need to describe "equal-status, cooperative contact" in detail. We would also review the research literature to help develop and clarify our definitions and our understanding of these concepts.

As our hypothesis takes shape, we begin the next step of the research process, during which we decide exactly how we will gather data. We must decide how cases will be selected and tested, how variables will be measured, and a host of related matters. Ultimately, these plans will lead to the observation phase (the bottom of the wheel of science), where we actually measure social reality. Before we can do this, we must have a very clear idea of what we are looking for and a well-defined strategy for conducting the search.

Making Observations and Using Statistics. To test Allport's theory of prejudice, we might begin with people from different racial or ethnic groups. We might place some subjects in cooperative situations and others in competitive situations. We would need to measure levels of prejudice before and after each type of contact. We might do this by administering a survey that asks subjects to agree or disagree with statements such as "Greater efforts must be made to racially integrate the public school system." Our goal would be to see if the people exposed to the cooperative contact situations actually become less prejudiced.

Now, finally, we come to the use of statistics in our research. During the observation phase, we will collect a great deal of numerical information or data. If our sample consists of 100 people, we will have 200 completed surveys measuring prejudice: 100 completed before the contact situation and 100 filled out afterward. Try to imagine dealing with 200 completed surveys. If we ask each respondent just five questions to measure their prejudice, we will have a total of 1000 separate pieces of information to deal with. What will we do? We'll have to organize and analyze this information, and statistics will become very valuable at this point. They will supply us with many ideas about "what to do" with the data. We will begin to look at some of the options in the next chapter; for now, let me stress two points about statistics.

First, statistics are crucial. Simply put, without statistics, quantitative research is impossible. We need statistics to analyze data and to shape and refine our theories of the social world. Second, and somewhat paradoxically, the role of statistics is limited. As Figure 1.1 makes clear, scientific research proceeds through several mutually interdependent stages. Statistics become directly relevant only at the end of the observation stage. Before any statistical analysis can be legitimately applied, however, the preceding phases of the process must have been successfully completed. If the researcher has asked poorly conceived questions or has made serious errors of design or method, then even the most sophisticated statistical analysis is valueless. As useful as they can be, statistics cannot substitute for rigorous conceptualization, detailed and careful planning, or creative use of theory. Statistics cannot salvage a poorly conceived or designed research project. They cannot make sense out of garbage. As the old saying goes: "Garbage in, garbage out."

On the other hand, inappropriate statistical applications can limit the usefulness of an otherwise carefully done project. Only by successfully completing *all* phases of the process can a quantitative research project hope to contribute to understanding. A reasonable knowledge of the uses and limitations of statistics is as essential to the education of the social scientist as is training in theory and methodology.

STATISTICS
IN EVERYDAY LIFE

Political campaigns sometimes use "push polls" to sway public opinion. These polls do not follow the scientific method. They are designed to influence opinions, sometimes by starting or circulating rumors and innuendo. For example, they may try to brand an opponent as untrustworthy by asking questions like "Would you still support the candidate if you found out that he is an alcoholic?" Even when completely fabricated, such a question can create negative associations in the minds of voters. Statistics may be used to "analyze" data gathered by push polls (or by marketing campaigns that use similar techniques), but if judged by scientific standards, the results are neither meaningful nor valid.

electorate"—which numbers over 130 million people) from a carefully drawn sample (usually about 1500 respondents). Inferential statistics will occupy our attention in Part II of this book. *(For practice in describing different statistical applications, see problems 1.3 and 1.7.)*

Discrete and Continuous Variables

In the next chapter, you will begin to encounter some of the broad array of statistics available to the social scientist. One aspect of using statistics that can be puzzling is deciding when to use which statistic. You will learn specific guidelines as you go along, but I will introduce some basic and general guidelines at this point. The first of these concerns discrete and continuous variables; the second, covered in the next section, concerns level of measurement.

DISCRETE VARIABLES

A variable is said to be **discrete** if it has a basic unit of measurement that cannot be subdivided. For example, number of people per household is a discrete variable. The basic unit is people, a variable that will always be measured in whole numbers; you'll never find 2.7 people living in a specific household. The scores of discrete variables will be 0, 1, 2, 3, or some other whole integer. Other examples of discrete variables include number of siblings, children, or cars. To measure these variables, we count the number of units (people, cars, siblings) for each case (household, person) and record results in whole numbers.

CONTINUOUS VARIABLES

A variable is **continuous** if it has scores that can be subdivided infinitely (at least theoretically). One example of a continuous variable would be time, which can be measured in minutes, seconds, milliseconds (thousands of a second), nanoseconds (billionths of a second), or even smaller units. In a sense, when we measure a continuous variable, we are always approximating and rounding off the scores. We could report somebody's time in the 100-yard dash as 10.7 seconds or 10.732451 seconds, but, since time can be infinitely subdivided, we will never be able to report the exact time elapsed. Since we cannot work with infinitely long numbers, we must report the scores on continuous variables as if they were discrete.

The scores of discrete and continuous variables may look the same even though we measure and process them differently. This distinction is one of the most basic in statistics and will be one of the criteria we use to choose among various statistics and graphic devices. *(For practice in distinguishing between discrete and continuous variables, see problems 1.4 through 1.8.)*

Level of Measurement

A second basic and very important guideline for the selection of statistics is the **level of measurement,** or the mathematical nature of the variables under consideration. Variables at the highest level of measurement have numerical scores and can be analyzed with a broad range of statistics. Variables at the lowest level of measurement have "scores" that are really just labels, not numbers at all. Statistics that require numerical variables are not appropriate and, often, completely meaningless when used with non-numerical variables. When selecting statistics, you must be sure that the level of measurement of the variable matches the mathematical operations required to compute the statistic.

For example, consider the variables age (measured in years) and income (measured in dollars). Both of these variables have numerical scores and could be summarized with a statistic such as the mean, or average (e.g., "The average income of this city is $43,000." "The average age of students on this campus is 19.7").

The average would be meaningless as a way of describing gender or zip codes, variables with non-numerical scores. Your personal zip code might *look* like a number, but it is merely an arbitrary label that happens to be expressed in digits. These "numbers" cannot be added or divided, and statistics like the average cannot be applied to this variable: The average zip code of a group of people is a meaningless statistic.

Determining the level at which a variable has been measured is one of the first steps in any statistical analysis, and we will consider this matter at some length. We will consider the level of measurement of variables throughout this text and each time a statistical technique is introduced.

There are three levels of measurement. In order of increasing sophistication, they are nominal, ordinal, and interval-ratio. Each is discussed separately.

THE NOMINAL LEVEL OF MEASUREMENT

Variables measured at the nominal level have "scores" or categories that are not numerical. Examples of variables at this level include gender, zip code, race, religious affiliation, and place of birth. At this lowest level of measurement, the only mathematical operation permitted is comparing the relative sizes of the categories (e.g., "There are more females than males in this dorm"). The categories or scores of nominal-level variables cannot be ranked with respect to each other and cannot be added, divided, or otherwise manipulated mathematically. Even when the scores or categories are expressed in digits (like zip codes and street addresses), all we can do is compare relative sizes of categories (e.g., "The most common zip code on this campus is 22033"). The categories themselves are not a mathematical scale; they are different from each other but not more or less or higher or lower than each other. Males and females differ in terms of gender, but neither category has more or less gender than the other. In the same way, a zip code of 54398 is different from but not "higher" than a zip code of 13427.

Nominal variables are rudimentary, but there are criteria and procedures that we need to observe in order to measure them adequately. In fact, these criteria apply to variables measured at all levels, not just nominal variables. First, the categories of nominal-level variables must be mutually exclusive of each other so that there is no ambiguity about the category or score of any given case. Each case must have one and only one score. Second, the categories must be exhaustive. In other words, there must be a category—at least an "other" or miscellaneous category—for every possible score that might be found.

Third, the categories of nominal variables should be relatively homogeneous. That is, our categories should include cases that are truly comparable, or, to put it another way, we need to avoid lumping apples with oranges. There are no hard-and-fast guidelines for judging if a set of categories is appropriately homogeneous. The researcher must make that decision in terms of the specific purpose of the research, and categories that are too broad for some purposes may be perfectly adequate for others.

Table 1.1 demonstrates some errors in measuring religious preference. Scale A in the table violates the criterion of mutual exclusivity because of overlap between the categories Protestant and Episcopalian. Scale B is not exhaustive, because it does not provide a category for people with no religious preference (None) or for people who belong to religions other than the three listed. Scale C uses a category (Non-Protestant, which would include Catholics, Jews, Buddhists, and so forth) that seems too broad for meaningful research. Scale D represents the way religious preference is often measured in North America, but these categories may be too general for some research projects and not compre-hensive enough for others. For example, an investigation of issues that have strong moral and religious content (assisted suicide, abortion, and capital punish-ment, for example) might need to distinguish between the various Protestant denominations, and an effort to document religious diversity would need to add categories for Buddhists, Muslims, and numerous other religious faiths.

As is the case with zip codes, numerical labels are sometimes used to iden-tify the categories of a variable measured at the nominal level. This practice is especially common when the data are being prepared for computer analysis. For example, the various religions might be labeled with a 1 indicating Protestant, a 2 signifying Catholic, and so on. Remember that these scores are merely labels or names and have no numerical quality to them. They cannot be added, subtracted, multiplied, or divided. The only mathematical operation per-missible with nominal variables is counting and comparing the number of cases in each category of the variable.

THE ORDINAL LEVEL OF MEASUREMENT

Variables measured at the ordinal level are more sophisticated than nominal-level variables. They have scores or categories that can be ranked from high to low, so, in addition to classifying cases into categories, we can describe the categories in terms of "more or less" with respect to each other. Thus, with ordinal-level vari-ables, not only can we say that one case is different from another; we can also say that one case is higher or lower, more or less than another.

TABLE 1.1 Four Scales for Measuring Religious Preference

Scale A (not mutually exclusive)	Scale B (not exhaustive)	Scale C (not homogeneous)	Scale D (an adequate scale)
Protestant	Protestant	Protestant	Protestant
Episcopalian	Catholic	Non-Protestant	Catholic
Catholic	Jew		Jew
Jew			None
None			Other
Other			

For example, the variable socioeconomic status (SES) is usually measured at the ordinal level. The categories of the variable are often ordered according to the following scheme:

4. Upper class

3. Middle class

2. Working class

1. Lower class

Individual cases can be compared in terms of the categories into which they are classified. Thus, an individual classified as a 4 (upper class) would be ranked higher than an individual classified as a 2 (working class) and a lower-class person (1) would rank lower than a middle-class person (3). Other variables that are usually measured at the ordinal level include attitude and opinion scales, such as those that measure prejudice, alienation, or political conservatism.

The major limitation of the ordinal level of measurement is that a particular score represents only a position with respect to some other score. We can distinguish between high and low scores, but the distance between the scores cannot be described in precise terms. Although we know that a score of 4 is more than a score of 2, we do not know if it is twice as much as 2.

Since we don't know what the exact distances are from score to score on an ordinal scale, our options for statistical analysis are limited. For example, addition (and most other mathematical operations) assumes that the intervals between scores are exactly equal. If the distances from score to score are not equal, 2 + 2 might equal 3 or 5 or even 15. Thus, strictly speaking, statistics such as the average, or mean (which requires that the scores be added together and then divided by the number of scores), are not permitted with ordinal-level variables. The most sophisticated mathematical operation fully justified with an ordinal variable is the ranking of categories and cases (although, as we will see, it is not unusual for social scientists to take liberties with this criterion).

THE INTERVAL-RATIO LEVEL OF MEASUREMENT[4]

The scores of variables measured at the interval-ratio level are actual numbers. This means, first of all, that there are equal intervals from score to score. For example, asking people how old they are will produce an interval-ratio–level variable (age) because the unit of measurement (years) has equal intervals (the distance from year to year is 365 days). Similarly, if we ask people how many siblings they have, we would produce a variable with equal intervals: Two siblings are one more than one and 13 is 1 more than 12.

Second, interval-ratio variables have true zero points. That is, the score of zero for these variables is not arbitrary; it indicates the absence or complete lack of whatever is being measured. For example, the variable "number of siblings"

---

[4]Many statisticians distinguish between the interval level (equal intervals) and the ratio level (true zero point). I find the distinction unnecessarily cumbersome in an introductory text and will treat these two levels as one.

has a true zero point because it is possible to have no siblings at all. Similarly, it is possible to have zero years of education, no income at all, a score of zero on a multiple-choice test, and to be zero years old (although not for very long). Other examples of interval-ratio variables include number of children, life expectancy, and years married. All mathematical operations are permitted for variables measured at the interval-ratio level.

COMPARING THE THREE LEVELS OF MEASUREMENT

Table 1.2 summarizes this discussion by presenting the basic characteristics of the three levels of measurement. Note that the number of permitted mathematical operations increases as we move from nominal to ordinal to interval-ratio levels of measurement. Ordinal-level variables are more sophisticated and flexible than nominal-level variables, and interval-ratio-level variables permit the broadest range of mathematical operations.

Let us end this section by making four points. The first stresses the importance of level of measurement, and the next three discuss some common points of confusion in applying this concept.

First, knowing the level of measurement of a variable is crucial because it tells us which statistics are appropriate and useful. Not all statistics can be used with all variables. As shown in Table 1.2, different statistics require different mathematical operations. For example, computing an average requires addition and division, and finding a median (or middle score) requires that the scores be ranked from high to low. Addition and division are appropriate only for interval-ratio–level variables, and ranking is possible only for variables that are at least ordinal in level of measurement. Your first step in dealing with a variable and selecting appropriate statistics is *always* to determine its level of measurement.

Second, the distinction made earlier between discrete and continuous variables is a concern only for interval-ratio-level variables. Nominal- and ordinal-level variables are almost always discrete. That is, researchers measure these variables by asking respondents to select the single category that best describes them, as

TABLE 1.2   Basic Characteristics of the Three Levels of Measurement

Levels	Examples	Measurement Procedures	Mathematical Operations Permitted
Nominal	Sex, race, religion, marital status	Classification into categories	Counting number in each category, comparing sizes of categories
Ordinal	Social class, attitude, and opinion scales	Classification into categories plus ranking of categories with respect to each other	All of the foregoing plus judgments of "greater than" and "less than"
Interval-ratio	Age, number of children, income	All of the foregoing plus description of distances between scores in terms of equal units	All of the foregoing plus all other mathematical operations (addition, subtraction, multiplication, division, square roots, etc.)

TABLE 1.3 **Measuring a Nominal Variable (Marital Status) and an Ordinal Variable (Support for Capital Punishment) as Discrete Variables**

What is your marital status? Are you presently:		Do you support the death penalty for persons convicted of homicide?	
Score	Category	Score	Category
1	Married	1	Strongly support
2	Divorced	2	Slightly support
3	Separated	3	Neither support or oppose
4	Widowed	4	Slightly oppose
5	Single	5	Strongly oppose

illustrated in Table 1.3. The scores of these survey items, as they are stated here, are discrete because respondents must select only one category and because these scores cannot be subdivided.

Interval-ratio variables can be either discrete (number of times you've been divorced, which must be a whole integer) or continuous (age, which could be measured to the nanosecond or more). Remember that, since we cannot work with

---

READING STATISTICS 1: Introduction

By this point in your education you have developed an impressive array of skills for reading words. Although you may sometimes struggle with a difficult idea or stumble over an obscure meaning, you can comprehend virtually any written work that you are likely to encounter.

As you continue your education in the social sciences, you must develop an analogous set of skills for reading numbers and statistics. To help you reach a reasonable level of literacy in statistics, I have included a series of boxed inserts in this text titled "Reading Statistics." These appear in most chapters and discuss how statistical results are typically presented in the professional literature. Each installment includes an extract or quotation from the professional literature so that we can analyze a realistic example.

As you will see, professional researchers use a reporting style that is quite different from the statistical language you will find in this text. Space in research journals is expensive, and the typical research project requires the analysis of many variables. Thus, a large volume of information must be summarized in very few words. Researchers may express in a word or two a result or an interpretation that will take us a paragraph or more to state.

Because this is an introductory textbook, I have been careful to break down the computation and logic of each statistic and to explain what we are doing when we use statistics. In this text we will never be concerned with more than a few variables at a time. We will be able to take pages or even entire chapters to develop a statistical idea or concept. Thus, a major theme of these boxed inserts will be to summarize how our comparatively long-winded (but more careful) vocabulary is translated into the concise language of the professional researcher.

When you have difficulty reading words, you consult reference books (especially dictionaries) to help you analyze the elements (words) of the passage. When you have difficulty reading statistics, you should do exactly the same thing. I hope you will find this text a valuable reference book, but if you learn enough to be able to use any source to help you read statistics, this text will have fulfilled one of its major goals.

TABLE 1.4   Measuring Income at the Ordinal Level

Score	Income Range
1	Less than $24,999
2	$25,000 to $49,999
3	$50,000 to $99,999
4	More than $100,000

infinitely long numbers, continuous interval-ratio variables have to be rounded off and reported as if they were discrete. The distinction relates more to our options for appropriate statistics or graphs than to the appearance of the variables.

Third, in determining level of measurement, always examine the way in which the scores of the variable are *actually stated*. This is particularly a problem with interval-ratio variables that have been measured at the ordinal level. To illustrate, consider income as a variable. If we asked respondents to list their exact income in dollars, we would generate scores that are interval-ratio in level of measurement. Measured in this way, the variable would have a true zero point (no income at all) and equal intervals from score to score (one dollar). It is more convenient for respondents, however, to ask them simply to check the appropriate category from a broad list, as in Table 1.4. The four scores or categories in Table 1.4 are ordinal in level of measurement because they are unequal in size. It is common for researchers to sacrifice precision (income in actual dollars) in this way for the convenience of the respondents. You should be careful to look at the way in which the variable is measured before making a decision about its level of measurement.

ONE STEP AT A TIME   **Determining the Level of Measurement of a Variable***

1: Inspect the scores of the variable *as they are actually stated*, keeping in mind the definition of the three levels of measurement (see Table 1.2).

2: Change the order of the scores. Do they still make sense? If the answer is *yes*, the variable is **nominal**. If the answer is *no*, proceed to step 3. To Illustrate: Gender is a nominal variable, and its scores can be stated in any order: (1) male, (2) female or (1) female, (2) male. Each statement of the scores is just as sensible as the other. On a nominal-level variable, no score is higher or lower than any other score and the order in which they are stated is arbitrary.

3: Is the distance between the scores unequal or undefined? If the answer is *yes*, the variable is **ordinal**. If the answer is *no*, proceed to step 4. To illustrate, consider the scale measuring support for capital punishment presented in Table 1.3. People who "strongly"

support the death penalty are more in favor than people who "slightly" support it, but the distance from one level of support to the next (from a score of 1 to a score of 2) is undefined. We do not have enough information to ascertain how much more or less one score is than another.

4: If you answered *no* in steps 2 and 3, the variable is **interval-ratio**. Variables at this level have scores that are actual numbers: They have an order with respect to each other and are a defined, equal distance apart. They also have a true zero point. Examples of interval-ratio variables include age, income, and number of siblings.

*This system for determining level of measurement was suggested by Michael G. Bisciglia, Louisiana State University.

Fourth, there is a mismatch between the variables that are usually of most interest to social scientists (race, sex, marital status, attitudes, and opinions) and the most powerful and interesting statistics (such as the mean). The former are typically nominal or, at best, ordinal in level of measurement, but the more sophisticated statistics require measurement at the interval-ratio level. This mismatch creates some very real difficulties for social science researchers. On one hand, researchers will want to measure variables at the highest, most precise level of measurement. If income is measured in exact dollars, for example, researchers can make very precise descriptive statements about the differences between people. For example: "Ms. Smith earns $12,547 more than Mr. Jones." If the same variable is measured in broad, unequal categories, such as those in Table 1.4, comparisons between individuals would be less precise and provide less information: "Ms. Smith earns more than Mr. Jones."

On the other hand, given the nature of the disparity, researchers are more likely to treat variables as if they were higher in level of measurement than they actually are. In particular, variables measured at the ordinal level, especially when they have many possible categories or scores, are often treated as if they were interval-ratio and analyzed with the more powerful, flexible, and interesting statistics available at the higher level. This practice is common, but researchers should be cautious in assessing statistical results and developing interpretations when the level of measurement criterion has been violated.

In conclusion, level of measurement is a very basic characteristic of a variable, and we will always consider it when presenting statistical procedures. Level of measurement is also a major organizing principle for the material that follows, and you should make sure that you are familiar with these guidelines. *(For practice in determining the level of measurement of a variable, see problems 1.4 through 1.8.)*

SUMMARY

1. The purpose of statistics is to organize, manipulate, and analyze data so that researchers can test theories and answer questions. Along with theory and methodology, statistics are a basic tool used by social scientists to enhance their understanding of the social world.

2. There are two general classes of statistics. Descriptive statistics are used to summarize the distribution of a single variable and the relationships between two or more variables. We use inferential statistics to generalize to populations from random samples.

3. Variables may be either discrete or continuous. They may be measured at any of three different levels. At the nominal level, we can compare category sizes. At the ordinal level, scores can be ranked from high to low. At the interval-ratio level, all mathematical operations are permitted. Interval-ratio-level variables can be either discrete or continuous. Variables at the nominal or ordinal level are almost always discrete.

GLOSSARY

Continuous variable. A variable with a unit of measurement that can be subdivided infinitely.

Data. Information expressed as numbers.

Data reduction. Summarizing many scores with a few statistics.

Dependent variable. A variable that is identified as an effect or outcome. The dependent variable is thought to be caused by the independent variable.

Descriptive statistics. The branch of statistics concerned with (1) summarizing the distribution of a single variable or (2) measuring the relationship between two or more variables.

Discrete variable. A variable with a basic unit of measurement that cannot be subdivided.

Hypothesis. A specific statement, derived from a theory, about the relationship between variables.

Independent variable. A variable that is identified as a cause. The independent variable is thought to cause the dependent variable.

Inferential statistics. The branch of statistics concerned with making generalizations from samples to populations.

Level of measurement. The mathematical characteristic of a variable and the major criterion for selecting statistical techniques. Variables can be measured at any of three levels, each permitting certain mathematical operations and statistical techniques. The characteristics of the three levels are summarized in Table 1.2.

Measures of association. Statistics that summarize the strength and direction of the relationship between variables.

Population. The total collection of all cases in which the researcher is interested.

Quantitative research. Research projects that collect data or information in the form of numbers.

Research. Any process of gathering information systematically and carefully to answer questions or test theories. Statistics are useful for research projects that collect numerical information or data.

Sample. A carefully chosen subset of a population. In inferential statistics, information is gathered from a sample and then generalized to a population.

Statistics. A set of mathematical techniques for organizing and analyzing data.

Theory. A generalized explanation of the relationship between two or more variables.

Variable. Any trait that can change values from case to case.

PROBLEMS

1.1 In your own words, describe the role of statistics in the research process. Using the "wheel of science" as a framework, explain how statistics link theory with research.

1.2 Find a research article in any social science journal. Choose an article on a subject of interest to you, and don't worry about being able to understand all of the statistics that are reported.
   a. How much of the article is devoted to statistics?
   b Is the research based on a sample from some population? How large is the sample? How were subjects or cases selected? Can the findings be generalized to some population?
   c. What variables are used? Which are independent and which are dependent? For each variable, determine the level of measurement and whether the variable is discrete or continuous.
   d. What statistical techniques are used? Try to follow the statistical analysis and see how much you can understand. Save the article and read it again after you finish this course and see if you do any better.

1.3 Distinguish between descriptive and inferential statistics. Describe a research situation that would use each type.

1.4 For each of the following items from a public-opinion survey, indicate the level of measurement and whether the variable will be discrete or continuous. *(HINT: Remember that only interval-ratio-level variables can be continuous.)*
   a. In what country were you born? _____
   b. What is your age? _____
   c. How many years of school have you completed? _____
   d. What is your occupation? _____
   e. If you were asked to use one of these four names for your social class, which would you say you belonged in?
   _____ Upper _____ Middle
   _____ Working _____ Lower
   f. What is your grade-point average? _____
   g. What is your major? _____
   h. The only way to deal with the drug problem is to legalize all drugs.
   _____ Strongly agree
   _____ Agree
   _____ Undecided
   _____ Disagree
   _____ Strongly disagree
   i. What is your astrological sign? _____
   j. How many brothers and sisters do you have? _____

1.5 Following are brief descriptions of how researchers measured a variable. For each situation, determine the level of measurement of the variable and whether it is continuous or discrete.

a. Race or ethnicity. Respondents were asked to check all that apply:

_____ Black

_____ White

_____ Hispanic

_____ Asian or Pacific Islander

_____ Native American

_____ Other (Please specify: _____)

b. Honesty. Subjects were observed as they passed by a spot where an apparently lost wallet was lying. The wallet contained money and complete identification. Subjects were classified into one of the following categories:

_____ returned the wallet with money;

_____ returned the wallet but kept the money;

_____ did not return the wallet.

c. Social class. Subjects were asked about their family situation when they were 16 years old. Was their family:

_____ Very well off compared to other families?

_____ About average?

_____ Not so well off?

d. Education. Subjects were asked how many years of schooling they had completed.

e. Racial integration on campus. Students were observed during lunchtime at the cafeteria for a month. The number of students sitting with students of other races was counted for each meal period.

f. Number of children. Subjects were asked: "How many children have you ever had? Please include any that may have passed away."

g. Student seating patterns in classrooms. On the first day of class, instructors noted where each student sat. Seating patterns were re-measured every two weeks until the end of the semester. Each student was classified as:

_____ same seat as last measurement;

_____ adjacent seat;

_____ different seat, not adjacent;

_____ absent.

h. Physicians per capita. The number of physicians was counted in each of 50 cities. The researchers used population data to compute the number of physicians per capita.

i. Physical attractiveness. A panel of 10 judges rated each of 50 photos of a mixed-race sample of males and females for physical attractiveness on a scale from 0 to 20, with 20 being the highest score.

j. Number of accidents. The number of traffic accidents per year for each of 20 intersections was recorded. Also, each accident was rated as:

_____ minor damage, no injuries;

_____ moderate damage, personal injury requiring hospitalization;

_____ severe damage and injury.

1.6 The following are descriptions of research projects. For each project, identify the independent and dependent variables, classify them in terms of level of measurement, and determine whether they are discrete or continuous.

a. For a research project in a political science course, a student has collected information about the quality of life and the degree of political democracy in 50 nations. Specifically, she used infant mortality rates to measure quality of life and the percentage of all adults who are permitted to vote in national elections as a measure of democratization. Her hypothesis is that quality of life is higher in more democratic nations.

b. A highway engineer wonders if a planned increase in the speed limit on a heavily traveled local avenue will result in any change in number of accidents. He plans to collect information on traffic volume, number of accidents, and number of fatalities for the six-month periods before and after the change.

c. Students are planning a program to promote "safe sex" and awareness of other health concerns for college students. To measure the effectiveness of the program, they plan to give a survey measuring knowledge to a random sample of students before and after the program.

d. A graduate student asks 500 female students if they have experienced any sexual harassment on campus. Each student is asked to estimate the frequency of these incidents as either "often, sometimes, rarely, or never." The researcher also gathers data on age and major to see if there is any connection between these variables and frequency of sexual harassment.

e. A supervisor in the Solid Waste Management Division of city government is assessing two different methods of trash collection. One area of the city is served by trucks with two-man crews who do "backyard" pickups, and the rest of the city is served by "hi-tech" single-person trucks with curbside pickup. The assessment measures include the number of complaints received from the two different areas over a six-month period, the amount of time per day required to service each area, and the cost per ton of trash collected.

f. Does tolerance for diversity vary by race or ethnicity? Samples of white, black, Asian, Hispanic, and Native Americans have been given a survey that measures their interest in and appreciation of cultures and groups other than their own.

g. States have drastically cut their budgets for mental health care. Will this increase the number of homeless people? A researcher contacts a number of agencies serving the homeless in each state and develops an estimate of the size of the population before and after the cuts.

h. The adult bookstore near campus has been raided and closed by the police. Your social research class has decided to poll a sample of students to find out: if he or she supports or opposes the closing of the store, how many times each has visited the store, and if he or she agrees or disagrees that "pornography causes sexual assaults on women." The class also collects information on the sex, age, religion, political philosophy, and major of the students to see if opinions are related to these characteristics.

1.7 For each of the following research situations, identify the level of measurement of all variables and indicate whether they are discrete or continuous. Also, decide which statistical applications are used: descriptive statistics (single variable), descriptive statistics (two or more variables), or inferential statistics. Remember that it is quite common for a given situation to require more than one type of application.

a. The administration of your university is proposing a change in parking policy. You select a random sample of students and ask each one if he or she favors or opposes the change. You then use the results to estimate the support for the change in the entire student body.

b. You ask everyone in your social research class for their highest grade in any math course and the grade on a recent statistics test. You compare the two sets of scores to see if there is any relationship.

c. Your aunt is running for mayor and hires you to question a sample of voters about their concerns. Specifically, she wants a profile of voters that will tell her what percentage belong to each political party, what percentage are male or female, and what percentage favor or oppose the widening of the main street in town.

d. Several years ago, a state reinstituted the death penalty for first-degree homicide. Did this reduce the homicide rate? A researcher has gathered information on the number of homicides in the state for the two-year periods before and after the change.

e. A local automobile dealer is concerned about customer satisfaction. He wants to mail a survey form to all customers for the past year and ask them if they are satisfied, very satisfied, or not satisfied with their purchases.

1.8 For each of the first 20 items in the General Social Survey (see Appendix G), indicate the level of measurement and whether the variable is continuous or discrete.

YOU ARE THE RESEARCHER

Introduction

The best way—maybe the only way—to learn statistics and to appreciate their importance is to apply and use them. This includes selecting the correct statistic for a given purpose, doing the calculations, and interpreting the result. I have included extensive end-of-chapter problems to give you multiple opportunities to select and calculate statistics and say what they mean. Most of these problems have been written so that they can be solved with just a simple hand calculator. I've purposely kept the number of cases unrealistically low so that the tedium of mere calculation would not interfere with the learning process. These problems present an important and useful opportunity to develop your statistical skills.

As important as they are, these end-of-chapter problems are simplified and several steps removed from the complex realities of social science research. To provide a more realistic statistical experience, I have included a feature called "You Are the Researcher" in which you will walk through many of the steps of a research project, making decisions about how to apply your growing knowledge of research and statistics and interpreting the statistical output you generate.

To conduct these research projects, you will analyze a shortened version of the 2008 General Social Survey (GSS). Please visit **www.cengagebrain.com** to download this database. At the CengageBrain.com home page, search for the ISBN of your title (from the back cover of your book) using the search box at the top of the page. This will take you to the product page where datasets and other free companion resources can be found. The GSS is a public-opinion poll that has been conducted on nationally representative samples of citizens of the United States since 1972. The full survey includes hundreds of questions covering a broad range of social and political issues. The version supplied with this text has a limited number of variables and cases but is still actual, "real-life" data, so you have the opportunity to practice your statistical skills in a more realistic context.

Even though the version of the GSS we use for this text is shortened, it is still a large data set, with almost 1500 respondents and 50 different variables, too large for even the most advanced hand calculator. To analyze the GSS, you will learn how to use a computerized statistical package called the Statistical Package for the Social Sciences (SPSS). A statistical package is a set of computer programs designed to analyze data. The advantage of these packages is that, since the programs are already written, you can capitalize on the power of the computer with minimal computer literacy and virtually no programming experience. Be sure to read Appendix F before attempting any data analysis.

In most of the research exercises, which begin in Chapter 2, you will make the same kinds of decisions as professional researchers and move through some of the steps of a research project. You will select variables and statistics, generate and analyze output, and express your conclusions. When you finish these exercises, you will be well prepared to conduct your own research project (within limits, of course) and perhaps make a contribution to the ever-growing social science research literature.

Part I   Descriptive Statistics

Part I consists of four chapters, each devoted to a different application of univariate descriptive statistics. Chapter 2 covers "basic" descriptive statistics, including percentages, ratios, rates, frequency distributions, and graphs. This material is relatively elementary and at least vaguely familiar to most people. Although these statistics are "basic," they are not necessarily simple or obvious, and the explanations and examples should be considered carefully before attempting the end-of-chapter problems or using them in actual research.

Chapters 3 and 4 cover measures of central tendency and dispersion, respectively. Measures of central tendency describe the typical case or average score (e.g., the mean), while measures of dispersion describe the amount of variety or diversity among the scores (e.g., the range, or the distance from the high score to the low score). These two types of statistics are presented in separate chapters to stress the point that centrality and dispersion are independent, separate characteristics of a variable. You should realize, however, that both measures are necessary and commonly reported together (along with some of the statistics presented in Chapter 2). To reinforce this idea, many of the problems at the end of Chapter 4 require the computation of a measure of central tendency from Chapter 3.

Chapter 5 is a pivotal chapter in the flow of the text. It takes some of the statistics from Chapters 2 through 4 and applies them to the normal curve, a concept of great importance in statistics. The normal curve is a line chart (see Chapter 2), which can be used to describe the position of scores using means (Chapter 3) and standard deviations (Chapter 4). Chapter 5 also uses proportions and percentages (Chapter 2).

In addition to its role in descriptive statistics, the normal curve is a central concept in inferential statistics, the topic of Part II of this text. Thus, Chapter 5 ends the presentation of univariate descriptive statistics and lays essential groundwork for the material to come.

2

Basic Descriptive Statistics
Percentages, Ratios and Rates,
Tables, Charts, and Graphs

By the end of this chapter, you will be able to

1. Explain the purpose of descriptive statistics in making data comprehensible.
2. Compute and interpret percentages, proportions, ratios, rates, and percentage change.
3. Construct and analyze frequency distributions for variables at each of the three levels of measurement.
4. Construct and analyze bar and pie charts, histograms, and line graphs.

R esearch results do not speak for themselves. They must be organized and manipulated so that whatever meaning they have can be quickly and easily understood by the researcher and by his or her readers. Researchers use statistics to clarify their results and communicate effectively. In this chapter, we consider some commonly used techniques for presenting research results, including percentages, rates, tables, and graphs. These univariate descriptive statistics are not mathematically complex (although they are not as simple as they might seem at first glance), but they are extremely useful for presenting research results clearly and concisely.

Percentages and Proportions

Consider the following statement: "Of the 269 cases handled by the court, 167 resulted in prison sentences of five years or more." While there is nothing wrong with this statement, the same fact could have been more clearly conveyed if it had been reported as a percentage: "About 62% of all cases resulted in prison sentences of five or more years."

Percentages supply a frame of reference for reporting research results, in the sense that they standardize the raw data to the base 100. The mathematical definition of a **percentage** is

FORMULA 2.1

$$\text{Percentage: } \% = \left(\frac{f}{N}\right) \times 100$$

where f = frequency, or the number of cases in any category
N = the number of cases in all categories

TABLE 2.1   Disposition of 269 Criminal Cases (fictitious data)*

Sentence	Frequency (f)	Percentage (%)
Five years or more	167	62.08
Less than five years	72	26.77
Suspended	20	7.44
Acquitted	10	3.72
	$N = 269$	100.01%

*The slight discrepancy in the totals of the percentage column is due to rounding error. See the Preface (Basic Mathematical Review) for more on rounding.

To illustrate the computation of percentages, consider the data presented in Table 2.1. How can we find the percentage of cases that resulted in sentences of five years or more? Note that there are 167 cases in the category ($f = 167$) and a total of 269 cases in all ($N = 269$). So

$$\text{Percentage } (\%) = \left(\frac{f}{N}\right) \times 100 = \left(\frac{167}{269}\right) \times 100$$

$$= (0.6208) \times 100 = 62.08\%$$

Using the same procedures, we can also find the percentage of cases that resulted in sentences of less than five years:

$$\text{Percentage } (\%) = \left(\frac{f}{N}\right) \times 100 = \left(\frac{72}{269}\right) \times 100$$

$$= (0.2677) \times 100 = 26.77\%$$

Percentages are easier to read and comprehend than raw frequencies. This advantage is particularly obvious when comparing groups of different sizes. For example, based on the information presented in Table 2.2, which college has the higher relative number of social science majors? Because the total enrollments are so different, comparisons are difficult to make from the raw frequencies. Percentages eliminate the difference in size of the two campuses by standardizing both distributions to the base of 100. The same data are presented in percentages in Table 2.3.

TABLE 2.2   Declared Major Fields on Two College Campuses (fictitious data)

Major	College A	College B
Business	103	3120
Natural sciences	82	2799
Social sciences	137	1884
Humanities	93	2176
	$N = 415$	$N = 9979$

TABLE 2.3 Declared Major Fields on Two College Campuses (fictitious data)

Major	College A	College B
Business	24.82%	31.27%
Natural sciences	19.76%	28.05%
Social sciences	33.01%	18.88%
Humanities	22.41%	21.81%
	100.00%	100.01%
	(415)	(9979)

Table 2.3 makes it easier to identify differences as well as similarities between the two colleges. College A has a much higher percentage of social science majors (even though the absolute number of social science majors is less than at College B) and about the same percentage of humanities majors. How would you describe the differences in the remaining two major fields?

Social scientists use **proportions** as well as percentages. Proportions vary from 0.00 to 1.00: They standardize results to a base of 1.00 instead of to the base of 100 for percentages. When we find a percentage, we divide the number of cases in a category (f) by the total number of all cases (N) and multiply the result by 100. To find a proportion, we do exactly the same division but do *not* multiply by 100.

FORMULA 2.2

$$\text{Proportion} = \frac{f}{N}$$

Percentages can be converted to proportions by dividing by 100 and, conversely, proportions can be converted to percentages by multiplying by 100. The two statistics are equivalent expressions of the same message and are interchangeable. For example, we could state the relative number of sentences of more than five years in Table 2.1 as a proportion:

$$\text{Proportion} = \frac{f}{N} = \frac{167}{269} = 0.62$$

How can we choose between these statistics? Percentages are easier for most people (including statisticians) to comprehend and generally would be preferred, especially when our goal is to express results clearly and quickly. Proportions are used less frequently, generally when we are working with probabilities (see Chapter 5). The preference for percentages is based solely on ease of communication. Proportions and percentages are equally valid ways of expressing results and you should be proficient in both. *(For practice in computing and interpreting percentages and proportions, see problems 2.1 and 2.2.)*

Here are some further guidelines on the use of percentages and proportions.

1. When the number of cases is small (say, fewer than 20), it is usually preferable to report the actual frequencies rather than percentages or proportions. With a small number of cases, the percentages can change drastically with relatively minor changes in the size of the data set. For example, if you begin

---

Applying Statistics 2.1: Communicating with Statistics

Not long ago, in a large social service agency, the following conversation took place between the executive director of the agency and a supervisor of one of the divisions.

> *Executive director:* Well, I don't want to seem abrupt, but I've only got a few minutes. Tell me, as briefly as you can, about this staffing problem you claim to be having.
>
> *Supervisor:* Ma'am, we just don't have enough people to handle our workload. Of the 177 full-time employees of the agency, only 50 are in my division. Yet 6231 of the 16,722 cases handled by the agency last year were handled by my division.
>
> *Executive director (smothering a yawn):* Very interesting. I'll certainly get back to you on this matter.

How could the supervisor have presented his case more effectively? Because he wants to compare two sets of numbers (his staff versus the total staff and the workload of his division versus the total workload of the agency), proportions or percentages would be a more forceful way of presenting results. What if the supervisor had said, "Only 28.25% of the staff is assigned to my division, but we handle 37.26% of the total workload of the agency"? Is this a clearer message?

The first percentage is found by

$$\% = \left(\frac{f}{N}\right) \times 100 = \frac{50}{177} \times 100$$

$$= (.2825) \times 100 = 28.25\%$$

and the second percentage is found by

$$\% = \left(\frac{f}{N}\right) \times 100 = \left(\frac{6231}{16,722}\right) \times 100$$

$$= (.3726) \times 100 = 37.26\%$$

with a group of 10 males and 10 females (that is, 50% of each gender) and then add another female, the percentage distributions will change to 52.38% female and 47.62% male. Of course, as the number of observations increases, each additional case will have a smaller impact. If we started with 500 males and females and then added one more female, the percentage of females would change by only a tenth of a percent (from 50% to 50.10%).

2. Always report the number of observations along with proportions and percentages. This permits the reader to judge the adequacy of the sample size and, conversely, helps to prevent the researcher from lying with statistics. Statements like "Two out of three people questioned prefer courses in statistics to any other course" might impress you, but the claim would lose its gloss if you learned that only three people were tested. *You should be extremely suspicious of reports that fail to mention the number of cases that were tested.*

3. Percentages and proportions can be calculated for variables at the ordinal and nominal levels of measurement, in spite of the fact that they require division. This is not a violation of the level-of-measurement guideline (see Table 1.2). Percentages and proportions do not require the division of the *scores* of the variable (as would be the case in computing the average score on a test, for example). Rather, the *number of cases* in a particular category (*f*) is divided by the total number of cases in the sample (*N*). When we make a statement like "43% of the sample is female," we are merely expressing the relative size of a category (female) of the variable (gender) in a convenient way.

ONE STEP AT A TIME **Finding Percentages and Proportions**

1: Determine the values for f (number of cases in a category) and N (number of cases in all categories). Remember that f will be the number of cases in a *specific category* (e.g., males on your campus) and N will be the number of cases in *all* categories (e.g., all students, males and females, on your campus) and that f will be smaller than N, except when the category

and the entire group are the same (e.g., when all students are male). Proportions cannot exceed 1.00, and percentages cannot exceed 100.00%.

2: For a proportion, divide f by N.

3: For a percentage, multiply the value you calculated in step 2 by 100.

Ratios, Rates, and Percentage Change

Ratios, rates, and percentage change provide some additional ways of summarizing results simply and clearly. Although they are similar to each other, each statistic has a specific application and purpose.

RATIOS

Ratios are especially useful for comparing the relative size of different categories of a variable and are determined by dividing the frequency in one category by the frequency in another. The formula for a ratio is

FORMULA 2.3

$$\text{Ratio} = \frac{f_1}{f_2}$$

where f_1 = the number of cases in the first category
f_2 = the number of cases in the second category

To illustrate the use of ratios, suppose that you were interested in the relative sizes of the various religious denominations and found that a particular community

---

Applying Statistics 2.2: Ratios

In Table 2.2, how many natural science majors are there compared to social science majors at College B? This question could be answered with frequencies, but a more easily understood way of expressing the answer would be with a ratio. The ratio of natural science to social science majors would be

$$\text{Ratio} = \frac{f_1}{f_2} = \frac{2799}{1884} = 1.49$$

For every social science major, there are 1.49 natural science majors at College B.

included 1370 Protestant families and 930 Catholic families. To find the ratio of Protestants (f_1) to Catholics (f_2), divide 1370 by 930:

$$\text{Ratio} = \frac{f_1}{f_2} = \frac{1370}{930} = 1.47$$

The ratio of 1.47 means that there are 1.47 Protestant families for every Catholic family.

Ratios can be very economical ways of expressing the relative size of two categories. In our example, it is obvious from the raw data that Protestants outnumber Catholics, but ratios tell us exactly how much one category outnumbers the other.

Ratios are often multiplied by some power of 10 to eliminate decimal points. For example, the ratio just computed might be multiplied by 100 and reported as 147 instead of 1.47. This would mean that there are 147 Protestant families for every 100 Catholic families in the community. To ensure clarity, the comparison units for the ratio are often expressed as well. Based on a unit of ones, the ratio of Protestants to Catholics would be expressed as 1.47:1. Based on hundreds, the same statistic might be expressed as 147:100. *(For practice in computing and interpreting ratios, see problems 2.1 and 2.2.)*

RATES

Rates provide still another way of summarizing the distribution of a single variable. Rates are defined as the number of actual occurrences of some phenomenon divided by the number of possible occurrences per some unit of time. Rates are usually multiplied by some power of 10 to eliminate decimal points. For example, the crude death rate for a population is defined as the number of deaths in that population (actual occurrences) divided by the number of people in the population (possible occurrences) per year. This quantity is then multiplied by 1000. The formula for the crude death rate can be expressed as

$$\text{Crude death rate} = \frac{\text{Number of Deaths}}{\text{Total population}} \times 1000$$

If there were 100 deaths during a given year in a town of 7000, the crude death rate for that year would be

$$\text{Crude death rate} = \frac{100}{7000} \times 1{,}000 = (0.01429) \times 1{,}000 = 14.29$$

Or, for every 1000 people, there were 14.29 deaths during this particular year. In the same way, if a city of 237,000 people experienced 120 auto thefts during a particular year, the auto theft rate would be

$$\text{Auto theft rate} = \frac{120}{237{,}000} \times 100{,}000 = (0.0005063) \times 100{,}000 = 50.63$$

Applying Statistics 2.3: Rates

In 2005, there were 2500 births in a city of 167,000. In 1965, when the population of the city was only 133,000, there were 2700 births. Is the birthrate rising or falling? Although this question can be answered from the preceding information, the trend in birthrates will be much more obvious if we compute birthrates for both years. Like crude death rates, crude birthrates are usually multiplied by 1000 to eliminate decimal points. For 1965:

$$\text{Crude birthrate} = \frac{2700}{133,000} \times 1000 = 20.30$$

In 1965, there were 20.30 births for every 1000 people in the city. For 2005:

$$\text{Crude birthrate} = \frac{2500}{167,000} \times 1000 = 14.97$$

In 2005, there were 14.97 births for every 1000 people in the city. With the help of these statistics, the decline in the birthrate is clearly expressed.

Or, for every 100,000 people, there were 50.63 auto thefts during the year in question. *(For practice in computing and interpreting rates, see problems 2.3 and 2.4a.)*

PERCENTAGE CHANGE

Measuring social change, in all its variety, is an important task for all social sciences. One very useful statistic for this purpose is the **percentage change**, which tells us how much a variable has increased or decreased over a certain span of time.

To compute this statistic, we need the scores of a variable at two different points in time. The scores could be in the form of frequencies, rates, or percentages. The percentage change will tell us how much the score has changed at the later time relative to the earlier time. Using death rates as an example once again, imagine a society suffering from a devastating outbreak of disease in which the death rate rose from 16 deaths per 1000 population in 1995 to 24 deaths per 1000 in 2005. Clearly, the death rate is higher in 2005, but by how much relative to 1995?

The formula for the percent change is

FORMULA 2.4
$$\text{Percent change} = \left(\frac{f_2 - f_1}{f_1} \right) \times 100$$

where f_1 = first score, frequency, or value

f_2 = second score, frequency, or value

In our example, f_1 is the death rate in 1995 ($f_1 = 16$) and f_2 is the death rate in 2005 ($f_2 = 24$). The formula tells us to subtract the earlier score from the later and then divide by the earlier score. The result expresses the size of the change in

Applying Statistics 2.4: Percentage Change

The American family has been changing rapidly over the past several decades. One major change has been an increase in the number of married women and mothers with young children who hold jobs outside the home. For example, in 1975, 36.7% of women with children under the age of 6 worked outside the home. In 2007, this percentage had risen to 72.2%.* How large is this change?

Obviously, the 2007 percentage is much higher. Calculating the percentage change will tell us the exact magnitude of the change. The 1975 percentage is f_1 and the 2007 figure is f_2, so

$$\text{Percent change} = \left(\frac{72.2 - 36.7}{36.7}\right)$$

$$= \left(\frac{35.5}{36.7}\right)$$

$$= 96.73\%$$

In the 30-year period between 1975 and 2005, the percentage of women with children younger than 6 who worked outside the home increased by 96.73%. This is an extremely large change (approaching 100%, or double the earlier percentage) in a short time frame and signals major changes in this social institution.

*U.S. Bureau of the Census. 2007. *Statistical Abstract of the United States, 2007.* Washington, DC: Government Printing Office. p. 380.

scores $(f_2 - f_1)$ relative to the score at the earlier time (f_1). The value is then multiplied by 100 to express the change in the form of a percentage:

$$\text{Percent change} = \left(\frac{24 - 16}{16}\right) \times 100 = \left(\frac{8}{16}\right) \times 100 = (.50) \times 100 = 50\%$$

The death rate in 2005 is 50% higher than in 1995. This means that the 2005 rate was equal to the 1995 rate *plus* half of the earlier score. If the rate had risen to 32 deaths per 1000, the percent change would have been 100% (the rate would have doubled), and if the death rate had fallen to 8 per 1000, the percent change would have been –50%. Note the negative sign: It means that the death rate has decreased by 50%. The 2005 rate would have been half the size of the 1995 rate.

An additional example should make the computation and interpretation of the percentage change clearer. Suppose we wanted to compare the projected population growth rates for various nations over a 50-year period. The necessary information is presented in Table 2.4, which shows the actual population for each nation in 2000 and the projected population for 2050. The "Increase/Decrease" column shows how many people will be added or lost over the 50-year time span.

Casual inspection will give us some information about population trends. For example, compare the "Increase/Decrease" column for Nigeria and the United States. These societies are projected to add roughly similar numbers of people (about 156 million for the United States, a little less for Nigeria), but, since the U.S. 2000 population is more than twice the size of the population of Nigeria, its percent change will be much lower (about 56% vs. about 115%).

TABLE 2.4   Projected Population Growth for Six Nations, 2000–2050

Nation	Population, 2000 (f_1)	Population, 2050 (f_2)	Increase/Decrease ($f_2 - f_1$)	Percent Change
China	1,263,638,000	1,303,723,000	40,085,000	3.17
United States	282,171,936	439,010,253	156,838,317	55.58
Canada	31,099,561	41,135,648	10,036,087	32.27
Mexico	99,926,620	147,907,650	47,981,030	48.02
Italy	57,719,337	50,389,841	−7,329,496	−12.70
Nigeria	123,178,818	264,262,405	141,083,587	114.54

Source: U.S. Bureau of the Census: http://www.census.gov/ipc/www/idb/informationGateway.php.

Calculating percent change will make comparisons more precise. The right-hand column shows the percent change in projected population for each nation. These values were computed by subtracting the 2000 population (f_1) from the 2050 population (f_2), dividing by the 2000 population, and multiplying by 100.

Although China has the largest population of these six nations, it will grow at the slowest rate (3.17%). The United States and Mexico will increase by about 50% (in 2050, their populations will be half again larger than in 2000) and Canada will grow by about one-third. Italy will actually lose people and its population will decline by over 12%. Nigeria has by far the highest growth rate: It will add the most people (relatively speaking) and its population will increase

---

ONE STEP AT A TIME   **Finding Ratios, Rates, and Percent Change**

To Find Ratios

1: Determine the values for f_1 and f_2. The value for f_1 will be the number of cases in the first category (e.g., the number of males on your campus), and the value for f_2 will be the number of cases in the second category (e.g., the number of females on your campus).

2: Divide the value of f_1 by the value of f_2.

3: You may multiply the value you calculated in step 2 by some power of 10 when reporting results.

To Find Rates

1: Determine the number of actual occurrences (e.g., births, deaths, homicides, assaults). This value will be the numerator of the formula.

2: Determine the number of possible occurrences. This value will usually be the total population for

the area in question and will be the denominator of the formula.

3: Divide the number of actual occurrences (step 1) by the number of possible occurrences (step 2).

4: Multiply the value you calculated in step 3 by some power of 10. Conventionally, birthrates and death rates are multiplied by 1000 and crime rates are multiplied by 100,000.

To Find Percent Change

1: Determine the values for f_1 and f_2. The former will be the score at time 1 (the earlier time) and the latter will be the score at time 2 (the later time).

2: Subtract f_1 from f_2.

3: Divide the quantity you found in step 2 by f_1.

4: Multiply the quantity you found in step 3 by 100.

STATISTICS	The Browns live in Montana and are preparing to send their daughter Jill to the University of California at Los Angeles (UCLA), the alma mater of both parents. Among the myriad decisions they must make is the question of transportation: Should Jill have a car with her at school? Mr. Brown thinks that California roads are much too dangerous: He points out that California suffered 3974 traffic fatalities in 2007 while Montana had only 277.
IN EVERYDAY LIFE	His daughter is ready for this argument: She points out that California's population (36,378,000) is many times larger than Montana's (957,000). How might a rate clarify the relative dangers of driving in the two states? Would a comparison of rates settle the argument? What sociological factors might account for the difference in the rates?

in size by over 100%. This means that in 2050 the population of Nigeria will be more than twice its 2000 size. *(For practice in computing and interpreting percent change, see problem 2.4b.)*

Frequency Distributions

Frequency distributions are tables that report the number of cases in each category of a variable. They are very helpful and commonly used ways of organizing and working with data. In fact, the construction of frequency distributions is almost always the first step in any statistical analysis.

To illustrate the usefulness of frequency distributions and to provide some data for examples, assume that the counseling center at a university is assessing its effectiveness. Any realistic evaluation research would collect a variety of information from a large group of students, but, for the sake of this example, we will confine our attention to just four variables and 20 students. The data are reported in Table 2.5.

Table 2.5 includes an unrealistically low number of cases, but it is still difficult to discern patterns. For example, try to ascertain the general level of satisfaction using this table. You may be able to do so with just 20 cases, but it will take some time and effort. Imagine the difficulty with 50 cases or 100 cases presented in this fashion. Clearly, the data need to be organized so that the researcher (and his or her audience) can easily understand the distribution of the variables. Frequency distributions are simple but extremely useful tools for organizing data and improving communication.

One general rule that applies to all frequency distributions, regardless of the level of measurement of the variable, is that the categories of the frequency distribution must be exhaustive and mutually exclusive. In other words, the categories must be stated so that each case to be sorted into one and only one category.

Beyond this rule, there are only guidelines to help you construct useful frequency distributions. The researcher has a fair amount of discretion in stating the categories of the frequency distribution (especially with variables measured at the interval-ratio level). I will identify the issues but, ultimately, the guidelines are only aids and helpful suggestions for decision making. As always,

TABLE 2.5   **Datat from Counseling Center Survey**

Student	Sex	Marital Status	Satisfaction with Services	Age
A	Male	Single	4	18
B	Male	Married	2	19
C	Female	Single	4	18
D	Female	Single	2	19
E	Male	Married	1	20
F	Male	Single	3	20
G	Female	Married	4	18
H	Female	Single	3	21
I	Male	Single	3	19
J	Female	Divorced	3	23
K	Female	Single	3	24
L	Male	Married	3	18
M	Female	Single	1	22
N	Female	Married	3	26
O	Male	Single	3	18
P	Male	Married	4	19
Q	Female	Married	2	19
R	Male	Divorced	1	19
S	Female	Divorced	3	21
T	Male	Single	2	20

the researcher has the final responsibility for making sensible decisions and presenting his or her data in a meaningful way.

FREQUENCY DISTRIBUTIONS FOR VARIABLES MEASURED AT THE NOMINAL LEVEL

Constructing frequency distributions for nominal-level variables is typically very straightforward. Count the number of times each category or score of the variable occurred and then display the frequencies in table format. Table 2.6 displays a frequency distribution for the variable "sex" from the counseling center survey. For purposes of illustration, a column for tallies has been included in this table to illustrate how the score could be counted. This column would *not* be included in the final form of the frequency distribution. Note that the table has a

TABLE 2.6   **Sex of Respondents, Counseling Center Survey**

Sex	Tallies	Frequency (*f*)
Male	### ###	10
Female	### ###	10
		$N = 20$

TABLE 2.7 Marital Status of Respondents, Counseling Center Survey

Status	Frequency (*f*)
Single	10
Married	7
Divorced	3
	N = 20

TABLE 2.8 Marital Status of Respondents, Counseling Center Survey

Status	Frequency (*f*)
Married	7
Not married	13
	N = 20

title, clearly labeled categories (male and female), and reports the total number of cases (*N*) at the bottom of the frequency column. These items must be included in *all* frequency distributions. The meaning of the table is quite clear. There are 10 males and 10 females in the sample, a fact that is much easier to comprehend from the frequency distribution than from the unorganized data presented in Table 2.5.

For some nominal variables, the researcher might have to make some choices about the number of categories included. For example, Table 2.7 presents the distribution of the variable "marital status" using the categories listed in Table 2.5 but, if the researcher wants to focus only on married and unmarried students, categories could be collapsed, as in Table 2.8. Notice that, when categories are grouped, information and detail will be lost. Table 2.8 would not allow the researcher to distinguish between the two unmarried states.

FREQUENCY DISTRIBUTIONS FOR VARIABLES MEASURED AT THE ORDINAL LEVEL

Frequency distributions for ordinal-level variables are constructed in the same way as for nominal-level variables. Table 2.9 reports the frequency distribution of "satisfaction" from the counseling center survey. Note that a column of percentages by category has been added to this table. Such columns heighten the clarity of the table (especially with larger samples) and are common adjuncts to the basic frequency distribution for variables measured at all levels. This table reports that most students were either satisfied or very satisfied with the services of the counseling center. The most common response (nearly half the sample) was "satisfied." If the researcher wanted to emphasize this major

TABLE 2.9 Satisfaction with Services, Couseling Center Survey

Satisfaction	Frequency (*f*)	Percentage (%)
(4) Very satisfied	4	20%
(3) Satisfied	9	45%
(2) Dissatisfied	4	20%
(1) Very dissatisfied	3	15%
	N = 20	100%

TABLE 2.10 Satisfaction with Services, Couseling Center Survey

Satisfaction	Frequency (f)	Percentage (%)
Satisfied	13	65%
Dissatisfied	7	35%
	N = 20	100%

trend, the categories could be collapsed, as in Table 2.10. Again, the price paid for this increased compactness is that some information (in this case, the exact breakdown of degrees of satisfaction and dissatisfaction) is lost. *(For practice in constructing and interpreting frequency distributions for nominal- and ordinal-level variables, see problem 2.5.)*

FREQUENCY DISTRIBUTIONS FOR VARIABLES MEASURED AT THE INTERVAL-RATIO LEVEL

In general, the construction of frequency distributions for variables measured at the interval-ratio level is more complex than for nominal and ordinal variables. Interval-ratio variables usually have a wide range of scores, and this means that the researcher must collapse or group categories to produce reasonably compact tables. To construct frequency distributions for interval-ratio—level variables, you must decide how many categories to use and how wide these categories should be.

For example, suppose you wished to report the distribution of the variable "age" for a sample drawn from a community. Unlike the college data reported in Table 2.5, a community sample would have a very broad range of ages. If you simply reported the number of times that each year of age (or score) occurred, you could easily wind up with a frequency distribution that contained 80, 90, or even more categories; such a table would be very difficult to read. The scores (years) must be grouped into larger categories to increase ease of comprehension. How large should these categories be? How many categories should be included in the table? Although there are no hard-and-fast rules, these decisions always involve a trade-off between more detail (a greater number of narrow categories) or more compactness (a smaller number of wide categories); this will be covered in the next section.

Constructing Frequency Distributions for Interval-Ratio—Level Variables

We will use the ages of the students in the counseling center survey to demonstrate the decision-making process for frequency distributions for interval-ratio level variables. Because of the narrow age range of a group of college students, we can use categories of only one year (these categories are often called **class intervals** when working with interval-ratio data). The frequency distribution is constructed by listing the ages in order, counting the number of times each score (year of age) occurs, and then totaling the number of scores for each category. Table 2.11 presents the information and reveals a concentration or clustering of scores in the 18 and 19 class intervals.

TABLE 2.11 **Age of Respondents, Counseling Center Survey (interval width = one year of age)**

Class Intervals	Frequency (f)
18	5
19	6
20	3
21	2
22	1
23	1
24	1
25	0
26	1
	$N = 20$

Even though this table is fairly clear, assume for the sake of illustration that you desire a more compact (less detailed) summary. To do this, you will have to group scores into wider class intervals. Increasing the interval width (say, to two years) will reduce the number of intervals and achieve a more compact expression. The grouping of scores in Table 2.12 clearly emphasizes the predominance of younger respondents. This trend in the data can be stressed even more by the addition of a column displaying the percentage of cases in each category.

STATED LIMITS

Note that the class intervals in Table 2.12 have been stated with an apparent gap between them (that is, the **stated class limits**, are separated by a distance of one unit). At first glance, these gaps may appear to violate the principle of exhaustiveness; but, because age has been measured in whole numbers, the gaps are actually not a problem. Given the level of precision of the measurement (in whole years, as opposed to, say, 10ths of a year), no case could have a score falling between these class intervals. For these data, the set of class intervals in Table 2.12 is exhaustive and mutually exclusive. Each of the 20 respondents in the sample can be sorted into one and only one age category.

However, consider the potential difficulties if age had been measured with greater precision. If age had been measured in 10ths of a year, into which class

TABLE 2.12 **Age of Respondents, Counseling Center Survey (interval width = one year of age)**

Class Intervals	Frequency (f)	Percentage (%)
18–19	11	55
20–21	5	25
22–23	2	10
24–25	1	5
26–27	1	5
	$N = 20$	100%

interval in Table 2.12 would a 19.4-year-old subject be placed? You can avoid this ambiguity by always stating the limits of the class intervals at the same level of precision as the data. Thus, if age were being measured in 10ths of a year, the limits of the class intervals in Table 2.12 would be stated in 10ths of a year. For example:

17.0–18.9

19.0–20.9

21.0–22.9

23.0–24.9

25.0–26.9

To maintain mutual exclusivity between categories, do not overlap the class intervals. If you state the limits of the class intervals at the same level of precision as the data (which might be in whole numbers, tenths, hundredths, etc.) and maintain a "gap" between intervals, you will always produce a frequency distribution where each case can be assigned to one and only one category.

MIDPOINTS

On occasion, when constructing certain graphs for example, you will need to work with the **midpoints** of the class intervals. A midpoint is the point exactly halfway between the upper and lower limits of a class interval; it can be found by dividing the sum of the upper and lower limits by 2.

Table 2.13 displays midpoints for two different sets of class intervals. *(For practice in finding midpoints, see problems 2.8b and 2.9b.)*

REAL LIMITS

For certain purposes, you must eliminate the "gap" between class intervals and treat a distribution as a continuous series of categories that border each other. This is necessary for some graphs (see Section 2.7) and for computing summary statistics for variables that have been grouped into frequency distributions.

ONE STEP AT A TIME	**Finding Midpoints**

1: Find the upper and lower limits of the lowest interval in the frequency distribution. For any interval, the upper limit is the highest score included in the interval and the lower limit is the lowest score included in the interval. For example, for the top set of intervals in Table 2.13, the lowest interval (0–2) includes scores of 0, 1, and 2. The upper limit of this interval is 2 and the lower limit is 0.

2: Add the upper and lower limits and divide by 2. For the interval 0–2: (0 + 2)/2 = 1. The midpoint for this interval is 1.

3: Midpoints for other intervals can be found by repeating steps 1 and 2 for each interval. As an alternative, you can find the midpoint for any interval by adding the value of the interval width to the midpoint of the next lower interval. For example, the lowest interval in Table 2.13 is 0–2 and the midpoint is 1. Intervals are 3 units wide (that is, they each include three scores), so the midpoint for the next higher interval (3–5) is 1 + 3, or 4. The midpoint for the interval 6–8 is 4 + 3, or 7, and so forth.

TABLE 2.13   Midpoints

Class Interval Width = 3	
Class Intervals	Midpoints
0–2	1.0
3–5	4.0
6–8	7.0
9–11	10.0

Class Interval Width = 6	
Class Intervals	Midpoints
100–105	102.5
106–111	108.5
112–117	114.5
118–123	120.5

We can use Table 2.12 to illustrate. The intervals in the table, stated with the apparent "gap," are called stated class limits and they organize the scores into a series of discrete, non-overlapping intervals. To treat the variable as continuous and find the **real class limits**, follow the instructions in the "One Step at a Time" box. The stated and real limits for Table 2.12 are:

Stated Limits	Real Limits
18–19	17.5–19.5
20–21	19.5–21.5
22–23	21.5–23.5
24–25	23.5–25.5
26–27	25.5–27.5

Table 2.14 presents additional illustrations of real limits for two different sets of class intervals. In both cases, the "gap" between the stated limits is 1. *(For practice in finding real limits, see problems 2.7c and 2.8d.)*

ONE STEP AT A TIME   **Finding Real Limits**

1: Find the distance (the "gap") between the stated class intervals. In Table 2.12, for example, this value is 1.

2: Divide the value found in step 1 in half.

3: Add the value found in step 2 to all upper stated limits and subtract it from all lower stated limits.

TABLE 2.14 Real Class Limits

Class Intervals (stated limits)	Real Class Limits
3–5	2.5–5.5
6–8	5.5–8.5
9–11	8.5–11.5

Class Intervals (stated limits)	Real Class Limits
100–105	99.5–105.5
106–111	105.5–111.5
112–117	111.5–117.5
118–123	117.5–123.5

CUMULATIVE FREQUENCY AND CUMULATIVE PERCENTAGE

Two commonly used adjuncts to the basic frequency distribution for interval-ratio data are the **cumulative frequency** and **cumulative percentage** columns. These additional columns allow us to tell at a glance how many cases fall below a given score or class interval in the distribution. To add these columns, follow the instructions in the One Step at a Time box. Table 2.15 shows a cumulative frequency column added to Table 2.12 and Table 2.16 displays the cumulative percentage column for the same distribution.

TABLE 2.15 Age of Respondents, Counseling Center Survey

Class Intervals	Frequency (f)	Cumulative Frequency
18–19	11	11
20–21	5	16
22–23	2	18
24–25	1	19
26–27	1	20
	$N = 20$	

TABLE 2.16 Age of Respondents, Counseling Center Survey

Class Intervals	Frequency (f)	Cumulative Frequency	Percentage	Cumulative Percentage
18–19	11	11	55%	55%
20–21	5	16	25%	80%
22–23	2	18	10%	90%
24–25	1	19	5%	95%
26–27	1	20	5%	100%
	$N = 20$		100%	

ONE STEP AT A TIME	Adding Cumulative Frequency and Percentage Columns to Frequency Distributions

To Add the Cumulative Frequency Column

1: Begin with the lowest class interval (the interval with the lowest scores). The entry in the cumulative frequency column will be the same as the number of cases in this interval.

2: Go to the next class interval. The cumulative frequency for this interval is the number of cases in the interval plus the number of cases in the lower interval.

3: Continue adding (or accumulating) cases from interval to interval until you reach the interval with the highest scores, which will have a cumulative frequency equal to N.

To Add the Cumulative Percentage Column

1: Compute the percentage of cases in each category one at a time, and then follow the pattern for the cumulative frequencies. The entry for the lowest class interval will be the same as the percentage of cases in the interval.

2: For the next higher interval, the cumulative percentage is the percentage of cases in the interval plus the percentage of cases in the lower interval.

3: Continue adding (or accumulating) percentages from interval to interval until you reach the interval with the highest scores, which will have a cumulative percentage of 100%.

These cumulative columns are quite useful in situations where the researcher wants to make a point about how cases are spread across the range of scores. For example, Tables 2.15 and 2.16 show clearly that most students in the counseling center survey are less than 21 years of age. If the researcher wishes to stress this fact, then these cumulative columns are quite handy. Most realistic research situations will be concerned with many more than 20 cases and/or many more categories than our tables have and the cumulative percentage column would usually be preferred to the cumulative frequencies column.

UNEQUAL CLASS INTERVALS

As a general rule, the class intervals of frequency distributions should be equal in size in order to maximize clarity and ease of comprehension. For example, note that all of the class intervals in Tables 2.15 and 2.16 are the same width (2 years). However, there are several other possibilities for stating class interval, and we will examine each situation separately.

Open-Ended Intervals. What would happen to the frequency distribution in Table 2.15 if we added one more student who was 47 years of age? We would now have 21 cases and there would be a large gap between the oldest respondent (now 47) and the second oldest (age 26). If we simply added the older student to the frequency distribution, we would have to include nine new class intervals (28–30, 31–32, 32–33, etc.) with zero cases in them before we got to the 46–47 interval. This would waste space and probably be unclear and confusing. An alternative way to handle the situation where we have a few very high or low scores would be to add an "open-ended" interval to the frequency distribution, as in Table 2.17.

TABLE 2.17 Age of Respondents, Counseling Center Survey ($N = 21$)

Class Intervals	Frequency (f)	Cumulative Frequency
18–19	11	11
20–21	5	16
22–23	2	18
24–25	1	19
26–27	1	20
28 and older	1	21
	$N = 21$	

The open-ended interval in Table 2.17 allows us to present the information more compactly than listing all of the empty intervals between "28–29" and "46–47." We could handle an extremely low score by adding an open-ended interval as the lowest class interval (e.g., "17 and younger"). There is a small price to pay for this efficiency (there is no information in Table 2.17 about the exact scores included in the open-ended interval), so this technique should not be used indiscriminately.

Intervals of Unequal Size. On some variables, most scores are tightly clustered together but others are strewn across a broad range of scores. Consider, as an example, the distribution of income in the United States. In 2008, most households (almost 56%) reported annual incomes between $25,000 and $100,000 and a sizeable grouping (about 23%) earned less than that. The problem (from a statistical point of view) comes with more affluent households, those with incomes above $100,000. The number of these households is quite small, of course, but we must still account for these cases.

If we tried to use a frequency distribution with equal intervals of, say, $10,000, we would need 30 or 40 or more intervals to include all of the more affluent households, and many of our intervals in the higher income ranges—especially those over $150,000—would have few or zero cases. In situations such as this, we can use intervals of unequal size to summarize the variable more efficiently, as in Table 2.18.

Some of the intervals in Table 2.18 are $10,000 wide, others are $25,000 or $50,000 wide, and two (the lowest and highest intervals) are open ended. Tables that use intervals of mixed widths might be a little confusing for the reader, but the trade-off in compactness and efficiency can be considerable. *(For practice in constructing and interpreting frequency distributions for interval-ratio level variables, see problems 2.5 to 2.9.)*

APPLYING FREQUENCY DISTRIBUTIONS FOR INTERVAL-RATIO VARIABLES: AN EXAMPLE

We covered a lot of ground in the preceding section, so let's pause and review these principles by considering a specific research situation. The following data represent the numbers of visits received over the past year by 90 residents of a retirement community.

0	52	21	20	21	24	1	12	16	12
16	50	40	28	36	12	47	1	20	7
9	26	46	52	27	10	3	0	24	50
24	19	22	26	26	50	23	12	22	26
23	51	18	22	17	24	17	8	28	52
20	50	25	50	18	52	46	47	27	0
32	0	24	12	0	35	48	50	27	12
28	20	30	0	16	49	42	6	28	2
16	24	33	12	15	23	18	6	16	50

Listed in this format, the data are a hopeless jumble from which no one could derive much meaning. The function of the frequency distribution is to arrange and organize these data so that their meanings will be made obvious.

First, we must decide how many class intervals to use in the frequency distribution. Following the guidelines presented in the One Step at a Time: Constructing Frequency Distributions for Interval-Ratio Variables box, let's use about 10 intervals ($k = 10$). By inspecting the data, we can see that the lowest score is 0 and the highest is 52. The range of these scores (R) is $52 - 0$, or 52. To find the approximate interval size (i), divide the range (52) by the number of intervals (10). Since $52/10 = 5.2$, we can set the interval size at 5.

The lowest score is 0, so the lowest class interval will be 0–4. The highest class interval will be 50–54, which will include the high score of 52. All that remains is to state the intervals in table format, count the number of scores that fall into each interval, and report the totals in a frequency column. These steps

TABLE 2.18   Distribution of Income by Household, United States, 2008

Income	Households (frequency)	Households (percent)
Less than $10,000	8,149,557	7.2%
$10,000 to $14,999	6,141,047	5.4%
$15,000 to $24,999	12,049,123	10.7%
$25,000 to $34,999	11,718,207	10.4%
$35,000 to $49,999	16,028,909	14.2%
$50,000 to $74,999	21,251,695	18.8%
$75,000 to $99,999	14,015,215	12.4%
$100,000 to $149,999	13,921,120	12.3%
$150,000 to $199,000	4,954,443	4.4%
$200,000 and above	4,872,013	4.3%
	113,101,329	100.1%

Source: U.S. Census Bureau, http://factfinder.census.gov/servlet/ADPTable?_bm = y&-geo_id = D&-qr_name = ACS_2008_1YR_G00_DP3&-ds_name = D&-_lang = en.

Applying Statistics 2.5: Frequency Distributions

The following list shows the ages of 50 prisoners enrolled in a work-release program. Is this group young or old? A frequency distribution will provide an accurate picture of the overall age structure.

18	60	57	27	19
20	32	62	26	20
25	35	75	25	21
30	45	67	41	30
37	47	65	42	25
18	51	22	52	30
22	18	27	53	38
27	23	32	35	42
32	37	32	40	45
55	42	45	50	47

We will use about 10 intervals to display these data. By inspection we see that the youngest prisoner is 18 and the oldest is 75. The range is thus 57. Interval size will be 57/10, or 5.7, which we can round off to either 5 or 6. Let's use a six-year interval beginning at 18. The limits of the lowest interval will be 18–23. Now we must state the limits of all the other intervals, count the number of cases in each interval, and display these counts in a frequency distribution.

Columns may be added for percentages, cumulative percentages, and/or cumulative frequency. The complete distribution, with a column added for percentages, is

Ages	Frequency	Percentages
18–23	10	20%
24–29	7	14%
30–35	9	18%
36–41	5	10%
42–47	8	16%
48–53	4	8%
54–59	2	4%
60–65	3	6%
66–71	1	2%
72–77	1	2%
	$N = 50$	100%

The prisoners seem to be fairly evenly spread across the age groups up to the 48–53 interval. There is a noticeable lack of prisoners in the oldest age groups and a concentration of prisoners in their 20s and 30s.

have been taken in Table 2.19, which also includes columns for the percentages and cumulative percentages. Note that this table is the product of several relatively arbitrary decisions. The researcher should remain aware of this fact and inspect the frequency distribution carefully. If the table is unsatisfactory for any reason, it can be reconstructed with a different number of categories and interval sizes.

Now, with the aid of the frequency distribution, some patterns in the data can be discerned. There are three distinct groupings of scores in the table. Ten residents were visited rarely, if at all (the 0–4 visits per year interval). The single largest interval, with 18 cases, is 20–24. Combined with the intervals immediately above and below, this represents quite a sizeable grouping of cases (42 out of 90, or 46.66% of all cases) and suggests that the dominant visiting rate is about twice a month, or approximately 24 visits per year. The third grouping, in the 50–54 class interval (12 cases), reflects a visiting rate of about once a week. The cumulative percentage column indicates that the majority of the resident (58.89%) were visited 24 or fewer times a year.

TABLE 2.19 Number of Visits per Year, 90 Rertirement Community Residents

Class Intervals	Frequency (f)	Cumulative Frequency	Percentage (%)	Cumulative Percentage
0–4	10	10	11.11%	11.11
5–9	5	15	5.56%	16.67
10–14	8	23	8.89%	25.26
15–19	12	35	13.33%	38.89
20–24	18	53	20.00%	58.89
25–29	12	65	13.33%	72.22
30–34	3	68	3.33%	75.55
35–39	2	70	2.22%	77.77
40–44	2	72	2.22%	79.99
45–49	6	78	6.67%	86.66
50–54	12	90	13.33%	99.99
	$N = 90$		99.99%*	

*Percentage columns will occasionally fail to total to 100% because of rounding error. If the total is between 99.90% and 100.10%, ignore the discrepancy. Discrepancies of greater than ±0.10% may indicate mathematical errors, and the entire column should be computed again.

ONE STEP AT A TIME | **Constructing Frequency Distributions for Interval-Ratio Variables**

1: Decide how many class intervals (k) you wish to use. One reasonable convention suggests that the number of intervals should be about 10. Many research situations may require fewer than 10 intervals ($k = 10$), and it is common to find frequency distributions with as many as 15 intervals. Only rarely will more than 15 intervals be used, since the resultant frequency distribution would be too large for easy comprehension.

2: Find the range (R) of the scores by subtracting the low score from the high score.

3: Find the size of the class intervals (i) by dividing R (from step 2) by k (from step 1):

$$i = R/k$$

Round the value of i to a convenient whole number. This will be the interval size or width.

4: State the lowest interval so that its lower limit is equal to or below the lowest score. By the same token, your highest interval will be the one that contains the highest score. Generally, intervals should be equal in size, but unequal and open-ended intervals may be used when convenient.

5: State the limits of the class intervals at the same level of precision as you have used to measure the data. Do not overlap intervals. You will thereby define the class intervals so that each case can be sorted into one and only one category.

6: Count the number of cases in each class interval, and report these subtotals in a column labeled "Frequency." Report the total number of cases (N) at the bottom of this column. The table may also include a column for percentages, cumulative frequencies, and cumulative percentages.

7: Inspect the frequency distribution carefully. Has too much detail been lost? If so, reconstruct the table with a greater number of class intervals (or use a smaller interval size). Is the table too detailed? If so, reconstruct the table with fewer class intervals (or use wider intervals). Are there too many intervals with no cases in them? If so, consider using open-ended intervals or intervals of unequal size. Remember that the frequency distribution results from a number of decisions you make in a rather arbitrary manner. If the appearance of the table seems less than optimal given the purpose of the research, redo the table until you are satisfied that you have struck the best balance between detail and conciseness.

8: Give your table a clear, concise title, and number the table if your report contains more than one. All categories and columns must also be clearly labeled.

Graphic Presentations of Data

Researchers frequently use graphing techniques to present their data in ways that are visually more dramatic than frequency distributions. These devices are particularly useful for communicating the overall shape of a distribution and for highlighting any clustering of cases in a particular range of scores. Many graphing techniques are available, but we will examine just four. The first two, pie and bar charts, are appropriate for discrete variables at any level of measurement. The last two, histograms and line charts (or frequency polygons), are used with both discrete and continuous interval-ratio variables but are particularly appropriate for the latter.

Here I explain how to construct graphs and charts "by hand." These days, however, computer programs are almost always used to produce graphic displays. Graphing software is sophisticated and flexible but also relatively easy to use, and you should learn to use these programs if they are available to you. The effort required to learn these programs will be repaid in the quality of the final product. The You Are the Researcher section at the end of this chapter includes a demonstration of how to produce bar charts and line charts.

PIE CHARTS

To construct a **pie chart**, begin by computing the percentage of all cases that fall into each category of the variable. Then divide a circle (the pie) into segments (slices) proportional to the percentage distribution. Be sure that the chart and all segments are clearly labeled.

Figure 2.1 is a pie chart that displays the distribution of "marital status" from the counseling center survey. The frequency distribution (Table 2.7) is reproduced as Table 2.20, with a column added for the percentage distribution. Since a circle's circumference is 360°, we will apportion 180° (or 50%) for the first category, 126° (35%) for the second, and 54° (15%) for the last category. The pie chart visually reinforces the relative preponderance of single respondents and the relative absence of divorced students in the counseling center survey.

BAR CHARTS

Like pie charts, **bar charts** are straightforward. Usually, the categories of the variable are arrayed along the horizontal axis (or *abscissa*) and frequencies, or percentages if you prefer, along the vertical axis (or *ordinate*). For each category of the variable, construct (or draw) a rectangle of constant width and with a height

FIGURE 2.1   Sample Pie Chart: Marital Status of Respondents ($N = 20$)

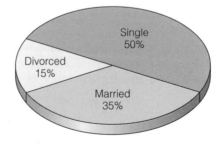

TABLE 2.20 **Marital Status of Respondents, Counseling Center Survey**

Status	Frequency (f)	Percentage (%)
Single	10	50
Married	7	35
Divorced	3	15
	$N = 20$	100%

FIGURE 2.2 **Sample Bar Chart: Marital Status of Respondents ($N = 20$)**

that corresponds to the number of cases in the category. The bar chart in Figure 2.2 reproduces the marital status data from Figure 2.1 and Table 2.20.

Figure 2.2 would be interpreted in exactly the same way as the pie chart in Figure 2.1, and researchers are free to choose between these two methods of displaying data. However, if a variable has more than four or five categories, the bar chart would be preferred, since the pie chart gets very crowded with many categories. To illustrate, Figure 2.3 uses a bar chart to display the data on visiting

FIGURE 2.3 **Sample Bar Chart for Visits per Year, Retirement Community Residents ($N = 90$)**

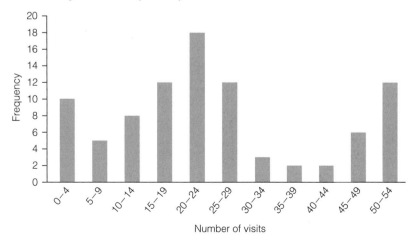

rates for the retirement community presented in Table 2.19. A pie chart for this same data would have had 11 different "slices," a more complex or "busier" picture than that presented by the bar chart. In Figure 2.3, the clustering of scores in the "20 to 24" range (approximately two visits a month) is readily apparent, as are the groupings in the "0 to 4" and "50 to 54" ranges.

Bar charts are particularly effective ways to display the relative frequencies for two or more categories of a variable when you want to emphasize some comparisons. Suppose, for example, that you wished to make a point about changing rates of homicide victimization for white males and females since 1955. Figure 2.4 displays the data in a dramatic and easily comprehended way. The bar chart shows that rates for males are higher than rates for females, that rates for both sexes were highest in 1975, and that rates declined after that time. *(For practice in constructing and interpreting pie and bar charts, see problems 2.5b and 2.10.)*

HISTOGRAMS

Histograms look a lot like bar charts and, in fact, are constructed in much the same way. However, histograms use real limits rather than stated limits, and the categories or scores of the variable border each other, as if they merged into each other in a continuous series. Therefore, these graphs are most appropriate for continuous interval-ratio–level variables, but they are commonly used for discrete interval-ratio–level variables as well. To construct a histogram from a frequency distribution, follow these steps.

1. Array the real limits of the class intervals or scores along the horizontal axis (abscissa).

2. Array frequencies along the vertical axis (ordinate).

3. For each category in the frequency distribution, construct a bar with height corresponding to the number of cases in the category and with width corresponding to the real limits of the class intervals.

4. Label each axis of the graph.

5. Title the graph.

FIGURE 2.4 **U.S. Homicide Victimization Rates by Gender, 1955–2006 (selected Years, Rates per 100,000 population, Whites only)**

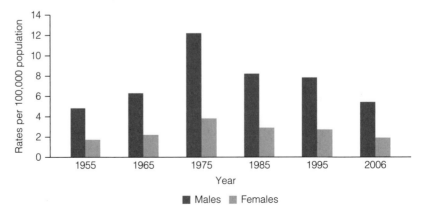

FIGURE 2.5 Age of Respondents, 2008 General Social Survey

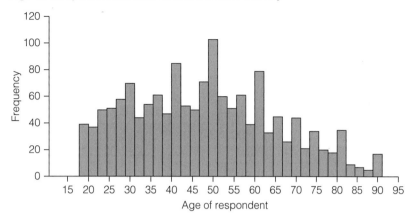

As an example, Figure 2.5 uses a histogram to display the distribution of ages for a sample of respondents to a national public-opinion poll. The bars in the graph are 5 years wide, and their uneven heights reflect the varying number of respondents for each 5-year group. The graph peaks around age 50, and the sample has more respondents (higher bars) who are younger than 50 and fewer respondents (lower bars) older than 50. Note also that there are no people in the sample younger than age 18, the usual cutoff point for respondents to public-opinion polls.

LINE CHARTS

Construction of a **line chart** (or **frequency polygon**) is similar to construction of a histogram. Instead of using bars to represent the frequencies, however, use a dot at the midpoint of each interval. Straight lines then connect the dots. Because the line is continuous from highest to lowest score, these graphs are especially appropriate for continuous interval-ratio—level variables but are frequently used with discrete interval-ratio—level variables as well. Figure 2.6

FIGURE 2.6 Number of Visits per Year, Retirement Community Residents (*N* = 90)

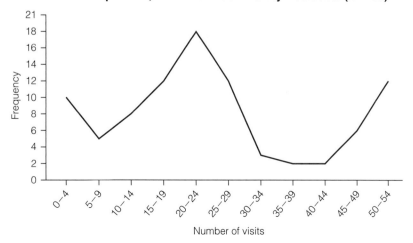

STATISTICS
IN EVERYDAY LIFE

There is a great deal of concern in U.S. society about the future of the American family and, more specifically, our high divorce rate. The U.S. divorce rate is higher than those of other Western industrialized nations (and almost twice the Canadian rate) but has been falling since the early 1980s. Why? One reason is that the marriage rate has also been falling—you can't get divorced unless you get married! Both trends are clearly displayed in the line chart in Figure 2.7.

FIGURE 2.7   U.S. Rates of Marriage and Divorce, 1950–2008 (rates per 100,000 population)

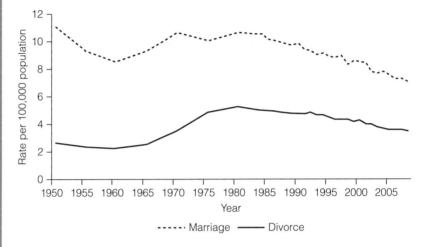

This information could have been presented in a table. What are the advantages of the "picture" presented in a graph? For example, can you see a difference in the rate of decline of the two lines? Is one rate falling faster than the other? What are the implications of this difference? What else does the graph say about family life in the U.S. that would not have been so clear from a table?

displays a line chart for the visiting data previously displayed in the bar chart in Figure 2.3.

Histograms and frequency polygons are alternative ways of displaying essentially the same message. Thus, the choice between the two techniques is left to the aesthetic pleasures of the researcher. *(For practice in constructing and interpreting histograms and line charts, see problems 2.7b, 2.8d, 2.9d, 2.11, and 2.12.)*

Interpreting Statistics: Changing Patterns of Workplace Surveillance

A sizeable volume of statistical material has been introduced in this chapter, and it will be useful to conclude by focusing on meaning and interpretation. What can you say after you have calculated percentages, built a frequency distribution, or constructed a graph or chart? Remember that statistics are tools to help us analyze information and answer questions. They never speak for themselves and they always have to be understood in the context of some research question or test of a hypothesis. This section provides an example of interpretation by posing and answering some questions from social science research. The interpretation (words) will be explicitly linked to the statistics (numbers) so that you will be able to see how and why conclusions are developed.

Your New Job and Workplace Surveillance. Congratulations! You have just landed a job with a major U.S. corporation and you now find yourself in the middle of the lunch hour in your cubicle. Should you log on to the Internet and spend a few minutes with your favorite role-playing game? Should you check your personal e-mail or contact your friends about plans for the weekend? Before making a decision, consider a series of reports issued by the American Management Association (AMA). These reports suggest that the chances are growing that you may be the subject of workplace surveillance and that your e-mail, telephone, and, more recently, your Internet use may be monitored by your employer.

Monitoring and Surveillance in 2007. Although the computer has become an important component of the workday for more and more people, your employer may feel compelled to monitor its use. Table 2.21 reports the percentage of companies that indicated they practiced a specific form of monitoring and surveillance.

Graphs are almost always a more effective method of presenting this type of information. The variable (type of monitoring and surveillance) is nominal level (the "types" are different from each other but do not form a scale), and, with nine possible scores, a bar chart would be preferred to a pie chart. Figure 2.8 shows that monitoring Internet connections was the most common form of surveillance, with about 66% of the companies practicing it. Storage and review of e-mail messages (about 43%) and telephone monitoring (about 45%) and were also very common. This graph clearly indicates that it would be unwise to surf the net on company time or to use the phone or e-mail for personal business.

TABLE 2.21   Monitornig and Surveillance, 2007

Type of Monitoring and Surveillance	% Yes
Monitoring Internet connections	66%
Storage and review of e-mail messages	43%
Telephone use (time spent, numbers called)	45%
Video surveillance for security purposes	47%
Storage and review of computer files	43%
Computer use (time logged on, keystroke counts, etc.)	45%
Recording and review of telephone conversations	16%
Video recording of employee job performance	7%
Storage and review of voice mail messages	10%

FIGURE 2.8   Monitoring and Surveillance of Employees, 2007

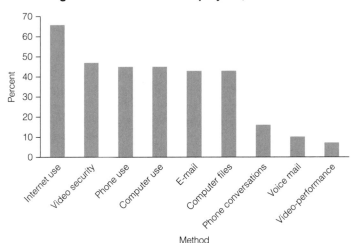

Monitoring and Surveillance over Time. What changes occurred in specific workplace monitoring practices between 1997 and 2007? Table 2.22 displays trends for five types of monitoring and surveillance. In 1997, the levels of monitoring were relatively low. Only about one-third of companies monitored telephone use, and no more than 16% practiced the other forms of surveillance. The levels rise quite dramatically over the period, and by 2007 four different forms of surveillance were being practiced by more than 40% of the companies.

Once again, these trends and patterns would be more clearly presented and appreciated in the form of a graph. Figure 2.9 presents the information in Table 2.22 in the form of a line graph. The graph clearly displays the general increase in monitoring and shows that monitoring of email messages, phone use, and computer files have become particularly common.

TABLE 2.22   Monitoring and Surveillance, 1997–2007

Type of Monitoring and Surveillance	Percent "Yes"					
	1997	1998	1999	2000	2001	2007
Storage and review of voice mail messages	5	5	6	7	8	10
Storage and review of computer files	14	20	21	31	36	43
Storage and review of e-mail messages	15	20	27	38	47	43
Telephone use (time spent, numbers called)	34	40	39	44	43	45
Computer use (time logged on, keystroke counts, etc.)	16	16	15	19	19	45

FIGURE 2.9 Monitoring and Surveillance, 1997–2007

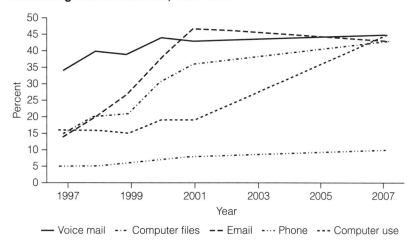

Computers may be ubiquitous features of employment in our information-age economy, but these data suggest that the potential for workplace monitoring and surveillance is also increasing. The computer on your desk is a double-edged sword. While it provides you with the necessary tools to do your job, employers are increasingly using the very same technology to watch you.

READING STATISTICS 2: Percentages, Rates, Tables, and Graphs

The statistics covered in this chapter are frequently used in the research literature of the social sciences—as well as in the popular press and the media—and one of the goals of this text is to help you develop your ability to understand and critically analyze them. Fortunately, this task is usually quite straightforward, but these statistical tools are sometimes not as simple as they appear and they can be misused. Here are some ideas to keep in mind when reading research reports that use these statistics.

First, there are many different formats for presenting results, and the tables and graphs you find in the research literature will not necessarily follow the conventions used in this text. Second, because of space limitations, tables and graphs may be presented with minimum detail. For example, the researcher may present a frequency distribution with only a percentage column.

Begin your analysis by examining the statistics carefully. If you are reading a table or graph, first read the title, all labels (that is, row and/or column headings), and any footnotes. These will tell you exactly what information is being presented. Inspect the body of the table or graph with the author's analysis in mind. See if you agree with the author's analysis. (You almost always will, but it never hurts to double-check and exercise your critical abilities.)

Finally, remember that most research projects analyze interrelationships among many variables. Because the tables and graphs covered in this chapter display variables one at a time, they are unlikely to be included in such research reports (or, perhaps, included only as background information). Even when they are not reported, you can be sure that the research began with an inspection of percentages,

(continued)

READING STATISTICS 2: (*continued*)

frequency distributions, or graphs for each variable. Univariate tables and graphs display a great deal of information about the variables in a compact, easily understood format and are almost universally used as descriptive devices.

STATISTICS IN THE PROFESSIONAL LITERATURE

Social scientists rely heavily on the U.S. census for information about the characteristics and trends of change in American society, including age composition, birthrates and death rates, residential patterns, educational levels, and a host of other variables. Census data is readily available (at www.census.gov), but since they represent information about the entire population, the numbers are often large, cumbersome, and awkward to use or understand. Thus, percentages, rates, and graphs are extremely useful statistical devices when analyzing or presenting census information.

Consider, for example, a recent report on the changing U.S. family.* The purpose of the report was to present information regarding the structure of the American family and to present and discuss recent changes and trends. Consider how this report might have read if the information had been given in words and raw numbers: "In 2007, there were 55,867,091 married-couple households, 19,947,474 female-headed households, and 37,258,717 nonfamily households." Can you distill any meaningful understandings about American family life from these sentences? Raw information simply does not speak for itself, and data have to be organized or placed in some context to reveal their meaning. Thus, social scientists almost always use percentages, rates, or graphs to present this kind of information so that they can understand it and convey their interpretations to others.

In contrast with raw information, consider the following table on family trends using percentages.

U.S. Households by Type, 1970 and 2007

	Percent of All Households	
	1970	2007
Family households:		
Married couples with children	40.3%	22.3%
Married couples without children	30.3%	28.3%
Other	10.6%	16.8%
Nonfamily households:		
Women living alone	5.6%	15.2%
Men living alone	11.5%	11.7%
Other	1.7%	5.6%
	100.0%	99.9%

A quick comparison of the two years reveals a dramatic decrease in the percentage of American households that consist of married couples living with children and an increase in the percentages of women living alone. What other trends can you see in the table?

*Kreider, Rose, and Elliot, Diana. 2009. "America's Families and Living Arrangements: 2007." U.S. Bureau of the Census: *Current Population Reports.* Available at http://www.census.gov/prod/2009pubs/p20-561.pdf.

SUMMARY

1. We considered several different ways of summarizing the distribution of a single variable and, more generally, reporting the results of our research. Our emphasis throughout was on the need to communicate our results clearly and concisely. You will often find that, as you strive to communicate statistical information to others, the meanings of the information will become clearer to you as well.

2. Percentages and proportions, ratios, rates, and percentage change represent several different ways to enhance

clarity by expressing our results in terms of relative frequency. Percentages and proportions report the relative occurrence of some category of a variable compared with the distribution as a whole. Ratios compare two categories with each other, and rates report the actual occurrences of some phenomenon compared with the number of possible occurrences per some unit of time. Percentage change shows the relative increase or decrease in a variable over time.

3. Frequency distributions are tables that summarize the entire distribution of some variable. Statistical analysis almost always starts with the construction and review of these tables for each variable. Columns for percentages, cumulative frequency, and/or cumulative percentages often enhance the readability of frequency distributions.

4. Pie and bar charts, histograms, and line charts (or frequency polygons) are graphic devices used to express the basic information contained in the frequency distribution in a compact and visually dramatic way.

SUMMARY OF FORMULAS

Percentages	2.1	$\% = \left(\dfrac{f}{N}\right) \times 100$
Proportions	2.2	$proportion = \left(\dfrac{f}{N}\right)$
Ratios	2.3	$ratio = \dfrac{f_1}{f_2}$
Percent Change	2.4	$Percent\ change = \left(\dfrac{f_2 - f_1}{f_1}\right) \times 100$

GLOSSARY

Bar chart. A graphic display device for discrete variables. Categories are represented by bars of equal width, the height of each corresponding to the number (or percentage) of cases in the category.

Class intervals. The categories used in the frequency distributions for interval-ratio variables.

Cumulative frequency. An optional column in a frequency distribution that displays the number of cases within an interval and all preceding intervals.

Cumulative percentage. An optional column in a frequency distribution that displays the percentage of cases within an interval and all preceding intervals.

Frequency distribution. A table that displays the number of cases in each category of a variable.

Frequency polygon. A graphic display device for interval-ratio variables. Class intervals are represented by dots placed over the midpoints, the height of each corresponding to the number (or percentage) of cases in the interval. All dots are connected by straight lines. Same as a **line chart**.

Histogram. A graphic display device for interval-ratio variables. Class intervals are represented by contiguous bars of equal width (equal to the class limits), the height of each corresponding to the number (or percentage) of cases in the interval.

Line chart. See **Frequency polygon**.

Midpoint. The point exactly halfway between the upper and lower limits of a class interval.

Percentage. The number of cases in a category of a variable divided by the number of cases in all categories of the variable, the entire quantity multiplied by 100.

Percent change. A statistic that expresses the magnitude of change in a variable from time 1 to time 2.

Pie chart. A graphic display device especially for discrete variables with only a few categories. A circle (the pie) is divided into segments proportional in size to the percentage of cases in each category of the variable.

Proportion. The number of cases in one category of a variable divided by the number of cases in all categories of the variable.

Rate. The number of actual occurrences of some phenomenon or trait divided by the number of possible occurrences per some unit of time.

Ratio. The number of cases in one category divided by the number of cases in some other category.

Real class limits. The class intervals of a frequency distribution when stated as continuous categories.

Stated class limits. The class intervals of a frequency distribution when stated as discrete categories.

PROBLEMS

2.1 SOC The tables that follow report the marital status of 20 respondents in two different apartment complexes. (*HINT: Make sure that you have the correct numbers in the numerator and denominator before answering these questions. For example, problem 2.1a asks for "the percentage of respondents in each complex who are married," and the denominators will be 20 for these two fractions. Problem 2.1d, on the other hand, asks for the "percentage of the single respondents who live in Complex B," and the denominator for this fraction will be 4 + 6, or 10.*)

Status	Complex A	Complex B
Married	5	10
Unmarried ("living together")	8	2
Single	4	6
Separated	2	1
Widowed	0	1
Divorced	1	0
	20	20

a. What percentage of the respondents in each complex are married?
b. What is the ratio of single to married respondents at each complex?
c. What proportion of each sample are widowed?
d. What percentage of the single respondents live in Complex B?
e. What is the ratio of the "unmarried/living together" to the "married" at each complex?

2.2 At St. Winnefred Hospital, the numbers of males and females in various job categories are as follows:

Job	Males	Females	Totals
Doctor	83	15	98
Nurse	62	116	178
Orderly	151	12	163
Lab Technician	32	30	62
Administrator	12	1	13
Clerk	72	141	213
Totals	412	315	727

Read each question carefully before constructing the fraction and solving for the answer. The formulas for these statistics are repeated here for your convenience:

$$\text{Percentage: } \% = \left(\frac{f}{N}\right) \times 100$$

$$\text{Proportion} = \frac{f}{N}$$

(*HINT: Be sure you place the proper number in the denominator of the fractions. For example, some problems use the total number of males or females as the denominator, whereas others use the total number of subjects.*)

a. What percentage of nurses are male?
b. What proportion of orderlies are female?
c. For doctors, what is the ratio of males to females?
d. What percentage of the total staff are males?
e. What is the ratio of males to females for the entire sample?
f. What proportion of the clerks are male?
g. What percentage of the sample are administrators?
h. What is the ratio of nurses to doctors?
i. What is the ratio of female lab technicians to male lab technicians?
j. What proportion of the males are clerks?

2.3 CJ The town of Shinbone, Kansas, has a population of 211,732 and experienced 47 bank robberies, 13 murders, and 23 auto thefts during the past year. Compute a rate for each type of crime per 100,000 population. (*HINT: Make sure that you set up the fraction with size of population in the denominator.*)

2.4 CJ The numbers of homicides in five U.S. states and five Canadian provinces for the years 1997 and 2008 were as follows on top of the next page.

a. Calculate the homicide rate per 100,000 population for each state and each province for each year. Relatively speaking, which state and which province had the highest homicide rates in each year? Which society seems to have the higher homicide rate? Write a paragraph describing these results.
b. Using the rates you calculated in part a, calculate the percent change between 1997 and 2005 for each state and each province. Which states and provinces had the largest increase and decrease? Which society seems to have the largest change in homicide rates? Summarize your results in a paragraph.

	1997		2008	
State	Homicides	Population	Homicides	Population
New Jersey	338	8,053,000	376	8,883,000
Iowa	52	2,852,000	76	3,003,000
Alabama	426	4,139,000	353	4,882,000
Texas	1327	19,439,000	1374	24,327,000
California	2579	32,268,000	2142	36,757,000

	1997		2008	
Province	Homicides	Population	Homicides	Population
Nova Scotia	24	936,100	12	936,600
Quebec	132	7,323,600	92	7,753,000
Ontario	178	11,387,400	176	12,936,300
Manitoba	31	1,137,900	54	1,206,000
British Columbia	116	3,997,100	117	4,383,800

Source: Statistics Canada, http://www.statcan.ca.s.

2.5 SOC The scores of 15 respondents on four variables are as reported next. These scores were taken from a public opinion survey called the General Social Survey, or the GSS. This data set is used for the computer exercises in this text. Small subsamples from the GSS will be used throughout the text to provide "real" data for problems. For the actual questions and other details, see Appendix G. The numerical codes for the variables are as follows:

Sex	Support for Gun Control	Level of Education	Age
1 = Male	1 = In favor	0 = Less than HS	Actual years
2 = Female	2 = Opposed	1 = HS	
		2 = Jr. college	
		3 = Bachelor's	
		4 = Graduate	

Case Number	Sex	Support for Gun Control	Level of Education	Age
1	2	1	1	45
2	1	2	1	48
3	2	1	3	55
4	1	1	2	32

(continued next column)

(continued)

Case Number	Sex	Support for Gun Control	Level of Education	Age
5	2	1	3	33
6	1	1	1	28
7	2	2	0	77
8	1	1	1	50
9	1	2	0	43
10	2	1	1	48
11	1	1	4	33
12	1	1	4	35
13	1	1	0	39
14	2	1	1	25
15	1	1	1	23

a. Construct a frequency distribution for each variable. Include a column for percentages.

b. Construct pie and bar charts to display the distributions of sex, support for gun control, and level of education.

2.6 SW A local youth service agency has begun a sex education program for teenage girls who have been referred by the juvenile courts. The girls were given a 20-item test for general knowledge about sex, contraception, and anatomy and physiology upon admission to the program and again after completing the program. The scores of the first 15 girls to complete the program are as follows.

Case	Pretest	Posttest		Case	Pretest	Posttest
A	8	12		I	5	7
B	7	13		J	15	12
C	10	12		K	13	20
D	15	19		L	4	5
E	10	8		M	10	15
F	10	17		N	8	11
G	3	12		O	12	20
H	10	11				

Construct frequency distributions for the pretest and posttest scores. Include a column for percentages. (*HINT: There were 20 items on the test, so the maximum range for these scores is 20. If you use 10 class intervals to display these scores, the interval size will be 2. Since there are no scores of 0 or 1 for either test, you may state the first interval as 2–3. To make comparisons easier, both frequency distributions should have the same intervals.*)

2.7 SOC Sixteen high school students completed a class to prepare them for the College Boards. Their scores were as follows.

420	345	560	650
459	499	500	657
467	480	505	555
480	520	530	589

These same 16 students were given a test of math and verbal ability to measure their readiness for college-level work. Scores are reported here in terms of the percentage of correct answers for each test.

Math Test			
67	45	68	70
72	85	90	99
50	73	77	78
52	66	89	75

Verbal Test			
89	90	78	77
75	70	56	60
77	78	80	92
98	72	77	82

a. Display each of these variables in a frequency distribution with columns for percentages and cumulative percentages.

b. Construct a histogram and frequency polygon for these data.

c. (*This problem is optional.*) Find the upper and lower real limits for the intervals you established.

2.8 GER Twenty-five residents of a community for senior citizens were asked to keep track of the number of hours they watched TV during the past week.

20	2	12	7	32
7	10	7	13	17
14	15	15	9	27
5	21	4	7	6
2	18	10	5	11

a. Construct a frequency distribution to display these data.

b. What are the midpoints of the class intervals?

c. Add columns to the table to display the percentage distribution, cumulative frequency, and cumulative percentages.

d. (*This problem is optional.*) Find the real limits for the intervals you selected.

e. Construct a histogram and a frequency polygon to display this distribution.

f. Write a paragraph summarizing this distribution of scores.

2.9 SOC Twenty-five students completed a questionnaire that measured their attitudes toward interpersonal violence. Respondents who scored high believed that in many situations a person could legitimately use physical force against another person. Respondents who scored low believed that in no situation (or very few situations) could the use of violence be justified.

52	47	17	8	92
53	23	28	9	90
17	63	17	17	23
19	66	10	20	47
20	66	5	25	17

a. Construct a frequency distribution to display these data.

b. What are the midpoints of the class intervals?

c. Add columns to the table to display the percentage distribution, cumulative frequency, and cumulative percentage.

d. Construct a histogram and a frequency polygon to display these data.

e. Write a paragraph summarizing this distribution of scores.

2.10 PA/CJ As part of an evaluation of the efficiency of your local police force, you have gathered the following data on police response time to calls for assistance during two different years. (Response times were rounded off to whole minutes.) Convert both frequency distributions into percentages, and construct pie charts and bar charts to display the data. Write a paragraph comparing the changes in response time between the two years.

Response Time, 2000	Frequency (f)
21 minutes or more	20
16–20 minutes	90
11–15 minutes	185
6–10 minutes	370
Less than 6 minutes	210
	875

(*continued next column*)

(*continued*)

Response Time, 2010	Frequency (f)
21 minutes or more	45
16–20 minutes	95
11–15 minutes	155
6–10 minutes	350
Less than 6 minutes	250
	895

2.11 SOC Figures 2.10 through 2.12 display trends in crime in the United States since 1984. Write a paragraph describing each of these graphs. What similarities and differences can you observe among the three graphs? (For example, do crime rates always change in the same direction?) Note the differences in the vertical axes from chart to chart—for

FIGURE 2.10 **U.S. Homicide Rates, 1984–2007 (per 100,000 population)**

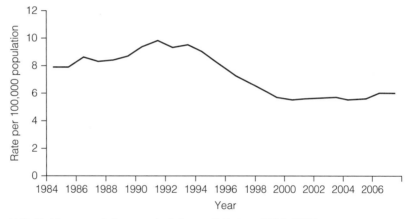

FIGURE 2.11 **U.S. Robbery and Aggravated Assault Rates, 1984–2007 (per 100,000 population)**

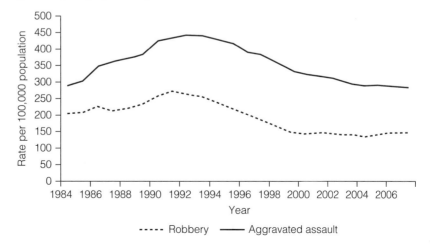

FIGURE 2.12 **U.S. Burglary and Car Theft Rates, 1984–2007**

homicide the axis ranges from 0 to 12, while for burglary and auto theft the range is from 0 to 1600. The latter crimes are far more common, and a scale with smaller intervals is needed to display the rates.

2.12 A hospital administrator has been keeping track of the number of times patients developed staph infections following surgery. Early in the year, the administrator required all medical staff to read a booklet titled *Fighting Staph*, and tripled the number of hand-sanitizer wall dispensers throughout the hospital. Data on the number of staph infections before and after the changes are presented here. Did the changes work? Is the hospital safer for post-surgical patients? Construct a line chart to display these two sets of data (use graphing software if available), and write a paragraph describing the changes.

Month	12 Months Before	12 Months After
January	23	25
February	25	21
March	20	18
April	19	12
May	15	9
June	17	10
July	24	11
August	28	15
September	23	17
October	20	14
November	21	18
December	22	20

YOU ARE THE RESEARCHER

Is There a "Culture War" in the United States?

One of the early steps in a research project is to produce frequency distributions and use them to inspect the variables. If nothing else, these tables will provide excellent background information and, sometimes, you can use them to begin to answer research questions. In this installment of "You Are the Researcher," you will use SPSS to produce summary tables for several variables that measure attitudes about controversial issues in U.S. society and that may map the battlefronts in what many call the American Culture Wars.

The United States seems to be divided on a number of religious, political, and cultural issues and values. We might characterize the opposing sides in terms of "liberal vs. conservative," "modern vs. traditional," or "progressive vs. old-school" and some of the most bitter debates across these lines include abortion, gay marriage, and gun control. As you know, debates over issues like these can be intense, bitter, and even violent: Adherents of one position may view their opponents with utter contempt, blast them with insults, and demonize them and dismiss their arguments. How deep is this fault line in U.S. society? How divided are the American people?

We can begin to investigate these questions by examining variables from the 2008 GSS. Pick three of these variables (see Appendix G or click **Utilities →** **Variables** on the menu bar of the Data Editor window of SPSS for a list of variables) that seem to differentiate the sides in the American culture war.

Before continuing, let's take a moment to consider the process of "picking variables." Technically, selecting a variable to represent a concept—like "culture war"—is called operationalization, and this can be one of the most difficult steps in a research project. Recall that we discussed operationalization in Chapter 1 when we reviewed the research process. When we move from the "Theory" stage of research to the "Hypothesis" stage (see Figure 1.1), we identify specific variables or measures (such as responses to survey items) that match our concepts (like prejudice). This can be challenging because our concepts are usually quite abstract and subject to a variety of perspectives. What exactly is a culture war and why are some positions liberal or traditional, conservative or progressive? In order to do research, we must use concrete, specific variables to represent our abstract and general concepts, but which variables relate to which concepts?

Any pairing we make between variables and concepts is bound to be at least a little arbitrary. In many cases, the best strategy is to use several variables to represent the concept: If our operationalizations are reasonable, our selected variables will "behave" similarly and each will behave as the abstract concept would if we could measure it directly. This is why I ask you to select three different variables to represent the culture wars. Each researcher may select different variables but, if everyone makes reasonable decisions, the chosen variables should be close representations of the concept.

After you have made your selections, complete the following steps.

Step 1: Identify Your Three Variables

Variable	SPSS Variable Name	Explain exactly what this variable measures
1		
2		
3		

Step 2: Operationalization

Explain why you selected each variable to represent an issue in the culture wars. How is the variable related to the debate? Which value or response of the variable is "liberal or progressive" and which is "conservative or traditional."

Variable	SPSS Variable Name	Explain how this variable exemplifies the culture war	Which value (e.g., "agree" or "support") is	
			Liberal	Conservative
1				
2				
3				

Step 3: Using SPSS for Windows to Produce Frequency Distributions

Now we are ready to generate output and get some background on the nature of-disagreements over values and issues among Americans. If necessary, click the SPSS icon on your monitor screen to start SPSS for Windows. Load the 2008 GSS by clicking the file name on the first screen or by clicking **File**, **Open**, and **Data** on the **SPSS Data Editor screen**. You may have to change the drive specification to locate the 2008 GSS data supplied with this text (probably named **GSS2008.sav**). Double click the file name to open the data set. When you see the message "SPSS Processor is Ready" on the bottom of the screen, you are ready to proceed.

Generating Frequency Distributions

We produced and examined a frequency distribution for the variable *sex* in Appendix F. Use the same procedures to produce frequency distributions for the three variables you used to represent the American culture wars. From the menu bar, click **Analyze**. From the menu that drops down, click **Descriptive Statistics** and **Frequencies**. The **Frequencies** window appears with the variables listed in alphabetical order in the left-hand box. The window may display variables by name (e.g., *abany, abnomore*) or by label (e.g., ABORTION IF WOMAN WANTS FOR ANY REASON). If labels are displayed, you may switch to variable names by clicking **Edit**, **Options**, and then making the appropriate selections on the "General" tab. See Appendix F and Table F.2 for further information.

Find the first of your three variables, click on its name to highlight it, and then click the arrow button in the middle of the screen to move it to the right-hand window. Find your other two variables and follow the same procedure to move their names to the right-hand window. SPSS will process all variables listed in the right-hand box together. Click **OK** in the upper-right-hand corner of the **Frequencies** window, and SPSS will rush off to create the frequency distributions you requested.

The tables will be in the **SPSS Viewer** window that will now be "closest" to you on the screen. The tables, along with other information, will be on the right-hand side of the window. To change the size of the output window, click the middle symbol (shaped like either a square or two intersecting squares) in the upper-right-hand corner of the **Output** window.

Reading SPSS Frequency Distributions

I will illustrate how to decipher the SPSS output using the variable *marital*, which measure current marital status. I chose *marital* so as not to duplicate any of the variables you selected in Step One. The output looks like this:

MARITAL STATUS

		Frequency	Percent	Valid Percent	Cumulative Percent
Valid	MARRIED	720	48.5	48.6	48.6
	WIDOWED	115	7.7	7.8	56.4
	DIVORCED	203	13.7	13.7	70.1
	SEPARATED	59	4.0	4.0	74.1
	NEVER MARRIED	384	25.9	25.9	100.0
	Total	1481	99.7	100.0	
Missing	NA	4	.3		
Total		1485	100.0		

Let's examine the elements of this table. The variable name and label are printed at the top of the output ("MARITAL STATUS"). The various categories are printed on the left. Moving one column to the right, we find the actual frequencies or the number of times each score of the variable occurred. We see that 720 of the respondents were married, 115 were widowed, and so forth. Next are two columns that report percentages. The entries in the "Percent" column are based on all respondents who were asked this question and include the scores "NA" (No Answer), "DK" (Don't Know), or "IAP" (Inapplicable). The "Valid Percent" column eliminates all cases with missing values. Since we almost always ignore missing values, we will pay attention only to the "Valid Percent" column (even though, in this case, only four scores are missing and the columns are virtually identical). The final column is a cumulative percentage column (see Table 2.16). For nominal-level variables like *marital*, this information is not meaningful, since the order in which the categories are stated is arbitrary.

The three frequency distributions you generated will have the same format and can be read in the same way. Use these tables—especially the "Valid Percents" column—to complete Step 4.

Step 4: Interpreting Results

Characterize your results by reporting the percentage (not the frequencies) of respondents who endorsed each response. How large are the divisions in American values? Is there consensus on the issue measured by your variable (do the great majority endorse the same response) or is there considerable disagreement? The lower the consensus, the greater the opportunity for the issue to be included in the culture war.

SPSS name of variable 1: _____

Summarize the frequency distribution in terms of the percentage of respondents who endorsed each position:

Are these results consistent with the idea that there is a "war" over Americans values? How?

SPSS name of variable 2: _____

Summarize the frequency distribution in terms of the percentage of respondents who endorsed each position:

Are these results consistent with the idea that there is a "war" over Americans values? How?

SPSS name of variable 3: _____

Summarize the frequency distribution in terms of the percentage of respondents who endorsed each position:

Are these results consistent with the idea that there is a "war" over Americans values? How?

3

Measures of Central Tendency

By the end of this chapter, you will be able to

1. Explain the purposes of measures of central tendency and interpret the information they convey.
2. Calculate, explain, and compare and contrast the mode, median, and mean.
3. Explain the mathematical characteristics of the mean.
4. Select an appropriate measure of central tendency according to level of measurement and skew.

One clear benefit of frequency distributions, graphs, and charts is that they summarize the overall shape of a distribution of scores in a way that can be quickly comprehended. Almost always, however, you will need to report more detailed information about the distribution. Specifically, two additional kinds of statistics are extremely useful: some idea of the typical or average case in the distribution (for example, "The average starting salary for social workers is $39,000 per year") and some idea of how much variety or heterogeneity there is in the distribution ("In this state, starting salaries for social workers range from $31,000 per year to $42,000 per year"). The first kind of statistic, called **measures of central tendency**, is the subject of this chapter. The second kind of statistic, measures of dispersion, is presented in Chapter 4.

The three commonly used measures of central tendency—the mode, median, and mean—are all probably familiar to you. All three summarize an entire distribution of scores by describing the most common score (the mode), the score of the middle case (the median), or the average score (the mean) of that distribution. These statistics are powerful because they can reduce huge arrays of data to a single, easily understood number. Remember that the central purpose of descriptive statistics is to summarize or "reduce" data.

Even though they share a common purpose, the three measures of central tendency are different statistics and they will have the same value only under certain conditions. As we shall see, they vary in terms of level of measurement, and, perhaps more importantly, they also vary in terms of how they define central tendency—they will not necessarily identify the same score or case as "typical." Thus, your choice of an appropriate measure of central tendency will depend in part on the way you measure the variable and in part on the purpose of the research.

The Mode

TABLE 3.1   Religious Preferences (fictitious data)

Denomination	Frequency
Protestant	128
Catholic	57
Jew	10
None	32
Other	15
	$N = 242$

The **mode** of any distribution of scores is the value that occurs most frequently. For example, in the set of scores 58, 82, 82, 90, 98, the mode is 82 because it occurs twice and the other scores occur only once.

The mode is a simple statistic, most useful when you are interested in the most common score and when you are working with nominal-level variables. In fact, the mode is the only measure of central tendency for nominal-level variables. For example, Table 3.1 reports the religious affiliations of a fictitious sample of 242 respondents. The mode of this distribution, the single largest category, is Protestant.

You should note that the mode has several limitations. First, some distributions have no mode at all (see Table 2.6). Other distributions have so many modes that the statistic becomes meaningless. For example, consider the distribution of test scores for a class of 100 students presented in Table 3.2. The distribution has modes at 55, 66, 78, 82, 90, and 97. Would reporting all six of these modes actually convey any useful information about central tendency in the distribution?

Second, the modal score of ordinal or interval-ratio variables may not be central to the distribution as a whole. That is, *most common* does not necessarily mean "typical" in the sense of identifying the center of the distribution. For example, consider the rather unusual (but not impossible) distribution of test scores in Table 3.3. The mode of the distribution is 93. Is this score very close to the majority of the scores? If the instructor summarized this distribution by reporting only the modal score, would he or she be conveying an accurate picture of the distribution as a whole? *(For practice in finding and interpreting the mode, see problems 3.1 to 3.7.)*

TABLE 3.2   A Distribution of Test Scores

Scores (% correct)	Frequency
97	10
91	7
90	10
86	8
82	10
80	3
78	10
77	6
75	9
70	7
66	10
55	10
	$N = 100$

TABLE 3.3   A Distribution of Test Scores

Scores (% correct)	Frequency
93	5
70	1
69	1
68	1
67	4
66	3
64	2
62	3
60	2
58	2
	$N = 24$

STATISTICS IN EVERYDAY LIFE	The modal (or most common) religious affiliation in the United States, by far, is Protestant. However, the nation is changing rapidly in this area of life, as in so many others. For example, since 1990, the percentage of Americans who identify as "Protestant" has dropped from 61% to 54%, while the percentage of Americans who profess no religious affiliation has almost doubled, from a little more than 8% to almost 16%.* Do these trends suggest that American society is growing stronger (more secular, more rational) or weaker (less moral, less guided by spiritual considerations)? Why? What additional information would you like to have before coming to a conclusion on this issue?

*Data are from the U.S. Bureau of the Census. 2009. *Statistical Abstract of the United States, 2009.* Table 75. Available at http://www.census.gov/prod/2009pub/10statab/pop.pdf.

The Median

Unlike the mode, the **median (Md)** is always at the exact center of a distribution of scores. The median is the score of the case that is at the middle of a distribution: Half the cases have scores higher than the median and half the cases have scores lower than the median. Thus, if the median family income for a community is $45,000, half the families earn more than $45,000 and half earn less.

Before finding the median, the cases must be placed in order from the highest to the lowest score—or from the lowest to the highest. Then you can determine the median by locating the case that divides the distribution into two equal halves. The median is the score of the middle case. If five students received grades of 93, 87, 80, 75, and 61 on a test, the median would be 80, the score that splits the distribution into two equal halves.

When the number of cases (*N*) is odd, the value of the median is unambiguous because there will always be a middle case. With an even number of cases, however, there will be two middle cases and, in this situation, the median is defined as the score exactly halfway between the scores of the two middle cases.

To illustrate, assume that seven students were asked to indicate their level of support for the intercollegiate athletic program at their universities on a scale ranging from 10 (indicating great support) to 0 (no support). After arranging their responses from high to low, you can find the median by locating the case that divides the distribution into two equal halves. With a total of seven cases, the middle case would be the fourth case, since there will be three cases above and three cases below this case. Table 3.4 lists the cases in order and identifies the median. With seven cases, the median is the score of the fourth case.

Now suppose we added one more student, whose support for athletics was measured as a 1 to the sample. This would make *N* an even number (8) and we would no longer have a single middle case. Table 3.5 presents the new distribution of scores, and, as you can see, any value between 7 and 5 would technically satisfy the definition of a median (that is, it would split the distribution into two equal halves of four cases each). We resolve this ambiguity by defining the median as the average of the scores of the two middle cases. In this example, the median would be defined as (7 + 5)/2, or 6.

TABLE 3.4 **Finding the Median with Seven Cases (*N* is odd)**

Case	Score	
1	10	
2	10	
3	8	
4	7	← Md
5	5	
6	4	
7	2	

TABLE 3.5 **Finding the Median with Eight Cases (*N* is even)**

Case	Score	
1	10	
2	10	
3	8	
4	7	
		Md = (7 + 5)/2 = 6
5	5	
6	4	
7	2	
8	1	

Remember that we follow different procedures to find the median depending whether *N* is odd or even. The procedures are stated in general terms in the One Step at a Time Box.

The median cannot be calculated for variables measured at the nominal level because it requires that scores be ranked from high to low. Remember that the scores of nominal-level variables cannot be ordered or ranked: The scores are different from each other but do not form a mathematical scale of any sort. The median can be found for either ordinal or interval-ratio data but is generally more appropriate for the former. *(The median may be found for any problem at the end of this chapter.)*

Other Measures of Position: Percentiles, Deciles, and Quartiles[1]

In addition to serving as a measure of central tendency, the median belongs to a class of statistics that measure position or location. The median identifies the exact middle of a distribution, but it is sometimes useful to locate other points as

---

[1]This section is optional.

ONE STEP AT A TIME   **Finding the Median**

1: Array the scores in order from high score to low score.

2: Count the number of scores to see if N is odd or even.

If N is ODD	If N is EVEN

3: The median will be the score of the middle case.

4: To find the middle case, add 1 to N and divide by 2.

5: The value you calculated in step 4 is the number of the middle case. The median is the score of this case. For example, if $N = 13$, the median will be the score of the $(13 + 1)/2$, or seventh, case.

3: The median is halfway between the scores of the two middle cases.

4: To find the first middle case, divide N by 2.

5: To find the second middle case, increase the value you computed in step 4 by 1.

6: Find the scores of the two middle cases. Add the scores together and divide by 2. The result is the median. For example, if $N = 14$, the median is the score halfway between the scores of the seventh and eighth cases.

7: If the middle cases have the same score, that score is defined as the median.

---

well. We may want to know, for example, the scores that split the distribution into thirds or fourths or the point below which a certain percentage of the cases fall. A familiar application of these measures would be scores on standardized tests, which are often reported in terms of location (for example, "A score of 476 is higher than 46% of the scores").

PERCENTILES

One commonly used statistic for reporting position is the **percentile**, which identifies the point below which a specific percentage of cases fall. If a score of 476 is reported as the 46th percentile, this means that 46% of the cases scored below this score.

To find a percentile, first arrange the scores in order. Next, multiply the number of cases (N) by the proportional value of the percentile. For example, the proportional value for the 46th percentile would be 0.46. The resultant value identifies the number of the case that marks the percentile. To find the 37th percentile for a sample of 78 cases, multiply 78 by 0.37. The result is 28.86, and the 37th percentile would be 86/100 of the distance between the scores of the 28th and 29th cases. In most cases, we would probably round off 28.86 to 29 and call the score of the 29th case the 37th percentile. The slight inaccuracy would be worth the savings in time and computational effort.

In terms of percentiles, the median is the 50th percentile, and in our example we would find the median by multiplying 78 by 0.50, finding the 39th case, and declaring the score of that case to be the 50th percentile. Notice again that we are cutting some corners here. Technically, the median would be the score halfway between the two middle cases (the 39th and 40th cases), but it is unlikely that this inaccuracy would be very significant.

DECILES AND QUARTILES

Some other commonly used measures of position are **deciles** and **quartiles**. Deciles divide the distribution of scores into tenths. So the first decile is the point below which 10% of the cases fall and is equivalent to the 10th percentile. The fifth decile is also the same as the 50th percentile, which is the same as (you guessed it) the median. Quartiles divide the distribution into quarters, and the first quartile is the same as the 25th percentile, the second quartile is the 50th percentile, and the third quartile is the 75th percentile. Any of these measures can be found by the method described earlier for percentiles. Remember that multiplying N by the proportional value of the percentile, decile, or quartile identifies the *case*, and it's the *score* of the case that actually marks the location. Also remember that this technique cuts some (probably minor) computational corners, so use it with caution. *(Quartiles, deciles, and percentiles may be found for any ordinal or interval-ratio variable in the problems at the end of this chapter.)*

The Mean

The **mean** ($\overline{X}$—read this as "ex-bar"),[2] or arithmetic average, is by far the most commonly used measure of central tendency. It reports the average score of a distribution, and is calculated by dividing the total sum of the scores by the number of scores (N).

To illustrate: A birth control clinic administered a 20-item test of general knowledge about contraception to 10 clients. The numbers of correct responses were 2, 10, 15, 11, 9, 16, 18, 10, 11, and 7. To find the mean of this distribution, add the scores (total = 109) and divide by the number of scores (10). The result (10.9) is the average score on the test.

The mathematical formula for the mean is

FORMULA 3.1

$$\overline{X} = \frac{\Sigma(X_i)}{N}$$

where $\overline{X}$ = the mean

$\Sigma(X_i)$ = the summation of the scores

N = the number of cases

Let's take a moment to consider the new symbols introduced in this formula. First, the symbol Σ (**uppercase Greek letter sigma**) is a mathematical operator

---

[2]This is the symbol for the mean of a sample. The mean of a population is symbolized with the Greek letter mu (μ—read this symbol as "mew").

just like the plus sign ($+$) or divide sign ($\div$). It stands for "the summation of" and directs us to add whatever quantities are stated immediately following it.[3] The second new symbol is X_i ("**X-sub-i** "), which refers to any single score—the "*i*th" score. If we wished to refer to the score of a particular case, the number of the case could replace the subscript. Thus, X_1 would refer to the score of the first case, X_2 to the score of the second case, X_{26} to the 26th case, and so forth.

The operation of adding all the scores is symbolized as $\Sigma(X_i)$. This combination of symbols directs us to sum the scores, beginning with the first score and ending with the last score in the distribution. Thus, Formula 3.1 states in symbols what has already been stated in words (to calculate the mean, add the scores and divide by the number of scores), but in a succinct and precise way. *(For practice in computing the mean, see any of the problems at the end of this chapter.)*

Because computation of the mean requires addition and division, it should be used with variables measured at the interval-ratio level. However, researchers do calculate the mean for variables measured at the ordinal level, because the mean is much more flexible than the median and is a central feature of many interesting and powerful advanced statistical techniques. Thus, if the researcher plans to do any more than merely describe his or her data, the mean will probably be the preferred measure of central tendency even for ordinal-level variables.

THREE CHARACTERISTICS OF THE MEAN

The mean is the most commonly used measure of central tendency, and we will consider its mathematical and statistical characteristics in some detail.

The Mean Balances All the Scores. First, the mean is an excellent measure of central tendency because it acts like a fulcrum that "balances" all the scores: The mean is the point around which all of the scores cancel out. We can express this property symbolically:

$$\Sigma(X_i - \overline{X}) = 0$$

This expression says that if we subtract the mean ($\overline{X}$) from each score (X_i) in a distribution and then sum the differences, the result will always be zero.

To illustrate, consider the test scores presented in Table 3.6. The mean of these five scores is 390/5, or 78. The difference between each score and the mean is listed in the right-hand column, and the sum of these differences is zero. The total of the negative differences (-19) is exactly equal to the total of the

---

ONE STEP AT A TIME	**Finding the Mean**

1: Add up the scores (X_i).   **2:** Divide the quantity you found in step 1 (ΣX_i) by N.

---

[3]See the Prologue (Basic Review of Math) for further information on the summation sign and on summation operations.

TABLE 3.6 **A Demonstration Showing That All Scores Cancel Out Around the Mean**

X_i	$(X_i - \overline{X})$
65	$65 - 78 = -13$
73	$73 - 78 = -5$
77	$77 - 78 = -1$
85	$85 - 78 = 7$
90	$90 - 78 = 12$
$\Sigma(X_i) = 390$	$\Sigma(X_i - \overline{X}) = 0$
$\overline{X} = 390/5 = 78$	

positive differences $(+19)$, as will always be the case. Thus, the mean "balances" the scores and is at the center of the distribution.

The Mean Minimizes the Variation of the Scores. A second characteristic of the mean is called the "least squares" principle, a characteristic that is expressed in the statement

$$\Sigma(X_i - \overline{X})^2 = \text{minimum}$$

or: The mean is the point in a distribution around which the variation of the scores (as indicated by the squared differences) is minimized. If the differences between the scores and the mean are squared and then added, the resultant sum will be less than the sum of the squared differences between the scores and any other point in the distribution.

To illustrate this principle, consider the distribution of five scores mentioned earlier: 65, 73, 77, 85, and 90. The differences between the scores and the mean have already been found. As illustrated in Table 3.7, if we square and sum these differences, we would get a total of 388. If we performed those same mathematical operations with any number other than the mean—say, the value 77—the result would be a sum greater than 388. Table 3.7 illustrates this point by showing that the sum of the squared differences around 77 is 393, a value greater than 388.

This least-squares principle underlines the fact that the mean is closer to all of the scores than the other measures of central tendency. Also, this characteristic of the mean is important for the statistical techniques of correlation and regression, topics we take up toward the end of this text.

The Mean Can Be Misleading If the Distribution Is Skewed. The final important characteristic of the mean is that it is affected by every score in the distribution. The mode (which is only the most common score) and the median (which deals only with the score of the middle case) are not so affected. This quality is both an advantage and a disadvantage. On one hand, the mean utilizes all the available information—every score in the distribution is included in the calculation of the mean. On the other hand, when a distribution has a few very

TABLE 3.7 A Demonstration Showing That the Mean Is the Point of Minimized Variation

X_i	$(X_i - \bar{X})$	$(X_i - \bar{X})^2$	$(X_i - 77)^2$
65	$65 - 78 = -13$	$(-13)^2 = 169$	$65 - 77 = (-12)^2 = 144$
73	$73 - 78 = -5$	$(-5)^2 = 25$	$73 - 77 = (-4)^2 = 16$
77	$77 - 78 = -1$	$(-1)^2 = 1$	$77 - 77 = (0)^2 = 0$
85	$85 - 78 = 7$	$(7)^2 = 49$	$85 - 77 = (8)^2 = 64$
90	$90 - 78 = 12$	$(12)^2 = 144$	$90 - 77 = (13)^2 = 169$
$\Sigma(X_i) = 390$	$\Sigma(X_i - \bar{X}) = 0$	$\Sigma(X_i - \bar{X})^2 = 388$	$\Sigma(X_i - 77)^2 = 393$

high or very low scores, the mean may be a very misleading measure of central-ity. When a distribution has some extremely high scores (this is called a *positive skew*), the mean will always have a greater numerical value than the median. If the distribution has some very low scores (a *negative skew*), the mean will be lower in value than the median.

To illustrate, consider Table 3.8. The five scores listed in column 1 have a mean and median of 25 (see column 2). In column 3, the scores are listed again with one score changed: 35 is changed to 3500. Look in column 4 and you will see that this change has no effect on the median; it remains at 25. This is because the median is based *only* on the score of the middle case and is not affected by changes in the scores of other cases in the distribution.

The mean, in contrast, is very much affected by the change because it takes *all* scores into account. The mean changes from 25 to 718 solely because of the one extreme score of 3500. Note also that the mean in column 4 is very different from four of the five scores listed in column 3. In this case, is the mean or the median a better representation of the scores? For distributions that have a few very high or very low scores, the mean may present a very misleading picture of the typical or central score. In these cases, the median may be the preferred measure of central tendency for interval-ratio variables. *(For practice in dealing with the effects of extreme scores on means and medians, see problems 3.11, 3.13, 3.14, and 3.15.)*

The general principle to remember is that, relative to the median, the mean is always pulled in the direction of extreme scores (i.e., scores that are much higher or lower than other scores). The mean and median will have the same

TABLE 3.8 A Demonstration Showing That the Mean Is Affected by Every Score

1 Scores	2 Measures of Central Tendency	3 Scores	4 Measures of Central Tendency
15	Mean = 125/5 = 25	15	Mean = 3590/5 = 718
20		20	
25	Median = 25	25	Median = 25
30		30	
35		3500	
$\Sigma = 125$		$\Sigma = 3590$	

FIGURE 3.1   A Positively Skewed Distribution (The mean is greater in value than the median)

FIGURE 3.2   A Negatively Skewed Distribution (The mean is less than the median)

FIGURE 3.3   An Unskewed, Symmetrical Distribution (The mean and median are equal)

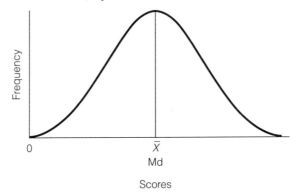

value when and only when a distribution is symmetrical. Figures 3.1 to 3.3 depict three different frequency polygons that demonstrate these relationships.

These relationships between medians and means also have a practical value. For one thing, a quick comparison of the median and mean will always tell you if a distribution is skewed and the direction of the skew. If the mean is less than the median, the distribution has a negative skew. If the mean is greater than the median, the distribution has a positive skew.

Second, these characteristics of the mean and median also provide a simple and effective way to "lie" with statistics. For example, if you want to maximize the average score of a positively skewed distribution, report the mean. Income data usually have a positive skew (there are only a few very wealthy people). If you want to impress someone with the general affluence of a mixed-income community, report the mean. If you want a lower figure, report the median.

STATISTICS IN EVERYDAY LIFE	The global reach of MacDonald's is well known and the near-ubiquity of American fast food provides a way of comparing cost-of-living around the globe. The Big Mac Index shows the local cost (in U.S. dollars) of this iconic burger. The latest data for 65 nations shows that prices range from $6.67 in Iceland to less than a dollar in Qatar, with an average of $2.45 and a median of $2.29.* What can you tell about the distribution of this variable from the fact that the mean is greater than the median?

*http://www.nationmaster.com/graph/eco_big_mac_ind-economy-big-mac-index.

Which measure is most appropriate for skewed distributions? This will depend on what point the researcher wishes to make but, as a rule, either both statistics or the median alone should be reported when the distribution has a few extreme scores.

Computing Measures of Central Tendency for Grouped Data[4]

Sometimes, data are presented in tables without measures of central tendency. In these situations, we can still compute means and medians if we are willing to make certain assumptions about how the scores are distributed. Usually, these assumptions will not be completely accurate, and the statistics we calculate based on them should be considered approximations of the true mean or median.

To illustrate, consider the following situation: As a new employee of a Youth Service Bureau, you are charged with preparing a report on truancy for a group of high school students. You have some truancy data left behind by the person you replaced, but the data are in the form of a frequency distribution (Table 3.9), and the ungrouped data are not available. What will you do?

TABLE 3.9   Truancy Data for 124 High School Students

Days Absent	Frequency
1–5	25
6–10	30
11–15	30
16–20	20
21–25	10
26–30	5
31–35	4
	$N = 124$

[4]This section is optional.

COMPUTING THE MEAN FOR GROUPED DATA

We need two values to compute a mean: the summation of the scores and the number of scores. For the distribution in Table 3.9, we know only the latter ($N = 124$). Because the data have been grouped, we do not know the original scores. We know from the table that 25 students were truant between one and five days, but we do not know exactly how many days. For all we know, these 25 students could have been truant for only one day each, or three or five days, or any combination of days between one and five, inclusive.

We can resolve this dilemma by assuming that all the scores are located at the midpoint of the interval. Then, by multiplying each midpoint by the number of cases in the interval, we can obtain an approximation of the sum of the original scores in the interval. To do this, add two columns to the original frequency distribution, one for the midpoints and one for the midpoints (m) multiplied by the frequencies (f). This is done for the truancy data in Table 3.10.

The summation of the last column—$\Sigma(fm)$—will be approximately the same value as the summation of the original scores, or $\Sigma(X_i)$.

So we have

$$\overline{X} = \frac{\Sigma(X_i)}{N} \cong \frac{\Sigma(fm)}{N}$$

where $\cong$ means "approximately equal to"

The final step in computing the mean for grouped data is to divide the total of the last column, $\Sigma(fm)$, by the number of cases:

FORMULA 3.2
$$\overline{X} = \frac{\Sigma(fm)}{N} = \frac{1567}{124} = 12.64$$

These 124 students were truant an average of 12.64 days over the year.

TABLE 3.10 Computing a Mean for Grouped Data

Days Absent	Frequency (f)	Midpoints (m)	Frequency × Midpoint ($f \times m$)
1–5	25	3	75
6–10	30	8	240
11–15	30	13	390
16–20	20	18	360
21–25	10	23	230
26–30	5	28	140
31–35	4	33	132
	$N = 124$		$\Sigma(fm) = 1567$

ONE STEP AT A TIME	Finding the Mean for Grouped Data

1: Find the midpoint (m) of each interval.

2: For each interval, multiply the midpoint (m) by the number of cases in the interval (f).

3: Find the sum of all the values calculated in step 2. This value is $\Sigma(fm)$.

4: Divide the quantity you found in step 3 by N. The result is the approximate mean.

COMPUTING THE MEDIAN FOR GROUPED DATA

To locate the median, we must first find the middle case of the distribution. With 124 cases in the truancy sample, the middle case is the ($N/2$), or 62nd, case. (Since N is even, we should find both middle cases but this shortcut will simplify calculations without much loss in accuracy.) We need to locate this case and then find the score associated with it. To do so, assume that the cases are evenly spaced throughout each interval in the frequency distribution. We then locate the interval that includes the middle case and, based on the assumption of equal spacing, find the score associated with that case.

To locate the interval that contains the median, add a cumulative frequency column to the original frequency distribution, as in Table 3.11. Looking at this column, we see that 55 cases are accumulated below the upper limit of the 6–10 interval and that 85 cases have been accumulated below the upper limit of the 11–15 interval. We know now that the middle, or 62nd, case is in the 11–15 interval and that the median is between 10.5 (the lower real limit of this interval) and 15.5 (the upper real limit), but we do not know the exact value.

To find the median, assume that all 30 of the cases in the 11–15 interval are evenly spaced throughout the interval, with case 56 located at the lower real limit (10.5) and case 85 at the upper real limit (15.5). Case 62 (our middle case) is the seventh case of the 30 in the interval. If the scores are evenly spaced, then the 62nd case will be 7/30 of the distance from 10.5 to 15.5. Now, 7/30 of 5 (the interval size) is 1.17, so we can approximate the median at 10.5 + 1.17, or 11.67.

TABLE 3.11 Computing a Median for Grouped Data

Days Absent	Frequency (f)	Cumulative Frequency (cf)
1–5	25	25
6–10	30	55
11–15	30	85
16–20	20	105
21–25	10	115
26–30	5	120
31–35	4	124
	$N = 124$	

Formula 3.3 summarizes these steps.

FORMULA 3.3

$$Md = rll + \left(\frac{N(.50) - cfb}{f} \right) i$$

where rll = real lower limit of the interval containing the median

cfb = cumulative frequency below the interval containing the median

f = number of cases in the interval containing the median

i = interval width

Applying this formula to our example, we would have

$$Md = 10.5 + \left(\frac{124(.50) - 55}{30} \right) 5 = 10.5 + \left(\frac{62 - 55}{30} \right) 5$$

$$= 10.5 + \left(\frac{7}{30} \right) 5 = 10.5 + 1.17 = 11.67$$

Thus, half of this sample was truant fewer than 11.67 days and half more than 11.67 days.

Because the value of the median (11.67) is almost a full day lower than the value of the mean (12.64), we can also conclude that this distribution has a positive skew, or a few very high scores. This skew is reflected in the frequency distribution itself, with most cases in the lower intervals. For purposes of description, the median may be the preferred measure of central tendency for these data, because the mean is affected by the relatively few very high scores.

This procedure may seem rather formidable the first time you attempt it. With a little practice, however, it will become routine. Of course, if you do not have access to the ungrouped data, there simply is no other way to find the median. *(For practice in finding means and medians with grouped data, see problems 3.5b and 3.6b.)*

ONE STEP AT A TIME　　**Finding the Median for Grouped Data**

1: Add a cumulative frequency column to the frequency distribution.

2: Find the middle case by multiplying N by 0.5.

3: Find the interval in the frequency distribution that contains the middle case.

4: Find the number of cases you will have to go into the interval to locate the middle case. This is given by N(0.5) – cfb (cfb = cumulative frequency below the interval that contains the median).

5: Divide the value you found in step 4 by f (the number of cases in the interval).

6: Multiply the number you found in step 5 by the interval size (i).

7: Add the number you found in step 6 to the lower real limit of the interval that contains the median. The result is an approximation of the median.

Choosing a Measure of Central Tendency

You should consider two main criteria when choosing a measure of central tendency. First, make sure that you know the level of measurement of the variable in question. This will generally tell you whether you should report the mode, median, or mean. Table 3.12 shows the relationship between the level of measurement and measures of central tendency. The capitalized, boldface "**YES**" identifies the most appropriate measure of central tendency for each level of measurement and the nonboldface "Yes" indicates the levels of measurement for which the measure is also permitted. An entry of "No" in the table means that the statistic cannot be computed for that level of measurement. Finally, the "Yes (?)" entry in the bottom row indicates that the mean is often used with ordinal-level variables even though, strictly speaking, this practice violates level-of-measurement guidelines.

Second, consider the definitions of the three measures of central tendency and remember that they provide different types of information. They will be the same value only under certain specific conditions (namely, for symmetrical distributions with one mode), and each has its own message to report. In many circumstances, you might want to report all three. The guidelines in Table 3.13

TABLE 3.12 **The Relationship Between Level of Measure and Measures of Central Tendency**

Measure of Central Tendency	Level of Measurement		
	Nominal	Ordinal	Interval-Ratio
Mode	**YES**	Yes	Yes
Median	No	**YES**	Yes
Mean	No	Yes (?)	**YES**

TABLE 3.13 **Choosing a Measure of Central Tendency**

Use the mode when:	1. The variable is measured at the nominal level.
	2. You want a quick and easy measure for ordinal and interval-ratio variables.
	3. You want to report the most common score.
Use the median when:	1. The variable is measured at the ordinal level.
	2. A variables measured at the interval-ratio level has a highly skewed distribution.
	3. You want to report the central score. The median always lies at the exact center of a distribution.
Use the mean when:	1. The variable is measured at the interval-ratio level (except when the variable is highly skewed).
	2. You want to report the typical score. The mean is "the fulcrum that exactly balances all of the scores."
	3. You anticipate additional statistical analysis.

stress both selection criteria and may be helpful when choosing a specific measure of central tendency.

Applying Statistics 3.1: The Mean and Median

Ten students have been asked how many hours they spent in the college library during the past week. What is the average "library time" for these students? The hours are reported in the following list, and we will find the mode, the median, and the mean for these data.

Student	Number of Visits to the Library Last Week (X_i)
1	0
2	2
3	5
4	5
5	7
6	10
7	14
8	14
9	20
10	30
$\Sigma(X_i) =$	107

By scanning the scores, we can see that two scores, 5 and 14, occurred twice, and no other score occurred more than once. This distribution has two modes, 5 and 14.

Because the number of cases is even, the median will be the average of the two middle cases after all cases have been ranked in order. With 10 cases, the first middle case will be the ($N/2$), or (10/2), or fifth case. The second middle case is the ($N/2$) + 1, or (10/2) + 1, or sixth case. The median will be the score halfway between the scores of the fifth and sixth cases. Counting down from the top, we find that the score of the fifth case is 7 and the score of the sixth case is 10. The median for these data is (7 + 10)/2, or (17/2), or 8.5. The median, the score that divides this distribution in half, is 8.5.

The mean is found by first adding all the scores and then dividing by the number of scores. The sum of the scores is 107, so the mean is

$$\overline{X} = \frac{\Sigma(X_i)}{N} = \frac{107}{10} = 10.7$$

These 10 students spent an average of 10.7 hours in the library during the week in question.

Note that the mean is higher than the median. This indicates a positive skew in the distribution (a few extremely high scores). By inspection we can see that the positive skew is caused by the two students who spent many more hours (20 hours and 30 hours) in the library than the other eight students.

SUMMARY

1. The three measures of central tendency presented in this chapter share a common purpose. Each reports some information about the most typical or representative value in a distribution. Appropriate use of these statistics permits the researcher to report important information about an entire distribution of scores in a single, easily understood number.

2. The mode reports the most common score and is used most appropriately with nominally measured variables.

3. The median (Md) reports the score that is the exact center of the distribution. It is most appropriately used with variables measured at the ordinal level and with variables measured at the interval-ratio level when the distribution is skewed.

4. The mean ($\overline{X}$), the most frequently used of the three measures, reports the most typical score. It is used most appropriately with variables measured at the interval-ratio level (except when the distribution is highly skewed).

5. The mean has a number of mathematical characteristics that are significant for statisticians. First, it is the point in a distribution of scores around which all other scores cancel out. Second, the mean is the point of minimized variation. Last, as distinct from the mode or median, the mean is affected by every score in the distribution and is therefore pulled in the direction of extreme scores.

SUMMARY OF FORMULAS

Mean	3.1	$\overline{X} = \dfrac{\Sigma(X_i)}{N}$
Mean for grouped data	3.2	$\overline{X} = \dfrac{\Sigma(fm)}{N}$
Median for grouped data	3.3	$Md = rll + \left(\dfrac{N(.50) - \text{cfb}}{f}\right)i$

GLOSSARY

Deciles. The points that divide a distribution of scores into 10ths.

Mean. The arithmetic average of the scores. $\overline{X}$ represents the mean of a sample, and μ, the mean of a population.

Measures of central tendency. Statistics that summarize a distribution of scores by reporting the most typical or representative value of the distribution.

Median (*Md*). The point in a distribution of scores above and below which exactly half of the cases fall.

Mode. The most common value in a distribution or the largest category of a variable.

Percentile. A point below which a specific percentage of the cases fall.

Quartiles. The points that divide a distribution into quarters.

Σ (uppercase Greek letter sigma). "The summation of."

Skew. The extent to which a distribution of scores has a few scores that are extremely high (positive skew) or extremely low (negative skew).

X_i ("*X-sub-i*"). Any score in a distribution.

PROBLEMS

3.1 $\boxed{\text{SOC}}$ A variety of information has been gathered from a sample of college freshmen and seniors, including their region of birth; the extent to which they support legalization of marijuana (measured on a scale on which 7 = strong support, 4 = neutral, and 1 = strong opposition); the amount of money they spend each week out-of-pocket for food, drinks, and entertainment; how many movies they watched in their dorm rooms last week; their opinion of cafeteria food (10 = excellent, 0 = very bad); and their religious affiliation. Some results are presented here. Find the *most appropriate* measure of central tendency for each variable for freshmen and then for seniors. Report both the measure you selected as well as its value for each variable (e.g., "Mode = 3" or "Median = 3.5"). (*HINT: Determine the level of measurement for each variable first. In general, this will tell you which measure of central tendency is appropriate. See the section "Choosing a Measure of Central Tendency" to review the relationship between measure of central tendency and level of measurement. Also, remember that the mode is the* most common *score, and especially remember to array scores from high to low before finding the median.*)

FRESHMEN

Student	Region of Birth	Legalization	Out-of-Pocket Expenses	Movies	Cafeteria Food	Religion
A	North	7	33	0	10	Protestant
B	North	4	39	14	7	Protestant
C	South	3	45	10	2	Catholic
D	Midwest	2	47	7	1	None
E	North	3	62	5	8	Protestant
F	North	5	48	1	6	Jew
G	South	1	52	0	10	Protestant
H	South	4	65	14	0	Other
I	Midwest	1	51	3	5	Other
J	West	2	43	4	6	Catholic

SENIORS

Student	Region of Birth	Legalization	Out-of-Pocket Expenses	Movies	Cafeteria Food	Religion
K	North	7	65	0	1	None
L	Midwest	6	62	5	2	Protestant
M	North	7	60	11	8	Protestant
N	North	5	90	3	4	Catholic
O	South	1	62	4	3	Protestant
P	South	5	57	14	6	Protestant
Q	West	6	40	0	2	Catholic
R	West	7	49	7	9	None
S	North	3	45	5	4	None
T	West	5	85	3	7	Other
U	North	4	78	5	4	None

3.2 A variety of information has been collected for all district high schools. Find the most appropriate measure of central tendency for each variable and summarize this information in a paragraph. (*HINT: The level of measurement of the variable will generally tell you which measure of central tendency is appropriate. Remember to organize the scores from high to low before finding the median.*)

High School	Enrollment	Largest Racial/ Ethnic Group	Percent College Bound	Most Popular Sport	Condition of Physical Plant (scale of 1–10 with 10 high)
1	1400	White	25	Football	10
2	1223	White	77	Baseball	7
3	876	Black	52	Football	5
4	1567	Hispanic	29	Football	8
5	778	White	43	Basketball	4
6	1690	Black	35	Basketball	5
7	1250	White	66	Soccer	6
8	970	White	54	Football	9

3.3 PS You have been observing the local Democratic Party in a large city and have compiled some information about a small sample of party regulars. Find the appropriate measure of central tendency for each variable.

Respondent	Sex	Social Class	No. of Years in Party	Education	Marital Status	Number of Children
A	M	High	32	High school	Married	5
B	M	Medium	17	High school	Married	0
C	M	Low	32	High school	Single	0
D	M	Low	50	Eighth grade	Widowed	7
E	M	Low	25	Fourth grade	Married	4
F	M	Medium	25	High school	Divorced	3
G	F	High	12	College	Divorced	3
H	F	High	10	College	Separated	2
I	F	Medium	21	College	Married	1
J	F	Medium	33	College	Married	5
K	M	Low	37	High school	Single	0
L	F	Low	15	High school	Divorced	0
M	F	Low	31	Eighth grade	Widowed	1

3.4 PA The following table presents the annual person-hours of time lost due to traffic congestion for a group of cities for 2007. This statistic is a measure of traffic congestion

City	Annual Person-Hours of Time Lost to Traffic Congestion per Year per Person
Baltimore	25
Boston	22
Buffalo	5
Chicago	22
Cleveland	7
Dallas	32
Detroit	29
Houston	32
Kansas City	8
Los Angeles	38
Miami	27
Minneapolis	22
New Orleans	10
New York	21
Philadelphia	21
Pittsburgh	8
Phoenix	23
San Antonio	21

(continued next column)

(continued)

City	Annual Person-Hours of Time Lost to Traffic Congestion per Year per Person
San Diego	29
San Francisco	29
Seattle	24
Washington, DC	31

Source: U.S. Bureau of the Census. 2009. *Statistical Abstract of the United States: 2009*. Table 1063. Available at http://www.census.gov/prod/2009pubs/10statab/trans.pdfhttp://www.census.gov/compendia/statab/.

a. Calculate the mean and median of this distribution.
b. Compare the mean and median. Which is the higher value? Why?
c. If you removed Los Angeles from this distribution and recalculated, what would happen to the mean? To the median? Why?
d. Report the mean and median as you would in a formal research report.

3.5 SOC For 15 respondents, data have been gathered on four variables (see the following table).
a. Find and report the appropriate measure of central tendency for each variable.
b. (*This problem is optional.*) For the variable age, construct a frequency distribution with interval

size = 5 and the first interval set at 15–19. Find the mean and median for the grouped variable and compare with the value for the ungrouped variable. How accurate is the estimate of the mean and median based on the grouped data?

Respondent	Marital Status	Racial/ Ethnic Group	Age	Attitude on Abortion Scale (high score = strong opposition)
A	Single	White	18	10
B	Single	Hispanic	20	9
C	Widowed	White	21	8
D	Married	White	30	10
E	Married	Hispanic	25	7
F	Married	White	26	7
G	Divorced	Black	19	9
H	Widowed	White	29	6
I	Divorced	White	31	10
J	Married	Black	55	5
K	Widowed	Asian American	32	4
L	Married	Native American	28	3
M	Divorced	White	23	2
N	Married	White	24	1
O	Divorced	Black	32	9

3.6 SOC The following table lists the median family incomes for 13 Canadian provinces and territories in 2000 and 2006. Compute the mean and median for each year and compare the two measures of central tendency. Which measure of central tendency is greater for each year? Are the distributions skewed? In which direction?

Province or Territory	2000	2006
Newfoundland and Labrador	38,800	50,500
Prince Edward Island	44,200	56,100
Nova Scotia	44,500	56,400
New Brunswick	43,200	54,000
Quebec	47,700	59,000
Ontario	55,700	66,600
Manitoba	47,300	58,700
Saskatchewan	45,800	60,500
Alberta	55,200	78,400
British Columbia	49,100	62,600
Yukon	56,000	76,000
Northwest Territories	61,000	88,800
Nunavut	37,600	54,300

3.7 Find the appropriate measure of central tendency for each variable displayed in problem 2.5. Report each statistic as you would in a formal research report.

3.8 Following are four variables for 30 cases from the General Social Survey. Age is reported in years. The variable *happiness* consists of answers to the question "Taken all together, would you say that you are (1) very happy (2) pretty happy, or (3) not too happy?" Respondents were asked how many sex partners they had over the past five years. Responses were measured on the following scale: 0–4 = actual numbers; 5 = 5–10 partners; 6 = 11–20 partners; 7 = 21–100 partners; 8 = more than 100.

a. For each variable, find the appropriate measure of central tendency and write a sentence reporting this statistical information as you would in a research report.

b. (*This problem is optional.*) For the variable age, construct a frequency distribution with interval size equal to 5 and the first interval set at 20–24. Compute the mean and median for the grouped data and compare with the values computed for the ungrouped data. How accurate are the estimates based on the grouped data?

Respondent	Age	Happiness	Number of Partners	Religion
1	20	1	2	Protestant
2	32	1	1	Protestant
3	31	1	1	Catholic

(*continued next page*)

(*continued*)

Respondent	Age	Happiness	Number of Partners	Religion
4	34	2	5	Protestant
5	34	2	3	Protestant
6	31	3	0	Jew
7	35	1	4	None
8	42	1	3	Protestant
9	48	1	1	Catholic
10	27	2	1	None
11	41	1	1	Protestant
12	42	2	0	Other
13	29	1	8	None
14	28	1	1	Jew
15	47	2	1	Protestant
16	69	2	2	Catholic
17	44	1	4	Other
18	21	3	1	Protestant
19	33	2	1	None
20	56	1	2	Protestant
21	73	2	0	Catholic
22	31	1	1	Catholic
23	53	2	3	None
24	78	1	0	Protestant
25	47	2	3	Protestant
26	88	3	0	Catholic
27	43	1	2	Protestant
28	24	1	1	None
29	24	2	3	None
30	60	1	1	Protestant

3.9 SOC The administration is considering a total ban on student automobiles. You have conducted a poll on this issue of 20 fellow students and 20 of the neighbors who live around the campus and have calculated scores for your respondents. On the scale you used, a high score indicates strong opposition to the proposed ban. The scores are presented here for both groups. Calculate an appropriate measure of central tendency and compare the two groups in a sentence or two.

Students		Neighbors	
10	11	0	7
10	9	1	6
10	8	0	0
10	11	1	3

(*continued next column*)

(*continued*)

Students		Neighbors	
9	8	7	4
10	11	11	0
9	7	0	0
5	1	1	10
5	2	10	9
0	10	10	0

3.10 SW As the head of a social services agency, you believe that your staff of 20 social workers is very much overworked compared to 10 years ago. The case loads for each worker are reported below for each of the two years in question. Has the average caseload increased? What measure of central tendency is most appropriate to answer this question? Why?

2000		2010	
52	55	42	82
50	49	75	50
57	50	69	52
49	52	65	50
45	59	58	55
65	60	64	65
60	65	69	60
55	68	60	60
42	60	50	60
50	42	60	60

3.11 SOC The following table lists the approximate number of cars per 1000 population for eight nations. Compute the mean and median for these data. Which is greater in value? Is there a positive skew in the data? How do you know?

Nation	Number of Cars per 1000 Population
United States	765
Canada	563
China	10
Russia	124
Japan	543
Mexico	138
Spain	471
United Kingdom	426

Source: http://www.nationmaster.com.

3.12 [SW] For the test scores first presented in problem 2.6 and reproduced here, compute a median and mean for both the pretest and posttest. Interpret these statistics.

Case	Pretest	Posttest
A	8	12
B	7	13
C	10	12
D	15	19
E	10	8
F	10	17
G	3	12
H	10	11
I	5	7
J	15	12
K	13	20
L	4	5
M	10	15
N	8	11
O	12	20

3.13 [SOC] A sample of 25 freshmen at a major university completed a survey that measured their degree of racial prejudice (the higher the score, the greater the prejudice).

a. Compute the median and mean scores for these data.

10	43	30	30	45
40	12	40	42	35
45	25	10	33	50
42	32	38	11	47
22	26	37	38	10

b. These same 25 students completed the same survey during their senior year. Compute the median and mean for this second set of scores, and compare them to the earlier set. What happened?

10	45	35	27	50
35	10	50	40	30
40	10	10	37	10
40	15	30	20	43
23	25	30	40	10

3.14 [SOC] You have compiled the following information on each of the graduates voted "most likely to succeed" by a local high school for a 10-year period. For each variable, find the appropriate measure of central tendency.

Case	Present Income	Marital Status	Owns a Home?	Years of Education Post–High School
A	104,000	Divorced	Yes	8
B	68,000	Divorced	No	4
C	54,000	Married	Yes	4
D	45,000	Married	No	4
E	40,000	Single	No	4
F	85,000	Separated	Yes	8
G	30,000	Married	No	3
H	27,000	Married	No	1
I	93,000	Married	Yes	6
J	48,000	Single	Yes	4

3.15 Professional athletes are threatening to strike because they claim that they are underpaid. The team owners have released a statement that says, in part, "The average salary for players was $1.2 million last year." The players counter by issuing their own statement that says, in part, "The average player earned only $753,000 last year." Is either side necessarily lying? If you were a sports reporter and had just read Chapter 3 of this text, what questions would you ask about these statistics?

YOU ARE THE RESEARCHER

The Typical American

Is there such a thing as a "typical" American? In this exercise, you will develop a profile of the average American based on measures of central tendency for 10 variables, chosen by you from the 2008 General Social Survey. Choose variables that you think are the most important in defining what it means to be a member of this society and then choose an appropriate measure of central tendency for each variable. Use this information to write a description of the typical American. We will also use this opportunity to introduce a new SPSS program.

Step 1: Choosing Your Variables

Scroll through the list of available variables either in Appendix G or by using the **Utilities → Variables** command in SPSS to get the online codebook. Select 10 variables that, in your view, are central to defining or describing the "typical American" and list them in the table below. *Select at least one variable from each of the three levels of measurement.*

Variable	SPSS Name	Explain exactly what this variable measures	Level of Measurement
1			
2			
3			
4			
5			
6			
7			
8			
9			
10			

Step 2: Getting the Statistics

Start *SPSS for Windows* by clicking the SPSS icon on your monitor screen. Load the 2008 GSS and when you see the message "SPSS Processor is Ready" on the bottom of the "closest" screen, you are ready to proceed.

Using the "Frequencies" Procedure for the Mode and Median

The only procedure in SPSS that will produce all three commonly used measures of central tendency (mode, median, and mean) is **Frequencies**. We used this procedure to produce frequency distributions in Chapter 2 and in Appendix F. Here you will use **Frequencies** to get modes and medians for any nominal- and ordinal-level variables you selected in Step 1.

Begin by clicking **Analyze** from the menu bar and then click **Descriptive Statistics** and **Frequencies**. In the **Frequencies** dialog box, find the names of your nominal and ordinal level variables in the list on the left and click the arrow button in the middle of the screen to move the names to the **Variables** box on the right.

To request specific statistics, click the **Statistics** button in the **Frequencies** dialog box, and the **Frequencies: Statistics** dialog box will open. Find the **Central Tendency** box on the right and click **Median** and **Mode**. Click **Continue**, and you will be returned to the **Frequencies** dialog box, where you might want to click the **Display Frequency Tables** box. When this box is *not* checked, SPSS will *not* produce frequency distribution tables, and only the statistics we request (mode and median) will appear in the **Output** window. Click **OK**, and SPSS will produce your output.

Record your results in the table below. Report the mode for all nominal-level variables and the median for ordinal-level variables. Use as many lines as needed.

Variable	SPSS Name	Level of Measurement	Mode	Median
1				
2				
3				
4				
5				
6				
7				
8				
9				

Using the "Descriptives" Procedure for the Mean

The **Descriptives** command in *SPSS for Windows* is designed to provide summary statistics for continuous interval-ratio–level variables. By default (i.e., unless you tell it otherwise), **Descriptives** produces the mean, the minimum and maximum scores (i.e., the lowest and highest scores, which can be used to compute the range), and the standard deviation. We will introduce the range and standard deviation in the next chapter.

To use **Descriptives**, click **Analyze**, **Descriptive Statistics**, and **Descriptives**. The **Descriptives** dialog box will open. This dialog box looks just like the **Frequencies** dialog box and works in the same way. Find the names of your interval-ratio–level variables in the list on the left and, once they are highlighted, click the arrow button in the middle of the screen to transfer them to the **Variables** box on the right. Click **OK** to produce your output and record your results in the table below, using as many lines as necessary.

Variable	SPSS Name	Level of Measurement	Mean
1			
2			
3			
4			
5			
6			
7			
8			
9			

Step 3: Interpreting Results

Examine the two tables displaying your results and write a summary paragraph or two describing the typical American. Be sure to report all ten variables and, as appropriate, describe the most common case (the mode), the typical case (the median), or the typical score (mean). Write as if you are reporting in a newspaper: Your goal should be clarity and accuracy. As an example, you might report that "The typical American is Protestant and voted for President Bush in 2004."

4

Measures of Dispersion

LEARNING OBJECTIVES   By the end of this chapter, you will be able to

1. Explain the purpose of measures of dispersion and the information they convey.
2. Compute and explain the index of qualitative variation (IQV), the range (R), the interquartile range (Q), the standard deviation (s), and the variance (s^2).
3. Select an appropriate measure of dispersion and correctly calculate and interpret the statistic.
4. Describe and explain the mathematical characteristics of the standard deviation.

In Chapters 2 and 3 you learned a variety of ways to describe a variable, including frequency distributions, graphs, and measures of central tendency. For a complete description of a distribution of scores, we must combine these with **measures of dispersion**, the subject of this chapter. While measures of central tendency describe the typical, average, or central score, measures of dispersion describe the variety, diversity, or heterogeneity of a distribution of scores.

The importance of the concept of **dispersion** might be easier to grasp if we consider a brief example. Suppose that the director of public safety wants to evaluate two ambulance services that have contracted with the city to provide emergency medical aid. As a part of the investigation, she has collected data on the response time of both services to calls for assistance. Data collected for the past year show that the mean response time is 7.4 minutes for Service A and 7.6 minutes for Service B. These averages, or means (calculated by adding up the response times to all calls and dividing by the number of calls), are very similar and provide no basis for judging one service as more or less efficient than the other. Measures of dispersion, however, can reveal substantial differences between distributions even when the measures of central tendency are the same. For example, consider Figure 4.1, which displays the distribution of response times for the two services in the form of frequency polygons, or line charts (see Chapter 2).

Compare the shapes of these two figures. Note that the line chart for Service B is much flatter than that for Service A. This is because the scores for Service B are more spread out, or more diverse, than the scores for Service A. In other words, Service B was much more variable in response time and had more scores in the high and low ranges and fewer in the middle. Service A was more consistent in its response time, and its scores are more clustered, or grouped, around the mean. Both distributions have essentially the same

FIGURE 4.1   Response Time for Two Ambulance Services

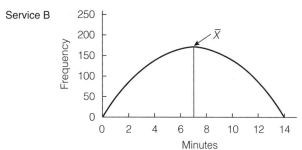

average response time, but there is considerably more *variation*, or dispersion, in the response times for Service B. If you were the director of public
safety, would you be more likely to select an ambulance service that was
always on the scene of an emergency in about the same amount of time
(Service A) or one that was sometimes very slow and sometimes very quick to
respond (Service B)? Note that if you had not considered dispersion, a possibly important difference in the performance of the two ambulance services
might have gone unnoticed.

Keep the two shapes in Figure 4.1 in mind as visual representations of the
concept of dispersion. The greater clustering of scores around the mean in
the distribution for Service A indicates *less* dispersion, and the flatter curve of
the distribution for Service B indicates *more* variety, or dispersion. Any of the
measures of dispersion discussed in this chapter will decrease in value as the
scores become less dispersed and the distribution becomes more peaked (that is,
as the distribution looks more and more like Service A's) and increase in value
as the scores become more dispersed and the distribution becomes flatter (that
is, as the distribution looks more and more like Service B's).

These ideas and Figure 4.1 may give you a general notion of what is meant
by dispersion, but the concept is not easily described in words alone. In this
chapter we introduce some of the more common measures of dispersion, each of
which provides a quantitative indication of the variety in a set of scores. We
begin with the index of qualitative variation, mention two measures—the range
and the interquartile range—briefly, and devote most of our attention to
the standard deviation and the variance.

The Index of Qualitative Variation (IQV)

We begin our consideration of measures of dispersion with the **index of qualitative variation (IQV)**. This statistic is rarely used in the professional research literature, but, since it is the only measure of dispersion available for nominal-level variables (although it can be used with any variable that has been grouped into a frequency distribution), it deserves some consideration.

The IQV is essentially the ratio of the amount of variation actually observed in a distribution of scores to the maximum variation that could exist in that distribution. The index varies from 0.00 (no variation) to 1.00 (maximum variation).

To illustrate the logic of this statistic, consider the increasing diversity of the United States population. Table 4.1 presents some information about the relative size of some groups for 2000 and some projections for 2025 and 2050 in three separate frequency distributions. Note that the values in the table are percentages instead of frequencies. This will greatly simplify computations. If there were no diversity in U.S. society (e.g., if everyone were white), the IQV would be 0.00. At the other extreme, if Americans were distributed equally across the five categories (i.e., if each group was 20% of the population), the IQV would achieve its maximum value of 1.00.

By inspection, you can see that the United States was quite diverse in 2000, not at all close to the extreme of zero diversity (or an IQV of 0.00). Whites comprised 71% of the population, but two other groups (blacks and Hispanics) were also quite sizeable at about 12% of the population. In 2025 and again in 2050, the percentage of whites declines and the proportional share of other groups, especially Hispanic Americans, increases. By 2050, the population will be much closer to maximum variation (an IQV of 1.00) than it was in 2000. Let us see how the IQV substantiates these observations. The computational formula for the IQV is

FORMULA 4.1

$$IQV = \frac{k(N^2 - \Sigma f^2)}{N^2(k - 1)}$$

where k = the number of categories

N = the number of cases

Σf^2 = the sum of the squared frequencies

TABLE 4.1   Racial and Ethnic Groups in the United States

Group	Percent of Total Population		
	2000	2025	2050
Non-Hispanic white Americans	71%	62%	53%
Black Americans	12%	13%	13%
Native Americans	1%	1%	1%
Asian Americans	4%	6%	9%
Hispanic Americans	12%	18%	24%
Totals =	100%	100%	100%

TABLE 4.2 **Finding the Sum of the Squared Frequencies**

Group	2000		2025		2050	
	f	f^2	f	f^2	f	f^2
Non-Hispanic white	71	5041	62	3844	53	2809
Black	12	144	13	169	13	169
Native American	1	1	1	1	1	1
Asian American	4	16	6	36	9	81
Hispanic	12	144	18	324	24	576
Total =	100	5346	100	4374	100	3636

This statistic uses the sum of the squared frequencies. It is most convenient to find this sum by adding a column to the frequency distribution for the squared frequencies and then summing this column. This procedure is illustrated in Table 4.2.

For each year, the sum of the frequency column is N and the sum of the squared frequency (Σf^2) is the total of the second column. Substituting these values into Formula 4.1 for the year 2000, we would have an IQV of 0.58:

$$\text{IQV} = \frac{k(N^2 - \Sigma f^2)}{N^2(k - 1)} = \frac{(5)(100^2 - 5346)}{(100^2)(4)} = \frac{(5)(10,000 - 5346)}{(10,000)(4)}$$

$$= \frac{(5)(4654)}{40,000} = \frac{23,270}{40,000} = 0.58$$

Since the values of k and N are the same for all three years, the IQV for the remaining years can be found by simply changing the values for (Σf^2). For the year 2025,

$$\text{IQV} = \frac{k(N^2 - \Sigma f^2)}{N^2(k - 1)} = \frac{(5)(100^2 - 4374)}{(100^2)(4)} = \frac{(5)(10,000 - 4374)}{(10,000)(4)}$$

$$= \frac{(5)(5626)}{40,000} = \frac{28,130}{40,000} = 0.70$$

and, similarly, for the year 2050,

$$\text{IQV} = \frac{k(N^2 - \Sigma f^2)}{N^2(k - 1)} = \frac{(5)(100^2 - 3636)}{(100^2)(4)} = \frac{(5)(10,000 - 3636)}{(10,000)(4)}$$

$$= \frac{(5)(6634)}{40,000} = \frac{31,820}{40,000} = 0.80$$

Thus, the IQV, in a quantitative and precise way, substantiates our earlier impressions.

ONE STEP AT A TIME | **Finding the Index of Qualitative Variation (IQV)**

If there is more than one column of frequencies (as in Table 4.1), complete these steps for each column in the table.

1: Add a column to the frequency distribution and label the new column "f^2."

2: Square each frequency and place the result in the new column you created in step 1.

3: Find the sum of the frequency (f) column and multiply this sum by itself. This value is N^2.

4: Find the sum of the squared frequency column. This value is Σf^2.

5: Subtract the sum of the squared frequencies (Σf^2) from N^2.

6: Multiply the value you found in step 5 by the number of categories (k).

7: Multiply N^2 by the number of categories minus 1 ($k - 1$).

8: Divide the quantity you found in step 6 by the quantity you found in step 7. The result is the IQV.

STATISTICS
IN EVERYDAY LIFE

In Chapter 3, we examined the changing religious profile of U.S. society, using percentages and the mode. Here, we present a table with more information about our increasing religious diversity and also note the IQV for the two years.

Religious Affiliation	1990	2008
Protestant	61.4%	53.6%
Catholic	26.8%	26.5%
Jewish	1.8%	1.2%
None	8.4%	15.8%
Other	1.6%	2.8%
TOTALS	100.0%	99.9%
IQV	0.68	0.77

What does the IQV add to the discussion in Chapter 3? How does it clarify the changes documented in the table?

The IQV of 0.58 for the year 2000 means that the distribution of frequencies shows almost 60% of the maximum variation possible. By 2050, the variation has increased to 80% of the maximum variation possible for this distribution of categories. The United States was already quite heterogeneous in 2000 and will grow increasingly diverse as the century progresses. *(For practice in calculating and interpreting the IQV, see problems 4.1 and 4.2.)*

The Range (*R*) and Interquartile Range (*Q*)

The **range (*R*)** is defined as the distance between the highest and lowest scores in a distribution:

FORMULA 4.2
$$R = \text{High score} - \text{Low score}$$

The range is quite easy to calculate and is perhaps most useful in gaining a quick and general notion of variability while scanning many distributions.

Unfortunately, the range has some important limitations related to the fact that it is based on only two scores (the highest and lowest scores). First, almost any sizeable distribution will contain some scores that are atypically high and/or low compared to most of the scores (for example, see Table 3.3). Thus, R might exaggerate the amount of dispersion for most of the scores in the distribution. Also, R yields no information about the variation of the scores between the highest and lowest scores.

The **interquartile range (Q)** is a kind of range. It avoids some of the problems associated with R by considering only the middle 50% of the cases in a distribution. To find Q, arrange the scores from highest to lowest and then divide the distribution into quarters (as distinct from halves when locating the median). The first quartile (Q_1) is the point below which 25% of the cases fall and above which 75% of the cases fall. The second quartile (Q_2) divides the distribution into halves (thus, Q_2 is equal in value to the median). The third quartile (Q_3) is the point below which 75% of the cases fall and above which 25% of the cases fall. Thus, if line LH represents a distribution of scores, the quartiles are located as shown:

The interquartile range is defined as the distance from the third to the first quartile, as stated in Formula 4.3:

FORMULA 4.3
$$Q = Q_3 - Q_1$$

The interquartile range essentially extracts the middle 50% of the distribution and, like R, is based on only two scores. Unlike the range, Q avoids the problem of being based on the most extreme scores, but it also fails to yield any information about the variation of the scores other than the two on which it is based.

COMPUTING THE RANGE AND INTERQUARTILE RANGE[1]

Table 4.3 presents a measure of educational attainment for 20 states. What are the range and interquartile range of these data? Note that the scores have already been ordered from high to low. This makes the range easy to calculate and is necessary for finding the interquartile range. Of these 20 states, Connecticut ranked the highest (34.7% of the population had a college degree) and West Virginia ranked the lowest (17.3%). The range is therefore $34.7 - 17.3$, or 17.4 ($R = 17.4$).

---

[1]This section is optional.

TABLE 4.3 Percent of Population Aged 25 and Older with a College Degree, 2007

Rank	State	Percent with College Degree
20 (highest)	Connecticut	34.7
19	New Hampshire	32.5
18	New York	31.7
17	California	29.5
16	Kansas	28.8
15	Montana	27.0
14	Maine	26.7
13	Florida	25.8
12	Pennsylvania	25.8
11	North Carolina	25.6
10	Arizona	25.3
9	Texas	25.2
8	Michigan	24.7
7	Idaho	24.5
6	Ohio	24.1
5	Indiana	22.1
4	Alabama	21.4
3	Louisiana	20.4
2	Mississippi	18.9
1 (lowest)	West Virginia	17.3

Source: U.S. Bureau of the Census, 2009. *Statistical Abstract of the United States, 2007.* Table 228. Available at http://www.census.gov/prod/2009pubs/10statab/educ.pdf.

To find Q, we must locate the first and third quartiles (Q_1 and Q_3). We will define these points in terms of the scores associated with certain cases, as we did when finding the median. Q_1 is determined by multiplying N by (.25). Since (20) $\times$ (.25) is 5, Q_1 is the score associated with the fifth case, counting up from the lowest score. The fifth case is Indiana, with a score of 22.1. So $Q_1 = 22.1$. The case that lies at the third quartile (Q_3) is given by multiplying N by (.75), and (20) $\times$ (.75) = 15th case. The 15th case, again counting up from the lowest score, is Montana, with a score of 27.0 ($Q_3 = 27.0$).[2] Therefore:

$$Q = Q_3 - Q_1$$

$$Q = 27.0 - 22.1$$

$$Q = 4.9$$

(For practice in finding and interpreting Q, see problem 4.5. The range may. be found for any of the problems at the end of this chapter except 4.1 and 4.2.)

[2]Note that, as we did in finding percentiles in Chapter 3, we are cutting some computational corners here. Technically, in this case, Q_1 would be halfway between the 5th and 6th scores and Q_3 would be halfway between the 15th and 16th scores. This slight inaccuracy is justified by considerable savings in time and computational effort, and similar shortcuts—defining quartiles in terms of the scores of the cases closest to Q_1 and Q_3—can be used routinely.

ONE STEP AT A TIME	Finding the Interquartile Range (*Q*)

1: Array the cases in order.

2: Find the case that lies at the first quartile (Q_1) by multiplying *N* by 0.25. If the result is not a whole number, round off to the nearest whole number. This is the number of the case that marks the first quartile. Note the *score* of this case.

3: Find the case that lies at the third quartile (Q_3) by multiplying *N* by 0.75. If the result is not a whole

number, round off to the nearest whole number. This is the number of the case that marks the third quartile. Note the *score* of this case.

4: Subtract the score of the case at the first quartile (Q_1)—see step 1—from the score of the case at the third quartile (Q_3)—see step 2. The result is the interquartile range (*Q*).

The Standard Deviation and Variance

A basic limitation of both *Q* and *R* is that they use only two scores and, thus, do not incorporate all the available information. Also, neither statistic provides any information on how far the scores are from each other or from some central point, such as the mean. How can we design a measure of dispersion that would correct these faults? We can begin with some specifications. A good measure of dispersion should:

1. Use all the scores in the distribution. The statistic should use all the information available.

2. Describe the average or typical deviation of the scores. The statistic should give us an idea about how far the scores are from each other or from the center of the distribution.

3. Increase in value as the scores became more diverse. This would be a very handy feature when comparing distributions because it would permit us to tell at a glance which was more variable: The higher the numerical value of the statistic, the greater the dispersion.

One way to develop a statistic to meet these criteria would be to start with the distances between each score and the mean, or the **deviations** ($X_i - \overline{X}$). The value of the deviations will increase as the differences between the scores and the mean increase. If the scores are more clustered around the mean (remember the graph for Service A in Figure 4.1), the deviations would be small. If the scores are more spread out, or more varied (like the scores for Service B in Figure 4.1), the deviations will be larger. How can we use the deviations to develop a useful statistic?

One course of action would be to use the sum of the deviations—$\Sigma(X_i - \overline{X})$—as the basis for a statistic, but, as we saw in Chapter 3, the sum of deviations will always be zero. To illustrate, consider a distribution of five scores: 10, 20, 30, 40, and 50. If we sum the deviations of the scores from the mean, we would always wind up with a total of zero, regardless of the amount of variety in the scores:

Scores (X_t)	Deviations ($X_i - \overline{X}$)
10	$(10 - 30) = -20$
20	$(20 - 30) = -10$
30	$(30 - 30) = 0$
40	$(40 - 30) = 10$
50	$(50 - 30) = 20$
$\Sigma(X_i) = 150$	$\Sigma(X_i - \overline{X}) = 0$
$\overline{X} = 150/5 = 30$	

Still, the sum of the deviations is a logical basis for a measure of dispersion, and statisticians have developed a way around the fact that the positive deviations always equal the negative deviations. If we square each of the deviations, all values will be positive because a negative number multiplied by itself becomes positive. For example: $(-20) \times (-20) = +400$. In our example, the sum of the squared deviations would be $(400 + 100 + 0 + 100 + 400)$, or 1000. Thus, a statistic based on the sum of the squared deviations will have the properties we want in a good measure of dispersion.

Before we finish designing our measure of dispersion, we must deal with another problem. The sum of the squared deviations will increase with sample size: The larger the *number* of scores, the greater the value of the measure. This would make it very difficult to compare the relative variability of distributions based on samples of different size. We can solve this problem by dividing the sum of the squared deviations by N (sample size) and thus standardizing for samples of different sizes.

These procedures yield a statistic known as the **variance**, which is symbolized as s^2. The variance is used primarily in inferential statistics, although it is a central concept in the design of some measures of association. For purposes of describing the dispersion of a distribution, a closely related statistic, called the **standard deviation** (symbolized as s), is typically used, and this statistic is our focus for the remainder of the chapter.

The formulas for the variance and standard deviation are

FORMULA 4.4

$$s^2 = \frac{\Sigma(X_i - \overline{X})^2}{N}$$

FORMULA 4.5

$$s = \sqrt{\frac{\Sigma(X_i - \overline{X})^2}{N}}$$

Strictly speaking, Formulas 4.4 and 4.5 are for the variance and standard deviation of a population. Slightly different formulas, with $N - 1$ instead of N in the denominator, should be used when we are working with random samples rather than entire populations. This is an important point because many of the

TABLE 4.4   Computing the Standard Deviation

Scores (X_i)	Deviations $(X_i - \overline{X})$	Deviations Squared $(X_i - \overline{X})^2$
10	$(10 - 30) = -20$	$(-20)^2 = 400$
20	$(20 - 30) = -10$	$(-10)^2 = 100$
30	$(30 - 30) = 0$	$(0)^2 = 0$
40	$(40 - 30) = 10$	$(10)^2 = 100$
50	$(50 - 30) = 20$	$(20)^2 = 400$
$\Sigma(X_i) = 150$	$\Sigma(X_i - \overline{X}) = 0$	$\Sigma(X_i - \overline{X})^2 = 1000$

electronic calculators and statistical software packages you might be using (including SPSS) use "$N - 1$" in the denominator and, thus, produce results that are at least slightly different from Formulas 4.4 and 4.5. The size of the difference will decrease as sample size increases, but the problems and examples in this chapter use small samples and the differences between using N and $N - 1$ in the denominator can be considerable in such cases. Some calculators offer the choice of "$N - 1$" or "N" in the denominator. If you use the latter, the values calculated for the standard deviation should match the values in this text.

To compute the standard deviation, it is advisable to construct a table such as Table 4.4 to organize computations. The five scores used in the previous example are listed in the left-hand column, the deviations are in the middle column, and the squared deviations are in the right-hand column.

The total of the last column in Table 4.4 is the sum of the squared deviations and should be substituted into the numerator of Formula 4.5. To finish solving the formula, divide the sum of the squared deviations by N and take the square root of the result. To find the variance, square the standard deviation. For our example problem, the variance is $s^2 = (14.14)^2 = 200$:

$$s = \sqrt{\frac{\Sigma(X_i - \overline{X})^2}{N}}$$

$$s = \sqrt{\frac{1000}{5}}$$

$$s = \sqrt{200}$$

$$s = 14.14$$

COMPUTING THE STANDARD DEVIATION: AN ADDITIONAL EXAMPLE

An additional example will help to clarify the procedures for computing and interpreting the standard deviation. A researcher is comparing the student bodies of two campuses. One college is located in a small town, and almost all the students reside on campus. The other is located in a large city, and the students are almost all part-time commuters. The researcher wishes to compare the ages of the students at the two campuses and has compiled the information presented

ONE STEP AT A TIME	**Finding the Standard Deviation (*s*) and the Variance (*s*²)**

To Begin

1: Construct a computing table like Table 4.4, with columns for the scores (X_i), the deviations ($X_i - \overline{X}$), and the deviations squared ($X_i - \overline{X}$)².

2: List the scores (X_i) in the left-hand column. Add up the scores and divide by N to find the mean. As a rule, state the mean in two places of accuracy or two digits to the right of the decimal point.

To Find the Values Needed to Solve Formula 4.5

1: Find the deviations ($X_i - \overline{X}$) by subtracting the mean from each score, one at a time. List the deviations in the second column. Generally speaking, you should state the deviations at the same level of accuracy (two places to the right of the decimal point) as the mean.

2: Add up the deviations. The sum must equal zero (within rounding error). If the sum of the deviations does not equal zero, you have made a computational

error and need to repeat step 1, perhaps at a higher level of accuracy.

3: Square each deviation and list the result in the third column.

4: Add up the squared deviations listed in the third column.

To Solve Formula 4.5

1: Transfer the sum of the squared deviations column to the numerator in Formula 4.5.

2: Divide the sum of the squared deviations (the numerator of the formula) by N.

3: Take the square root of the quantity you computed in the previous step. This is the standard deviation.

To Find the Variance (*s*²)

1: Square the value of the standard deviation (*s*).

in Table 4.5. Which student body is older and more diverse on age? (Needless to say, these very small groups are much too small for serious research and are used here only to simplify computations.)

We see from the means that the students from the residential campus are quite a bit younger than the students from the urban campus (19 vs. 23 years of age). Which group is more diverse in age? Computing the standard deviation will answer this question. To solve Formula 4.5, substitute the sum of the right-hand column ("Deviations Squared") in the numerator and N (5 in this case) in the denominator:

$$\text{Residential Campus } s = \sqrt{\frac{\Sigma(X_i - \overline{X})^2}{N}} = \sqrt{\frac{4}{5}} = \sqrt{0.8} = 0.89$$

$$\text{Urban Campus } \quad s = \sqrt{\frac{\Sigma(X_i - \overline{X})^2}{N}} = \sqrt{\frac{88}{5}} = \sqrt{17.6} = 4.20$$

The higher value of the standard deviation for the urban campus means that it is more diverse. As you can see by scanning the scores, the students at the residential college are within a narrow age range ($R = 20 - 18 = 2$), whereas the students at the urban campus are more mixed and include students of age 25 and 30 ($R = 30 - 18 = 12$). *(For practice in computing and*

TABLE 4.5 Computing the Standard Deviation for Two Campuses

	Residential Campus	
Ages (X_i)	Deviations $(X_i - \bar{X})$	Deviations Squared $(X_i - \bar{X})^2$
18	$(18 - 19) = -1$	$(-1)^2 = 1$
19	$(19 - 19) = 0$	$(0)^2 = 0$
20	$(20 - 19) = 1$	$(1)^2 = 1$
18	$(18 - 19) = -1$	$(-1)^2 = 1$
20	$(20 - 19) = 1$	$(1)^2 = 1$
$\Sigma(X_i) = 95$	$\Sigma(X_i - \bar{X}) = 0$	$\Sigma(X_i - \bar{X})^2 = 4$

$$\bar{X} = \frac{\Sigma(X_i)}{N} = \frac{95}{5} = 19$$

	Urban Campus	
Ages (X_i)	Deviations $(X_i - \bar{X})$	Deviations Squared $(X_i - \bar{X})^2$
20	$(20 - 23) = -3$	$(-3)^2 = 9$
22	$(22 - 23) = -1$	$(-1)^2 = 1$
18	$(18 - 23) = -5$	$(-5)^2 = 25$
25	$(25 - 23) = 2$	$(2)^2 = 4$
30	$(30 - 23) = 7$	$(7)^2 = 49$
$\Sigma(X_i) = 115$	$\Sigma(X_i - \bar{X}) = 0$	$\Sigma(X_i - \bar{X})^2 = 88$

$$\bar{X} = \frac{\Sigma(X_i)}{N} = \frac{115}{5} = 23$$

interpreting the standard deviation, see any of the problems at the end of this chapter except 4.1 and 4.2. Problems with smaller data sets, such as 4.3 to 4.5, are recommended for practicing computations until you are comfortable with these procedures.)

Computing the Standard Deviation from Grouped Data[3]

When data have been grouped into a frequency distribution, we face the same problem in computing the standard deviation that we faced with the mean and median: The exact value of the scores is no longer known. We resolve this problem by using the midpoints of the intervals as approximations of the original scores. In other words, instead of subtracting the scores from the mean to find the deviations, we'll subtract the midpoints from the mean. This procedure assumes that the scores in each interval of the frequency distribution are

---

[3]This section is optional.

TABLE 4.6   Computing the Mean for Grouped Data

Correct	Frequency (f)	Midpoints (m)	Frequency × Midpoint (f × m)
0–2	1	1	1
3–5	2	4	8
6–8	3	7	21
9–11	4	10	40
12–14	3	13	39
15–17	2	16	32
18–20	2	19	38
	17		$\Sigma(fm) = 179$

$$\overline{X} = \frac{\Sigma(X_i)}{N} = \frac{179}{17} = 10.53$$

clustered at the midpoint and the value we compute for s will be an approximation of the actual standard deviation.

To illustrate this procedure, we will use the data presented in Table 4.6, which reports the number of quiz questions missed in a class of 17 students. As you can see from the frequency column, some students got only a few items correct, two students got 18–20 items correct, and most got between 6 and 14 items correct. Table 4.6 demonstrates the computation of the mean for grouped data (see Section 3.7). The mean is 10.53, and we will use this value to compute the standard deviation.

The formula for computing the standard deviation for grouped data is

FORMULA 4.6

$$s = \sqrt{\frac{\Sigma f(m - \overline{X})^2}{N}}$$

where f = the number of cases in an interval

m = the midpoint of that interval

$\overline{X}$ = the mean

N = the total number of cases

Formula 4.6 tells us to subtract the value of the mean $(\overline{X})$ from the midpoint (m) of each interval, square the result, and then multiply that quantity by the number of cases in the interval (f). Next, divide by N, and, finally, find the square root of the resultant quantity. It will be helpful to organize computations in table format, as in Table 4.7.

To solve Formula 4.6, substitute the sum of the far right-hand column—$\Sigma f(m - \overline{X})^2$—into the numerator under the square root sign, divide by N, and take the square root of the resultant sum:

$$s = \sqrt{\frac{\Sigma f(m - \overline{X})^2}{N}} = \sqrt{\frac{436.22}{17}} = \sqrt{25.66} = 5.07$$

TABLE 4.7 Computing the Standard Deviation for Grouped Data

Correct	f	m	$m - \overline{X}$	$(m - \overline{X})^2$	$f(m - \overline{X})^2$
0–2	1	1	$1 - 10.53 = -9.53$	90.82	90.82
3–5	2	4	$4 - 10.53 = -6.53$	42.64	85.28
6–8	3	7	$7 - 10.53 = -3.53$	12.46	37.38
9–11	4	10	$10 - 10.53 = -0.53$	0.28	1.12
12–14	3	13	$13 - 10.53 = 2.47$	6.10	18.30
15–17	2	16	$16 - 10.53 = 5.47$	29.92	59.84
18–20	2	19	$19 - 10.53 = 8.47$	71.74	143.48
	17				$\Sigma f(m - \overline{X})^2 = 436.22$

ONE STEP AT A TIME	**Finding the Standard Deviation (s) and the Variance (s^2) from Grouped Data**

1: If necessary, construct a computing table like Table 4.6 and find the mean by dividing the sum of the frequency (f) column by N.

2: Construct a computing table like Table 4.7.

3: Subtract the mean from the midpoint of each interval. Note the result in a column labeled "$m - \overline{X}$"

4: Square the value you found in step 3 and note the result in a column labeled "$(m - \overline{X})^2$."

5: Multiply the value you found in step 4 by the frequency in that interval. Note the result in a column labeled "$f(m - \overline{X})^2$."

6: Add up the values in the "$f(m - \overline{X})^2$" column and transfer this sum to the numerator in Formula 4.6.

7: Divide the sum of the "$f(m - \overline{X})^2$" column (see step 6) by N.

8: Take the square root of the quantity you computed in step 7. This is the standard deviation (s).

To Find the Variance (s^2) for Grouped Data

1: Square the value of the standard deviation (s).

These 17 students averaged 10.53 items correct, and the distribution has a standard deviation of 5.07. Remember that these are approximations of the actual mean and standard deviations, based on the assumption that the scores in each interval are clustered at the midpoint. Of course, it would be preferable to have the exact values. But if the data have been grouped into a frequency distribution, the procedures presented here and in in Chapter 3 are the only ways to approximate the mean and standard deviation. *(For practice in computing descriptive statistics with grouped data, see problems 4.8b and 4.14b.)*

Interpreting the Standard Deviation

It is very possible that the meaning of the standard deviation (i.e., why we calculate it) is not completely obvious to you at this point. You might be asking: "Once I've gone to the trouble of calculating the standard deviation, what do

I have?" The meaning of this measure of dispersion can be expressed in three ways. The first and most important involves the normal curve, and we will defer this interpretation until the next chapter.

A second way of thinking about the standard deviation is as an index of variability that increases in value as the distribution becomes more variable. In other words, the standard deviation is higher for more diverse distributions and lower for less diverse distributions. The lowest value the standard deviation can have is zero, and this would occur for distributions with no dispersion (i.e., if every single case in the sample had exactly the same score). Thus, 0 is the lowest value possible for the standard deviation (although there is no upper limit).

A third way to get a feel for the meaning of the standard deviation is by comparing one distribution with another. We already did this when comparing the two ambulance services in Figure 4.1 and the residential and urban campuses in Table 4.5. You might also do this when comparing one group against another (e.g., men vs. women, blacks vs. whites) or the same variable at two different times. For example, suppose we found that the ages of the students on a particular campus had changed over time as indicated by the following summary statistics:

2000	2010
$\bar{X} = 21$	$\bar{X} = 25$
$s = 1$	$s = 3$

In 2000, students were, on the average, 21 years of age. By 2010, the average age had risen to 25. Clearly, the student body has grown older, and, according to the

---

STATISTICS
IN EVERYDAY LIFE

The citizens of more affluent societies can routinely expect to live into their 70s, 80s, or even older. This is not the case, of course, for nations with lower standards of living, nutrition, and health care. The table below presents statistical information on life expectancy for a group of 94 nations at three different income levels. Note that average life expectancy increases as income level increases but the measures of dispersion (standard deviation and range) do just the opposite. Why are lower-income nations so much more diverse on this variable? (*Hint: Note the ranges: Are people in less affluent societies more vulnerable across the life course?*)

Summary Statistics on Life Expectancy by Income Level for 94 Nations*

Income Level	Mean	Standard Deviation	High	Low	Range	N
High	80.3	1.2	83	78	5	27
Medium	71.6	4.4	79	52	27	41
Low	54.7	8.1	74	41	33	26

*Data are from Population Reference Bureau. 2009. *World Population Data Sheet*. Available at http://www.prb.org/Publications/Datasheets/2009/2009wpds.aspx.

Applying Statistics 4.1: The Standard Deviation

At a local preschool, 10 children were observed for 1 hour, and the number of aggressive acts committed by each was recorded in the following list. What is the standard deviation of this distribution? We will use Formula 4.5 to compute the standard deviation.

Number of Aggressive Acts

(X_i)	$(X_i - \overline{X})$	$(X_i - \overline{X})^2$
1	$1 - 4 = -3$	9
3	$3 - 4 = -1$	1
5	$5 - 4 = 1$	1
2	$2 - 4 = -2$	4
7	$7 - 4 = 3$	9
11	$11 - 4 = 7$	49

(continued next column)

(continued)

(X_i)	$(X_i - \overline{X})$	$(X_i - \overline{X})^2$
1	$1 - 4 = -3$	9
8	$8 - 4 = 4$	16
2	$2 - 4 = -2$	4
0	$0 - 4 = -4$	16

$\Sigma(X_i) = 40 \quad \Sigma(X_i - \overline{X}) = 0 \quad \Sigma(X_i - \overline{X})^2 = 118$

$$\overline{X} = \frac{\Sigma(X_i)}{N} = \frac{40}{10} = 4.0$$

Substituting into Formula 4.5, we have

$$s = \sqrt{\frac{\Sigma(X_i - \overline{X})^2}{N}} = \sqrt{\frac{118}{10}} = \sqrt{11.8} = 3.44$$

The standard deviation for these data is 3.44.

Applying Statistics 4.2: Describing Dispersion

The homicide rates of five western and five New England states are listed in the tables below. Which group of states has the higher rate? Which is more variable?

Homicide Rate for Five New England States, 2007 (homicides per 100,000 population)

State	Homicides (X_i)	Deviations $(X_i - \overline{X})$	Deviations Squared $(X_i - \overline{X})^2$
Connecticut	3.0	0.88	0.77
Massachusetts	2.8	0.68	0.46
Vermont	1.9	−0.22	0.05
Rhode Island	1.8	−0.32	0.10
New Hampshire	1.1	−1.02	1.04
	$\Sigma(X_i) = 10.6$	$\Sigma(X_i - \overline{X}) = 0.00$	$\Sigma(X_i - \overline{X})^2 = 2.42$

Source: U.S. Bureau of the Census, 2009. *Statistical Abstract of the United States, 2009.* Table 297. Washington, DC: U.S. Government Printing Office.

(continued)

Applying Statistics 4.2: (*continued*)

$$\overline{X} = \frac{\Sigma X_i}{N} = \frac{10.6}{5} = 2.12$$

$$s = \sqrt{\frac{\Sigma(X_i - \overline{X})^2}{N}} = \sqrt{\frac{2.42}{5}} = \sqrt{0.48} = 0.69$$

Homicide Rate for Five Western States, 2007 (homicides per 100,000 population)

State	Homicides (X_i)	Deviations ($X_i - \overline{X}$)	Deviations Squared ($X_i - \overline{X}$)^2
Nevada	7.5	3.46	11.97
California	6.2	2.16	4.67
Wyoming	3.1	−0.94	0.88
Oregon	1.9	−2.14	4.58
Montana	1.5	−2.54	6.45
	$\Sigma(X_i) = 20.2$	$\Sigma(X_i - \overline{X}) = 0.00$	$\Sigma(X_i - \overline{X})^2 = 28.55$

Source: U.S. Bureau of the Census, 2009. *Statistical Abstract of the United States, 2009*. Table 297. Washington, DC: U.S. Government Printing Office.

$$\overline{X} = \frac{\Sigma X_i}{N} = \frac{20.2}{5} = 4.04$$

$$s = \sqrt{\frac{\Sigma(X_i - \overline{X})^2}{N}} = \sqrt{\frac{28.55}{5}} = \sqrt{5.71} = 2.39$$

With such small groups, you can tell by inspection that the western states have higher homicide rates. This impression is confirmed by both the median, which is 1.9 for the New England states (Vermont is the middle case) and 3.1 for the western states (Wyoming is the middle case), and the mean (2.12 for the New England states and 4.04 for the western states). For both groups, the mean is greater than the median, indicating a positive skew. Note also that the skew is greater for the western states (that is, the difference between mean and median is greater). This is caused by Nevada, which has a much higher score than any other state. However, even with Nevada removed, the western states still have higher rates. With Wyoming removed, the mean of the remaining four western states is 3.18.

The five western states are also much more variable. The range for the western states is 6 ($R = 7.5 - 1.5 = 6$), much higher than the range for the New England states of 1.9 ($R = 3.0 - 1.1 = 1.9$). Similarly, the standard deviation for the western states (2.39) is more than three times greater than the standard deviation for the eastern states (.69). The greater dispersion in the western states is also largely a reflection of Nevada's extremely high score.

In summary, the five western states average higher homicide rates and are also more variable than the five eastern states.

standard deviation, it has also grown more diverse in age. The lower standard deviation for 2000 shows that ages would be more clustered around the mean in that year (remember the distribution for Service A in Figure 4.1). In 2010, in contrast, the larger standard deviation means that the distribution is flatter and more spread out, like the distribution for Service B in Figure 4.1. In other words, compared to 2010, the students in 2000 were more similar to each other and more clustered in a narrower age range. The standard deviation is extremely useful for making comparisons of this sort between distributions of scores.

READING STATISTICS 3: Measures of Central Tendency and Dispersion

As was the case with frequency distributions, measures of central tendency and dispersion may not be presented in the professional research literature. These reports focus on the relationships between variables, not on describing them one by one. Univariate descriptive statistics are usually just the first step in the data analysis, not the ultimate concern, and probably will not be included in the final publication. On the other hand, some statistics (for example, the mean and standard deviation) serve a dual function. They not only are valuable descriptive statistics but also form the basis for many analytical techniques. Thus, they may be reported in the latter role if not in the former.

When included in research reports, measures of central tendency and dispersion will most often be presented in some summary form such as a table. A fragment of such a summary table might look like this:

Variable	$\bar{X}$	s	N
Age	33.2	1.3	1078
Number of children	2.3	0.7	1078
Years married	7.8	1.5	1052
Income	55,786	1500	987

These tables describe the overall characteristics of the sample succinctly and, if you inspect the table carefully, you will have a good sense of the nature of the sample on the traits relevant to the project. Note that the number of cases varies from variable to variable in the preceding table. This is normal in social science research and is caused by missing data or incomplete information on some of the cases.

STATISTICS IN THE PROFESSIONAL LITERATURE

Professors Fiona Kay and Jean Wallace were concerned with the effects of mentors on the careers of lawyers. Specifically, they wondered about the effects of specific types of mentoring relationships on career advancement and rewards. They also were concerned with gender differences: Do male and female lawyers benefit in different ways or at different levels from having a mentor? The researchers studied the careers of almost 1600 lawyers in Ontario, Canada.

In this installment of "Reading Statistics," we review some of the descriptive statistics reported by Kay and Wallace. As is usually the case, they reported these statistics as background information and to help identify differences and important patterns. The actual hypotheses are tested with more advanced statistics.

Kay and Wallace present means and standard deviations for about 25 different variables, six of which are reproduced below. These results show little difference between the genders on the variables measuring relationships with mentors and some of the variables measuring the progress of and respondents' satisfaction with their careers. While they did find large differences in earnings, there was little gender difference in other variables such as satisfaction with career.

Want to find out more? See the following source.

Adapted from Kay, Fiona and Wallace, Jean, 2009. "Mentors as Social Capital: Gender, Mentors, and Career Rewards in Law Practice," Table 1, *Sociological Inquiry*, 79: 418–452. Used by permission.

(continued)

Means and Standard Deviations for Selected Variables: Males vs. Females

Variable	Males		Females	
	$\overline{X}$	s	$\overline{X}$	s
Closeness of relation with mentor (*1 = not at all close* to *4 = very close*)	2.85	0.69	2.85	0.73
Quality of mentorship (*1 = very poor* to *5 = very good*)	3.53	0.62	3.51	0.62
Earnings	205,502	143,417	134,577	109,730
Progress toward meeting career goals	3.41	1.01	3.25	0.96
Value of work to respondent	3.52	0.92	3.47	1.00
Work satisfaction	3.69	0.62	3.64	0.58

Interpreting Statistics: The Changing Distribution of Income in the United States

In this installment of Interpreting Statistics, we examine the changing distribution of income in the United States. We will use the latest information from the Census Bureau to answer several questions: Is average income rising or falling? Does the average American receive more or less income than in the past? Is the distribution of income becoming more unequal (i.e., are the rich getting richer)? We can answer these questions by looking at changes in measures of central tendency and dispersion. We will use the mean to look at changes in average income and the median to observe changes in income for the average American. The standard deviation would be the preferred measure of dispersion for an interval-ratio–level variable such as income, but this statistics is not available. Instead, we will use a variation of the interquartile range to measure dispersion.

Before considering the data, we should keep in mind that virtually any distribution of income will be positively skewed (see Figure 3.1). That is, in any group, community, state, or nation, the incomes of most people will be grouped around the mean or median but some people—the wealthiest members of the group—will have incomes far higher than the average. Of course, some people will also have incomes far below the mean or median, and their incomes may even approach zero. However, low incomes will have less effect on the mean than very high incomes. The lowest income cannot be less than zero, but the highest incomes can be in the millions or even billions. Since the mean uses *all* the scores, including the very highest, and is pulled in the direction of extreme scores, mean income will always be greater than median income.

Figure 4.2 shows the changes in mean and median income for the United States over almost four decades. Income is expressed in 2008 dollars to eliminate the effect of inflation over the years. Without this adjustment, recent incomes would appear to be much higher than older incomes, not because of increased

FIGURE 4.2   Mean and Median Household Incomes, U.S., 1967–2008 (2008 dollars)

buying power and well-being, but rather because of the changing value of the dollar. Also, Figure 4.2 is based on total income for entire households, not individual income.

The line labeled "Median" in Figure 4.2 shows that the income of the average American household was about $40,000 in 1967. This value gradually trends upward, with some declines in the mid-1970s, early 1980s, and early 1990s, all periods of recession. There is a pronounced increase in median income during the boom economy of the 1990s, but then we see a decline and leveling off as the economy once again fell into recession. By 2008, median income for the average American household had reached $50,303, about $10,000 higher than in 1967. This indicates an increase in standard of living for the average American household over this time period.

The pattern of the mean income is almost identical to that of the median. It rises and falls in almost exactly the same ways and at the same times. Notice, however, that the mean is always higher than the median, a reflection of the characteristic positive skew of income data. Also notice that the size of the gap between the mean and the median increases over time, indicating that the degree of skew is increasing. In other words, the difference between the median and mean increases because the underlying distribution is becoming more skewed—the income of the rich is increasing relative to the income of the typical American household.

An increasing skew means that the scores are becoming more diverse, and this means, in turn, that all measures of dispersion computed on the distribution of income between 1967 and 2008 will increase in value. The Census Bureau provides statistics similar to the interquartile range, and these are presented in Figure 4.3. Each line in the figure marks a certain percentile. For example, the 10th percentile is the income that splits the distribution into two groups: 10% of households earned less than this value and 90% earned more. The 80th percentile also splits the distribution into two groups: 20% earned more than this income and 80% earned less.

We can measure dispersion by comparing the incomes associated with the various percentiles: the 80th with the 20th and the 95th with the 10th. These will produce ranges similar to Q. Figure 4.3 shows that the lower percentiles

FIGURE 4.3 Percentiles of Household Income, U.S., 1967–2008 (2008 dollars)

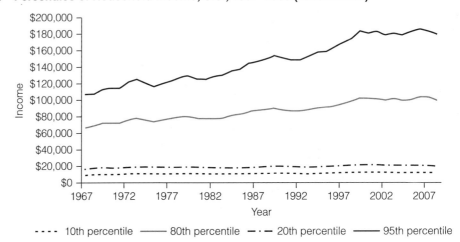

generally don't change in value but that the higher percentiles rise, at least until the early years of the 21st century. While the financial situation of lower-income Americans has remained virtually unchanged over this time period, the income of higher-income Americans has generally risen. This means that more affluent Americans received a much higher level of income at the end of the time period, which is very consistent with the increasing positive skew in Figure 4.2.

Taken together, these graphs show increases in the income of the average American household (median) and the average income for all American households (mean). However, the increase was not shared equally by all. It was virtually nonexistent for the less affluent (10th and 20th percentiles) and quite robust for those with higher incomes. Thus, dispersion increased. People with modest incomes continued to have modest incomes, and, consistent with the ancient folk wisdom, the rich got richer.

SUMMARY

1. Measures of dispersion summarize information about the heterogeneity, or variety, in a distribution of scores. When combined with an appropriate measure of central tendency, these statistics convey a large volume of information in just a few numbers. While measures of central tendency locate the central points of the distribution, measures of dispersion indicate the amount of diversity in the distribution.

2. The index of qualitative variation (IQV) can be computed for any variable that has been organized into a frequency distribution. It is the ratio of the amount of variation observed in the distribution to the maximum variation possible in the distribution. The IQV is most appropriate for variables measured at the nominal level.

3. The range (R) is the distance from the highest to the lowest score in the distribution. The interquartile range

(Q) is the distance from the third to the first quartile (the "range" of the middle 50% of the scores). These two ranges can be used with variables measured at either the ordinal or interval-ratio level.

4. The standard deviation (s) is the most important measure of dispersion because of its central role in many more advanced statistical applications. The standard deviation has a minimum value of zero (indicating no variation in the distribution) and increases in value as the variability of the distribution increases. It is used most appropriately with variables measured at the interval-ratio level.

5. The variance (s^2) is used primarily in inferential statistics and in the design of some measures of association.

SUMMARY OF FORMULAS

Index of Qualitative Variation	4.1	$IQV = \dfrac{k(N^2 - \Sigma f^2)}{N^2(k-1)}$	Standard Deviation	4.5	$s = \sqrt{\dfrac{\Sigma(X_i - \overline{X})^2}{N}}$
Range	4.2	$R = \text{High score} - \text{Low score}$			
Interquartile Range	4.3	$Q = Q_3 - Q_1$	Standard Deviation, Grouped Data	4.6	$s = \sqrt{\dfrac{\Sigma f(m - \overline{X})^2}{N}}$
Variance	4.4	$s^2 = \dfrac{\Sigma(X_i - \overline{X})^2}{N}$			

GLOSSARY

Deviations. The distances between the scores and the mean.

Dispersion. The amount of variety, or heterogeneity, in a distribution of scores.

Index of qualitative variation (IQV). A measure of dispersion for variables that have been organized into frequency distributions.

Interquartile range (Q). The distance from the third quartile to the first quartile.

Measures of dispersion. Statistics that indicate the amount of variety, or heterogeneity, in a distribution of scores.

Range (R). The highest score minus the lowest score.

Standard deviation. The square root of the squared deviations of the scores around the mean, divided by N. The most important and useful descriptive measure of dispersion; s represents the deviation of a sample; σ represents the standard deviation of a population.

Variance. The squared deviations of the scores around the mean divided by N. A measure of dispersion used primarily in inferential statistics and also in correlation and regression techniques; s^2 represents the variance of a sample; σ^2 represents the variance of a population.

PROBLEMS

4.1 SOC The marital status of residents of four apartment complexes is reported here. Compute the index of qualitative variation (IQV) for each neighborhood. Which is the most heterogeneous of the four? Which is the least? (*HINT: It may be helpful to organize your computations as in Table 4.2.*)

Complex A

Marital Status	Frequency
Single	26
Married	31
Divorced	12
Widowed	5
	$N = 74$

Complex B

Marital Status	Frequency
Single	10
Married	12
Divorced	8
Widowed	7
	$N = 37$

Complex C

Marital Status	Frequency
Single	20
Married	30
Divorced	2
Widowed	1
	$N = 53$

Complex D

Marital Status	Frequency
Single	52
Married	3
Divorced	20
Widowed	10
	$N = 85$

4.2 The following table shows the major areas of study selected by the first-year college students living in two small dormitories. Compute the index of qualitative variation for the table. Which dorm is more diverse in terms of this variable?

Major	Dorm A	Dorm B
Education	20	10
Social Science	7	17
Computer Science	3	4
Business	10	7
Other	7	10
	47	48

4.3 Compute the range and standard deviation of the following 10 scores. (*HINT: It will be helpful to organize your computations as in Tables 4.4.*)

10, 12, 15, 20, 25, 30, 32, 35, 40, 50

4.4 Compute the range and standard deviation of the following 10 test scores.

66, 75, 69, 72, 84, 90, 96, 70, 55, 45

4.5 SOC In problem 3.8, you computed mean and median income for 13 Canadian provinces and territories in two separate years. Now compute the standard deviation and range for each year, and, taking account of the two measures of central tendency and the two measures of dispersion, write a paragraph summarizing the distributions. What do the measures of dispersion add to what you already knew about central tendency? Did the provinces become more or less variable over the period? The scores are reproduced here.

Province or Territory	2000	2006
Newfoundland and Labrador	38,800	50,500
Prince Edward Island	44,200	56,100
Nova Scotia	44,500	56,400
New Brunswick	43,200	54,000
Quebec	47,700	59,000
Ontario	55,700	66,600
Manitoba	47,300	58,700
Saskatchewan	45,800	60,500
Alberta	55,200	78,400
British Columbia	49,100	62,600
Yukon	56,000	76,000
Northwest Territories	61,000	88,800
Nunavut	37,600	54,300

4.6 a. Compute the standard deviation for the pretest and posttest scores that were used in problems 2.6 and 3.12. The scores are reproduced here. Taking into account all of the information you have on these variables, write a paragraph describing how the sample changed from test to test. What does

the standard deviation add to the information you already had?

Case	Pretest	Posttest
A	8	12
B	7	13
C	10	12
D	15	19
E	10	8
F	10	17
G	3	12
H	10	11
I	5	7
J	15	12
K	13	20
L	4	5
M	10	15
N	8	11
O	12	20

b. Frequency distributions for both pretest and posttest scores are presented here. Use this information to compute a mean and standard deviation for each group, and compare these values with the actual mean and standard deviation you computed earlier. How accurate are the approximate mean and standard deviation as compared to the actual statistics?

Pretest

Scores	Frequency (f)
0–4	2
5–9	4
10–14	7
15–19	2
20–24	0
	N = 15

Posttest

Scores	Frequency (f)
0–4	0
5–9	3
10–14	7
15–19	3
20–24	2
	N = 15

4.7 SOC Labor force participation rates (percent employed), percent high school graduates, and mean income for males and females in 10 states are reported here. Calculate a mean and a standard deviation for both groups for each variable and describe the differences. Are males and females unequal on any of these variables? How great is the gender inequality?

State	Labor Force Participation		% High School Graduates		Mean Income	
	Male	Female	Male	Female	Male	Female
A	74	54	65	67	35,623	27,345
B	81	63	57	60	32,345	28,134
C	81	59	72	76	35,789	30,546
D	77	60	77	75	38,907	31,788
E	80	61	75	74	42,023	35,560
F	74	52	70	72	34,000	35,980
G	74	51	68	66	25,800	19,001
H	78	55	70	71	29,000	26,603
I	77	54	66	66	31,145	30,550
J	80	75	72	75	34,334	29,117

4.8 Data on several variables measuring overall heath and well-being for five nations are reported here for 2005, with projections to 2010. Are nations becoming more or less diverse on these variables? Calculate the mean, range, and standard deviation for each year for each variable. Summarize the results in a paragraph.

	Life Expectancy (years)		Infant Mortality Rate*		Fertility Rate[†]	
	2005	2010	2005	2010	2005	2010
Canada	80	81	4.8	5.0	1.6	1.6
United States	78	78	6.5	6.1	2.1	2.1
Mexico	75	76	20.9	17.8	2.5	6.6
Columbia	72	74	21.0	16.9	2.6	2.2
Japan	81	82	3.3	2.8	1.4	1.2
China	72	74	24.2	19.4	1.7	1.8
Sudan	59	53	62.5	78.1	4.9	4.5
Kenya	53	59	60.7	53.5	5.0	4.6
Italy	80	80	5.9	5.4	1.3	1.3
Germany	79	79	4.2	4.0	1.4	1.4

Sources: For 2005: U.S. Bureau of the Census, 2007. *Statistical Abstract of the United States, 2007.* p. 837. Available at http://www.census.gov/compendia /statab/. For 2010: U.S. Bureau of the Census, 2009. *Statistical Abstract of the United States, 2009.* Table 1303. Available at http://www.census.gov/prod/2009pubs/10statab/intlstat.pdf.
*Number of deaths of children under one year of age per 1000 live births.
[†]Average number of children per female.

4.9 In problem 3.11, you computed measures of central tendency for the number of cars per 1000 population for eight nations. The scores are reproduced here. Compute the standard deviation for this variable, and write a paragraph summarizing the mean, median, and standard deviation.

Nation	Number of Cars per 1000 Population
United States	765
Canada	563
China	10
Russia	124
Japan	543
Mexico	138
Spain	471
United Kingdom	426

Source: http://www.nationmaster.com.

4.10 Per capita expenditures for public transportation for 20 cities are reported here for 2000 and 2010. Compute a mean and standard deviation for each year, and describe the differences in expenditures for the five-year period.

City	2000	2010
A	52	197
B	87	124
C	101	131
D	52	197
E	115	119
F	88	87
G	100	150
H	101	209
I	95	110
J	78	140
K	103	178
L	107	94
M	85	125
N	117	200
O	167	225
P	180	210
Q	120	201
R	78	141
S	55	248
T	92	131
U	99	152

4.11 Compute the range and standard deviation for the data presented in problem 3.14. The data are reproduced here. What would happen to the value of the standard deviation if you removed Los Angeles from this distribution and recalculated? Why?

City	Annual Person-Hours of Time Lost to Traffic Congestion per Year per Person
Baltimore	25
Boston	22
Buffalo	5
Chicago	22
Cleveland	7
Dallas	32
Detroit	29
Houston	32
Kansas City	8
Los Angeles	38
Miami	27
Minneapolis	22
New Orleans	10
New York	21
Philadelphia	21
Pittsburgh	8
Phoenix	23
San Antonio	21
San Diego	29
San Francisco	29
Seattle	24
Washington, DC	31

Source: U.S. Bureau of the Census. 2009. *Statistical Abstract of the United States: 2009.* Table 1063. Available at http://www.census.gov/prod/2009pubs/10statab/trans.pdf.

4.12 SOC Listed here are the rates of abortion per 100,000 women for 20 states in 1973 and 1975. Describe what happened to these distributions over the two-year period. Did the average rate increase or decrease? What happened to the dispersion of this distribution? What happened between 1973 and 1975 that might explain these changes in central tendency and dispersion? (*HINT: It was a Supreme Court decision.*)

State	1973	1975
Maine	3.5	9.5
Massachusetts	10.0	25.7
New York	53.5	40.7
Pennsylvania	12.1	18.5

(continued next column)

(continued)

State	1973	1975
Ohio	7.3	17.9
Michigan	18.7	20.3
Iowa	8.8	14.7
Nebraska	7.3	14.3
Virginia	7.8	18.0
South Carolina	3.8	10.3
Florida	15.8	30.5
Tennessee	4.2	19.2
Mississippi	0.2	0.6
Arkansas	2.9	6.3
Texas	6.8	19.1
Montana	3.1	9.9
Colorado	14.4	24.6
Arizona	6.9	15.8
California	30.8	33.6
Hawaii	26.3	31.6

Source: United States Bureau of the Census, *Statistical Abstract of the United States: 1977* (98th edition). Washington, DC: Government Printing Office, 1977.

4.13 SW One of your goals as the new chief administrator of a large social service bureau is to equalize work loads within the various divisions of the agency. You have gathered data on caseloads per worker within each division. Which division comes closest to the ideal of an equalized workload? Which is farthest away?

A	B	C	D
50	60	60	75
51	59	61	80
55	58	58	74
60	55	59	70
68	56	59	69
59	61	60	82
60	62	61	85
57	63	60	83
50	60	59	65
55	59	58	60

4.14 You're the governor of the state and must decide which of four metropolitan police departments will win the annual award for efficiency. The performance of each department is summarized in monthly arrest statistics as reported here. Which department will win the award? Why?

	Departments			
	A	B	C	D
$\bar{X} =$	601.30	633.17	592.70	599.99
$s =$	2.30	27.32	40.17	60.23

4.15 At St. Algebra College, the math department ran some special sections of the freshman math course using a variety of innovative teaching techniques. Students were randomly assigned to either the traditional sections or the experimental sections, and all students were given the same final exam. The results of the final are summarized here. What was the effect of the experimental course?

Traditional	Experimental
$\bar{X} =$ 77.8	$\bar{X} =$ 76.8
$s =$ 12.3	$s =$ 6.2
$N =$ 478	$N =$ 465

4.16 a. Compute the standard deviation for both sets of data presented in problem 3.13 and reproduced here. Compare the standard deviation computed for freshmen with the standard deviation computed for seniors. What happened? Why? Does this change relate at all to what happened to the mean over the four-year period? How? What happened to the shapes of the underlying distributions?

Freshmen				
10	43	30	30	45
40	12	40	42	35
45	25	10	33	50
42	32	38	11	47
22	26	37	38	10

Seniors				
10	45	35	27	50
35	10	50	40	30
40	10	10	37	10
40	15	30	20	43
23	25	30	40	10

b. Frequency distributions for both freshmen and seniors are presented here. Use this information to compute a mean and standard deviation for each group, and compare these values with the actual mean and standard deviation you computed earlier. How accurate are the approximate mean and standard deviation as compared to the actual statistics?

Freshmen	
Scores	Frequency (f)
10–19	5
20–29	3
30–39	8
40–49	8
50–59	1
	$N =$ 25

Seniors	
Scores	Frequency (f)
10–19	7
20–29	4
30–39	6
40–49	6
50–59	2
	$N =$ 25

YOU ARE THE RESEARCHER

Following Up on Earlier Projects

Two projects are presented below. The first is a follow-up on the project presented at the end of Chapter 3. The second is a follow-up on the project presented at the end of Chapter 2. In the second project, you will be introduced to a new SPSS command. You are urged to complete both.

PROJECT 1: The Typical American Revisited

In Chapter 3, you described the typical American using 10 variables selected from the 2008 GSS. Now you will examine variation or dispersion on some of the variables you selected.

Step 1: Choosing the Variables

Select at least five of the ordinal and interval-ratio variables you used in Chapter 3. Add more variables if you had fewer than five variables at this level of measurement.

Step 2: Getting the Statistics

Use **Descriptives** to find the range and standard deviation for each of your selected variables. With the 2008 GSS loaded, click **Analyze**, **Descriptive Statistics**, and **Descriptives** from the Main Menu of SPSS. The **Descriptives** dialog box will open. Find the names of your variables in the list on the left and click the right arrow button to transfer them to the **Variables** box. Click **OK**, and SPSS will produce the same output you analyzed in Chapter 3. Now, however, we will consider dispersion rather than central tendency.

The standard deviation for each variable is reported in the column labeled "Std. Deviation," and the range can be computed from the values given in the Minimum and Maximum columns. At this point, the range is probably easier to understand and interpret than the standard deviation. As we have seen, the latter is more meaningful when we have a point of comparison. For example, suppose we were interested in the variable *tvhours* and how television viewing habits have changed over the years. The **Descriptives** output for 2008 shows that people watched an average of 3.03 hours a day with a standard deviation of 2.77. Suppose that data from 1978 showed an average of 3.70 hours of television viewing a day with a standard deviation of 1.1. You could conclude that television watching had, on the average, decreased over the 30-year period but that Americans had also become much more diverse in their viewing habits. Record your results in the table below.

Variable	SPSS Name	Mean	Range	Standard Deviation
1				
2				
3				
4				
5				

Step 3: Interpreting Results

Start with the descriptions of the variables you wrote in Chapter 3 and add information about dispersion, referring both to the range and the standard deviation. Your job now is both to describe the "typical" American *and* to suggest the amount of variation around this central point.

PROJECT 2: The Culture Wars Revisited

In Chapter 2, you examined some of the dimensions of the "culture wars" in U.S. society. In this project, you will reexamine the topic and learn how to use a new SPSS command to combine existing variables into a new summary variable. This computed variable can be used to summarize feelings and attitudes "in general" and to explore new dimensions of the issue.

Step 1: Creating a Scale to Summarize Attitudes toward Abortion

One of the most controversial issues in the American culture wars is abortion: Under what conditions, if any, should abortion be legal? The GSS data set supplied with this text includes two variables that measure support for legal abortion. The variables differ in context: One asks specifically if an abortion should be available if the mother is married and wants no more children (*abnomore*). The other asks if an abortion should be available when the woman wants it "for any reason" (*abany*). Since these two situations are distinct, each item should be analyzed in its own right. Suppose, however, that you wanted to create a summary scale that indicated a person's *overall* feelings about abortion.

One way to do this would be to add the scores on the two variables together. This would create a new variable, which we will call *abscale*, with three possible scores. If a respondent was consistently "pro-abortion" and answered "yes" (coded as "1") to both items, the respondent's score on the summary variable would be 1 + 1, or 2. A score of 3 would occur when a respondent answered "yes" (1) to one item and "no" (2) to the other. This might be labeled an "intermediate" or "moderate" position. The final possibility would be a score of 4 if the respondent answered "no" (coded as "2") to both items. This would be a consistent "anti-abortion" position. The table below summarizes the scoring possibilities.

If response on *abany* is:	and	response on *abnomore* is:	score on *abscale* will be:
1 (Yes)		1 (Yes)	2 (pro-abortion)
1 (Yes)		2 (No)	3 (moderate)
2 (No)		1 (Yes)	3 (moderate)
2 (No)		2 (No)	4 (anti-abortion)

The new variable, *abscale*, summarizes each respondent's overall position on the issue. Once created, *abscale* could be analyzed, transformed, and manipulated exactly like a variable actually recorded in the data file.

Using Compute. To use the **Compute** command, click **Transform** and then **Compute** from the main menu. The **Compute Variable** window will appear. Find the **Target Variable** box in the upper left-hand corner of this window. The first thing we need to do is assign a name (*abscale*) to the new variable we are about to compute and type that name in this box. Next, we need to tell SPSS how to compute the new variable. In this case, *abscale* will be computed by adding the scores of *abany* and *abnomore*. Find *abany* in the variable list on the left and click the arrow button in the middle of the screen to transfer the variable name to the **Numeric Expression** box. Next, click the plus sign (+) on the calculator pad under the **Numeric Expression** box, and the sign will appear next to *abany*. Finally, highlight *abnomore* in the variable list and click the arrow button to transfer the variable name to the **Numeric Expression** box.

The expression in the **Numeric Expression** box should now read

abany + abnomore

Click **OK**, and *abscale* will be created and added to the data set. If you want to keep this new variable permanently, click **Save** from the **File** menu, and the updated data set with *abscale* added will be saved to disk. (If you are using the student version of SPSS, remember that your data set is limited to 50 variables and you will not be allowed to save the data set with *abscale* added.)

Examining the Variables. We now have three variables that measure attitudes toward abortion—two items referring to specific situations, and a more general, summary item. It is always a good idea to check the frequency distribution for computed variables to make sure that the computations were carried out as we intended. Use the **Frequencies** procedure (click **Analyze, Descriptive Statistics,** and **Frequencies**) to get tables for *abany, abnomore,* and *abscale.* Your output will look like this:

ABORTION IF WOMAN WANTS FOR ANY REASON* (*abany*)

		Frequency	Percent	Valid Percent	Cumulative Percent
Valid	YES	397	26.7	41.8	41.8
	NO	552	37.2	58.2	100.0
	Total	949	63.9	100.0	
Missing	IAP	494	33.3		
	DK	31	2.1		
	NA	11	.7		
	Total	536	36.1		
Total		1485	100.0		

*This table has been slightly edited.

MARRIED–WANTS NO MORE CHILDREN* (*abnomore*)

		Frequency	Percent	Valid Percent	Cumulative Percent
Valid	YES	424	28.6	44.7	44.7
	NO	525	35.4	55.3	100.0
	Total	949	63.9	100.0	
Missing	IAP	494	33.3		
	DK	28	1.9		
	NA	14	.9		
	Total	536	36.1		
Total		1485	100.0		

*This table has been slightly edited.

Abscale*

		Frequency	Percent	Valid Percent	Cumulative Percent
Valid	2.00	351	23.6	37.7	37.7
	3.00	111	7.5	11.9	49.7
	4.00	468	31.5	50.3	100.0
	Total	930	62.6	100.0	
Missing	System	555	37.4		
Total		1485	100.0		

*This table has been slightly edited.

Missing Cases. Note that only 949 of the 1485 total respondents to the 2008 GSS answered the two specific abortion items. Remember that no respondent is given the entire GSS, and the vast majority of the "missing cases" received a form of the GSS that did not include these two items. These cases are coded as "IAP" in the tables above.

Now look at *abscale* and note that even fewer cases (930) are included in the summary scale than in the two original items. When SPSS executes a **Compute** statement, it automatically eliminates any cases that are missing scores on any of the constituent items. If these cases were not eliminated, a variety of errors and misclassifications could result. For example, if cases with missing scores were included, a person who scored a 2 ("anti-abortion") on *abany* and then failed to respond to *abnomore* would have a total score of 2 on *abscale*. Thus, this case would be treated as "pro-abortion" when the only information we have indicates that this respondent is "anti-abortion." Cases with missing scores are deleted to avoid this type of error.

Step 2: Interpreting the Results

Compare the distributions of these three variables with each other and write a report in which you answer the following questions:

1. How does the level of approval for abortion differ from one specific situation to another?
2. What percent of the respondents approved and disapproved of abortion in both circumstances? What percentage approved in one situation but not the other? (*Use the distribution of abscale to answer this question.*)
3. What do these patterns reveal about value consensus in American society? Are Americans generally in agreement about the issue of abortion?

5

The Normal Curve

LEARNING OBJECTIVES

By the end of this chapter, you will be able to

1. Define and explain the concept of the normal curve.
2. Convert empirical scores to *Z* scores and use *Z* scores and the normal curve table (Appendix A) to find areas above, below, and between points on the curve.
3. Express areas under the curve in terms of probabilities.

The **normal curve** is a concept of great importance in statistics. In combination with the mean and standard deviation, it is used to make precise descriptive statements about empirical distributions. Also, the normal curve is central to the theory that underlies inferential statistics. Thus, this chapter concludes our treatment of Part I (Descriptive Statistics) and lays important groundwork for Part II (Inferential Statistics).

The normal curve is sometimes used to assign grades on tests, and you may be familiar with this application. In this guise, the normal curve is often called the "bell curve," and grading "on the curve" means that the instructor wants the scores to follow a specific pattern: The modal grade is a C and there will be equal numbers of As and Fs, Bs and Ds. In other words, the distribution of grades should look like a bell or a mound.

Properties of the Normal Curve

The normal curve is a theoretical model, a kind of frequency polygon, or line chart, that is unimodal (i.e., has a single mode, or peak), perfectly smooth, and symmetrical (unskewed), so its mean, median, and mode are all exactly the same value. It is bell-shaped, and its tails extend infinitely in both directions. Of course, no empirical distribution matches this ideal model perfectly, but some variables (e.g., test results from large classes, standardized test scores such as the GRE) are close enough to permit the assumption of normality. In turn, this assumption makes possible one of the most important uses of the normal curve—the description of empirical distributions based on our knowledge of the theoretical normal curve.

The crucial point about the normal curve is that distances along the horizontal axis, when measured in standard deviations from the mean, always encompass the same proportion of the total area under the curve. In other

FIGURE 5.1   IQ Scores for a Group of Males

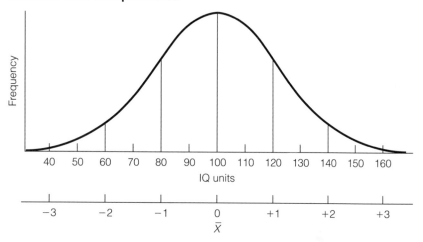

FIGURE 5.2   IQ Scores for a Group of Females

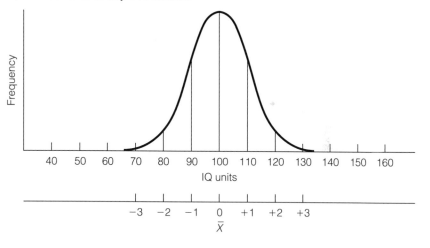

words, the distance from any point to the mean—when measured in standard deviations—will cut off exactly the same part of the area under the curve.

To illustrate, Figures 5.1 and 5.2 present two hypothetical distributions of IQ scores for fictional groups of males and females, both normally distributed, such that:

Males	Females
$\bar{X} = 100$	$\bar{X} = 100$
$s = 20$	$s = 10$
$N = 1000$	$N = 1000$

Figures 5.1 and 5.2 are drawn with two scales on the horizontal axis of the graph. The upper scale is stated in "IQ units" and the lower scale in standard deviations from the mean. These scales are interchangeable and we can easily

shift from one to the other. For example, for the males, an IQ score of 120 is one standard deviation (remember that, for the male group, $s = 20$) above the mean and an IQ of 140 is two standard deviations above (to the right of) the mean. Scores to the left of the mean are marked as negative values because they are less than the mean. An IQ of 80 is one standard deviation below the mean, an IQ score of 60 is two standard deviations less than the mean, and so forth. Figure 5.2 is marked in a similar way, except that, since its standard deviation is a different value ($s = 10$), the markings occur at different points. For the female sample, one standard deviation above the mean is an IQ of 110, one standard deviation below the mean is an IQ of 90, and so forth.

Recall that, on any normal curve, distances along the horizontal axis, when measured in standard deviations, always encompass exactly the same proportion of the total area under the curve. Specifically, the distance between one standard deviation above the mean and one standard deviation below the mean (or ± 1 standard deviation) encompasses exactly 68.26% of the total area under the curve. This means that in Figure 5.1, 68.26% of the total area lies between the score of 80 (-1 standard deviation) and 120 ($+1$ standard deviation). The standard deviation for females is 10, so the same percentage of the area (68.26%) lies between the scores of 90 and 110. On any normal distribution, 68.26% of the total area will always fall between ± 1 standard deviation, regardless of the trait being measured and the number values of the mean and standard deviation.

It will be useful to familiarize yourself with the following relationships between distances from the mean and areas under the curve:

Between	Lies
±1 standard deviation	68.26% of the area
±2 standard deviations	95.44% of the area
±3 standard deviations	99.72% of the area

These relationships are displayed graphically in Figure 5.3.

We can describe empirical distributions that are at least approximately normal using these relationships between distance from the mean and area.

FIGURE 5.3   Areas Under the Theoretical Normal Curve

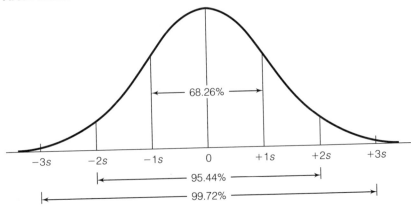

STATISTICS	Tests of intelligence are designed to generate distributions of scores that are approximately normal. That is, the tests mix easier and harder questions so that the mean will be about 100 and the final distribution of scores will be bell-shaped and have a mode of about 100 and roughly equal numbers of high and low scores.
IN EVERYDAY LIFE	Intelligence testing can be controversial and, although we cannot go into detail here, you should be aware that there is considerable debate about the meaning of IQ scores. Tests of intelligence may have cultural and other biases and there is considerable disagreement about whether or not they actually measure native mental ability. In particular, you should be aware that the more or less normal distribution of IQ scores does not necessarily mean that the underlying and elusive quality called intelligence—however that might be defined—is also normally distributed.

The position of individual scores can be described with respect to the mean, the distribution as a whole, or any other score in the distribution.

The areas between scores can also be expressed in numbers of cases rather than percentage of total area. For example, a normal distribution of 1000 cases will contain about 683 cases (68.26% of 1000 cases) between ±1 standard deviation of the mean, about 954 between ±2 standard deviations, and about 997 between ±3 standard deviations. Thus, for any normal distribution, only a few cases will be farther away from the mean than ±3 standard deviations.

Using the Normal Curve

We have seen that we can find areas under the normal curve for scores that are 1, 2, or 3 standard deviations above or below the mean. To work with values that are not exact multiples of the standard deviation, we must express the original scores in units of the standard deviation or convert them into **Z scores**. The original scores could be in any unit of measurement (feet, IQ, dollars), but Z scores always have the same values for their mean (0) and standard deviation (1).

COMPUTING Z SCORES

Think of converting the original scores into Z scores as a process of changing value scales—similar to changing from meters to yards, kilometers to miles, or gallons to liters. These units are different but equally valid ways of expressing distance, length, or volume. For example, a mile is equal to 1.61 kilometers, so two towns that are 10 miles apart are also 16.1 kilometers apart and a "5k" race covers about 3.10 miles. Although you may be more familiar with miles than kilometers, either unit works perfectly well as a way of expressing distance.

In the same way, the original (or "raw") scores and Z scores are two equally valid but different ways of measuring distances under the normal curve. In Figure 5.1, for example, we could describe a particular score in terms of IQ units ("John's score was 120") or standard deviations ("John scored one standard deviation above the mean").

When we compute Z scores, we convert the original units of measurement (IQ scores, inches, dollars, etc.) to Z scores and, thus, "standardize" the normal curve to a distribution that has a mean of 0 and a standard deviation of 1. The mean of the empirical normal distribution will be converted to 0, its standard deviation to 1, and all values will be expressed in Z-score form. The formula for computing Z scores is

FORMULA 5.1

$$z = \frac{X_i - \overline{X}}{s}$$

This formula will convert any score (X_i) from an empirical normal distribution into the equivalent Z score. To illustrate with the men's IQ data (Figure 5.1), the Z-score equivalent of a raw score of 120 would be

$$z = \frac{120 - 100}{20} = +1.00$$

The Z score of positive 1.00 indicates that the original score lies one standard deviation unit above (to the right of) the mean. A negative score would fall below (to the left of) the mean. *(For practice in computing Z scores, see any of the problems at the end of this chapter.)*

THE NORMAL CURVE TABLE

The theoretical normal curve has been very thoroughly described by statisticians. The areas related to any Z score have been precisely determined and organized into a table format. This **normal curve table**, or Z-score table, is presented as Appendix A in this text; for purposes of illustration, a small portion of it is reproduced here as Table 5.1.

The normal curve table consists of three columns, with Z scores in the left-hand column a, areas between the Z score and the mean in the middle column b, and areas beyond the Z score in the right-hand column c. To find the area between any Z score and the mean, go down the column labeled "Z" until you find the Z score. For example, go down column a either in Appendix A or in Table 5.1 until you find a Z score of +1.00. The entry in column b shows that the "area between mean and Z" is 0.3413. The table presents all areas in the form of proportions, but we can easily translate these into percentages by multiplying them by 100 (see Chapter 2). We could say either "a proportion of 0.3413 of the total area

ONE STEP AT A TIME	Finding Z Scores

1: Subtract the value of the mean ($\overline{X}$) from the value of the score (X_i).	**2:** Divide the quantity found in step 1 by the value of the standard deviation (s). The result is the Z-score equivalent for this raw score.

TABLE 5.1   **An Illustration of How to Find Areas Under the Normal Curve Using Appendix A**

(a) Z	(b) Area Between Mean and Z	(c) Area Beyond Z
0.00	0.0000	0.5000
0.01	0.0040	0.4960
0.02	0.0080	0.4920
0.03	0.0120	0.4880
⋮	⋮	⋮
1.00	0.3413	0.1587
1.01	0.3438	0.1562
1.02	0.3461	0.1539
1.03	0.3485	0.1515
⋮	⋮	⋮
1.50	0.4332	0.0668
1.51	0.4345	0.0655
1.52	0.4357	0.0643
1.53	0.4370	0.0630
⋮	⋮	⋮

under the curve lies between a Z score of 1.00 and the mean" or "34.13% of the total area lies between a score of 1.00 and the mean."

To illustrate further, find the Z score of 1.50 either in column a of Appendix A or the abbreviated table presented in Table 5.1. This score is 1½ standard deviations to the right of the mean and corresponds to an IQ of 130 for the men's IQ data. The area in column b for this score is 0.4332. This means that a proportion of 0.4332—or a percentage of 43.32%—of all the area under the curve lies between this score and the mean.

The third column in the table presents "Areas Beyond Z." These are areas above positive scores or below negative scores. This column will be used when we want to find an area above or below certain Z scores, an application that will be explained later in this chapter.

To conserve space, the normal curve table in Appendix A includes only positive Z scores. Since the normal curve is perfectly symmetrical, however, the area between the score and the mean—column b—for a negative score will be exactly the same as for a positive score of the same numerical value. For example, the area between a Z score of -1.00 and the mean will also be 34.13%, exactly the same as the area we found previously for a score of $+1.00$. As will be repeatedly demonstrated later, however, the sign of the Z score is extremely important and should be carefully noted.

For practice in using Appendix A to describe areas under an empirical normal curve, verify that the following Z scores and areas are correct for the men's IQ distribution. For each IQ score, the equivalent Z score is computed using Formula 5.1, and then Appendix A is consulted to find areas between the score and the mean. ($\overline{X} = 100$, $s = 20$ throughout.)

IQ Score	Z Score	Area Between Z and the Mean
110	+0.50	19.15%
125	+1.25	39.44%
133	+1.65	45.05%
138	+1.90	47.13%

The same procedures apply when the Z-score equivalent of an actual score happens to be a minus value (that is, when the raw score lies below the mean).

IQ Score	Z Score	Area Between Z and the Mean
93	−0.35	13.68%
85	−0.75	27.34%
67	−1.65	45.05%
62	−1.90	47.13%

Remember that the areas in Appendix A will be the same for Z scores of the same numerical value regardless of sign. For the men's IQ data, the area between the score of 138 (-1.90) and the mean is the same the area between 62 ($+1.90$) and the mean. *(For practice in using the normal curve table, see any of the problems at the end of this chapter.)*

FINDING THE TOTAL AREA ABOVE AND BELOW A SCORE

To this point, we have seen how the normal curve table can be used to find areas between a Z score and the mean. The information presented in the table can also be used to find other kinds of areas in empirical distributions that are at least approximately normal in shape. For example, suppose you need to determine the total area below the scores of two male subjects in the distribution described in Figure 5.1. The first subject has a score of 117 ($X_1 = 117$), which is equivalent to a Z score of $+0.85$:

$$z_1 = \frac{X_i - \overline{X}}{s} = \frac{117 - 100}{20} = \frac{17}{20} = +0.85$$

The plus sign of the Z score indicates that the score should be placed above (to the right of) the mean. To find the area below a positive Z score, the area between the score and the mean—given in column b—must be added to the area below the mean. As we noted earlier, the normal curve is symmetrical (unskewed) and its mean will be equal to its median. Therefore, the area below the mean (just like the median) will be 50%. Study Figure 5.4 carefully. We are interested in the shaded area.

By consulting the normal curve table, we find that the area between the score and the mean in column b is 30.23% of the total area. The area below a Z score of $+0.85$ is therefore 80.23% (50.00% + 30.23%). This subject scored higher than 80.23% of the persons tested.

FIGURE 5.4   Finding the Area Below a Positive Z Score

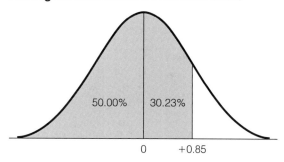

The second subject has an IQ score of 73 ($X_2 = 73$), which is equivalent to a Z score of -1.35:

$$z_2 = \frac{X_i - \overline{X}}{s} = \frac{73 - 100}{20} = -\frac{27}{20} = -1.35$$

To find the area below a negative score, we use column c, labeled "Area Beyond Z." The area of interest is depicted in Figure 5.5, and we must determine the size of the shaded area. The area beyond a score of -1.35 is given as 0.0885, which we can express as 8.85%. The second subject ($X_2 = 73$) scored higher than 8.85% of the tested group.

In the foregoing examples, we found the area below a score. Essentially the same techniques are used to find the area above a score. If we need to determine the area above an IQ score of 108, for example, we would first convert to a Z score,

$$Z = \frac{X_i - \overline{X}}{s} = \frac{108 - 100}{20} = \frac{8}{20} = +0.40$$

and then proceed to Appendix A. The shaded area in Figure 5.6 represents the area in which we are interested. The area above a positive score is found in the "Area Beyond Z" column, and, in this case, the area is 0.3446, or 34.46%.

FIGURE 5.5   Finding the Area Below a Negative Z Score

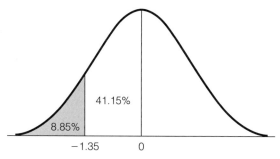

FIGURE 5.6 Finding the Area Above a Positive Z Score

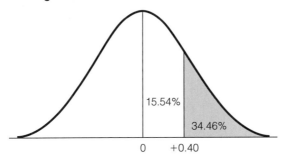

15.54%

34.46%

0   +0.40

TABLE 5.2 Finding Areas Above and Below Positive and Negative Scores

	When the Z Score Is	
To Find Area:	Positive	Negative
Above Z	Look in column c	Add column b area to .5000 or to 50.00%
Below Z	Add column b area to .5000 or to 50.00%	Look in column c

These procedures are summarized in Table 5.2 and in the One Step at a Time box. They might be confusing at first, and you should *always* draw the curve and shade in the areas in which you are interested. *(For practice in finding areas above or below Z scores, see problems 5.1 to 5.7.)*

ONE STEP AT A TIME **Finding Areas Above and Below Positive and Negative Z Scores**

1: Compute the Z score. Note whether the score is positive or negative.

2: Find the Z score in column a of the normal curve table (Appendix A).

3: To find the total area:

Below a Positive Z Score:

Add the column b area for this Z score to .5000, or multiply the column b area by 100 and add it to 50.00%.

or

Above a Positive Z Score:

Look in column c for this Z score. This value is the area above the score expressed as a proportion.

To express the area as a percentage, multiply the column c area by 100.

or

Below a Negative Z Score:

Look in column c for this Z score. This value is the area below the score expressed as a proportion. To express the area as a percentage, multiply the column c area by 100.

or

Above a Negative Z Score:

Add the column b area for this Z score to .5000, or multiply the column b area by 100 and add it to 50.00%. The result is the total area above this score.

FINDING AREAS BETWEEN TWO SCORES

On occasion, you will need to determine the area between two scores. When the scores are on opposite sides of the mean, the area between them can be found by adding the areas between each score and the mean. Using the men's IQ data as an example, if we wished to know the area between the IQ scores of 93 and 112, we would convert both scores to Z scores, find the area between each score and the mean in column b of Appendix A, and add these two areas together. The first IQ score of 93 converts to a Z score of -0.35:

$$Z_1 = \frac{X_i - \overline{X}}{s} = \frac{93 - 100}{20} = -\frac{7}{20} = -0.35$$

The second IQ score (112) converts to $+0.60$:

$$Z_2 = \frac{X_i - \overline{X}}{s} = \frac{112 - 100}{20} = \frac{12}{20} = 0.60$$

Both scores are placed on Figure 5.7. We are interested in the total shaded area. The total area between these two scores is 13.68% + 22.57%, or 36.25%. Therefore, 36.25% of the total area (or about 363 of the 1000 cases) lies between the IQ scores of 93 and 112.

When the scores of interest are on the same side of the mean, a different procedure must be followed to determine the area between them. For example, if we were interested in the area between the scores of 113 and 121, we would begin by converting these scores into Z scores:

$$Z_1 = \frac{X_i - \overline{X}}{s} = \frac{113 - 100}{20} = \frac{13}{20} = +0.65$$

$$Z_2 = \frac{X_i - \overline{X}}{s} = \frac{121 - 100}{20} = \frac{21}{20} = +1.05$$

The scores are noted in Figure 5.8; we are interested in the shaded area. To find the area between two scores on the same side of the mean, find the area between each score and the mean (given in column b of Appendix A) and then subtract the smaller area from the larger. Between the Z score of $+0.65$

FIGURE 5.7 Finding the Area Between Two Scores

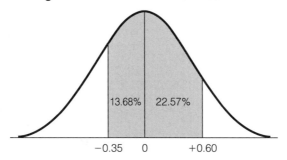

FIGURE 5.8 Finding the Area Between Two Scores

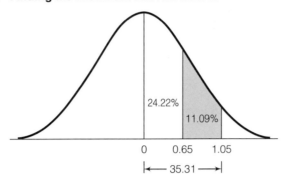

and the mean lies 24.22% of the total area. Between +1.05 and the mean lies 35.31% of the total area. Therefore, the area between these two scores is 35.31% − 24.22%, or 11.09% of the total area. The same technique would be followed if both scores had been below the mean. The procedures for finding areas between two scores are summarized in Table 5.3 and in the One Step at a Time box. *(For practice in finding areas between two scores, see problems 5.3, 5.4, and 5.6 to 5.9.)*

TABLE 5.3 Finding Areas Between Scores

Situation	Procedure
Scores are on the SAME side of the mean	Find areas between each score and the mean in column b. Subtract the smaller area from the larger area.
Scores are on OPPOSITE sides of the mean	Find areas between each score and the mean in column b. Add the two areas together.

ONE STEP AT A TIME   Finding Areas Between *Z* Scores

1: Compute the *Z* scores for both raw scores. Note whether the scores are positive or negative.

2: Find the areas between each score and the mean in column b.

If the Scores Are on the *Same* Side of the Mean	If the Scores Are on *Opposite* Sides of the Mean
Subtract the smaller area from the larger area. Multiply this value by 100 to express it as a percentage.	Add the two areas together to get the total area between the scores. Multiply this value by 100 to express it as a percentage.

Applying Statistics 5.1: Finding Z Scores and Areas

You have just received your score on a test of intelligence. If your score was 78 and you know that the mean score on the test was 67 with a standard deviation of 5, how does your score compare with the distribution of all test scores?

If you can assume that the test scores are normally distributed, you can compute a Z score and find the area below or above your score. The Z-score equivalent of your raw score would be

$$Z = \frac{X_i - \overline{X}}{s} = \frac{78 - 67}{5} = \frac{11}{5} = +2.20$$

Turning to Appendix A, we find that the "Area Between Mean and Z" for a Z score of 2.20 is 0.4861, which could also be expressed as 48.61%. Since this is a positive Z score, we need to add this area to 50.00% to find the total area below. Your score is higher than $48.61 + 50.00$, or 98.61%, of all the test scores. You did pretty well!

Applying Statistics 5.2: Finding Z Scores and Areas

All sections of Biology 101 at a large university were given the same final exam. Test scores were distributed normally, with a mean of 72 and a standard deviation of 8. What percentage of students scored between 60 and 69 (a grade of D) and what percentage scored between 70 and 79 (a grade of C)? The first two scores are both below the mean. Using Table 5.2 as a guide, we must first compute Z scores, find areas between each score and the mean, and then subtract the smaller area from the larger.

$$Z_1 = \frac{X_i - \overline{X}}{s} = \frac{60 - 72}{8} = -\frac{12}{8} = -1.50$$

$$Z_2 = \frac{X_i - \overline{X}}{s} = \frac{69 - 72}{8} = -\frac{3}{8} = -0.38$$

Using column b, we see that the area between $Z = -1.50$ and the mean is .4332 and the area between $Z = -0.38$ and the mean is .1480. Subtracting the smaller from the larger

$(.4332 - .1480)$ gives .2852. Changing to percentage format, we can say that 28.52% of the students earned a D on the test.

To find the percentage of students who earned a C, we must add column b areas together, since the scores (70 and 79) are on opposite sides of the mean (see Table 5.2):

$$Z_1 = \frac{X_i - \overline{X}}{s} = \frac{70 - 72}{8} = -\frac{2}{8} = -0.25$$

$$Z_2 = \frac{X_i - \overline{X}}{s} = \frac{79 - 72}{8} = \frac{7}{8} = 0.88$$

Using column b, we see that the area between $Z = -0.25$ and the mean is .0987 and the area between $Z = 0.88$ and the mean is .3106. Therefore, the total area between these two scores is $.0987 + .3106$, or 0.4093. Translating to percentages again, we can say that 40.93% of the students earned a C on this test.

Using the Normal Curve to Estimate Probabilities

To this point, we have thought of the theoretical normal curve as a way of describing the percentage of total area above, below, and between scores in an empirical distribution. We have also seen that these areas can be converted into the number of cases above, below, and between scores. In this section, we introduce the idea that the theoretical normal curve may also be thought of as a distribution of probabilities. Specifically, we may use the properties of the theoretical normal curve (Appendix A) to estimate the probability that a randomly selected case will have a score that falls in a certain range. In terms of techniques, these probabilities will be found in exactly the same way as areas were found. Before we consider these mechanics, however, let us examine what is meant by the concept of probability.

Although we are rarely systematic or rigorous about it, we all attempt to deal with probabilities every day, and, indeed, we base our behavior on our estimates of the likelihood that certain events will occur. We constantly ask (and answer) questions such as: What is the probability of rain? Of drawing to an inside straight in poker? Of the worn-out tires on my car going flat? Of passing a test if I don't study?

To estimate the probability of an event, we must first be able to define what would constitute a "success." The preceding examples contain several different definitions of a success (rain, drawing a certain card, flat tires, and passing grades). To determine a probability, a fraction must be established, with the numerator equaling the number of events that would constitute a success and the denominator equaling the total number of possible events where a success could theoretically occur:

$$\text{Probability} = \frac{\#\ successes}{\#\ events}$$

To illustrate, assume that we wish to know the probability of selecting a specific card—say, the king of hearts—in one draw from a well-shuffled deck of cards. Our definition of a success is specific (drawing the king of hearts); and with the information given, we can establish a fraction. Only one card satisfies our definition of success, so the number of events that would constitute a success is 1; this value will be the numerator of the fraction. There are 52 possible events (that is, 52 cards in the deck), so the denominator will be 52. The fraction is thus 1/52, which represents the probability of selecting the king of hearts on one draw from a well-shuffled deck of cards. Our probability of success is 1 out of 52.

We can leave this fraction as it is, or we can express it in several other ways. For example, we can express it as an odds ratio by inverting the fraction, showing that the odds of selecting the king of hearts on a single draw are 52:1 (or a probability of fifty-two to one). We can express the fraction as a proportion by dividing the numerator by the denominator. For our example, the corresponding proportion is .0192, which is the proportion of all possible events that would satisfy our definition of a success. In the social sciences, probabilities are usually expressed as proportions, and we will follow this convention throughout the remainder of this section. Using p to represent "probability," the probability of drawing the king of hearts (or any specific card) can be expressed as

$$p \text{ (king of hearts)} = \frac{\# \, successes}{\# \, events} = \frac{1}{52} = .0192$$

As conceptualized here, probabilities have an exact meaning: Over the long run, the events that we define as successes will bear a certain proportional relationship to the total number of events. The probability of .0192 for selecting the king of hearts in a single draw really means that, over thousands of selections of one card at a time from a full deck of 52 cards, the proportion of successful draws would be .0192. Or, for every 10,000 draws, 192 would be the king of hearts, and the remaining 9808 selections would be other cards. Thus, when we say that the probability of drawing the king of hearts in one draw is .0192, we are essentially applying to a single draw our knowledge of what would happen over thousands of draws.

Like proportions, probabilities range from 0.00 (meaning that the event has absolutely no chance of occurrence) to 1.00 (a certainty). As the value of the probability increases, the likelihood that the defined event will occur also increases. A probability of .0192 is close to zero, and this means that the event (drawing the king of hearts) is unlikely, or improbable.

These techniques can be used to establish simple probabilities in any situation in which we can specify the number of successes and the total number of events. For example, a single die has six sides, or faces, each with a different value, ranging from 1 to 6. The probability of getting any specific number (say, a 4) in a single roll of a die is therefore

$$p \text{ (rolling a 4)} = \frac{1}{6} = .1667$$

Combining this way of thinking about probability with our knowledge of the theoretical normal curve allows us to estimate the likelihood of selecting a case that has a score within a certain range. For example, suppose we wished to estimate the probability that a randomly chosen subject from the distribution of men's IQ scores would have an IQ score between 95 and the mean score of 100. Our definition of a success here would be the selection of any subject with a score in the specified range. Normally, we would next establish a fraction with the numerator equal to the number of subjects with scores in the defined range and the denominator equal to the total number of subjects. However, if the empirical distribution is normal in form, we can skip this step, since the probabilities, in proportion form, are already stated in Appendix A. That is, the areas in Appendix A can be interpreted as probabilities.

To determine the probability that a randomly selected case will have a score between 95 and the mean, we would convert the original score to a Z score:

$$Z = \frac{X_i - \overline{X}}{s} = \frac{95 - 100}{20} = -\frac{5}{20} = -0.25$$

Using Appendix A, we see that the area between this score and the mean is 0.0987. This is the probability we are seeking. The probability that a randomly selected case will have a score between 95 and 100 is 0.0987 (or, rounded off, 0.1, or 1 out of 10). In the same fashion, the probability of selecting a subject from any range of scores can be estimated. Note that the techniques for estimating probabilities are exactly the same as those for finding areas. The only new information introduced in this section is the idea that the areas in the normal curve table can also be thought of as probabilities.

To consider an additional example, what is the probability that a randomly selected male will have an IQ less than 123? We will find probabilities in exactly the same way we found areas. The score (X_i) is above the mean and, following the directions in Table 5.2, we will find the probability we are seeking by adding the area in column b to 0.5000. First, we find the Z score:

$$Z = \frac{X_i - \overline{X}}{s} = \frac{123 - 100}{20} = \frac{23}{20} = +1.15$$

Next, look in column b of Appendix A to find the area between this score and the mean. Then add the area (0.3749) to 0.5000. The probability of selecting a male with an IQ of less than 123 is 0.3749 + 0.5000, or 0.8749. Rounding this value to .88, we can say that the odds are .88 (very high) that we will select a male with an IQ score in this range. Technically, remember that this probability expresses what would happen over the long run: For every 100 males selected

---

Applying Statistics 5.3: Finding Probabilities

The distribution of scores on a biology final exam used in Applying Statistics 5.2 had a mean of 72 and a standard deviation of 8. What is the probability that a student selected at random will have a score less than 61? More than 80? Less than 98? To answer these questions, we must first calculate Z scores and then consult Appendix A. We are looking for probabilities, so we will leave the areas in proportion form. The Z score for a score of 61 is

$$Z_1 = \frac{X_i - \overline{X}}{s} = \frac{61 - 72}{8} = -\frac{11}{8} = -1.38$$

This score is a negative value (below, or to the left of, the mean) and we are looking for the area below. Using Table 5.1 as a guide, we see that we must use column c to find the area below a negative score. This area is .0838. Rounding off, we can say that the odds of selecting a student with a score less than 61 are only 8 out of 100. This low value tells us this would be an unlikely event.

The Z score for the score of 80 is

$$Z_2 = \frac{X_i - \overline{X}}{s} = \frac{80 - 72}{8} = \frac{8}{8} = 1.00$$

The Z score is positive, and to find the area above (greater than) 80, we look in column c (see Table 5.1). This value is .1587. The odds of selecting a student with a score greater than 80 is roughly 16 out of 100, about twice as likely as selecting a student with a score of less than 61.

The Z score for the score of 98 is

$$Z_1 = \frac{X_i - \overline{X}}{s} = \frac{98 - 72}{8} = \frac{26}{8} = 3.25$$

To find the area below a positive Z score, we add the area between the score and the mean (column b) to .5000 (see Table 5.1). This value is .4994 + .5000, or .9994. It is extremely likely that a randomly selected student will have a score less than 98. Remember that scores more than ±3 standard deviations from the mean are very rare.

Finding Probabilities

1: Compute the Z score (or scores). Note whether the score is positive or negative.

2: Find the Z score (or scores) in column a of the normal curve table (Appendix A).

3: Find the area above or below the score (or between the scores) as you would normally (see the three previous One Step at a Time boxes in this chapter) and express the result as a proportion. Typically, probabilities are expressed as a value between 0.00 and 1.00 rounded to two digits beyond the decimal point.

from this group over an infinite number of trials, 88 would have IQ scores less than 123 and 12 would not.

Let me close by stressing a very important point about probabilities and the normal curve. The probability is very high that any case randomly selected from a normal distribution will have a score close in value to that of the mean. The shape of the normal curve is such that most cases are clustered around the mean and decline in frequency as we move farther away—either to the right or to the left—from the mean value. In fact, given what we know about the normal curve, the probability that a randomly selected case will have a score within ±1 standard deviation of the mean is 0.6826. Rounding off, we can say that 68 out of 100 cases—or about two-thirds of all cases—selected over the long run will have a score between ±1 standard deviation, or Z score, of the mean. The probabilities are that any randomly selected case will have a score close in value to the mean.

In contrast, the probability of the case having a score beyond 3 standard deviations from the mean is very small. Look in column c ("Area Beyond Z") for a Z score of 3.00 and you will find the value .0014. Adding the areas in the upper tail (beyond ±3.00) to the area in the lower tail (beyond ±3.00), gives us .0014 + .0014, for a total of .0028. The probability of selecting a case with a very high score or a very low score is .0028. If we randomly select cases from a normally distributed variable, we would select cases with Z scores beyond ±3.00 only 28 times out of every 10,000 trials.

The general point to remember is that cases with scores close to the mean are common and cases with scores that have scores far above or below the mean are rare. This relationship is central for an understanding of inferential statistics in Part II. *(For practice in using the normal curve table to find probabilities, see problems 5.8 to 5.10 and 5.13.)*

STATISTICS
IN EVERYDAY LIFE

If you enjoy games of chance—bingo, card games, the state lottery, or board games like Parcheesi—you can use your knowledge of the laws of probability to improve your performance and your likelihood of winning. This point was illustrated in the recent best seller *Bringing Down the House*, later made into a movie called *21*. The book describes the adventures of a group of MIT students who used their understanding of probability to develop a system that almost guaranteed they could beat the house in the game of blackjack. They won hundreds of thousands of dollars before they were finally stopped by casino security.

SUMMARY

1. The normal curve, in combination with the mean and standard deviation, can be used to construct precise descriptive statements about empirical distributions that are normally distributed. This chapter also lays some important groundwork for Part II.

2. To work with the theoretical normal curve, raw scores must be transformed into their equivalent Z scores. Z scores allow us to find areas under the theoretical normal curve (Appendix A).

3. We considered three uses of the theoretical normal curve: finding total areas above and below a score, finding areas between two scores, and expressing these areas as probabilities. This last use of the normal curve is especially germane because inferential statistics are centrally concerned with estimating the probabilities of defined events in a fashion very similar to the process introduced in this chapter.

SUMMARY OF FORMULAS

Z Scores   5.1 $Z = \dfrac{X_i - \overline{X}}{s}$

GLOSSARY

Normal curve. A theoretical distribution of scores that is symmetrical, unimodal, and bell-shaped. The standard normal curve always has a mean of 0 and a standard deviation of 1.

Normal curve table. Appendix A; a detailed description of the area between a Z score and the mean of any standardized normal distribution.

Z scores. Standard scores; the way scores are expressed after they have been standardized to the theoretical normal curve.

PROBLEMS

5.1 Scores on a quiz were normally distributed and had a mean of 10 and a standard deviation of 3. For each of the following scores, find the Z score and the percentage of area above and below the score.

X_i	Z Score	% Area Above	% Area Below
5			
6			
7			
8			
9			
11			
12			
14			
15			
16			
18			

5.2 Assume that the distribution of a college entrance exam is normal, with a mean of 500 and a standard deviation of 100. For each of the following scores, find the equivalent Z score, the percentage of the area above the score, and the percentage of the area below the score.

X_i	Z Score	% Area Above	% Area Below
650			
400			
375			
586			
437			
526			
621			
498			
517			
398			

5.3 The senior class has been given a comprehensive examination to assess educational experience. The mean on the test was 74 and the standard deviation was 10. What percentage of the students had scores

  a. between 75 and 85? _____

  b. between 80 and 85? _____

  c. above 80? _____

  d. above 83? _____

  e. between 80 and 70? _____

  f. between 75 and 70? _____

  g. below 75? _____

  h. below 77? _____

  i. below 80? _____

  j. below 85? _____

5.4 For a normal distribution where the mean is 50 and the standard deviation is 10, what percentage of the area is

  a. between the scores of 40 and 47? _____

  b. above a score of 47? _____

  c. below a score of 53? _____

  d. between the scores of 35 and 65? _____

  e. above a score of 72? _____

  f. below a score of 31 and above a score of 69? _____

  g. between the scores of 55 and 62? _____

  h. between the scores of 32 and 47? _____

5.5 At St. Algebra College, the 200 freshmen enrolled in introductory biology took a final exam on which their mean score was 72 and their standard deviation was 6. The following table presents the grades of 10 students. Convert each into a Z score and determine the *number of people* who scored higher or lower than each of the 10 students. (*HINT: Multiply the appropriate proportion by N and round the result.*)

X_i	Z Score	Number of Students Above	Number of Students Below
60			
57			
55			
67			
70			
72			
78			
82			
90			
95			

5.6 If a distribution of test scores is normal, with a mean of 78 and a standard deviation of 11, what percentage of the area lies

  a. below 60? _____

  b. below 70? _____

  c. below 80? _____

  d. below 90? _____

  e. between 60 and 65? _____

  f. between 65 and 79? _____

  g. between 70 and 95? _____

  h. between 80 and 90? _____

  i. above 99? _____

  j. above 89? _____

  k. above 75? _____

  l. above 65? _____

5.7 A scale measuring prejudice has been administered to a large sample of respondents. The distribution of scores is approximately normal, with a mean of 31 and a standard deviation of 5. What percentage of the sample had scores

  a. below 20? _____

  b. below 40? _____

  c. between 30 and 40? _____

  d. between 35 and 45? _____

  e. above 25? _____

  f. above 35? _____

5.8 $\boxed{\text{SOC}}$ On the scale mentioned in problem 5.7, if a score of 40 or more is considered "highly prejudiced," what is the probability that a person selected at random will have a score in that range?

5.9 For a math test on which the mean was 59 and the standard deviation was 4, what is the probability that a student randomly selected from this class will have a score

  a. between 55 and 65? _____

  b. between 60 and 65? _____

  c. above 65? _____

  d. between 60 and 50? _____

  e. between 55 and 50? _____

  f. below 55? _____

5.10 The average number of drop-outs for a school district has been 305 per year with a standard deviation of 50. What is the probability that next year the number of drop-outs will be

  a. less than 250? _____

  b. less than 300? _____

  c. more than 350? _____

  d. more than 400? _____

  e. between 250 and 350? _____

  f. between 300 and 350? _____

  g. between 350 and 375? _____

5.11 The local police force gives all applicants an entrance exam and accepts only those applicants who score in the top 15% on this test. If the mean score this year is 87 and the standard deviation is 8, would an individual with a score of 110 be accepted?

5.12 To be accepted into an honor society, students must have GPAs in the top 10% of the school. If the mean GPA is 2.78 and the standard deviation is .33, which of the following GPAs would qualify?

3.20, 3.21, 3.25, 3.30, 3.35

5.13 In a distribution of scores with a mean of 35 and a standard deviation of 4, which event is more likely: that a randomly selected score will be between 29 and 31 or that a randomly selected score will be between 40 and 42?

5.14 After taking the state merit examinations for the positions of school counselor and social worker, you receive the following information on the tests and on your performance. On which of the tests did you do better?

School Counselor	Social Worker
$\bar{X} = 118$	$\bar{X} = 27$
$s = 17$	$s = 3$
Your score = 127	Your score = 29

PART I CUMULATIVE EXERCISES

1. To what extent do people apply religion to the problems of everyday living? Fifteen people have responded to a series of questions including the following:

1. What is your religion?
   1. Protestant
   2. Catholic
   3. Jewish
   4. None
   5. Other

2. On a scale of 1 to 10 (with 10 being the highest), how strong would you say your faith is?

3. How many times a day do you pray?

4. "When things aren't going well, my religion is my major source of comfort."
   1. Strongly agree
   2. Slightly agree
   3. Neither agree nor disagree
   4. Slightly disagree
   5. Strongly disagree

5. How old are you?

Case	Religion	Strength	Pray	Comfort	Age
1	1	2	0	2	30
2	2	8	1	1	67
3	1	8	3	1	45
4	1	6	0	1	43
5	3	9	3	1	32
6	3	3	0	3	18
7	4	0	0	5	52
8	5	9	6	1	37
9	1	5	0	2	54
10	2	8	2	1	55
11	2	3	0	5	33
12	1	6	1	3	45
13	1	8	2	2	37
14	1	7	2	1	50
15	2	9	1	1	25

For each variable, construct a frequency distribution and calculate appropriate measures of central tendency and dispersion. Write a sentence summarizing each variable.

2. A survey measuring attitudes toward interracial dating was administered to 1000 people. The survey asked the following questions:

1. What is your age?

2. What is your sex?
   1. Male
   2. Female

3. "Marriages between people of different racial groups just don't work out and should be banned by law."
   1. Strongly agree
   2. Slightly agree
   3. Neither agree nor disagree
   4. Slightly disagree
   5. Strongly disagree

4. How many years of schooling have you completed?

5. Which of the following categories best describes the place where you grew up?
   1. Large city
   2. Medium-size city
   3. Suburbs of a city
   4. Small town
   5. Rural area

6. What is your marital status?
   1. Married
   2. Separated or divorced
   3. Widowed
   4. Never married

The scores of 20 respondents are reproduced here.

Case	Age	Sex	Attitude on Interracial Dating	Years of School	Area	Marital Status
1	17	1	5	12	1	4
2	25	2	3	12	2	1
3	55	2	3	14	2	1
4	45	1	1	12	3	1
5	38	2	1	10	3	1
6	21	1	1	16	5	1
7	29	2	2	16	2	2
8	30	2	1	12	4	1
9	37	1	1	12	2	1
10	42	2	3	18	5	4
11	57	2	4	12	2	3
12	24	2	2	12	4	1
13	27	1	2	18	3	2
14	44	1	1	15	1	1
15	37	1	1	10	5	4
16	35	1	1	12	4	1
17	41	2	2	15	3	1
18	42	2	1	10	2	4
19	20	2	1	16	1	4
20	21	2	1	16	1	4

a. For each variable, construct a frequency distribution and select and calculate an appropriate measure of central tendency and a measure of dispersion. Summarize each variable in a sentence.

b. For all 1000 respondents, the mean age was 34.70 with a standard deviation of 3.4 years. Assuming the distribution of age is approximately normal, compute Z scores for each of the first 10 respondents and determine the percentage of the area below (younger than) each respondent.

3. The following data set is taken from the General Social Survey. Abbreviated versions of the questions along with the meanings of the codes are also presented. See Appendix G for the codes and the complete question wordings. For each variable, construct a frequency distribution and select and calculate an appropriate measure of central tendency and a measure of dispersion. Summarize each variable in a sentence.

1. How many children have you ever had? (Values are actual numbers.)

2. Respondent's educational level:
   0. Less than HS
   1. HS
   2. Jr. college
   3. Bachelor's degree
   4. Graduate school

3. Race:
   1. White
   2. Black
   3. Other

4. "It is sometimes necessary to discipline a child with a good, hard spanking."
   1. Strongly agree
   2. Agree
   3. Disagree
   4. Strongly Disagree

5. Number of hours of TV watched per day. (Values are actual numbers of hours.)

6. What is your religious preference?
   1. Protestant
   2. Catholic
   3. Jewish
   4. None
   5. Other

Case	Number of Children	Years of School	Race	Attitude on Spanking	TV Hours	Religion
1	3	1	1	3	3	1
2	2	0	1	4	1	1
3	4	2	1	2	3	1
4	0	3	1	1	2	1
5	5	1	1	3	2	1
6	1	1	1	3	3	1
7	9	0	1	1	6	1
8	6	1	2	3	4	1

(*continued on next page*)

(continued)

Case	Number of Children	Years of School	Race	Attitude on Spanking	TV Hours	Religion
9	4	3	1	1	2	4
10	2	1	3	1	1	1
11	2	0	1	2	4	1
12	4	1	2	1	5	2
13	0	1	1	3	2	2
14	2	1	1	4	2	1
15	3	1	2	3	4	1
16	2	0	1	2	2	1
17	2	1	1	2	2	1
18	0	3	1	3	2	1
19	3	0	1	3	5	2
20	2	1	2	1	10	1
21	2	1	1	3	4	1
22	1	0	1	3	5	1
23	0	2	1	1	2	2
24	0	1	1	2	0	4
25	2	4	1	1	1	2

Part II — Inferential Statistics

The six chapters in this part cover the techniques and concepts of inferential statistics. Generally speaking, these applications allow us to learn about large groups (populations) from smaller, carefully selected subgroups (samples). These statistical techniques are powerful and extremely useful. They are used to poll public opinion, to research the potential market for new consumer products, to project the winners of elections, to test the effects of new drugs, and in hundreds of other ways both inside and outside the social sciences.

Chapter 6 includes a brief description of the technology of sampling, or how subgroups are selected so as to justify making inferences to populations. This section is intended to give you a general overview of the process and some insight into the actual selection of cases for the sample, not as a comprehensive or detailed treatment of the subject. The most important part of this chapter, however, concerns the sampling distribution, the single most important concept in inferential statistics. The sampling distribution is normal in shape, and it is the key link between populations and samples.

There are two main applications in inferential statistics, and Chapter 7 covers the first: using statistical information from a sample (e.g., a mean or a proportion) to estimate the characteristics of a population. The technique, called *estimation*, is most commonly used in public opinion polling and election projection.

Chapters 8 through 11 cover the second application of inferential statistics: hypothesis testing. Most of the relevant concepts for this material are introduced in Chapter 8, and each chapter covers a different situation in which hypothesis testing is done. For example, Chapter 9 presents the techniques used when we are comparing information from two different samples or groups (e.g., men vs. women), while Chapter 10 covers applications involving more than two groups or samples (e.g., Republicans vs. Democrats vs. Independents).

Hypothesis testing is one of the more challenging aspects of statistics for beginning students, and I have included an abundance of learning aids to ease the chore of assimilating this material. Hypothesis testing is also one of the most common and important statistical applications to be found in social science research. Mastery of this material is essential for developing your ability to read the professional literature.

6

Introduction to Inferential Statistics
Sampling and the Sampling Distribution

LEARNING OBJECTIVES By the end of this chapter, you will be able to

1. Explain the purpose of inferential statistics in terms of generalizing from a sample to a population.
2. Define and explain the basic techniques of random sampling.
3. Explain and define these key terms: population, sample, parameter, statistic, representative, EPSEM.
4. Differentiate between the sampling distribution, the sample, and the population.
5. Explain the two theorems presented.

One of the goals of social science research is to test our theories using many different people, groups, societies, and historical eras. Obviously, we can have the greatest confidence in theories that have stood up to testing against the greatest variety of cases and social settings. However, a major problem in social science research is that the most appropriate populations for our tests are very large. For example, a theory concerning political party preference among U.S. citizens would be most suitably tested using the entire electorate, but it is impossible to interview every member of this group (about 130 million people). Indeed, even for theories that reasonably could be tested with smaller populations—such as a local community or the student body at a university—the logistics of gathering data from every single case in the population are staggering to contemplate.

How can we test our theories if it is too difficult or expensive to research entire populations? To deal with this problem, social scientists select samples, or subsets of cases, from the populations of interest. Our goal in inferential statistics is to learn about the characteristics of a population (often called **parameters**) based on what we can learn from the sample. Two applications of inferential statistics are covered in this text. In estimation procedures, covered in Chapter 7, an "educated guess" of the population parameter is made, based on what is known about the sample. In hypothesis testing, covered in Chapters 8 through 11, the validity of a hypothesis about the population is tested against sample outcomes. In this chapter, we look briefly at sampling (the techniques for selecting cases for a sample) and then introduce a key concept in inferential statistics: the sampling distribution.

Probability Sampling: Basic Concepts

Social scientists have developed a variety of sampling techniques. In this chapter, we review the basic procedures for selecting probability samples, the only type of sample that fully supports the use of inferential statistics to generalize to populations. These samples are often described as *random*, and you may be more familiar with this terminology. Because of its greater familiarity, I will often use the term *random sample* in the following chapters. The term *probability sample* is preferred, however, because, in everyday language, *random* is often used to mean "by coincidence" or to give a connotation of unpredictability. As you will see, probability samples are selected carefully and methodically, in ways that leave no room for haphazardness. Interviewing the people you happen to meet in a mall one afternoon may be "random" in some sense of the word, but this technique will not result in a sample that could support inferential statistics.

Before considering probability sampling, let me point out that social scientists often use nonprobability samples. For example, social scientists studying small-group dynamics or the structure of attitudes or personal values might use the students enrolled in their classes as subjects. These "convenience" samples are very useful for a number of purposes (e.g., exploring ideas or pretesting survey forms before embarking on a more ambitious project) and are typically cheaper and easier to assemble. The major limitation of these samples is that results cannot be generalized beyond the group being tested. If a theory of prejudice, for example, has been tested only on the students in a particular section of introductory sociology at a particular university, we cannot conclude that results would apply to other types of people. Therefore, we cannot place a lot of confidence in theories tested on nonprobability samples, even when they produce very strong evidence.

When constructing a probability sample, our goal is to select cases so that the final sample is **representative** of the population from which it was drawn. A sample is representative if it reproduces the important characteristics of the population. For example, if the population consists of 60% females and 40% males, the sample should have about the same proportional makeup. In other words, a representative sample is very much like the population—only smaller. It is crucial for inferential statistics that samples be representative. How can we assure ourselves that our samples are representative? Unfortunately, it is not possible to guarantee that samples will be representative. However, we can maximize the chances of a representative sample by following the principle of EPSEM (the "**E**qual **P**robability of **SE**lection **M**ethod"), the fundamental

---

STATISTICS

IN EVERYDAY LIFE

Advertisers sometimes phrase their appeals in language that might seem scientific. Claims such as "80% of the 2000 people surveyed prefer Brand X" might sound convincing, and a sample size of 2000 certainly seems impressive. What if the sample consisted solely of people leaving a promotional event for Brand X during which free samples of the product were distributed? Does the claim still have legitimacy? Only properly selected probability samples can be used to make conclusions about larger populations.

principle of probability sampling. To follow EPSEM and maximize the probability that our sample will be representative, we select the sample so that every element or case in the population has an equal probability of being selected. Remember that EPSEM and representativeness are two different things. In other words, the fact that a sample is selected according to EPSEM does not guarantee that it will be an exact representation of the population. The probability is very high that an EPSEM sample will be representative; but, just as a perfectly honest coin will sometimes show 10 heads in a row when flipped, an EPSEM sample will occasionally present an inaccurate picture of the population. One of the great strengths of inferential statistics is that they allow the researcher to estimate the probability of this type of error and interpret results accordingly.

EPSEM Sampling Techniques

This section presents four techniques for selecting random samples. The simplest technique is presented first; the remaining techniques are refinements or variations on the basic technique.

SIMPLE RANDOM SAMPLING

The most basic EPSEM sampling technique produces a **simple random sample**. To use this technique, we need a list of all cases in the population and a system of selection that will guarantee that every case has an equal chance of being chosen for the sample. The selection process could be based on a variety of operations (for example, flipping coins, throwing dice, drawing numbers from a hat). Cases are often selected by using tables of random numbers. These tables are lists of numbers that have no pattern to them (that is, they are random); an example of such a table is available at the Web site for this text. Alternately, computer programs can be written to produce lists of random numbers.

To use random numbers to select a sample, first assign a unique identification number to each case in the population. Then pick numbers from the table or the computer-generated list one at a time and select the cases whose identification numbers match the selected numbers. Using random numbers will produce an EPSEM sample because the numbers in the table are in random order and any number is just as likely as any other number. Stop selecting cases when you have reached your desired sample size; if an identification number is selected more than once, ignore the repeats.[1]

SYSTEMATIC RANDOM SAMPLING

Following the foregoing procedures will produce random samples as long as you select from a complete list of the population. However, this technique can be very cumbersome when the list is long. Consider the situation when your

---

[1]Ignoring identification numbers when they repeat is called "sampling without replacement" and, technically, this practice compromises the randomness of the selection process. However, if the sample is a small fraction of the population, we will be unlikely to select the same case twice and ignoring repeats will not bias our conclusions.

population numbers 10,000 cases. It is perfectly possible that the first case you select will come from the front of the list, the second from the back, the third from the front again, and so on. To save time in such a situation, researchers often use a technique called **systematic sampling**, where only the first case is randomly selected. Thereafter, every kth case is selected. For example, if you are drawing from a list of 10,000 and desire a sample of 200, select the first case randomly and every 10,000/200th, or 50th, case thereafter. If you randomly start with case 13, then your second case will be case 63, your third case 113, and so on, until you reach the end of the list.

Note that systematic sampling does not strictly conform to the criterion of EPSEM. That is, once the first case has been selected, the other cases no longer have an equal probability of being chosen. In our example, cases other than the 13th, the 63rd, the 113th, and so on will not be selected for the sample. In general, this increased probability of error is very slight as long as the list from which cases are chosen is random, or at least noncyclical with respect to the traits you wish to measure. For example, there might be a problem if you are concerned with ethnicity and are drawing your sample from an alphabetical list, because there is a tendency for certain ethnic names to begin with the same letter (for example, the Irish prefix *O*). Therefore, when using systematic sampling, pay careful attention to the nature of the population list as well as your sampling technique.

STRATIFIED RANDOM SAMPLING

A third type of EPSEM sampling produces a **stratified sample**. This technique is very desirable because it guarantees that the sample will be representative on the selected traits. To apply this technique, you first *stratify* (or divide) the population list into sublists according to some relevant trait and then sample from the sublists. If you select a number of cases from each sublist proportional to the numbers for that characteristic in the population, the sample will be representative of the population.

For example, suppose you are drawing a sample of 300 of your classmates and you wish to have proportional representation from every major field on campus. If only 10% of the student body is majoring in zoology, random and systematic sampling could result in a sample with very few (or even no) zoologists. If, however, you first divide the population into sublists by major, you can use the EPSEM technique to select exactly 30 zoologists from the appropriate sublist. Following the same procedure with other majors will create a sample that is, by definition, representative of the population on this characteristic. Thus, stratified samples are guaranteed to meet the all-important criterion of representativeness (at least for the traits that are used to stratify the samples). The major limitation of stratified random sampling is that we often lack detailed information about the exact composition of the population. Without information about the population, we will be unable to establish a scheme for stratification and determine how many cases should be taken from each sublist.

CLUSTER SAMPLING

To this point, sampling techniques have been presented as straightforward processes of randomly selecting cases from a list or sublists of the population. However, sampling is rarely so uncomplicated, and the major difficulty almost always centers on what might appear, at first glance, to be the easiest part: finding a list of the population. For many populations of interest to the social sciences, there are no complete, up-to-date lists. There is no list of United States citizens, no list of the residents of any given state, and no complete, up-to-date list of residents of your local community. Telephone books might appear to be complete lists of local residents, but they omit unlisted numbers and are very likely to be outdated.

Social scientists have devised several ways of dealing with the limitations imposed by the scarcity of lists. Probably the most significant of these is **cluster sampling**, which involves selecting groups of cases (clusters) rather than single cases. The clusters are often based on geography, and the selection of clusters often proceeds in stages. For example, you might draw a cluster sample of your city or town by first numbering all voting precincts. Next, you would use an EPSEM technique to select a sample of precincts. The second stage of selection would involve numbering the blocks within each of the selected precincts and, following EPSEM, selecting a sample of blocks. A third stage might involve the selection of households within each selected block. When these stages are completed, you would have a sample that had a very high probability of being representative of the entire city without ever using a list of city residents.

Cluster samples are somewhat less likely to be representative of the population than simple random samples of comparable size. In part, this lower accuracy is a result of the multiple selection stages. With a simple random sample, the sample is drawn in one selection from the list of the population. In a multistage cluster sample, each stage in the selection process (e.g., first the precincts, then the blocks, and then the households) has a probability of error. That is, each time we sample, we run the risk of selecting an unrepresentative sample. In simple random sampling, we run this risk once; with cluster sampling we will run the risk anew at each stage.

Although we have to treat inferences about populations based on cluster samples with some additional caution, we usually have no alternative method of sampling. While it may be extremely difficult (or even impossible) to construct an accurate list of an entire city population, all you need to compile a cluster sample is a map (or a list of voting precincts, census tracts, and so forth).

Before we leave this discussion of sampling techniques, let me stress a major point. The purpose of inferential statistics is to acquire knowledge about populations based on the information derived from samples of that population. Each of the statistics to be presented in the following chapters in Part II requires that samples be selected according to EPSEM. While even the most painstaking and sophisticated sampling techniques are no guarantee, the probability is high that EPSEM samples will be representative of the populations from which they are selected.

| **STATISTICS** IN EVERYDAY LIFE | The sampling techniques presented in this chapter are also used for commercial purposes and for mass marketing. The next time you are annoyed by a telemarketer calling during dinnertime, remember that your phone number may have been selected by a sophisticated, complex application of random and cluster sampling. This probably will not lessen your annoyance, but it will remind you that sampling technology is a part of daily life and in many areas outside of the social sciences. |

The Sampling Distribution

Once we have selected a probability sample according to some EPSEM procedure, what do we know? On one hand, we can gather a great deal of information from the cases in the sample. On the other hand, we know nothing about the population. Indeed, if we had information about the population, we probably wouldn't need the sample. Remember that we use inferential statistics to learn more about populations, and information from the sample is important primarily insofar as it allows us to generalize to the population.

When we use inferential statistics, we generally measure some variable (e.g., age, political party preference, or opinions about abortion) in the sample and then use the information from the sample to learn more about that variable in the population. In Part I of this text, you learned that three types of information are generally necessary to adequately characterize a variable: (1) the shape of its distribution, (2) some measure of central tendency, and (3) some measure of dispersion. Clearly, all three kinds of information can be gathered (or computed) on a variable from the cases in the sample. Just as clearly, none of the information is available for the population. Except in rare situations (for example, IQ tests are designed so that the scores of a population will be approximately normal in distribution, and the distribution of income is almost always positively skewed), nothing can be known about the exact shape of the distribution of a variable in the population. The means and standard deviations of variables in the population are also unknown. Let me remind you that if we had this information for the population, inferential statistics would be unnecessary.

In statistics, we link information from the sample to the population with the **sampling distribution**: *the theoretical, probabilistic distribution of a statistic for all possible samples of a certain size (N).* That is, the sampling distribution is the distribution of a statistic (e.g., a mean or a proportion) based on every conceivable combination of cases from the population. A crucial point about the sampling distribution is that its characteristics are based on the laws of probability, not on empirical information, and are very well known. In fact, the sampling distribution is the central concept in inferential statistics, and a prolonged examination of its characteristics is certainly in order.

THE THREE DISTRIBUTIONS USED IN INFERENTIAL STATISTICS

FIGURE 6.1 Relationships Between the Sample, the Sampling Distribution, and the Population

As illustrated by Figure 6.1, all applications of inferential statistics move between the sample and the population via the sampling distribution. Thus, three separate and distinct distributions of a variable are involved in every application of inferential statistics:

1. The population distribution of the variable, which, while empirical, is unknown. Amassing information about or making inferences to the population is the sole purpose of inferential statistics.

2. The sampling distribution of the variable, which is nonempirical, or theoretical. Because of the laws of probability, a great deal is known about this distribution. Specifically, the shape, central tendency, and dispersion of the distribution can be deduced and, therefore, the distribution can be adequately characterized.

3. The sample distribution of the variable, which is empirical (i.e., it exists in reality) and is known. The shape, central tendency, and dispersion of the variable can be ascertained for the sample. Remember, however, that information from the sample is important primarily insofar as it allows the researcher to learn about the population.

The utility of the sampling distribution is implied by its definition. Because it includes the statistics from all possible sample outcomes, the sampling distribution enables us to estimate the probability of any particular sample outcome, a process that will occupy our attention for the next five chapters.

CONSTRUCTING A SAMPLING DISTRIBUTION

The sampling distribution is theoretical, which means that it is never obtained in reality by the researcher. However, to understand better the structure and function of the distribution, let's consider an example of how one might be constructed. Suppose we wanted to gather some information about the age of a particular community of 10,000 individuals. We draw an EPSEM sample of 100 residents, ask all 100 respondents their age, and use those individual scores to compute a mean age of 27. This score is noted on the graph in Figure 6.2. Note that this sample is one of countless possible combinations of 100 people taken from this population of 10,000, and the statistic (the mean of 27) is one of millions of possible sample outcomes.

Now replace the first 100 respondents and draw another sample of the same size ($N = 100$) and again compute the average age. Assume that the mean for the second sample is 30 and note this sample outcome on Figure 6.2. This second sample is another of the countless combinations of 100 people taken from this population of 10,000, and the sample mean of 30 is another of the millions of possible sample statistics. Replace these respondents and draw still another sample, calculate and note the mean, replace this third sample, and draw a fourth sample, continuing these operations an infinite number of times, calculating and noting the mean of each sample. Now try to imagine what Figure 6.2 would look like after tens of thousands of individual samples had been collected and the mean had been computed for each sample. What shape, mean, and standard deviation would this distribution of sample means have after we had collected all possible combinations of 100 respondents from the population of 10,000?

For one thing, we know that each sample will be at least slightly different from every other sample, since it is very unlikely that we will sample exactly the same 100 people twice. Since each sample will be unique, each sample mean will be at least slightly different in value. We also know that not all the samples will be representative, even if we follow EPSEM closely. For example, if we continue taking samples of 100 people long enough, we will eventually choose

FIGURE 6.2 Constructing a Sample Distribution

a sample that includes only the very youngest residents. Such a sample would have a mean much lower than the true population mean. Likewise, some of our samples will include only senior citizens and will have means that are much higher than the population mean. Common sense suggests, however, that such nonrepresentative samples will be rare and that most sample means will cluster around the true population value.

To illustrate further, assume that we somehow come to know that the true mean age of the population of 10,000 individuals is 30. Since, as we have just seen, most of the sample means will also be approximately 30, the sampling distribution of these sample means should peak at 30. Some of the samples will be unrepresentative and their means will "miss the mark," but the frequency of such misses should decline as we get farther away from the population mean of 30. That is, the distribution should slope to the base as we move away from the population value—sample means of 29 or 31 should be common; means of 20 or 40 should be rare. Since the samples are random, their means should miss an equal number of times on either side of the population value, and the distribution itself should therefore be roughly symmetrical. In other words, the sampling distribution of all possible sample means should be approximately normal and will resemble the distribution presented in Figure 6.3. Recall from Chapter 5 that, on any normal curve, cases close to the mean (say, within ± 1 standard deviation) are common and cases far away from the mean (say, beyond 3 standard deviations) are rare.

TWO THEOREMS

These commonsense notions about the shape of the sampling distribution and other very important information about central tendency and dispersion are stated in two theorems. The first of these theorems states:

> If repeated random samples of size N are drawn from a normal population with mean μ and standard deviation σ, then the sampling distribution of sample means will be normal, with a mean μ and a standard deviation of $\sigma/\sqrt{N}$.

To translate: If we begin with a trait that is normally distributed across a population (IQ scores, for example) and take an infinite number of equal-sized random samples from that population, then the sampling distribution of sample means will be normal. If we know that the variable is distributed normally in the population, we can assume that the sampling distribution will be normal.

FIGURE 6.3   A Sampling Distribution of Sample Means

The theorem tells us more than the shape of the sampling distribution, however. It also defines its mean and standard deviation. In fact, it says that the mean of the sampling distribution will be exactly the same value as the mean of the population. That is, if we know that the mean IQ of the entire population is 100, then we know that the mean of any sampling distribution of sample mean IQ scores will also be 100. Exactly why this should be so cannot be fully explained at this level. Recall, however, that most sample means will cluster around the population value over the long run. Thus, the fact that these two values are equal should have intuitive appeal. As for dispersion, the theorem says that the standard deviation of the sampling distribution, also called the **standard error of the mean**, will be equal to the standard deviation of the population divided by the square root of N (symbolically: $\sigma/\sqrt{N}$).

If the mean and standard deviation of a normally distributed variable in a population are known, the theorem allows us to compute the mean and standard deviation of the sampling distribution.[2] Thus, we will know exactly as much about the sampling distribution (shape, central tendency, and dispersion) as we ever knew about any empirical distribution.

The first theorem requires that the variable be normally distributed in the population. What happens when the distribution of the variable in question is unknown or is known *not* to be normal in shape (like income, which always has a positive skew)? These eventualities (very common, in fact) are covered by a second theorem, called the **Central Limit Theorem**:

> If repeated random samples of size N are drawn from any population with mean μ and standard deviation σ, then, as N becomes large, the sampling distribution of sample means will approach normality, with mean μ and standard deviation $\sigma/\sqrt{N}$.

To translate: The sampling distribution of sample means will become normal in shape as sample size increases for *any* variable, even when the variable is not normally distributed across the population. When N is large, the mean of the sampling distribution will equal the population mean, and its standard deviation (or the standard error of the mean) will be equal to $\sigma/\sqrt{N}$.

The Central Limit Theorem is important because it removes the condition that the variable be normally distributed in the population. Whenever sample size is large, we can assume that the sampling distribution is normal, with a mean equal to the population mean and a standard deviation equal to $\sigma/\sqrt{N}$. Thus, even if we are working with a variable that is known to have a skewed distribution (like income), we can still assume a normal sampling distribution.

The issue remaining, of course, is to define what is meant by a large sample. A good rule of thumb is that if sample size (N) is 100 or more, the Central Limit Theorem applies, and you can assume that the sampling distribution of sample statistics is normal in shape. When N is less than 100, you must have good evidence of a normal population distribution before you can assume that the sampling distribution is normal. Thus, a normal sampling distribution can be ensured by the expedient of using fairly large samples.

---

[2]In the typical research situation, the values of the population mean and standard deviation are, of course, unknown. However, these values can be estimated from sample statistics, as we shall see in the chapters that follow.

THE SAMPLING DISTRIBUTION: AN ADDITIONAL EXAMPLE

Developing an understanding of the sampling distribution—what it is and why it's important—can be a challenging task. It may be helpful to list briefly the most important points about the sampling distribution:

1. Its definition: *The sampling distribution is the distribution of a statistic (such as a mean or a proportion) for all possible sample outcomes of a certain size.*

2. Its shape: *normal* (see Chapter 5 and Appendix A).

3. Its central tendency and dispersion: *The mean of the sampling distribution is the same value as the mean of the population. The standard deviation of the sampling distribution—or the standard error—is equal to the population standard deviation divided by the square root of N (see the theorems).*

4. The role of the sampling distribution in inferential statistics: *It links the sample with the population* (see Figure 6.1)

To reinforce these points, let's consider an additional example using the General Social Survey (GSS), the database used for SPSS exercises in this text. The GSS has been administered to randomly selected samples of adult Americans since 1972 and explores a broad range of characteristics and issues, including confidence in the Supreme Court, attitudes about assisted suicide, number of siblings, and level of education. The GSS has its limits, of course, but it is a very valuable resource for testing theory and for learning more about American society. Focusing on this survey, let's review the roles played by the population, the sample, and the sampling distribution when we use this database.

We'll start with the population, or the group we want to learn more about. In the case of the GSS, the population consists of all adult (older than 18) Americans, which includes 230 million people. Clearly, we can never interview all of these people and learn what they are like or what they are thinking about abortion, gun control, sex education in the public schools, or any other issue. We should also note that this information is worth having. It could help inform public debates, provide some basis in fact for the discussion of many controversial issues (e.g., the polls show consistently that the majority of Americans favor some form of gun control), and assist people in clarifying their personal beliefs. If the information is valuable, how can we learn more about this huge population?

This brings us to the sample, a carefully chosen subset of the population. The General Social Survey is administered to about 3000 people, each of whom is chosen by a sophisticated technology based on the principle of EPSEM. A key point to remember is that samples chosen by this method are very likely to be representative of the populations from which they were selected. Whatever is true of the sample will also be true of the population (with some limits and qualifications, of course).

The respondents are contacted at home and asked for background information (religion, gender, years of education, and so on) as well as their opinions and attitudes. When all of this information is collated, the GSS database includes information (shape, central tendency, dispersion) on hundreds of variables (age, level of prejudice, marital status) for the people in the sample. So we have a lot of information about the variables for the *sample* (the people who actually responded to the survey), but no information about these variables for the *population* (the 230 million adult Americans). How do we get from the known sample to the unknown population? This is the central question of inferential statistics; the answer, as you hopefully realize by now, is "by using the sampling distribution."

Remember that, unlike the sample and the population, the sampling distribution is theoretical, and, because of the theorems presented earlier in this chapter, we know its shape, central tendency, and dispersion. For any variable from the GSS, the theorems tell us that:

- The sampling distribution will be normal in shape because the sample is "large" (N is much greater than 100). This will be true regardless of the shape of the variable in the population.
- The mean of the sampling distribution will be the same value as the mean of the population. If *all* adult Americans have completed an average of 13.5 years of schooling ($\mu = 13.5$), the mean of the sampling distribution will also be 13.5.
- The standard deviation (or standard error) of the sampling distribution is equal to the population standard deviation (σ) divided by the square root of N.

Thus, the theorems tell us the statistical characteristics of this distribution (shape, central tendency, and dispersion), and this information allows us to link the sample to the population.

How does the sampling distribution link the sample to the population? The fact that the sampling distribution will be normal when N is large is crucial. This means that more than two-thirds (68%) of all samples will be within ± 1 Z score of the mean of the sampling distribution (which is the same value as the mean of the population), about 95% are within ± 2 Z scores, and so forth. We do not (and cannot) know the actual value of the mean of the sampling distribution, but we do know that the probabilities are very high that our sample statistic is approximately equal to this parameter. Similarly, the theorems give us crucial information about the mean and standard error of the sampling distribution that we can use, as you will see in the chapters that follow, to link information from the sample to the population.

To summarize, our goal is to infer information about the population (in the case of the GSS, all adult Americans). When populations are too large to test, we use information from randomly selected samples, carefully drawn from the population of interest, to estimate the characteristics of the population. In the case of the GSS, the full sample consists of about 3000 adult Americans who have responded to the questions on the survey. The sampling distribution, the theoretical distribution whose characteristics are defined by the theorems, links the known sample to the unknown population.

STATISTICS
IN EVERYDAY LIFE

The General Social Survey has been administered since 1972, and this permits us to track change in our society over time. As one illustration, data from the GSS demonstrates that—contrary to widely shared beliefs—U.S. society is not growing more permissive or liberal on all moral issues. Attitudes about extramarital sex, for example, have grown more disapproving. In the late 1970s, 72% of respondents said that extramarital sex was "always wrong." By 2008, this percentage had grown to 82%.

TABLE 6.1   Symbols for Means and Standard Deviations of Three Distributions

	Mean	Standard Deviation	Proportion
1. Samples	$\overline{X}$	s	P_s
2. Populations	μ	σ	P_u
3. Sampling distributions			
of means	$\mu_{\overline{X}}$	$\sigma_{\overline{X}}$	
of proportions	μ_p	σ_p	

Symbols and Terminology

In the following chapters, we will be working with three entirely different distributions. Furthermore, we will be concerned with several different kinds of sampling distributions—including the sampling distribution of sample means and the sampling distribution of sample proportions.

To distinguish among these various distributions, we will often use symbols. The symbols for the means and standard deviations of samples and populations have already been introduced in Chapters 3 and 4. For quick reference, Table 6.1 presents the symbols that will be used for the sampling distribution. Basically, the sampling distribution is denoted with Greek letters that are subscripted according to the sample statistic of interest.

Note that the mean and standard deviation of a sample are denoted with English letters ($\overline{X}$ and s), while the mean and standard deviation of a population are denoted with Greek-letter equivalents (μ and σ). Proportions calculated on samples are symbolized as P_s (P-sub-s; s for sample), while population proportions are denoted as P_u (P-sub-u; u for "universe" or population). The symbols for the sampling distribution are Greek letters with English-letter subscripts. The mean and standard deviation of a sampling distribution of sample means are $\mu_{\overline{X}}$ ("mu-sub-x-bar") and $\sigma_{\overline{X}}$ ("sigma-sub-x-bar"). The mean and standard deviation of a sampling distribution of sample proportions are μ_p ("mu-sub-p") and σ_p ("sigma-sub-p").

SUMMARY

1. Since populations are almost always too large to test, a fundamental strategy of social science research is to select a sample from the defined population and then to use information from the sample to generalize to the population. This is done either by estimation or by hypothesis testing.

2. Several techniques are commonly used for selecting random samples. Each of these techniques involves selecting cases for the sample according to EPSEM. Even the most rigorous technique, however, cannot guarantee that the sample will be representative. One of the great strengths of inferential statistics is that the probability that the sample is nonrepresentative can be estimated.

3. The sampling distribution, the central concept in inferential statistics, is a theoretical distribution of all possible sample outcomes. Since its overall shape, mean, and standard deviation are known (under the conditions specified in the two theorems), the sampling distribution can be adequately characterized and utilized by researchers.

4. The two theorems that were introduced in this chapter state that when the variable of interest is normally distributed in the population or when sample size is large, the sampling distribution will be normal in shape, its mean will be equal to the population mean, and its standard deviation (or standard error) will be equal to the population standard deviation divided by the square root of N.

5. All applications of inferential statistics involve generalizing from the sample to the population by means of the sampling distribution. Both estimation procedures and hypothesis testing incorporate the three distributions, and it is crucial that you develop a clear understanding of each distribution and its role in inferential statistics

GLOSSARY

Central Limit Theorem. A theorem that specifies the mean, standard deviation, and shape of the sampling distribution, given that the sample is large.

Cluster sampling. A method of sampling by which geographical units are randomly selected and all cases within each selected unit are tested.

EPSEM. The **E**qual **P**robability of **SE**lection **M**ethod for selecting samples. Every element or case in the population must have an equal probability of selection for the sample.

μ. The mean of a population.

μ_p. The mean of a sampling distribution of sample proportions.

$\mu_{\bar{X}}$. The mean of a sampling distribution of sample means.

P_s. (P-sub-s) Any sample proportion.

P_u. (P-sub-u) Any population proportion.

Parameter. A characteristic of a population.

Representative. The quality a sample is said to have if it reproduces the major characteristics of the population from which it was drawn.

Sampling distribution. The distribution of a statistic for all possible sample outcomes of a certain size. Under conditions specified in two theorems, the sampling distribution will be normal in shape, with a mean equal to the population value and a standard deviation equal to the population standard deviation divided by the square root of N.

Simple random sample. A method for choosing cases from a population by which every case and every combination of cases has an equal chance of being included.

Standard error of the mean. The standard deviation of a sampling distribution of sample means.

Stratified sample. A method of sampling by which cases are selected from sublists of the population.

Systematic sampling. A method of sampling by which the first case from a list of the population is randomly selected. Thereafter, every kth case is selected.

PROBLEMS

6.1 Imagine that you had to gather a random sample ($N = 300$) of the student body at your school. How would you acquire a list of the population? Would the list be complete and accurate? What procedure would you follow in selecting cases (that is, simple or systematic random sampling)? Would cluster sampling be an appropriate technique (assuming that no list was available)? Describe in detail how you would construct a cluster sample.

6.2 This exercise is tedious but, if you want a more concrete and tangible understanding of sampling distributions and the two theorems presented in this chapter, then it may have a significant payoff. At the end of this problem are listed the ages of a population of college students ($N = 50$). By a random method

(such as a table of random numbers), draw at least 50 samples of size 2 (that is, 50 pairs of cases), compute a mean for each sample, and plot the means on a frequency polygon. (Incidentally, this exercise will work better if you draw 100 or 200 samples and/or use larger samples than $N = 2$.)

a. The curve you've just produced is a sampling distribution. Observe its shape; after 50 samples, it should be approaching normality. What is your estimate of the population mean (μ) based on the shape of the curve?

b. Calculate the mean of the sampling distribution ($\mu_{\bar{X}}$). Be careful to do this by summing the sample means (not the scores) and dividing by the number of samples you've drawn. Now compute the population mean (μ). These two means should be very close in value because $\mu_{\overline{X}} = \mu$ by the Central Limit Theorem.

c. Calculate the standard deviation of the sampling distribution (use the means as scores) and the standard deviation of the population. Compare these two values. You should find that $\sigma_{\overline{X}} = \sigma/\sqrt{N}$.

d. If none of the preceding exercises turned out as they should have, it is for one or more of the following reasons:

   1. You didn't take enough samples. You may need as many as 100 or 200 (or more) samples to see the curve begin to look "normal."

   2. Sample size (2) is too small. An N of 5 or 10 would work much better.

6.3 Your sampling method is not truly random and/or the population is not arranged in random fashion.

17	20	20	19	20
18	21	19	20	19
19	22	19	23	19
20	23	18	20	20
22	19	19	20	20
23	17	18	21	20
20	18	20	19	20
22	17	21	21	21
21	20	20	20	22
18	21	20	22	21

YOU ARE THE RESEARCHER

Using SPSS to Draw Random Samples, Compute Statistics, and Estimate a Parameter

SPSS for Windows has a procedure for drawing random samples, which we can use to illustrate some points about sampling and to demonstrate that random samples will produce statistics that are close approximations of the corresponding population values or parameters. For this demonstration, the 2008 GSS sample will be treated as a population and its characteristics will be treated as parameters.

In this exercise, you will draw three random samples of different sizes from the 2008 GSS sample and use the sample means to estimate the "population" mean. The average age of the full GSS 2008 sample (48.16) will serve as the parameter or μ. The samples you will draw will be approximately 10%, 25%, and 50% of the population size, and SPSS will select them by an EPSEM procedure. These samples may be considered "simple random samples."

As a part of this procedure you will also request the "standard error of the mean," or S.E. MEAN. This is the standard deviation of the sampling distribution ($\sigma_{\overline{X}}$). This statistic will be of interest because we can expect our sample means to be within a standard error of the population value or parameter.

With the 2008 GSS loaded, click **Data** from the menu bar of the Data Editor and then click **Select Cases**. The **Select Cases** window appears and presents a number of different options. To select random samples, click the button next to "Random sample of cases" and then click on the **Sample** button. The **Select Cases: Random Sample** window will open. We can specify the size of the sample in two different ways. If we use the first option, we can specify that the sample will include a certain percentage of cases in the database. The second option allows us to specify the exact number of cases in the sample. Let's use the first option. Choose a value between 8 and 12 and type that value into the box on the first line. Click **Continue**, and then click **OK** on the **Select Cases** window. The sample will be selected and can now be processed.

To find the mean age for this sample, click **Analyze, Descriptive Statistics**, and then **Descriptives**. Find *age* in the variable list and transfer it to the **Variable(s):** window.

On the **Descriptives** menu, click the **Options** button and select S.E. MEAN in addition to the usual statistics. Click **Continue** and then **OK** and the requested statistics will appear in the output window.

Write your results on line 1 of the table below. Write in the value that you chose for your sample (between 8% and 12%) and, in the next column, write in the actual sample size (N). Next, write in the values for the sample mean and standard error in the appropriate columns. In column 5, write in the value of the sample mean (column 3) to the left of the ± sign and the value of the standard error to the right of the ± sign. This interval should include the value of the population mean (48.16). Complete column 6 by subtracting the population mean (48.16) from the sample mean (column 3) for this sample. To produce the next sample, return to the **Select Cases** window by clicking **Data** and **Select Cases**. Click the **Reset** button at the bottom of the window and then click **OK** and the full data set ($N = 1477$) will be restored. Repeat the procedure we followed previously. Click the button next to "Random sample of cases" and then click on the **Sample** button. The **Select Cases: Random Sample** window will open. Choose a number between 23 and 27 and type that value in the box, click **Continue** and **OK**, and the new sample will be selected.

Run the **Descriptives** procedure for this new sample (don't forget S.E. MEAN) and note the results in the table as we did with the first sample. Write the values in on line 2. Finally, repeat these steps for a third sample. This time, choose a value between 48 and 52 to enter into the box next to "Percent of all cases" in the "Select Cases: Random Sample" box. Note results for this sample on line 3 in the table.

	1 Sample %	2 Sample Size (N)	3 Sample Mean	4 Standard Error	5 Sample Mean ± Standard Error	6 Sample Mean − Population Mean
1	___ %				___ ± ___	
2	___ %				___ ± ___	
3	___ %				___ ± ___	

What do you notice about the table? Look at column 4. What happens to the standard error as sample size increases? Does this mean that larger samples will provide more accurate estimates of population values? How?

Look at column 5. If these samples are representative, the intervals stated in this column should include the value of the population mean (48.16). Do they? Finally, look at column 6, which shows the distance between each sample mean and the population mean. The mean of the smallest sample is probably furthest from the population mean and the distance should decrease as sample size increases, but remember that we are dealing with chance and probabilities here so the patterns you observe may not match these expectations. At any rate, the most important point is that all three sample means should be within 1 standard error of the population mean.

7

Estimation Procedures

LEARNING OBJECTIVES   By the end of this chapter, you will be able to

1. Explain the logic of estimation and the role of the sample, sampling distribution, and the population.
2. Define and explain the concepts of bias and efficiency.
3. Construct and interpret confidence intervals for sample means and sample proportions.
4. Explain the relationships between confidence level, sample size, and the width of the confidence interval.

The object of this branch of inferential statistics is to estimate population values or parameters from statistics computed from samples. Although the techniques presented in this chapter may be new to you, you are familiar with public opinion polls and election projections, their most common applications. Polls and surveys on every conceivable issue—from the sublime to the trivial—have become a staple of the mass media and popular culture. The statistical techniques you will learn in this chapter are the same as those used by the most reputable, sophisticated, and scientific pollsters.

The standard procedure for estimating population values is to construct a **confidence interval**, a mathematical statement that says that the parameter lies within a certain interval or range of values. For example, a confidence interval might say "66% ± 3%—or, between 63% and 69%—of Americans approve of capital punishment."[1] In the media, the central value of the interval (66% in this case) is usually stressed, but it is important to realize that the population parameter (the percentage of *all* Americans who approve of capital punishment) could be anywhere in the interval between 63% and 69%.

Estimation Selection Criteria

Estimation procedures are based on sample statistics. Which of the many available sample statistics should be used? Estimators can be selected according to two criteria: **bias** and **efficiency**. Estimates should be based on sample statistics that are unbiased and relatively efficient. We cover each of these criteria separately.

---

[1]This estimate is based on the General Social Survey, 2008, which was administered to a representative sample of adult residents of the United States.

BIAS

An estimator is unbiased if the mean of its sampling distribution is equal to the population value of interest. We know from the theorems presented in Chapter 6 that sample means meet this criterion. The mean of the sampling distribution of sample means (which we will note symbolically as $\mu_{\overline{X}}$) is the same as the population mean (μ).

Sample proportions (P_s) are also unbiased. That is, if we calculate sample proportions from repeated random samples of size N and then array them in a line chart, the sampling distribution of sample proportions will have a mean (μ_p) equal to the population proportion (P_u). Thus, if we are concerned with coin flips and sample honest coins 10 at a time ($N = 10$), the sampling distribution will have a mean equal to 0.5, which is the probability that an honest coin will show heads (or tails) when flipped. All statistics other than sample means and sample proportions are biased (that is, have sampling distributions with means not equal to the population value).

Since sample means and proportions are unbiased, we can determine the probability that they lie within a given distance of the population values we are trying to estimate. To illustrate, consider a specific problem. Assume that we wish to estimate the average income of a community. A random sample of 500 households is taken ($N = 500$), and a sample mean of \$45,000 is computed. In this example, the population mean is the average income of *all* households in the community and the sample mean is the average income for the 500 households that happened to be selected for our sample. Note that we do not know the value of the population mean (μ)—if we did, we wouldn't need the sample—but it is μ that we are interested in. The sample mean of \$45,000 is important primarily insofar as it can give us information about the population mean.

The two theorems presented in Chapter 6 give us a great deal of information about the sampling distribution of all possible sample means in this situation. Because N is large ($N > 100$), we know that the sampling distribution is normal and that its mean is equal to the population mean. We also know that all normal curves contain about 68% of the cases (the cases here are sample means) within ± 1 Z, 95% of the cases within ± 2 Z's, and more than 99% of the cases within ± 3 Z's of the mean. Remember that we are discussing the sampling distribution here—the distribution of all possible sample outcomes, or, in this instance, sample means. Thus, the probabilities are very good (approximately 68 out of 100 chances) that our sample mean of \$45,000 is within ± 1 Z, excellent (95 out of 100) that it is within ± 2 Z's, and overwhelming (99 out of 100) that it is within ± 3 Z's of the mean of the sampling distribution (which is the same value as the population mean). These relationships are graphically depicted in Figure 7.1.

If an estimator is unbiased, it is probably an accurate estimate of the population parameter (μ in this case). However, in less than 1% of the cases, a sample mean will be more than ± 3 Z's away from the mean of the sampling distribution (very inaccurate) by random chance alone. We literally have no idea if our particular sample mean of \$45,000 is in this small minority. We do know, however, that the odds are high that our sample mean is considerably closer than 3 Z's to the mean of the sampling distribution and, thus, to the population mean.

FIGURE 7.1 **Areas Under the Sampling Distribution of Sample Means**

EFFICIENCY

The second desirable characteristic of an estimator is efficiency, which is the extent to which the sampling distribution is clustered about its mean. Efficiency, or clustering, is essentially a matter of dispersion, as we saw in Chapter 4 (see Figure 4.1). The smaller the standard deviation of a sampling distribution, the greater the clustering and the higher the efficiency. Remember that the standard deviation of the sampling distribution of sample means, or the standard error of the mean, is equal to the population standard deviation divided by the square root of N. Therefore, the standard deviation of the sampling distribution is an inverse function of $N (\sigma_{\bar{X}} = \sigma/\sqrt{N})$. As sample size increases, $\sigma_{\bar{X}}$ will decrease. We can improve the efficiency (or decrease the standard deviation of the sampling distribution) for any estimator by increasing sample size.

An example should make this clearer. Consider two samples of different sizes:

Sample 1	Sample 2
$\bar{X}$ = \$45,000	$\bar{X}$ = \$45,000
N_1 = 100	N_2 = 1,000

Both sample means are unbiased, but which is the more efficient estimator? Consider sample 1 and assume, for the sake of illustration, that the population standard deviation (σ) is \$500.[2] In this case, the standard deviation of the sampling distribution of all possible sample means with an N of 100 would be $\sigma/\sqrt{N}$, or $500/\sqrt{100}$, or \$50.00. For sample 2, the standard deviation of all possible sample means with an N of 1000 would be much smaller. Specifically, it would be equal to $500/\sqrt{1000}$, or \$15.81.

The sampling distribution based on sample 2 is much more clustered than the sampling distribution based on sample 1. In fact, sampling distribution

---

[2]In reality, of course, the value of σ would be unknown.

FIGURE 7.2 A Sampling Distribution with $N = 100$ and $\sigma_{\bar{X}} = \$50.00$

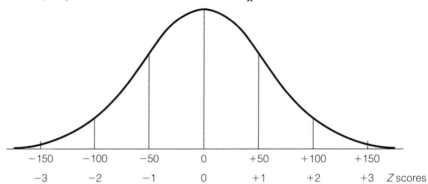

-150	-100	-50	0	$+50$	$+100$	$+150$	
-3	-2	-1	0	$+1$	$+2$	$+3$	Z scores

FIGURE 7.3 A Sampling Distribution with $N = 1000$ and $\sigma_{\bar{X}} = \$15.81$

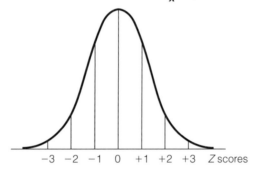

-3	-2	-1	0	$+1$	$+2$	$+3$	Z scores

2 contains 68% of all possible sample means within ±15.81 of μ, while sampling distribution 1 requires a much broader interval of ±50.00 to do the same. The estimate based on a sample with 1000 cases is much more likely to be close in value to the population parameter than is an estimate based on a sample of 100 cases. Figures 7.2 and 7.3 illustrate these relationships graphically.

The key point to remember is that the standard deviation of all sampling distributions is an inverse function of N: the larger the sample, the greater the clustering and the higher the efficiency. In part, these relationships between sample size and the standard deviation of the sampling distribution do nothing more than underscore our commonsense notion that more confidence can be placed in large samples than in small (as long as both have been randomly selected).

Interval Estimation Procedures

The first step in constructing an interval estimate is to decide on the risk you are willing to take of being wrong. An interval estimate is wrong if it does *not* include the population parameter. This probability of error is called **alpha** (symbolized by Greek letter α). The exact value of alpha will depend on the nature of the research situation, but a 0.05 probability is commonly used.

FIGURE 7.4 **The Sampling Distribution with Alpha (α) Equal to 0.05**

.0250

.4750 — .4750

.0250

.9500

Setting alpha equal to 0.05, also called using the 95% **confidence level**, means that over the long run the researcher is willing to be wrong only 5% of the time. Or, to put it another way, if an infinite number of intervals were constructed at this alpha level (and with all other things being equal), 95% of them would contain the population value and 5% would not. In reality, of course, only one interval is constructed, and, by setting the probability of error very low, we are setting the odds in our favor that the interval will include the population value.

The second step is to picture the sampling distribution, divide the probability of error equally into the upper and lower tails of the distribution, and then find the corresponding Z score. For example, if we decided to set alpha equal to 0.05, we would place half (0.025) of this probability in the lower tail and half in the upper tail of the distribution. The sampling distribution would thus be divided as illustrated in Figure 7.4.

We need to find the Z score that marks the beginnings of the shaded areas. In Chapter 5, we learned how to calculate Z scores and find areas under the normal curve. Here, we will reverse that process. We need to find the Z score beyond which lies a proportion of .0250 of the total area. To do this, go down column c of Appendix A until you find this proportion (.0250). The associated Z score is 1.96. Since the curve is symmetrical and we are interested in both the upper and lower tails, we designate the Z score that corresponds to an alpha of .05 as ±1.96 (see Figure 7.5).

We now know that 95% of all possible sample outcomes fall within ±1.96 Z scores of the population value. In reality, of course, there is only one sample

FIGURE 7.5 **Finding the *Z* Score That Corresponds to an Alpha (α) of 0.05**

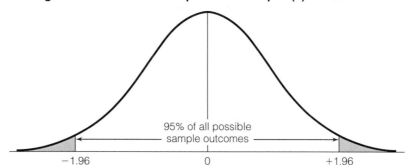

95% of all possible
sample outcomes

−1.96   0   +1.96

TABLE 7.1 *Z* Scores for Various Levels of Alpha (α)

Confidence Level	Alpha (α)	α/2	Z Score
90%	0.100	0.0500	±1.65
95%	0.050	0.0250	±1.96
99%	0.010	0.0050	±2.58
99.9%	0.001	0.0005	±3.32
99.99%	0.0001	0.00005	±3.90

outcome, but, if we construct an interval estimate based on ±1.96 *Z*'s, the probabilities are that 95% of all such intervals will trap the population value. Thus, we can be 95% confident that our interval contains the population value.

Besides the 95% level, there are four other commonly used confidence levels: the 90% level (α = .10), the 99% level (α = .01), the 99.9% level (α = .001), and the 99.99% level (α = .0001). To find the corresponding *Z* scores for these levels, follow the procedures outlined earlier for an alpha of 0.05. Table 7.1 summarizes all the information you will need.

You should turn to Appendix A and confirm for yourself that the *Z* scores in Table 7.1 correspond to these alpha levels. As you do, note that, in the cases where alpha is set at 0.10 and 0.01, the precise areas we seek are not in the table. For example, with an alpha of 0.10, we would look in column c ("Area Beyond Z") for the area .0500. Instead we find an area of .0505 (*Z* = ±1.64) and an area of .0495 (*Z* = ±1.65). The *Z* score we are seeking is somewhere between these two other scores. When this occurs, take the larger of the two scores as Z. This will make the interval as wide as possible under the circumstances and is thus the most conservative course of action. For the lowest values of alpha, we encounter the same problem (the exact areas are not in the table) and resolve it the same way, taking the largest score as Z. *(For practice in finding Z scores for various levels of confidence, see problem 7.3.)*

The third step is actually to construct the confidence interval. In the sections that follow, we construct interval estimates, first for sample means and then for sample proportions.

INTERVAL ESTIMATION PROCEDURES FOR SAMPLE MEANS (LARGE SAMPLES)

The formula for constructing a confidence interval based on sample means is given in Formula 7.1.

FORMULA 7.1
$$\text{c.i.} = \overline{X} \pm Z\left(\frac{\sigma}{\sqrt{N}}\right)$$

where c.i. = confidence interval

$\overline{X}$ = the sample mean

Z = the *Z* score as determined by the alpha level

$\dfrac{\sigma}{\sqrt{N}}$ = the sample deviation of the sampling distribution or the standard error of the mean

As an example, suppose you wanted to estimate the average IQ of a community and had randomly selected a sample of 200 residents, with a sample mean IQ of 105. Assume that the population standard deviation for IQ scores is about 15, so we can set σ equal to 15. If we are willing to run a 5% chance of being wrong and set alpha at 0.05, the corresponding Z score will be ±1.96. These values can be substituted directly into Formula 7.1, and an interval can be constructed:

$$\text{c.i.} = \overline{X} \pm Z\left(\frac{\sigma}{\sqrt{N}}\right)$$

$$\text{c.i.} = 105 \pm 1.96\left(\frac{15}{\sqrt{200}}\right)$$

$$\text{c.i.} = 105 \pm 1.96\left(\frac{15}{14.14}\right)$$

$$\text{c.i.} = 105 \pm (1.96)(1.06)$$

$$\text{c.i.} = 105 \pm 2.08$$

That is, our estimate is that the average IQ for the population in question is somewhere between 102.92 (105 − 2.08) and 107.08 (105 + 2.08). Since 95% of all possible sample means are within ±1.96 Z's (or 2.08 IQ units in this case) of the mean of the sampling distribution, the odds are very high that our interval will contain the population mean. In fact, even if the sample mean is as far off as ±1.96 Z's (which is unlikely), our interval will still contain $\mu_{\overline{X}}$ and, thus, μ. Only if our sample mean is one of the few that is more than ±1.96 Z's from the mean of the sampling distribution will we have failed to include the population mean.

Note that the value of the population standard deviation was supplied in our example. Needless to say, it is unusual to have such information about a population. In the great majority of cases, we will have no knowledge of σ, but we can estimate σ with s, the sample standard deviation. Unfortunately, s is a biased estimator of σ, and the formula must be changed slightly to correct for the bias. For larger samples, the bias of s will not affect the interval very much. The revised formula for cases in which σ is unknown is

FORMULA 7.2

$$\text{c.i.} = \overline{X} \pm Z\left(\frac{s}{\sqrt{N-1}}\right)$$

In comparing this formula with Formula 7.1, note that there are two changes. First, σ is replaced by s, and, second, the denominator of the last term is the square root of $N - 1$ rather than the square root of N. The latter change is the correction for the fact that s is biased.

Let me stress here that the substitution of s for σ is permitted only for large samples (that is, samples with 100 or more cases). For smaller samples, when the value of the population standard deviation is unknown, the standardized normal distribution summarized in Appendix A cannot be used in the estimation process. To construct confidence intervals from sample means with samples

---

Applying Statistics 7.1: Estimating a Population Mean

A study of the leisure activities of Americans was conducted on a sample of 1000 households. The respondents identified television viewing as a major form of recreation. If the sample reported an average of 6.2 hours of television viewing a day, what is the estimate of the population mean? The information from the sample is

$$\overline{X} = 6.2$$
$$s = 0.7$$
$$N = 1000$$

If we set alpha at 0.05, the corresponding Z score will be ± 1.96, and the 95% confidence interval will be

$$\text{c.i.} = \overline{X} \pm Z\left(\frac{s}{\sqrt{N-1}}\right)$$

$$\text{c.i.} = 6.2 \pm 1.96\left(\frac{0.7}{\sqrt{1000-1}}\right)$$

$$\text{c.i.} = 6.2 \pm 1.96\left(\frac{0.7}{31.61}\right)$$

$$\text{c.i.} = 6.2 \pm 1.96(0.02)$$

$$\text{c.i.} = 6.2 \pm 0.04$$

Based on this result, we would estimate that the population spends an average of 6.2 ± 0.04 hours per day viewing television. The lower limit of our interval estimate $(6.2 - 0.04)$ is 6.16, and the upper limit $(6.2 + 0.04)$ is 6.24. Thus, another way to state the interval would be

$$6.16 \leq \mu \geq 6.24$$

The population mean is greater than or equal to 6.16 and less than or equal to 6.24. Because alpha was set at the 0.05 level, this estimate has a 5% chance of being wrong (that is, of not containing the population mean).

---

smaller than 100, we must use a different theoretical distribution, called the Student's t distribution, to find areas under the sampling distribution. We defer the presentation of the t distribution until Chapter 8 and confine our attention here to estimation procedures for large samples only.

Let us close this section by working through a sample problem with Formula 7.2. Average income for a random sample of a particular community is $45,000, with a standard deviation of $200. What is the 95% interval estimate of the population mean, μ?

Given that

$$\overline{X} = \$45,000$$
$$s = \$200$$
$$N = 500$$

and using an alpha of .05, the interval can be constructed:

$$\text{c.i.} = \overline{X} \pm Z\left(\frac{s}{\sqrt{N-1}}\right)$$

$$\text{c.i.} = 45,000 \pm 1.96\left(\frac{200}{\sqrt{499}}\right)$$

$$\text{c.i.} = 45,000 \pm 17.55$$

ONE STEP AT A TIME	Constructing Confidence Intervals for Sample Means Using Formula 7.2

1: Select an alpha level and find the associated Z score in Table 7.1. If you use the conventional alpha level of 0.05, the Z score is ±1.96.

2: Substitute the sample values into the proper formula. The value of the population standard deviation (σ) will almost always be unknown, so we will almost always use Formula 7.2 to construct the confidence interval.

To Solve Formula 7.2

1: Find the square root of $N - 1$.

2: Divide the value you found in step 1 into s, the sample standard deviation.

3: Multiply the value you found in step 2 by the value of Z.

4: The value you found in step 3 is the width of the confidence interval. To find the lower and upper limits of the interval, subtract and add this value to the sample mean.

Interpreting the Confidence Interval

1: Express the confidence interval in a sentence or two that identifies each of the these elements:

a. The sample statistic (a mean in this case)
b. The confidence interval.
c. Sample size (N)
d. The population to which you are estimating
e. The confidence level (e.g., 95%)

The average income for the community as a whole is between \$44,982.45 (45,000 − 17.55) and \$45,017.55 (45,000 + 17.55). Remember that this interval has only a 5% chance of being wrong (that is, of not containing the population mean). *(For practice with confidence intervals for sample means, see problems 7.1, 7.4–7.7, 7.18, and 7.19a–7.19c.)*

INTERVAL ESTIMATION PROCEDURES FOR SAMPLE PROPORTIONS (LARGE SAMPLES)

Estimation procedures for sample proportions are essentially the same as those for sample means. The major difference is that, since proportions are different statistics, we must use a different sampling distribution. In fact, again based on the Central Limit Theorem (see Chapter 6), we know that sample proportions have sampling distributions that are normal in shape with means (μ_p) equal to the population value (P_u) and standard deviations (σ_p) equal to $\sqrt{\dfrac{P_u(1 - P_u)}{N}}$.

The formula for constructing confidence intervals based on sample proportions is

FORMULA 7.3

$$\text{c.i.} = P_s \pm Z\sqrt{\frac{P_u(1 - P_u)}{N}}$$

The values for P_s and N come directly from the sample, and the value of Z is determined by the confidence level, as was the case with sample means. This leaves one unknown in the formula, P_u—the same value we are trying to estimate. This dilemma can be resolved by setting the value of P_u at 0.5. Since the second term in the numerator under the radical $(1 - P_u)$ is the reciprocal of P_u, the entire expression will always have a value of 0.5×0.5, or 0.25, which is the maximum value this expression can attain. That is, if we set P_u at any value other than 0.5, the expression $P_u(1 - P_u)$ will decrease in value. If we set P_u at 0.4, for example, the second term $(1 - P_u)$ would be 0.6, and the value of the entire expression would decrease to 0.24. Setting P_u at 0.5 ensures that the expression $P_u(1 - P_u)$ will be at its maximum possible value and, consequently, the interval will be at maximum width. This is the most conservative solution possible to the dilemma posed by having to assign a value to P_u in the estimation equation.

To illustrate these procedures, assume that you wish to estimate the proportion of students at your university who missed at least one day of classes because of illness last semester. Out of a random sample of 200 students, 60 reported that they had been sick enough to miss classes. The sample proportion on which we will base our estimate is thus 60/200, or 0.30. At the 95% level, the interval estimate will be

$$\text{c.i.} = P_s \pm Z \sqrt{\frac{P_u(1 - P_u)}{N}}$$

$$\text{c.i.} = 0.30 \pm 1.96 \sqrt{\frac{(0.5)(0.5)}{200}}$$

$$\text{c.i.} = 0.30 \pm 1.96 \sqrt{\frac{0.25}{200}}$$

$$\text{c.i.} = 0.30 \pm (1.96)(0.04)$$

$$\text{c.i.} = 0.30 \pm 0.08$$

Based on this sample proportion of 0.30, you would estimate that the proportion of students who missed at least one day of classes because of illness was between 0.22 and 0.38. The estimate could, of course, also be phrased in percentages by reporting that between 22% and 38% of the student body was affected by illness at least once during the past semester. *(For practice with confidence intervals for sample proportions, see problems 7.2, 7.8–7.12, and 7.19d–7.19g.)*

STATISTICS

IN EVERYDAY LIFE

What is the happiest day of the year for Americans? According to Gallup's Daily Mood Index, in 2009, the three days in which most Americans felt "happiness and enjoyment without stress and worry" were Easter (64%), Flag Day (63%), and Mother's Day (62%). These results, which are based on daily samples of about 1000 Americans, are accurate to within ±3%, and have a 95% confidence level.

How could we explain these rankings? What is it about these holidays that make them more enjoyable and less stressful than, say, Christmas or Thanksgiving?

Data from Gallup: "Americans Less Happy, More Stressed in 2009." http://www.gallup.com/poll/124904/Americans-Less-Happy-Stressed-2009.aspx.

Applying Statistics 7.2: Estimating Population Proportions

If 45% of a random sample of 1000 Americans reports that walking is their major physical activity, what is the estimate of the population value? The sample information is

$$P_s = 0.45$$
$$N = 1000$$

Note that the percentage of "walkers" has been stated as a proportion. If we set alpha at 0.05, the corresponding Z score will be ± 1.96, and the interval estimate of the population proportion will be

$$\text{c.i.} = P_s \pm Z\sqrt{\frac{P_u(1 - P_u)}{N}}$$

$$\text{c.i.} = 0.45 \pm 1.96\sqrt{\frac{(0.5)(0.5)}{1000}}$$

$$\text{c.i.} = 0.45 \pm 1.96\sqrt{0.00025}$$

$$\text{c.i.} = 0.45 \pm (1.96)(0.016)$$

$$\text{c.i.} = 0.45 \pm 0.03$$

We estimate that the proportion of the population for which walking is the major form of physical exercise is between 0.42 and 0.48. That is, the lower limit of the interval estimate is $(0.45 - 0.03)$ or 0.42, and the upper limit is $(0.45 + 0.03)$ or 0.48. We may also express this result in percentages: Between 42% and 48% of the population walk as their major form of physical exercise. This interval has a 5% chance of not containing the population value.

ONE STEP AT A TIME   Constructing Confidence Intervals for Sample Proportions

Begin by selecting an alpha level and finding the associated Z score in Table 7.1. If you use the conventional alpha level of 0.05, the Z score is ± 1.96.

To Solve Formula 7.3

1: Substitute the value 0.25 for the expression $P_u(1 - P_u)$ in the numerator of the fraction under the square root sign.

2: Divide N into 0.25.

3: Find the square root of the value you found in step 2.

4: Multiply the value you found in step 3 by the value of Z.

5: The value you found in step 4 is the width of the confidence interval. To find the lower and upper limits of the interval, subtract and add this value to the sample proportion.

Interpreting the Confidence Interval

1: Express the confidence interval in a sentence or two that identifies each of the these elements:

a. The sample statistic (a proportion in this case)
b. The confidence interval
c. Sample size (N)
d. The population to which you are estimating
e. The confidence level (e.g., 95%)

COMPUTING CONFIDENCE INTERVALS: A SUMMARY

To this point, we have covered the construction of confidence intervals for sample means and sample proportions. In both cases, the procedures assume large samples (N greater than 100). The procedures for constructing confidence

---

Applying Statistics 7.3: Estimating Population Proportions

A total of 1609 adult Canadians were randomly selected to participate in a study of attitudes toward homosexuality and same-sex marriages. Some results are reported here. What is the level of support in the population? The sample information, expressed in terms of the proportion agreeing, is

"Gays and lesbians should have the same rights as heterosexuals."

$$P_s = 0.72$$
$$N = 1609$$

"Marriage should be expanded to include same-sex unions."

$$P_s = 0.60$$
$$N = 1609$$

For the first item, the confidence interval estimate to the population at the 95% confidence level is

$$\text{c.i.} = P_s \pm Z\sqrt{\frac{P_u(1 - P_u)}{N}}$$

$$\text{c.i.} = 0.72 \pm 1.96\sqrt{\frac{(0.5)(0.5)}{1609}}$$

$$\text{c.i.} = 0.72 \pm 1.96\sqrt{0.00016}$$

$$\text{c.i.} = 0.72 \pm (1.96)(0.013)$$

$$\text{c.i.} = 0.72 \pm 0.03$$

Expressing these results in terms of percentages, we can conclude that between 69% and 75% of adult Canadians support equal rights for gays and lesbians. This estimate is based on a sample of 1609 and is constructed at the 95% confidence level.

For the second survey item, the confidence interval estimate to the population at the 95% confidence level is

$$\text{c.i.} = P_s \pm Z\sqrt{\frac{P_u(1 - P_u)}{N}}$$

$$\text{c.i.} = 0.60 \pm 1.96\sqrt{\frac{(0.5)(0.5)}{1609}}$$

$$\text{c.i.} = 0.60 \pm 1.96\sqrt{0.00016}$$

$$\text{c.i.} = 0.60 \pm (1.96)(0.013)$$

$$\text{c.i.} = 0.60 \pm 0.03$$

Again, expressing results in terms of percentages, we can conclude that between 57% and 63% of adult Canadians support same-sex marriages. This estimate is based on a sample of 1609 and is constructed at the 95% confidence level. (Note that the width of the second confidence interval is exactly the same as the first. This is because we are using the same values for Z score and sample size in both estimates.)

---

intervals for small samples are not covered in this text. Table 7.2 presents the three formulas for confidence intervals organized by the situations in which they are used. For sample means when the population standard deviation is known, use Formula 7.1. When the population standard deviation is unknown (which is the usual case), use Formula 7.2. For sample proportions, always use Formula 7.3.

CONTROLLING THE WIDTH OF INTERVAL ESTIMATES

The width of a confidence interval for either sample means or sample proportions can be partly controlled by manipulating two terms in the equation. First,

TABLE 7.2   Choosing Formulas for Confidence Intervals

If the Sample Statistic Is a	and	Use Formula
mean	the population standard deviation is known	7.1 c.i. $= \bar{X} \pm Z\left(\dfrac{\sigma}{\sqrt{N}}\right)$
mean	the population standard deviation is unknown	7.2 c.i. $= \bar{X} \pm Z\left(\dfrac{s}{\sqrt{N-1}}\right)$
proportion		7.3 c.i. $= P_s \pm Z\sqrt{\dfrac{P_u(1-P_u)}{N}}$

the confidence level can be raised or lowered and, second, the interval can be widened or narrowed by gathering samples of different size. The researcher alone determines the risk he or she is willing to take of being wrong (that is, of not including the population value in the interval estimate). The exact confidence level (or alpha level) will depend, in part, on the purpose of the research. For example, if potentially harmful drugs were being tested, the researcher would naturally demand very high levels of confidence (99.99% or even 99.999%). On the other hand, if intervals are being constructed only for loose "guesstimates," then much lower confidence levels can be tolerated (such as 90%).

The relationship between interval size and confidence level is that intervals widen as confidence levels increase. This should make intuitive sense. Wider intervals are more likely to include the population value; hence, more confidence can be placed in them.

To illustrate this relationship, let us return to the example where we estimated the average income for a community. In this problem, we were working with a sample of 500 residents, and the average income for this sample was $45,000, with a standard deviation of $200. We constructed the 95% confidence interval and found that it extended 17.55 around the sample mean (that is, the interval was $45,000 ± 17.55).

If we had constructed the 90% confidence interval (a lower confidence level) for these sample data, the Z score in the formula would have decreased to 1.65, and the interval would have been narrower:

$$\text{c.i.} = \bar{X} \pm Z\left(\frac{s}{\sqrt{N-1}}\right)$$

$$\text{c.i.} = 45,000 \pm 1.65\left(\frac{200}{\sqrt{499}}\right)$$

$$\text{c.i.} = 45,000 \pm 1.65(8.95)$$

$$\text{c.i.} = 45,000 \pm 14.77$$

On the other hand, if we had constructed the 99% confidence interval, the Z score would have increased to ± 2.58, and the interval would have been wider:

$$\text{c.i.} = \overline{X} \pm Z\left(\frac{s}{\sqrt{N-1}}\right)$$

$$\text{c.i.} = 45{,}000 \pm 2.58\left(\frac{200}{\sqrt{499}}\right)$$

$$\text{c.i.} = 45{,}000 \pm 2.58(8.95)$$

$$\text{c.i.} = 45{,}000 \pm 23.09$$

At the 99.9% confidence level, the Z score would be ± 3.32, and the interval would be wider still:

$$\text{c.i.} = \overline{X} \pm Z\left(\frac{s}{\sqrt{N-1}}\right)$$

$$\text{c.i.} = 45{,}000 \pm 3.32\left(\frac{200}{\sqrt{499}}\right)$$

$$\text{c.i.} = 45{,}000 \pm 3.32(8.95)$$

$$\text{c.i.} = 45{,}000 \pm 29.71$$

We can readily see the increase in interval size in Table 7.3. Although I have used sample means to illustrate, exactly the same relationships apply to sample proportions. *(To explore further the relationship between alpha and interval width, see problem 7.13.)*

Sample size bears the opposite relationship to interval width. As sample size increases, interval width decreases. Larger samples give more precise (narrower) estimates. Again, an example should make this clearer. In Table 7.4, confidence intervals for four samples of various sizes are constructed and then grouped together for purposes of comparison. The sample data are the same as in Table 7.3, and the confidence level is 95% throughout. The relationships illustrated in Table 7.4 also hold true, of course, for sample proportions. *(To explore further the relationship between sample size and interval width, see problem 7.14.)*

TABLE 7.3 Interval Estimates for Four Confidence Levels ($\overline{X}$ = $45,000, s = $200, N = 500 throughout)

Alpha	Confidence Level	Interval	Interval Width
0.10	90%	$45,000 ± 14.77	$29.54
0.05	95%	$45,000 ± 17.55	$35.10
0.01	99%	$45,000 ± 23.09	$46.18
0.001	99.9%	$45,000 ± 29.71	$59.42

TABLE 7.4 **Interval Estimates for Four Different Samples ($\overline{X}$ = \$45,000, s = \$200, alpha = 0.05 throughout)**

Sample 1 (N = 100)	Sample 2 (N = 500)
c.i. = \$45,000 ± 1.96(200/$\sqrt{99}$)	c.i. = 45,000 ± 1.96(200/$\sqrt{499}$)
c.i. = \$45,000 ± 39.40	c.i. = 45,000 ± 17.55

Sample 3 (N = 1000)	Sample 4 (N = 10,000)
c.i. = \$45,000 ± 1.96(200/$\sqrt{999}$)	c.i. = 45,000 ± 1.96(200/$\sqrt{9,999}$)
c.i. = \$45,000 ± 12.40	c.i. = 45,000 ± 3.92

Sample	N	Interval Width
1	100	\$78.80
2	500	\$35.10
3	1000	\$24.80
4	10,000	\$7.84

Notice that the decrease in interval width (or increase in precision) does not bear a constant, or linear, relationship with sample size. For example, sample 2 is five times larger than sample 1, but the interval is not five times as narrow. This is an important relationship because it means that N might have to be increased many times over to improve the accuracy of an estimate appreciably. Since the cost of a research project is a direct function of sample size, this relationship implies a point of diminishing returns in estimation procedures. A sample of 10,000 will cost about twice as much as a sample of 5000, but estimates based on the larger sample will not be twice as precise.

Interpreting Statistics: Predicting Election Outcomes and Judging the Performance of the President

The statistical techniques covered in this chapter have become a part of everyday life in the United States and in many other societies. In politics, for example, estimation techniques are used to track public sentiment, measure how citizens perceive the performance of our leaders, and project the likely winners of upcoming elections. We should acknowledge, before proceeding, that these applications of estimation techniques can be controversial. Many people wonder if the easy availability of polls makes politicians overly sensitive to the whims of public sentiment. Others are concerned that election projections might decrease people's readiness to participate in the democratic process and, in particular, cast their votes on Election Day. These are serious concerns, of course, but we can do little more than acknowledge them in this text and hope that you have the opportunity to pursue them in other, more appropriate settings.

TABLE 7.5   **Polls and Actual Results, Presidential Race, 2008**

Date/Poll	Margin of Error	Obama	McCain
11-3-2008, CSPAN	±3%	54%	43%
11-2-2008, Gallup	±2%	53%	42%
11-2-2008, Fox News	±3%	50%	43%
11-1-2008, NBC News	±3%	51%	43%
11-1-2008, CNN	±4%	53%	46%
11-1-2008, ABC News	±2%	54%	43%
11-1-2008, CBS News	±4%	54%	41%
	ACTUAL VOTE	55%	44%

Data from: http://www.daytodaypolitics.com/polls/presidential_election_Obama_vs
_McCain_2008.htm.

In this installment of Interpreting Statistics, we examine the accuracy of election projections for the 2008, 2004, and 2000 presidential elections, and we'll also examine the polls that have measured the approval ratings of incumbent U.S. presidents since the middle of the 20th century. Both kinds of polls use the same formulas introduced in this chapter to construct confidence intervals (although the random samples were assembled by a complex and sophisticated technology that is beyond the scope of this text).

PREDICTING THE 2008 PRESIDENTIAL ELECTION

Pollsters are very accurate in predicting the outcomes of presidential elections, as illustrated by the 2008 race between Barack Obama and John McCain. Table 7.5 presents the predictions of seven different polls just before Election Day. The actual vote is also noted in the table. The predictions differ from each other because pollsters use different sampling techniques and different definitions of "likely voter." The table shows that virtually all predictions for both candidates were within their respective margins of error—which were ±2%, 3%, or 4%, depending on sample size. The two predictions that were inaccurate were for President Obama: Both Fox and NBC News were too low, the former by 2 percentage points and the latter by 1.

"TOO CLOSE TO CALL"

Remember that 95% confidence intervals are accurate only to within 2–4 percentage points, depending on sample size. This means that the polls cannot identify the likely victor in very close races. Unlike the 2008 election, the 2004 and 2000 elections were very close, and the polls indicated a statistical dead heat right up to the end of the campaigns. Table 7.6 shows CNN's "poll of polls"—or the averages of polls from various sources—for the final weeks of the 2004 campaign in which Republican George Bush ran against Democrat John Kerry and the final breakdown of votes for these two candidates. The polls included by CNN were based on sample sizes of about 1000.

TABLE 7.6   Polling Results and Actual Vote, 2004 Presidential Election (The margin of error for these projections is about ±3%)

Date of Poll	Bush	Kerry
November 1	48%	46%
October 25	49%	46%
October 18	50%	45%
ACTUAL VOTE	51%	48%

The final polls in October and November show that the race was very close and that the difference between the candidates was so small that a winner could not be projected. For example, in the November 1 poll, Bush's support could have been as low as 45% (48% − 3%) and Kerry's could have been as high as 49% (46% + 3%). When the confidence intervals overlap, the race is said to be "too close to call" and "a statistical dead heat."

The 2000 presidential election between George Bush and Vice President Al Gore was even closer than the 2004 election. When the ballots were finally counted (and recounted), the candidates were in a virtual tie for the popular vote, separated by only a few thousand votes out of more than 100 million votes cast. Democrat Al Gore received 48.3% of the popular vote, while Republican George Bush garnered 48.1%. After a lengthy and intense battle over the electoral votes of Florida, the Supreme Court decided the election in favor of Bush.

THE UPS AND DOWNS OF PRESIDENTIAL POPULARITY

Once you get to be president, you are not free of public-opinion surveys. Since the middle of the 20th century, pollsters have tracked the president's popularity by asking samples of adult Americans if they approve or disapprove of the way the president is handling the job. Approval ratings for Presidents Bush and Obama are presented in Figure 7.6. For purposes of clarity, only the sample percentages are used for this graph, but remember that the confidence interval estimate would range about ±3% (at the 95% confidence level) around these points.

Perhaps the single most dramatic feature of Figure 7.6 is the huge increase in approval for President Bush that followed the 9/11 terrorist attacks on the World Trade Center and the Pentagon in 2001. This burst of support reflects the increased solidarity, strong emotions, and high levels of patriotism with which Americans responded to the attacks. The president's approval rating reached an astounding 90% shortly after the attacks but then, inevitably, began to trend down as the society (and politics) gradually returned to the daily routine of everyday business. President Bush's popularity received another boost with the invasion of Iraq in March 2003, but then began to sink to prewar levels and below, reflecting the increasing problems faced by the Bush administration and a growing reluctance of many Americans to support the war effort.

Americans responded to President Obama's inauguration with great enthusiasm and high levels of approval. His initial ratings were in the high 70s—much

FIGURE 7.6 **Presidential Approval, 2001–2010**

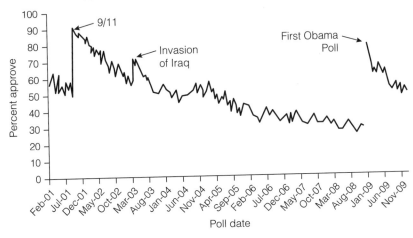

higher than President Bush's first ratings and more than double his last—but then gradually declined throughout his first year in office as the nation's economic woes continued and the administration was caught up in a lengthy, acrimonious debate over health care.

Returning to the controversies surrounding public opinion polls, many would argue that it is harmful to judge the president so continuously. To some extent, these approval ratings expose the president (and other leaders and politicians whose performance is similarly measured) to the whims of public sentiment and, at some level, tend to pressure him or her to cater to popular opinion and shape ("spin") his or her image to maintain support. On the other hand, information like that presented in Figure 7.6 supplies some interesting and useful insights into the affairs of the society as a whole, not only of the political institution. For example, the approval ratings in Figure 7.6 provide convincing evidence that Americans responded to the attacks of 9/11 with strong solidarity and cohesion, contrary to the expectations of some. The numbers also show the debilitating effects of controversial wars and a sputtering economy on public morale and confidence in the leadership of the nation.

STATISTICS

IN EVERYDAY LIFE

"Dewey Defeats Truman," read the banner headline of the morning edition of the *Chicago Tribune* the day after the 1948 presidential election. The headline became famous because the final results were just the reverse: Not only had Truman won his bid for reelection, but he won by a handy margin. What happened? Because the political "experts" were certain that Truman would lose, the pollsters had stopped taking polls early in the fall of 1948 and missed a late surge of support for the incumbent president. This major embarrassment for the *Chicago Tribune* (and many other newspapers) resulted in the current practice of polling voters right up until Election Day.

What kinds of opinions or attitudes would be more likely to change rapidly, as did support for President Truman? Can you think of attitudes or opinions that would be less likely to fluctuate?

READING STATISTICS 4: Public-Opinion Polls

You are most likely to encounter the estimation techniques covered in this chapter in the mass media in the form of public-opinion polls, election projections, and the like. Responsible reporting by the media will usually emphasize the estimate itself (for example, "In a survey of the American public, 57% of the respondents said that they approve of the president's performance") but also will report the width of the interval ("This estimate is accurate to within ±3%" or "Figures from this poll are subject to a sampling error of ±3%"), the alpha level (usually as the confidence level of 95%), and the size of the sample ("1458 households were surveyed").

Election projections and voter analyses, at least at the presidential level, have been common since the middle of the 20th century and are discussed further in the "Interpreting Statistics" section of this chapter. More recently, public opinion polling has been used to gauge reactions to everything from the newest movies to the hottest gossip to the president's conduct during the latest crisis, and polls are routinely included in news stories in the media. I would include an example or two of these applications here, but polls have become so pervasive that you can find your own example. Just pick up a newsmagazine or newspaper and leaf through it casually, and I bet you'll find at least one poll. Read the story and see if you can identify the population, the confidence interval width, the sample size, and the confidence level. Bring the news item to class and dazzle your instructor.

As a citizen as well as a social scientist, you should be extremely suspicious of polls that do not include such vital information as sample size or interval width. You should also check to see how the sample was assembled. Nonrandom samples cannot be regarded as representative of the American public, or for that matter, of any population other than the sample itself. You might encounter samples such these when news or sports programs ask viewers to call in and register their opinions about some current controversy. The "samples" assembled in this way are not representative, and the results should be regarded as "for entertainment only" or, perhaps, as a basis for discussion. They should not be regarded as reflections of public opinion in general. You should, of course, read all polls and surveys critically and analytically, but you should place confidence only in polls that are based on samples selected according to the rule of EPSEM (see Chapter 6) from some defined population.

Advertisements, commercials, and reports published by partisan groups sometimes report statistics that seem to be estimates of the population values. Often, however, such estimates are based on woefully inadequate sample sizes and biased sampling procedures, and the data are collected under circumstances that evoke a desired response. "Person in the street" (or shopper in the grocery store) interviews have a certain folksy appeal but must not be accorded the same credibility as surveys conducted by reputable polling firms.

PUBLIC-OPINION SURVEYS IN THE PROFESSIONAL LITERATURE

For the social sciences, probably the single most important consequence of the growth in opinion polling is that many nationally representative databases are now available for research purposes. These high-quality databases are often available at no cost or for a nominal fee, and they make it possible to conduct "state-of-the-art" research without the expense and difficulty of collecting data yourself. This is an important development because we can now test our theories against very high-quality data, and our conclusions can have a stronger empirical basis. Our research efforts can have greater credibility with our colleagues, with policy makers, and with the public at large.

One of the more important and widely used databases of this sort is called the General Social Survey, or the GSS. Since 1972, the National Opinion Research Council has questioned a nationally representative sample of Americans about a wide variety of issues and concerns. Since many of the questions are asked every year, the GSS offers a longitudinal record of American sentiment and opinion about a large variety of topics for more than three decades. Each year, new topics of current concern are added and explored, and the variety of information available continues to expand. Like other nationally representative samples, the GSS sample is chosen by a complex probability design that resembles cluster sampling (see Chapter 6). Sample size varies from 1400 to over 4000, and estimates based on samples this large will be accurate to within about ±3%. The computer exercises in this text are based on the 2008 GSS; this database is described more fully in Appendix F.

SUMMARY

1. Population values can be estimated with sample values. Confidence intervals estimate that the population value falls within a certain range of values and has a known probability of being incorrect.
2. Estimates based on sample statistics must be unbiased and relatively efficient. Of all the sample statistics, only means and proportions are unbiased. The means of the sampling distributions of these statistics are equal to the respective population values. Efficiency is largely a matter of sample size. The greater the sample size, the lower the value of the standard deviation of the sampling distribution, the more tightly clustered the sample outcomes will be around the mean of the sampling distribution, and the more efficient the estimate.
3. The width of a confidence interval is a function of the risk we are willing to take of being wrong (the alpha level) and the sample size. The interval widens as our probability of being wrong decreases and as sample size decreases.

SUMMARY OF FORMULAS

Confidence interval for a sample mean, large samples, population standard deviation known:

$$7.1 \quad \text{c.i.} = \overline{X} \pm Z\left(\frac{\sigma}{\sqrt{N}}\right)$$

Confidence interval for a sample mean, large samples, population standard deviation unknown:

$$7.2 \quad \text{c.i.} = \overline{X} \pm Z\left(\frac{s}{\sqrt{N-1}}\right)$$

Confidence interval for a sample proportion, large samples:

$$7.3 \quad \text{c.i.} = P_s \pm Z\sqrt{\frac{P_u(1 - P_u)}{N}}$$

GLOSSARY

Alpha (α). The probability of error, or the probability that a confidence interval does not contain the population value. Alpha levels are usually set at 0.10, 0.05, 0.01, or 0.001.

Bias. A criterion used to select sample statistics as estimators. A statistic is unbiased if the mean of its sampling distribution is equal to the population value of interest.

Confidence interval. An estimate of a population value in which a range of values is specified.

Confidence level. A frequently used alternate way of expressing alpha, the probability that an interval estimate will not contain the population value. Confidence levels of 90%, 95%, 99%, and 99.9% correspond to alphas of 0.10, 0.05, 0.01, and 0.001, respectively.

Efficiency. The extent to which the sample outcomes are clustered around the mean of the sampling distribution.

PROBLEMS

7.1 For each set of sample outcomes below, construct the 95% confidence interval for estimating μ, the population mean.

   a. $\overline{X} = 5.2$
      $s = 0.7$
      $N = 157$

   b. $\overline{X} = 100$
      $s = 9$
      $N = 620$

   c. $\overline{X} = 20$
      $s = 3$
      $N = 220$

   d. $\overline{X} = 1020$
      $s = 50$
      $N = 329$

   e. $\overline{X} = 7.3$
      $s = 1.2$
      $N = 105$

   f. $\overline{X} = 33$
      $s = 6$
      $N = 220$

7.2 For each of the following sets of sample outcomes, construct the 99% confidence interval for estimating P_u.

   a. $P_s = 0.40$
      $N = 548$

   b. $P_s = 0.37$
      $N = 522$

   c. $P_s = 0.79$
      $N = 121$

   d. $P_s = 0.14$
      $N = 100$

   e. $P_s = 0.43$
      $N = 1049$

   f. $P_s = 0.63$
      $N = 300$

7.3 For each of the following confidence levels, determine the corresponding Z score.

Confidence Level	Alpha	Area Beyond Z	Z Score
95%	0.05	0.0250	±1.96
94%			
92%			
97%			
98%			

7.4 SW You have developed a series of questions to measure job satisfaction of bus drivers in New York City. A random sample of 100 drivers has an average score of 10.6, with a standard deviation of 2.8. What is your estimate of the average job satisfaction score for the population as a whole? Use the 95% confidence level.

7.5 SOC A researcher has gathered information from a random sample of 178 households. For each of the following variables, construct confidence intervals to estimate the population mean. Use the 90% level.
  a. An average of 2.3 people resides in each household. Standard deviation is 0.35.
  b. There was an average of 2.1 television sets ($s = 0.10$) and 0.78 telephones ($s = 0.55$) per household.
  c. The households averaged 6.0 hours of television viewing per day ($s = 3.0$).

7.6 SOC A random sample of 100 HBO movies contained an average of 3.38 acts of physical violence per program. At the 99% level, what is your estimate of the population value?
$$\overline{X} = 3.38$$
$$s = 0.30$$
$$N = 100$$

7.7 SOC A random sample of 429 college students was interviewed about a number of matters.
  a. They reported that they had spent an average of $345.23 on textbooks during the previous semester. If the sample standard deviation for these data is $15.78, construct an estimate of the population mean at the 99% level.
  b. They also reported that they had visited the health clinic an average of 1.5 times a semester. If the sample standard deviation is 0.3, construct an estimate of the population mean at the 99% level.
  c. On the average, the sample had missed 2.8 days of classes per semester because of illness. If the sample standard deviation is 1.0, construct an estimate of the population mean at the 99% level.

  d. On the average, the sample had missed 3.5 days of classes per semester for reasons other than illness. If the sample standard deviation is 1.5, construct an estimate of the population mean at the 99% level.

7.8 CJ A random sample of 100 inmates at a maximum security prison shows that exactly 10 of the respondents had been the victims of violent crime during their incarceration. Estimate the proportion of victims for the population as a whole, using the 90% confidence level. (*HINT: Calculate the sample proportion P_s before using Formula 7.3. Remember that proportions are equal to frequency divided by N.*)

7.9 SOC The survey mentioned in problem 7.5 found that 25 of the 178 households consisted of unmarried couples who were living together. What is your estimate of the population proportion? Use the 95% level.

7.10 PA A random sample of 260 workers in a high-rise office building revealed that 30% were very satisfied with the quality of elevator service. At the 99% level, what is your estimate of the population value?

7.11 SOC A random sample of 1496 respondents of a major metropolitan area was questioned about a number of issues. Construct estimates to the population at the 95% level for each of the results reported next. Express the final confidence interval in percentages (e.g., "between 40 and 45% agreed that premarital sex was always wrong").
  a. When asked to agree or disagree with the statement "Explicit sexual books and magazines lead to rape and other sex crimes," 823 agreed.
  b. When asked to agree or disagree with the statement "Hand guns should be outlawed," 650 agreed.
  c. 375 of the sample agreed that marijuana should be legalized.
  d. 1023 of the sample said that they had attended a church, synagogue, mosque, or other place of worship at least once within the past month.
  e. 800 agreed that public elementary schools should have sex education programs starting in the fifth grade.

7.12 SW A random sample of 100 patients treated in a program for alcoholism and drug dependency over the past 10 years was selected. It was determined that 53 of the patients had been readmitted to the program at least once. At the 95% level, construct an estimate of the population proportion.

7.13 For the following sample data, construct five different interval estimates of the population mean, one each for the 90%, 95%, 99%, 99.9%, and 99.99% levels. What happens to the interval width as confidence level increases? Why?

$$\overline{X} = 100$$
$$s = 10$$
$$N = 500$$

7.14 For each of the following three sample sizes, construct the 95% confidence interval. Use a sample proportion of 0.40 throughout. What happens to interval width as sample size increases? Why?

$$P_s = 0.40$$
Sample A: $N = 100$
Sample B: $N = 1000$
Sample C: $N = 10,000$

7.15 PS Two individuals are running for mayor of Shinbone. You conduct an election survey a week before the election and find that 51% of the respondents prefer candidate A. Can you predict a winner? Use the 99% level. (*HINT: In a two-candidate race, what percentage of the vote would the winner need? Does the confidence interval indicate that candidate A has a sure margin of victory? Remember that while the population parameter is probably α = 0.01 in the confidence interval, it may be anywhere in the interval.*)

$$P_s = 0.51$$
$$N = 578$$

7.16 SOC The World Values Survey is administered periodically to random samples from societies around the globe. Listed here are the number of respondents in each nation who said that they are "very happy." Compute sample proportions, and construct confidence interval estimates for each nation at the 95% level.

Nation	Year	Number "Very Happy"	Sample Size	Confidence Interval
Great Britain	2006	528	1041	
Japan	2005	311	1096	
Brazil	2006	510	1500	
China	2007	424	2015	
Malaysia	2006	428	1201	
Mali	2007	598	1534	
Russia	2006	214	2033	

7.17 SOC The fraternities and sororities at St. Algebra College have been plagued by declining membership over the past several years and want to know if the incoming freshman class will be a fertile recruiting ground. Not having enough money to survey all 1600 freshmen, they commission you to survey the interests of a random sample. You find that 35 of your 150 respondents are "extremely" interested in social clubs. At the 95% level, what is your estimate of the number of freshmen who would be extremely interested? (*HINT: The high and low values of your final confidence interval are proportions. How can proportions also be expressed as numbers?*)

7.18 SOC You are the automotive specialist for the monthly magazine *Consumer Beware*. Part of your job is to investigate the claims made by automakers in their TV commercials. You are particularly suspicious of a new economy car that the automaker claims will get 78 miles per gallon. After checking the mileage figures for a random sample of 125 owners of this car, you find an average miles per gallon of 75.5, with a standard deviation of 3.7. At the 99% level, do your results tend to confirm or refute the manufacturer's claims?

7.19 SOC The results listed next are from a survey given to a random sample of the American public. For each sample statistic, construct a confidence interval estimate of the population parameter at the 95% confidence level. Sample size (N) is 2987 throughout.
   a. The average occupational prestige score was 43.87, with a standard deviation of 13.52.
   b. The respondents reported watching an average of 2.86 hours of TV per day, with a standard deviation of 2.20.
   c. The average number of children was 1.81, with a standard deviation of 1.67.
   d. Of the 2987 respondents, 876 identified themselves as Catholic.
   e. Five hundred thirty-five of the respondents said that they had never married.
   f. The proportion of respondents who said they voted for Bush in the 2004 presidential election was 0.52.
   g. When asked about capital punishment, 2425 of the respondents said that they favored the death penalty for murder.

Estimating the Characteristics of the Typical American

SPSS does not provide a program specifically for constructing confidence intervals, although some of the procedures we'll cover in future chapters do include confidence intervals as part of the output. Rather than make use of these programs, we will use SPSS to produce sample statistics from the 2008 GSS, which you can then use as the basis for interval estimates to the population (U.S. society as of 2008).

Step 1: Choosing the Variables

As a framework for this exercise, we will return to the task of describing the "typical American" begun in Chapter 4 and continued in Chapter 5. Select four of the variables you used in the earlier exercises and add four more that you did NOT use earlier. Make sure that you have at least one variable from the nominal level and at least one measured at the ordinal or interval-ratio level. List the variables, along with level of measurement, here.

Variable	SPSS Name	Explain exactly what this variable measures	Levels of Measurement
1			
2			
3			
4			
5			
6			
7			
8			

Step 2: Getting Sample Statistics

For interval-ratio variables and ordinal level variables with 4 or more scores, use **Descriptives** to get sample means, standard deviations, and sample sizes. For nominal-level variables and for ordinal-level variables with 3 or fewer scores, use **Frequencies** to produce frequency distributions. For either procedure, click **Analyze → Descriptive Statistics** and then select either **Frequencies** or **Descriptives**. Select your variables from the box on the left and click the arrow to move the variable name to the window on the right. SPSS will process all variables listed at the same time.

Step 3: Constructing Confidence Intervals

Once you have the SPSS output, use the results to construct 95% confidence intervals around the sample statistics. For nominal-level variables and ordinal-level variables with 3 or fewer scores, select one category (e.g., female for *sex* or Catholic for *relig*) and look in the **Valid Percent** column of the frequency distribution to get the percentage of cases in the sample in that category. Change this value to a proportion (divide by 100). This value is P_s and can be substituted directly into Formula 7.3. Remember to estimate P_u at 0.5. After you have found the confidence interval, change your proportion back to a percentage and record your results in the table below.

For ordinal-level variables with 4 or more scores and for interval-ratio variables, find the sample mean, standard deviation, and sample size in the output of the **Descriptives** procedure. Substitute these values into Formula 7.2 and record your results in the table below.

Step 4: Recording Results

Use the table to summarize your confidence intervals.

Variable	SPSS Name	Sample Statistic $(\bar{X}$ or $P_s)$	95% Confidence Interval	N
1				
2				
3				
4				
5				
6				
7				
8				

NOTE: Some ordinal-level variables, like *news* and *attend*, use scores to stand for broad categories (see Appendix G for examples) and sample means should be interpreted in terms of these categories, not as absolute scores. For example, if the mean of *news* is 3.7, this would indicate that the average score for this variable is somewhere in between "once a week" (a score of 3) and "less than once a week" (a score of 4).

Step 5: Reporting Results

For each statistic, express the confidence interval in words, as if you were reporting results in a newspaper story. Be sure to include each of the following: the sample statistic, sample size, the upper and lower limits of the confidence interval, and the confidence level (95%), and identify the population. A sample sentence might read "I estimate that 45% of all Americans support candidate X for President. This estimate is accurate to within ±3% and has a 95% chance of being accurate and is based on a sample of 1362 adult Americans.

8

Hypothesis Testing I
The One-Sample Case

By the end of this chapter, you will be able to

1. Explain the logic of hypothesis testing.
2. Define and explain the conceptual elements involved in hypothesis testing, especially the null hypothesis, the sampling distribution, the alpha level, and the test statistic.
3. Explain what it means to "reject the null hypothesis" or "fail to reject the null hypothesis."
4. Identify and cite examples of situations in which one-sample tests of hypotheses are appropriate.
5. Test the significance of single-sample means and proportions using the five-step model and correctly interpret the results.
6. Explain the difference between one- and two-tailed tests and specify when each is appropriate.
7. Define and explain Type I and Type II errors and relate each to the selection of an alpha level.
8. Use the Student's *t* distribution to test the significance of a sample mean for a small sample.

In Chapter 7 you were introduced to inferential statistical techniques for estimating population parameters from sample statistics. In Chapters 8 through 11, we investigate a second application of inferential statistics, called **hypothesis testing** or **significance testing**. In this chapter, the techniques for hypothesis testing in the one-sample case are introduced. These procedures could be used in situations such as the following:

1. A researcher working for a senior citizens advocacy group has asked a sample of 789 older residents of a particular state if they have been victimized by crime over the past year. He has obtained the percentage of the entire population of the state that was victimized by crime from the state government Web site and wants to test his hypothesis that senior citizens, as represented by this sample, are more likely to be victimized than the population in general.

2. The board of trustees of a small college wants to expand its athletic program but is concerned about the academic progress of student athletes. They have asked the dean to gather information about the academic performance of athletes at a nearby university, which has a large athletics program. The dean will compare the GPAs of a random sample of 235 student athletes with the GPA of all students at the university.

3. A sociologist is assessing the effectiveness of a rehabilitation program for alcoholics in her city. The program serves a large area and she cannot test every single client. Instead, she draws a random sample of 127 people from the list of all clients and questions them on a variety of issues. She notices that, on the average, the people in her sample miss fewer days of work each year than workers in the city as a whole. On the basis of this unexpected finding, she decides to conduct a test to see if the workers in her sample really are more reliable than workers in the community as a whole.

In each of these situations, we have randomly selected samples (of senior citizens, athletes, or treated alcoholics) that we want to compare to a population (the entire state, student body, or city). Note that we are not interested in the sample per se but in the larger group from which it was selected (*all* senior citizens in the state, *all* athletes on this campus, or *all* people who have completed the treatment program). Specifically, we want to know if the groups represented by the samples differ from population parameters on a specific statistic (victimization rates, GPAs, or absenteeism).

Of course, it would be better if we could test all senior citizens, athletes, or treated alcoholics rather than these smaller samples. However, as we have seen, researchers usually do not have the resources to test everyone in a large group and must use random samples instead. In these situations, conclusions will be based on a comparison of a single sample (representing the larger group) and the population. For example, if we found that the rate of victimization for the sample of senior citizens was higher than the rate for the state as a whole, we might conclude that "senior citizens are significantly more likely to be crime victims." The word *significantly* is a key word: It means that the difference between the sample's victimization rate and the population's rate is very unlikely to be caused by random chance alone. In other words, it is very likely that *all* senior citizens (not just those in the sample) have a higher victimization rate than the state as a whole. On the other hand, if we found little difference between the GPAs of the sample of athletes and the whole student body, we might conclude that athletes (*all* athletes on this campus, not just the athletes in the sample) are essentially the same as other students in terms of academic achievement.

Thus, we can use samples to represent larger groups (senior citizens, athletes, or treated alcoholics) and compare and contrast the characteristics of the sample with those of the population and be extremely confident in our conclusions. Remember, however, that there is no guarantee that random samples will be representative; thus, there will always be a small amount of uncertainty in our conclusions. One of the great advantages of inferential statistics is that we will be able to estimate the probability of error and evaluate our decisions accordingly.

An Overview of Hypothesis Testing

We'll begin with a general overview of hypothesis testing, using the third research situation mentioned earlier as an example, and introduce the more technical considerations and proper terminology throughout the remainder of the chapter. Let's examine this situation in some detail. First, the main question here is "Are people treated in this program more reliable workers than people in the community in general?" In other words, what the researcher would really like to do is compare *all* clients (the *population* of alcoholics treated in the program) with the entire metropolitan area. If she had information for both of these groups, she could answer the question easily, completely, and finally.

The problem is that the researcher does not have the time or money to gather information on the thousands of people who have been treated by the program. Instead, she has drawn a random sample of 127 clients. The absentee rates for the sample and the community are:

Community	Sample of Treated Alcoholics
μ = 7.2 days per year	$\overline{X}$ = 6.8 days per year
σ = 1.43	N = 127

We can see that there is a difference: The average rate of absenteeism for the sample is lower than the rate for the community. Although it's tempting, we can't make any conclusions yet because we are working with a random sample of the population we are interested in, not the population itself (*all* people treated in the program).

Figure 8.1 should clarify these relationships. The community is symbolized by the largest circle because it is the largest group. The population of all treated alcoholics is also symbolized by a large circle because it is a sizeable group, although only a fraction of the community as a whole. The random sample of 127, the smallest of the three groups, is symbolized by the smallest circle.

The labels on the arrows connecting the circles summarize the major questions and connections in this research situation. As we noted earlier, the main question is "Does the population of *all* treated alcoholics have different absentee rates than the community as a whole?" The population of treated alcoholics, too large to test, is represented by the random sample of 127.

FIGURE 8.1 A Test of Hypothesis for Single-Sample Means

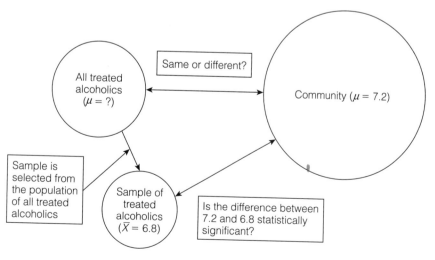

We are interested in what caused the observed difference between the sample mean of 6.8 and the community mean of 7.2. There are two possible explanations for this difference, and we consider them one at a time.

The first explanation is that the difference between the community mean (7.2) and the sample mean (6.8) reflects a real difference in absentee rates between the population of all treated alcoholics and the community. The difference is "statistically significant," which means that it is very unlikely to have occurred by random chance alone. If this explanation is true, the population of all treated alcoholics is different from the community and the sample did *not* come from a population with a mean absentee rate of 7.2 days.

The second explanation is called the **null hypothesis** (symbolized as H_0, or *H*-sub-zero). It states that the observed difference between sample and community means was caused by mere random chance: There is no important difference between treated alcoholics and the community, and the difference between the sample mean and the community mean is trivial and due to random chance. If the null hypothesis is true, treated alcoholics are just like everyone else and have a mean absentee rate of 7.2 days.

Which explanation is correct? We cannot answer this question with absolute certainty as long as we are working with a sample rather than the entire group. We can, however, set up a decision-making procedure so conservative that one of the explanations can be chosen, with the knowledge that the probability of choosing the wrong explanation is very low.

This decision-making process begins with the assumption that the second explanation—the null hypothesis—is correct. Symbolically, the assumption that the mean absenteeism rate for all treated alcoholics is the same as the rate for the community as a whole can be stated as

$$H_0: \mu = 7.2 \text{ days per year}$$

Remember that μ refers to the mean for *all* treated alcoholics, not just the 127 in the sample. This assumption, $\mu = 7.2$, can be tested statistically.

If the null hypothesis ("*The population of treated alcoholics is not different from the community as a whole and has a μ of 7.2*") is true, then the probability of getting the observed sample outcome ($\overline{X} = 6.8$) can be found. Let us add an objective decision rule in advance. If the odds of getting the observed difference are less than 0.05 (5 out of 100, or 1 in 20), we will reject the null hypothesis. If this explanation were true, a difference of this size (7.2 days vs. 6.8 days) would be a very rare event, and in hypothesis testing we always bet *against* rare events.

How can we estimate the probability of the observed sample outcome ($\overline{X} = 6.8$) if the null hypothesis is correct? This probability can be determined by using our knowledge of the sampling distribution of all possible sample outcomes. Looking back at the information we have and applying the Central Limit Theorem (see Chapter 6), we can assume that the sampling distribution is normal in shape and has a mean of 7.2 (because $\mu_{\overline{X}} = \mu$) and a standard deviation of $1.43/\sqrt{127}$ because $\sigma_{\overline{X}} = \sigma/\sqrt{N}$. We also know that the standard normal distribution can be interpreted as a distribution of probabilities (see Chapter 5) and that this sample outcome ($\overline{X} = 6.8$) is one of thousands of possible sample outcomes. The sampling distribution, with the sample outcome noted, is depicted in Figure 8.2.

FIGURE 8.2   The Sampling Distribution of All Possible Sample Means

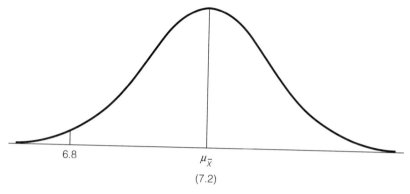

Using our knowledge of the standardized normal distribution, we can add further useful information to this sampling distribution of sample means. Specifically, we can depict our decision rule ("Any sample outcome with probability less than 0.05 will cause us to reject the null hypothesis") with Z scores. The probability of 0.05 can be translated into an area and divided equally into the upper and lower tails of the sampling distribution. Using Appendix A, we find that the Z-score equivalent of this area is ±1.96. (To review finding Z scores from areas or probabilities, see Chapter 7). The areas and Z scores are depicted in Figure 8.3.

The decision rule can now be rephrased. Any sample outcome falling in the shaded areas depicted in Figure 8.3 has a probability of occurrence of less than 0.05. Such an outcome would be a rare event and would cause us to reject the null hypothesis.

All that remains is to translate our sample outcome into a Z score to see where it falls on the curve. To do this, we use the standard formula for locating any score under a normal distribution. When we use known or empirical distributions, this formula is expressed as

$$Z = \frac{X_i - \overline{X}}{s}$$

FIGURE 8.3   The Sampling Distribution of All Possible Sample Means

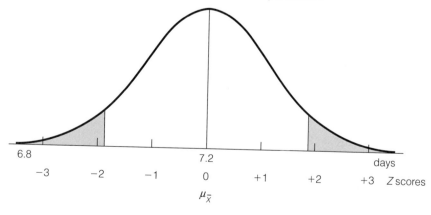

Or, to find the equivalent Z score for any raw score, subtract the mean of the distribution from the raw score and divide by the standard deviation of the distribution.

Since we are now concerned with the sampling distribution of all sample means rather than an empirical distribution, the symbols in the formula will change, but the form remains exactly the same:

FORMULA 8.1

$$Z = \frac{\overline{X} - \mu}{\sigma/\sqrt{N}}$$

Or, to find the equivalent Z score for any sample mean, subtract the mean of the sampling distribution (which is equal to the population mean, or μ) from the sample mean and divide by the standard deviation of the sampling distribution.

Recalling the data given on this problem, we can now find the Z-score equivalent of the sample mean:

$$Z = \frac{6.8 - 7.2}{1.43/\sqrt{127}}$$

$$Z = \frac{-0.40}{0.127}$$

$$Z = -3.15$$

In Figure 8.4, this Z score of -3.15 is noted on the distribution of all possible sample means, and we see that the sample outcome does fall in the shaded area. If the null hypothesis is true, this sample outcome has a probability of occurrence of less than 0.05. The sample outcome ($\overline{X} = 6.8$, or $Z = -3.15$) would be rare if the null hypothesis was true, and the researcher may therefore reject the null hypothesis. The sample of 127 treated alcoholics comes from a population that is significantly different from the community on absenteeism. Or, to put it another way, the sample does not come from a population that has a mean of 7.2 days of absences.

Keep in mind that our decisions in significance testing are based on random samples. On rare occasions, an EPSEM sample may not be representative of the population from which it was selected. The decision-making process just outlined has a very high probability of resulting in correct decisions, but we always

FIGURE 8.4 **The Sampling Distribution of Sample Means, with the Sample Outcome ($\overline{X} = 6.8$) Noted in Z Scores**

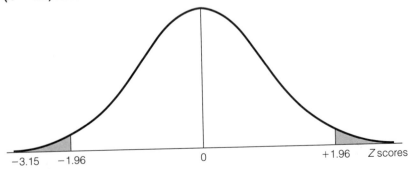

The federal Food and Drug Administration (FDA) estimates that it takes an average of eight years for a new drug to be approved for use in the United States. Before the manufacturer of a drug can apply for FDA approval, the drug must be tested first on animals, then on samples of healthy volunteer humans, and finally on samples of humans who have been diagnosed with the specific disease or condition that the manufacturer claims the drug was designed to treat. The results of these studies must show that the drug is safe and effective and that its benefits outweigh its risks. All the tests are based on null hypotheses: that the drug is ineffective in the treatment of the target disease or condition.

face an element of risk when we must work with samples rather than populations. The decision to reject the null hypothesis might be incorrect if this sample happens to be one of the few that is unrepresentative of the population of alcoholics treated in this program. One important strength of hypothesis testing is that we can estimate the probability of making an incorrect decision. In the example at hand, the null hypothesis was rejected and the probability that this decision was incorrect is 0.05—the decision rule established at the beginning of the process. To say that the probability of rejecting the null hypothesis incorrectly is 0.05 means that, if we repeated this same test an infinite number of times, we would incorrectly reject the null hypothesis only 5 times out of every 100.

The Five-Step Model for Hypothesis Testing

The previous section introduced the concepts and procedures used in hypothesis testing. Next, we will examine a **five-step model** for organizing all tests of hypothesis.

Step 1. Making assumptions and meeting test requirements

Step 2. Stating the null hypothesis

Step 3. Selecting the sampling distribution and establishing the critical region

Step 4. Computing the test statistic

Step 5. Making a decision and interpreting the results of the test

We now look at each step individually, using the problem from the previous section as an example throughout.

Step 1. Making Assumptions and Meeting Test Requirements. Any application of statistics requires that we make certain assumptions. Specifically, three assumptions about the testing situation and the variables involved have to be satisfied when conducting hypothesis tests with a mean. First, we must be sure that we are working with a random sample, one that has been selected according to the rules of EPSEM. Second, we must assume that the variable being tested is interval-ratio in level of measurement to justify the computation of a mean. Finally, we must assume that the sampling distribution of all possible sample means is normal in shape so that we may use the standardized normal distribution to find areas under the sampling distribution. We can be sure that this assumption is satisfied by using large samples (see the Central Limit Theorem in Chapter 6).

Usually, we will state these assumptions in abbreviated form as a mathematical model for the test. For example:

> Model: Random sampling
> Level of measurement is interval-ratio
> Sampling distribution is normal

Step 2. Stating the Null Hypothesis. The null hypothesis is always a statement of "no difference," but its exact form will vary depending on the test being conducted. In the single-sample case, the null hypothesis states that the sample comes from a population with a certain characteristic. In our example, the null hypothesis is that the population of treated alcoholics is "no different" from the community as a whole, that their average days of absenteeism is also 7.2, and that the difference between 7.2 and the sample mean of 6.8 is caused by random chance. As we saw previously, the null hypothesis would be stated as

$$H_0: \mu = 7.2$$

where μ refers to the mean of the population of treated alcoholics. The null hypothesis is the central element in any test of hypothesis because the entire process is aimed at rejecting or failing to reject the H_0.

Usually, the researcher believes there is a significant difference and desires to reject the null hypothesis. The researcher's belief is stated in a **research hypothesis (H_1)**, a statement that directly contradicts the null hypothesis. Thus, the researcher's goal in hypothesis testing is often to support the research hypothesis by rejecting the null hypothesis.

The research hypothesis can be stated in several ways. One form would state that the population from which the sample was selected did *not* have a certain characteristic or, in terms of our example, had a mean that was *not* equal to a specific value:

$$(H_1: \mu \neq 7.2)$$

where $\neq$ means "not equal to"

Symbolically, this statement asserts that the population of all treated alcoholics is different from the community as a whole.

The research hypothesis is enclosed in parentheses to emphasize that it has no formal standing in the hypothesis-testing process (except, as we shall see in the next section, in choosing between one-tailed and two-tailed tests). It serves as a reminder of what the researcher believes to be the truth.

Step 3. Selecting the Sampling Distribution and Establishing the Critical Region. The sampling distribution is the probabilistic yardstick against which a particular sample outcome is measured. By assuming that the null hypothesis is true (and *only* by this assumption), we can attach values to the mean and standard deviation of the sampling distribution and thus measure the probability of any specific sample outcome. There are several different sampling distributions, but for now we will confine our attention to the sampling distribution described by the standard normal curve, as summarized in Appendix A.

The **critical region** consists of the areas under the sampling distribution that include unlikely sample outcomes. We must define what we mean by "unlikely" prior to the test of hypothesis. That is, we must specify in advance those sample outcomes so unlikely that they will lead us to reject the H_0. This decision rule will establish the critical region, or **region of rejection**: the areas under the sampling distribution that contain unlikely sample outcomes. In our earlier example, this area began at a Z score of ± 1.96, called **Z(critical)**, that was graphically displayed in Figure 8.3. The shaded area is the critical region. Any sample outcome for which the Z-score equivalent fell in this area (that is, below -1.96 or above $+1.96$) would have caused us to reject the null hypothesis.

By convention, the size of the critical region is reported as alpha (α), the proportion of all of the area included in the critical region. In our example, our **alpha level** was 0.05. Other commonly used alphas are 0.10, 0.01, 0.001, and 0.0001.

These decisions can be stated in abbreviated form. The critical region is noted by the Z scores that mark its beginnings.

$$\text{Sampling distribution} = Z \text{ distribution}$$
$$\alpha = 0.05$$
$$Z(\text{critical}) = \pm 1.96$$

(For practice in finding Z(critical) scores, see problem 8.1a.)

Step 4. Computing the Test Statistic. To evaluate the probability of the sample outcome, the sample value must be converted into a Z score. Finding this Z score is called computing the **test statistic**, and the resultant value will

---

ONE STEP AT A TIME	**Completing Step 4 of the Five-Step Model: Computing** *Z***(obtained)**

Use these procedures if the population standard deviation (σ) is known or sample size (N) is greater than 100. See the section entitled "The Student's t Distribution" for procedures when s is unknown and N is less than 100.

To Compute the Test Statistic Using Formula 8.1

1: Find the square root of N.

2: Divide the square root of N into the population standard deviation (σ).

3: Subtract the population mean (μ) from the sample mean ($\bar{X}$).

4: Divide the quantity you found in step 3 by the quantity you found in step 2. This value is Z(obtained).

TABLE 8.1 Making a Decision in Step 5 and Interpreting the Results of the Test

Situation	Decision	Interpretation
The test statistic is in the critical region	Reject the null hypothesis (H_0)	The difference is statistically significant
The test statistic is not in the critical region	Fail to reject the null hypothesis (H_0)	The difference is not statistically significant

be referred to as **Z(obtained)** in order to differentiate the test statistic from the score that marks the beginning of the critical region. In our example, we found a Z(obtained) of 3.15. *(For practice in computing obtained Z scores for means, see problems 8.1c, 8.2 to 8.7, and 8.15e and f.)*

Step 5. Making a Decision and Interpreting the Results of the Test. Finally, the test statistic is compared with the critical region. If the test statistic falls into the critical region, our decision will be to reject the null hypothesis. If the test statistic does not fall into the critical region, we fail to reject the null hypothesis. In our example, the two values were

$$Z(\text{critical}) = \pm 1.96$$

$$Z(\text{obtained}) = -3.15$$

and we saw that the Z(obtained) fell in the critical region (see Figure 8.4). Our decision was to reject the null hypothesis, which stated: "Treated alcoholics have a mean absentee rate of 7.2 days." When we reject this null hypothesis, we are saying that treated alcoholics do *not* have a mean absentee rate of 7.2 days and that there *is* a difference between them and the community. We can also say that the difference between the sample mean of 6.8 and the community mean of 7.2 is statistically significant, or unlikely to be caused by random chance alone.

Note that, in order to complete step 5, you have to do two things. First you reject or fail to reject the null hypothesis (see Table 8.1). Second, you need to say what this decision means. In this case, we rejected the null hypothesis, which means that there is a significant difference between the mean of the sample and the mean for the entire community and, therefore, we can conclude that treated alcoholics are different from the community as a whole.

ONE STEP AT A TIME	**Completing Step 5 of the Five-Step Model: Making a Decision and Interpreting Results**

1: Compare the Z(obtained) to the Z(critical). If Z(obtained) is *in* the critical region, *reject* the null hypothesis. If Z(obtained) is *not in* the critical region, *fail to reject* the null hypothesis.

2: Interpret your decision in the terms of the original question. For example, our conclusion for the example problem was "Treated alcoholics miss significantly fewer days of work than the community as a whole."

Think of hypothesis testing as analogous to gambling. Imagine that you have been invited to participate in a game of chance involving coin flips: heads you win, tails your opponent wins. You would agree to participate in such a game only if you could *assume* that the coin was honest and that the probabilities of heads and tails were both 0.5. What if your opponent flips 10 tails in a row? Think of these flips as a *test* of your original assumption that the coin is honest. At some point, as the coin shows tail after tail, you must compare the outcomes with your original assumption of honesty and make a *decision*: Either the coin is weighted toward tails or you have just witnessed a very rare series of events. If you conclude that the game is rigged, you have rejected the null hypothesis that there is no difference in the probabilities of heads and tails. While no one could blame you for walking away, note that there is a slight chance that your decision was wrong: It is possible (though very unlikely) that the game is not rigged and the coin is not weighted toward tails. In hypothesis testing, we also *make assumptions* (steps 1 through 3), *test* these assumptions in step 4, and make a *decision*, based on probabilities, in step 5.

This five-step model will serve us as a framework for decision making throughout the hypothesis-testing chapters. The exact nature and method of expression for our decisions will be different for different situations. However, the five-step model will provide a common frame of reference for all significance testing.

Choosing a One-Tailed or Two-Tailed Test

The five-step model for hypothesis testing is fairly rigid, and the researcher has little room for making choices. Nonetheless, the researcher must still deal with two options. First, he or she must decide between a one-tailed and a two-tailed test. Second, an alpha level must be selected. In this section we discuss the former decision; we discuss the latter in the next section.

The choice between a one- and two-tailed test is based on the researcher's expectations about the population from which the sample was selected. These expectations are reflected in the research hypothesis (H_1), which is contradictory to the null hypothesis and usually states what the researcher believes to be "the truth." In most situations, the researcher will wish to support the research hypothesis by rejecting the null hypothesis.

The format for the research hypothesis may take either of two forms, depending on the relationship between what the null hypothesis states and what the researcher believes to be the truth. The null hypothesis states that the population has a specific characteristic. In the example that has served us throughout this chapter, the null stated, in symbols, "All treated alcoholics have the *same* absentee rate (7.2 days) as the community." The researcher might believe that the population of treated alcoholics actually has *less* absenteeism (their population mean is *lower than* the value stated in the null hypothesis) or *more* absenteeism (their population mean is *greater than* the value stated in the null hypothesis), or he or she might be unsure about the direction of the difference.

If the researcher is unsure about the direction, the research hypothesis would state only that the population mean is "not equal" to the value stated in the null hypothesis. The research hypothesis stated in our example ($\mu \neq 7.2$) was in this format. This is called a **two-tailed test** of significance because it means that the researcher will be equally concerned with the possibility that the true population value is greater than *or* less than the value specified in the null hypothesis.

In other situations, the researcher might be concerned only with differences in a specific direction. If the direction of the difference can be predicted or if the researcher is concerned only with differences in one direction, a **one-tailed test** can be used. A one-tailed test may take one of two forms, depending on the researcher's expectations about the direction of the difference. If the researcher believes that the true population value is greater than the value specified in the null hypothesis, the research hypothesis would use the ">" (or, "greater than") symbol. In our example, if we had predicted that treated alcoholics had *higher* absentee rates than the community (or averaged *more* days of absenteeism), our research hypothesis would have been

$$(H_1: \mu > 7.2)$$

where $>$ signifies "greater than"

If we predicted that treated alcoholics had lower absentee rates than the community (or averaged *fewer* days of absenteeism than 7.2), our research hypothesis would have been

$$(H_1: \mu < 7.2)$$

where $<$ signifies "less than"

One-tailed tests are often appropriate when programs designed to solve a problem or improve a situation are being evaluated. If the program for treating alcoholics made them *less* reliable workers, for example, the program would be a failure, at least on that criterion. In this situation, the researcher may focus only on outcomes that would indicate that the program is a success (i.e., when treated alcoholics have lower rates) and conduct a one-tailed test with a research hypothesis in the form: $H_1: \mu < 7.2$. Or consider the evaluation of a program designed to reduce unemployment. The evaluators would be concerned only with outcomes that show a decrease in the unemployment rate. If the rate shows no change or if unemployment increases, the program is a failure, and both of these outcomes might be considered equally negative by the researchers. Thus, the researchers could legitimately use a one-tailed test that stated that unemployment rates for graduates of the program would be less than $(<)$ rates in the community.

ONE- VERSUS TWO-TAILED TESTS AND THE FIVE-STEP MODEL

In terms of the five-step model, the choice of a one-tailed or two-tailed test determines what we do with the critical region in step 3. As you recall, in a two-tailed test, we split the critical region equally into the upper and lower tails of the sampling distribution. In a one-tailed test, we place the entire critical area in one tail of the sampling distribution. If we believe that the population characteristic is greater than the value stated in the null hypothesis (if the H_1 includes the $>$ symbol), we place the entire critical region in the upper tail. If we believe that the characteristic is less than the value stated in the null hypothesis (if the H_1 includes the $<$ symbol), the entire critical region goes in the lower tail.

TABLE 8.2   One- vs. Two-Tailed Tests, $\alpha = 0.05$

If the Research Hypothesis Uses	The Test Is	And Concern Is with	Z(critical) =
$\neq$	Two-tailed	Both tails	±1.96
$>$	One-tailed	Upper tail	+1.65
$<$	One-tailed	Lower tail	−1.65

TABLE 8.3   Finding Critical Z Scores for One-Tailed Tests

Alpha	Two-Tailed Value	One-Tailed Value Upper Tail	One-Tailed Value Lower Tail
0.10	±1.65	+1.29	−1.29
0.05	±1.96	+1.65	−1.65
0.01	±2.58	+2.33	−2.33
0.001	±3.32	+3.10	−3.10
0.0001	±3.90	−3.70	+3.70

For example, in a two-tailed test with alpha equal to 0.05, the critical region begins at Z(critical) = ±1.96. In a one-tailed test at the same alpha level, the Z(critical) is +1.65 if the upper tail is specified and −1.65 if the lower tail is specified. Table 8.2 summarizes the procedures to follow in terms of the nature of the research hypothesis. The difference in placing the critical region is graphically summarized in Figure 8.5, and the critical Z scores for the most common alpha levels are given in Table 8.3 for both one- and two-tailed tests.

Note that the critical Z values for one-tailed tests at all values of alpha are closer to the mean of the sampling distribution. Thus, a one-tailed test is more likely to reject the H_0 without changing the alpha level (assuming that we have specified the correct tail). One-tailed tests are a way of statistically both having and eating your cake and should be used whenever (1) the direction of the difference can be confidently predicted or (2) the researcher is concerned only with differences in one tail of the sampling distribution. An example should clarify these procedures.

USING A ONE-TAILED TEST

A sociologist has noted that sociology majors seem more sophisticated, charming, and cosmopolitan than the rest of the student body. A "Sophistication Scale" test has been administered to the entire student body and to a random sample of 100 sociology majors, and these results have been obtained:

Student Body	Sociology Majors
$\mu = 17.3$	$\bar{X} = 19.2$
$\sigma = 7.4$	$N = 100$

FIGURE 8.5 **Establishing the Critical Region, One-Tailed Tests Versus Two-Tailed Tests ($\alpha = 0.05$)**

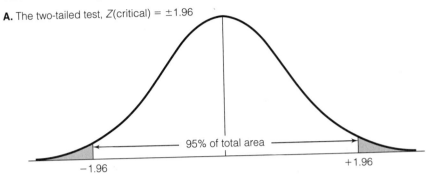

A. The two-tailed test, Z(critical) = ± 1.96

95% of total area

-1.96 $+1.96$

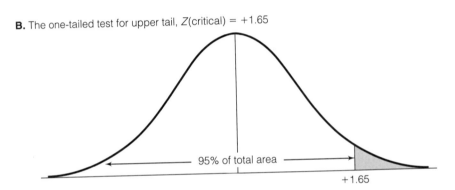

B. The one-tailed test for upper tail, Z(critical) = $+1.65$

95% of total area

$+1.65$

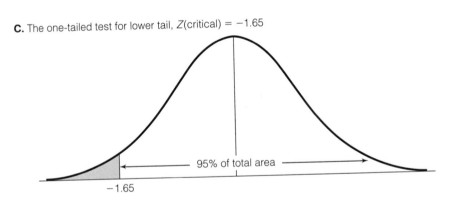

C. The one-tailed test for lower tail, Z(critical) = -1.65

95% of total area

-1.65

We will use the five-step model to test the H_0 of no difference between sociology majors and the general student body.

Step 1. Making Assumptions and Meeting Test Requirements. Since we are using a mean to summarize the sample outcome, we must assume that the Sophistication Scale generates interval-ratio–level data. With a sample size of 100, the Central Limit Theorem applies, and we can assume that the sampling distribution is normal in shape.

Model: Random sampling
Level of measurement is interval-ratio
Sampling distribution is normal

Step 2. Stating the Null Hypothesis (H_0). The null hypothesis states that there is no difference between sociology majors and the general student body. The research hypothesis (H_1) will also be stated at this point. The researcher has predicted a direction for the difference ("Sociology majors are *more* sophisticated"), so a one-tailed test is justified. The one-tailed research hypothesis asserts that sociology majors have a higher ($>$) score on the Sophistication Scale. The two hypotheses may be stated as

$$H_0: \mu = 17.3$$
$$(H_1: \mu > 17.3)$$

Step 3. Selecting the Sampling Distribution and Establishing the Critical Region. We will use Appendix A to find areas under the sampling distribution. If alpha is set at 0.05, the critical region will begin at the Z score $+1.65$. That is, the researcher has predicted that sociology majors are *more* sophisticated and that this sample comes from a population that has a mean *greater than* 17.3, so he or she will be concerned only with sample outcomes in the upper tail of the sampling distribution. If sociology majors are *the same as* other students in terms of sophistication (if the H_0 is true) or if they are *less* sophisticated (and come from a population with a mean less than 17.3), the theory is not supported. These decisions may be summarized as

$$\text{Sampling distribution} = Z \text{ distribution}$$
$$\alpha = 0.05$$
$$Z(\text{critical}) = +1.65$$

Step 4. Computing the Test Statistic.

$$Z(\text{obtained}) = \frac{\overline{X} - \mu}{\sigma/\sqrt{N}}$$

$$Z(\text{obtained}) = \frac{19.2 - 17.3}{7.4/\sqrt{100}}$$

$$Z(\text{obtained}) = +2.57$$

Step 5. Making a Decision and Interpreting Test Results. In this step, we will compare the $Z(\text{obtained})$ with the $Z(\text{critical})$:

$$Z(\text{critical}) = +1.65$$
$$Z(\text{obtained}) = +2.57$$

The test statistic falls into the critical region and this outcome is depicted in Figure 8.6. We will reject the null hypothesis because, if the H_0 were true, a difference of

FIGURE 8.6 Z(obtained) Versus Z(critical) ($\alpha = 0.05$, one tailed test)

this size would be very unlikely. There is a significant difference between sociology majors and the general student body in terms of sophistication. Since the null hypothesis has been rejected, the research hypothesis (sociology majors are more sophisticated) is supported. *(For practice in dealing with tests of significance for means that may call for one-tailed tests, see problems 8.2, 8.3, 8.6, 8.8, and 8.17.)*

Selecting an Alpha Level

In addition to deciding between one-tailed and two-tailed tests, the researcher must select an alpha level. We have seen that the alpha level plays a crucial role in hypothesis testing. When we assign a value to alpha, we define what we mean by an "unlikely" sample outcome. If the probability of the observed sample outcome is lower than the alpha level (if the test statistic falls into the critical region), then we reject the null hypothesis.

How can we make reasonable decisions about the value of alpha? Recall that, in addition to defining what will be meant by *unlikely*, the alpha level is the probability that the decision to reject the null hypothesis if the test statistic falls into the critical region will be incorrect. In hypothesis testing, the error of incorrectly rejecting the null hypothesis—or rejecting a null hypothesis that is actually true—is called **Type I error**, or **alpha error**. To minimize this type of error, use very small values for alpha.

To elaborate: When an alpha level is specified, the sampling distribution is divided into two sets of possible sample outcomes. The critical region includes all unlikely or rare sample outcomes. The remainder of the area consists of all sample outcomes that are not rare. The lower the level of alpha, the smaller the critical region and the greater the distance between the mean of the sampling distribution and the beginnings of the critical region. Compare, for the sake of illustration, the alpha levels and values for Z(critical) for two-tailed tests presented in Table 8.4. As you may recall, Table 7.1 also presented this information.

As alpha goes down, the critical region becomes smaller and moves farther away from the mean of the sampling distribution. The lower the alpha level, the harder it will be to reject the null hypothesis and, because a Type I error can

TABLE 8.4 **The Relationship Between Alpha and Z(critical) for a Two-Tailed Test**

If Alpha Equals:	The Two-Tailed Critical Region Will Begin at Z(critical) Equal to
0.100	±1.65
0.050	±1.96
0.010	±2.58
0.001	±3.32

occur only if our decision in step 5 is to reject the null hypothesis, the lower the probability of Type I error. To minimize the probability of rejecting a null hypothesis that is in fact true, use very low alpha levels.

However, there is a complication. As the critical region decreases in size (as alpha levels decrease), the noncritical region—the area between the two Z(critical) scores in a two-tailed test—becomes larger. All other things being equal, the lower the alpha level, the less likely that the sample outcome will fall into the critical region. This raises the possibility of a second type of incorrect decision, called **Type II error**, or **beta error**: failing to reject a null that is, in fact, false. The probability of a Type I error decreases as the alpha level decreases, but the probability of a Type II error increases. Thus, the two types of error are inversely related, and it is not possible to minimize both in the same test. As the probability of one type of error decreases, the probability of the other increases, and vice versa.

It may be helpful to clarify in table format the relationships between decision making and errors. Table 8.5 lists the two decisions we can make in step 5 of the five-step model: We either reject or fail to reject the null hypothesis. Table 8.4 also lists the two possible conditions of the null hypothesis: It is either actually true or actually false. The table combines these possibilities into a total of four possible outcomes, two of which are desirable ("OK") and two of which indicate that an error has been made.

Let's consider the two desirable ("OK") outcomes first. We want to reject false null hypotheses and fail to reject true null hypotheses. The goal of any scientific investigation is to verify true statements and reject false statements.

The remaining two combinations are errors or situations we wish to avoid. If we reject a null hypothesis that is actually true, we are saying that a true statement is false. Likewise, if we fail to reject a null hypothesis that is actually false, we are saying that a false statement is true. Obviously, we would always prefer to wind up in one of the boxes labeled "OK"—and always reject false statements and accept the truth when we find it. Remember, however, that

TABLE 8.5 **Decision Making and the Five-Step Model**

The H_0 Is Actually:	Decision	
	Reject	Fail to Reject
True	Type I, or α, error	OK
False	OK	Type II, or β, error

STATISTICS IN EVERYDAY LIFE	In social science research, the 0.05 alpha level has become the standard indicator of a significant difference. This alpha level means that we will incorrectly reject the null hypothesis only five times out of every 100 tests. These might seem like excellent odds but research that involves potentially harmful drugs would call for even lower alpha levels (0.001, 0.0001, or even lower) to minimize the possibility of making incorrect decisions and endangering health.

hypothesis testing always carries an element of risk and that it is not possible to minimize the chances of both Type I and Type II errors simultaneously.

What all of this means, finally, is that you must think of selecting an alpha level as an attempt to balance the two types of error. Higher alpha levels will minimize the probability of Type II error (saying that false statements are true), and lower alpha levels will minimize the probability of Type I error (saying that true statements are false). Normally, in social science research, we will want to minimize Type I error, and lower alpha levels (0.05, 0.01, 0.001, or lower) will be used. The 0.05 level in particular has emerged as a generally recognized indicator of a significant result. However, the widespread use of the 0.05 level is simply a convention, and there is no reason that alpha cannot be set at virtually any sensible level (such as 0.04, 0.027, 0.083). The researcher has the responsibility of selecting the alpha level that seems most reasonable in terms of the goals of the research project.

The Student's *t* Distribution

To this point, we have considered only situations involving single-sample means where the value of the population standard deviation (σ) was known. Needless to say, in most research situations, the value of σ will be unknown. However, a value for σ is required in order to compute the standard error of the mean (σ/N), convert our sample outcome into a Z score, and place the Z(obtained) on the sampling distribution (step 4). How can we reasonably obtain a value for the population standard deviation?

It might seem sensible to estimate σ with s, the sample standard deviation. As we noted in Chapter 7, s is a biased estimator of σ but the degree of bias decreases as sample size increases. For large samples (that is, samples with 100 or more cases), the sample standard deviation yields an adequate estimate of σ. Thus, for large samples, we simply substitute σ for s in the formula for Z(obtained) in step 4 and continue to use the standard normal curve to find areas under the sampling distribution.[1]

For smaller samples, however, when σ is unknown, an alternative distribution called the **Student's *t* distribution** must be used to find areas under the sampling distribution and establish the critical region. The shape of the *t* distribution varies as a function of sample size. The relative shapes of the *t* and Z distributions are

---

[1] Even though its effect will be minor and will decrease with sample size, we will always correct for the bias in s by using the term $N - 1$ rather than N in the computation for the standard deviation of the sampling distribution when σ is unknown.

FIGURE 8.7 The *t* Distribution and the *Z* Distribution

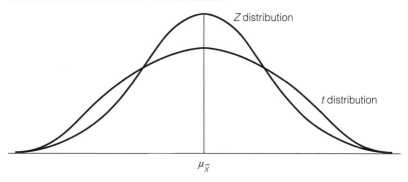

depicted in Figure 8.7. For small samples, the *t* distribution is much flatter than the *Z* distribution, but, as sample size increases, the *t* distribution comes to resemble the *Z* distribution more and more until the two are essentially identical for sample sizes greater than 120. As *N* increases, the sample standard deviation (*s*) becomes a more and more adequate estimator of the population standard deviation (*s*), and the *t* distribution becomes more and more like the *Z* distribution.

THE DISTRIBUTION OF *t*: USING APPENDIX B

The *t* distribution is summarized in Appendix B. The *t* table differs from the *Z* table in several ways. First, there is a column at the left of the table labeled *df* for "degrees of freedom."[2] Since the exact shape of the *t* distribution varies by sample size, the exact location of the critical region varies as well. Degrees of freedom, which are equal to $N - 1$ in the case of a single-sample mean, must be computed first, before the critical region for any alpha can be located. Second, alpha levels are arrayed across the top of Appendix B in two rows, one row for the one-tailed tests and one for two-tailed tests. To use the table, begin by locating the selected alpha level in the appropriate row.

The third difference is that the entries in the table are the actual scores, called ***t*(critical)**, that mark the beginnings of the critical regions and not areas under the sampling distribution. To illustrate the use of this table with single-sample means, find the critical region for alpha equal to 0.05, two-tailed test, for $N = 30$. The degrees of freedom will be $N - 1$, or 29; reading down the proper column, you should find a value of 2.045. Thus, the critical region for this test will begin at $t(\text{critical}) = \pm 2.045$.

Notice that this *t*(critical) is larger in value than the comparable *Z*(critical), which for a two-tailed test at an alpha of 0.05 would be ±1.96. This is because the *t* distribution is flatter than the *Z* distribution (see Figure 8.7). On the *t* distribution, the critical regions will begin farther away from the mean of the sampling distribution; therefore, the null hypothesis will be harder to reject.

[2]Degrees of freedom refers to the number of values in a distribution that are free to vary. For a sample mean, a distribution has $N - 1$ degrees of freedom. This means that, for a specific value of the mean and of N, $N - 1$ scores are free to vary. For example, if the mean is 3 and $N = 5$, the distribution of five scores would have $5 - 1 = 4$ degrees of freedom. When the values of four of the scores are known, the value of the fifth is fixed. If four scores are 1, 2, 3, and 4, the fifth must be 5 and no other value.

Applying Statistics 8.1: Testing a Sample Mean for Significance

For a random sample of 152 felony cases tried in a local court, the average prison sentence was 27.3 months. Is this significantly different from the average prison term for felons statewide ($\mu = 28.7$)? We will use the five-step model to organize the decision-making process.

Step 1. Making Assumptions and Meeting Test Requirements.

> Model: Random sampling
> Level of measurement is interval-ratio
> Sampling distribution is normal

Since this is a large sample ($N > 100$) and length of sentence is an interval-ratio variable, we can conclude that the model assumptions are satisfied

Step 2. Stating the Null Hypothesis (H_0). The null hypothesis would say that the average sentence locally (for *all* felony cases) is equal to the statewide average. In symbols:

$$H_0: \mu = 28.7$$

The research question does not specify a direction; it only asks if the local sentences are "different from" (not higher or lower than) the statewide average. This suggests a two-tailed test:

$$(H_1: \mu \neq 28.7)$$

Step 3. Selecting the Sampling Distribution and Establishing the Critical Region. Since this is a large sample, we can use Appendix A to establish the critical region and state the critical scores as Z scores (as opposed to *t* scores).

> Sampling distribution = Z distribution
> $\alpha = 0.05$
> Z(critical) = ± 1.96

Step 4. Computing the Test Statistic. The necessary information for conducting a test of the null hypothesis is

$$\overline{X} = 27.3 \qquad \mu = 28.7$$
$$s = 3.7$$
$$N = 152$$

The test statistic, Z(obtained), would be

$$Z(\text{obtained}) = \frac{\overline{X} - \mu}{s/\sqrt{N - 1}}$$

$$Z(\text{obtained}) = \frac{27.3 - 28.7}{3.7/\sqrt{152 - 1}}$$

$$Z(\text{obtained}) = \frac{-1.40}{3.7/\sqrt{151}}$$

$$Z(\text{obtained}) = \frac{-1.40}{0.30}$$

$$Z(\text{obtained}) = -4.67$$

Note that, even though sample size is large, we have used "$N - 1$" rather than "N" under the square root sign. We will do this routinely whenever σ is unknown, even though the effects on the value of Z(obtained) will be very small.

Step 5. Making a Decision and Interpreting the Test Results. With alpha set at 0.05, the critical region begins at Z(critical) = ± 1.96. With an obtained Z score of -4.67, the null hypothesis would be rejected. The difference between the prison sentences of felons convicted in the local court and felons convicted statewide is statistically significant. The difference is so large that we may conclude that it did not occur by random chance. The decision to reject the null hypothesis has a 0.05 probability of being wrong.

Furthermore, the smaller the sample size (the lower the degrees of freedom), the larger the value of **t(obtained)** necessary to reject the H_0.

Also note that the values of t(critical) decrease as degrees of freedom increase. For one degree of freedom, the t(critical) is 12.706 for an alpha of 0.05, two-tailed test, but this score grows smaller for larger samples. For degrees of freedom greater than 120, the value of t(critical) is the same as the comparable value of Z(critical), or ±1.96. As sample size increases, the t distribution resembles the Z distribution more and more until, with sample sizes greater than 120, the two distributions are essentially identical.[3]

USING THE *t* DISTRIBUTION IN A TEST OF HYPOTHESIS

To demonstrate the uses of the t distribution in more detail, we will work through an example problem. Note that, in terms of the five-step model, the changes occur mostly in steps 3 and 4. In step 3, the sampling distribution will be the t distribution, and degrees of freedom (df) must be computed before locating the critical region or the t(critical) score. In step 4, a slightly different formula for computing the test statistic, t(obtained), will be used. As compared with the formula for Z(obtained), s will replace σ and N − 1 will replace N. Specifically,

FORMULA 8.2
$$t(\text{obtained}) = \frac{\overline{X} - \mu}{s/\sqrt{N-1}}$$

A researcher wonders if commuter students are different from the general student body in terms of academic achievement. She has gathered a random sample of 30 commuter students and has learned from the registrar that the mean grade-point average for all students is 2.50 ($\mu = 2.50$), but the standard deviation of the population (σ) has never been computed. Sample data are reported here. Is the sample from a population that has a mean of 2.50?

Student Body	Commuter Students
$\mu = 2.50 \ (= \mu_{\overline{X}})$	$\overline{X} = 2.78$
$\sigma = ?$	$s = 1.23$
	$N = 30$

Step 1. Making Assumptions and Meeting Test Requirements.

Model: Random sampling
Level of measurement is interval-ratio
Sampling distribution is normal

Step 2. Stating the Null Hypothesis.

$$H_0: \mu = 2.50$$
$$(H_1: \mu \neq 2.50)$$

You can see from the research hypothesis that the researcher has not predicted a direction for the difference. This will be a two-tailed test.

---

[3]Appendix B abbreviates the t distribution by presenting a limited number of critical t scores for degrees of freedom between 31 and 120. If the degrees of freedom for a specific problem equal 77 and alpha equals 0.05, two-tailed, we have a choice between a t(critical) of 2.000 (df = 60) and a t(critical) of 1.980 (df = 20). In situations such as these, take the larger table value as t(critical). This will make rejection of H_0 less likely and is therefore the more conservative course of action.

Step 3. Selecting the Sampling Distribution and Establishing the Critical Region. Since σ is unknown and the sample size is small, the t distribution will be used to find the critical region. Alpha will be set at 0.01.

$$\text{Sampling distribution} = t \text{ distribution}$$
$$\alpha = 0.01, \text{ two-tailed test}$$
$$\text{df} = (N - 1) = 29$$
$$t(\text{critical}) = \pm 2.756$$

Step 4. Computing the Test Statistic.

$$t(\text{obtained}) = \frac{\overline{X} - \mu}{s/\sqrt{N - 1}}$$

$$t(\text{obtained}) = \frac{2.78 - 2.50}{1.23/\sqrt{29}}$$

$$t(\text{obtained}) = \frac{0.28}{0.23}$$

$$t(\text{obtained}) = +1.22$$

Step 5. Making a Decision and Interpreting Test Results. The test statistic does not fall into the critical region. Therefore, the researcher fails to reject the H_0. The difference between the sample mean (2.78) and the population mean (2.50) is not statistically significant. The difference is no greater than what would be expected if only random chance were operating. The test statistic and critical regions are displayed in Figure 8.8.

To summarize, when testing single-sample means, we must make a choice regarding the theoretical distribution we will use to establish the critical region. The choice is straightforward. If the population standard deviation (σ) is known or sample size is large, the Z distribution (summarized in Appendix A) will be used. If σ is unknown and the sample is small, the t distribution (summarized in Appendix B) will be used. These decisions are summarized in Table 8.6. *(For practice in using the t distribution in a test of hypothesis, see problems 8.8 to 8.10 and 8.17.)*

ONE STEP AT A TIME

Completing Step 4 of the Five-Step Model: Computing t(obtained)

Follow these procedures when using the Student's t distribution.

To Compute the Test Statistic Using Formula 8.2

1: Find the square root of $N - 1$.

2: Divide the quantity you found in step 1 into the sample standard deviation (s).

3: Subtract the population mean (μ) from the sample mean ($\overline{X}$).

4: Divide the quantity you found in step 3 by the quantity you found in step 2. This value is t(obtained).

FIGURE 8.8 **Sampling Distribution Showing *t*(obtained) Versus *t*(critical) ($\alpha = 0.05$, two-tailed test, df = 29)**

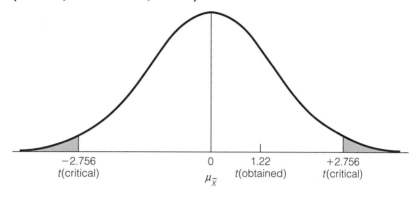

TABLE 8.6 **Choosing a Sampling Distribution When Testing Single-Sample Means for Significance**

If Population Standard Deviation (σ) is	Sampling Distribution
Known	*Z* distribution
Unknown and sample size (*N*) is large	*Z* distribution
Unknown and sample size (*N*) is small	*t* distribution

ONE STEP AT A TIME	**Completing Step 5 of the Five-Step Model: Making a Decision and Interpreting Results**

1: Compare the *t*(obtained) to the *t*(critical). If *t*(obtained) is *in* the critical region, *reject* the null hypothesis. If *t*(obtained) is *not in* the critical region, *fail to reject* the null hypothesis.

2: Interpret your decision in terms of the original question. For example, our conclusion for the example problem used in this section was "There is no significant difference between the average GPAs of commuter students and the general student body."

Tests of Hypothesis for Single-Sample Proportions (Large Samples)

In many cases, we work with variables that are not interval-ratio in level of measurement. One alternative in this situation would be to use a sample proportion (P_s), rather than a sample mean, as the test statistic. As we shall see, the overall procedures for testing single-sample proportions are the same as those for testing means. The central question is still "Does the population from which the sample was drawn have a certain characteristic?" We still conduct the test based on the assumption that the null hypothesis is true, and we still evaluate the probability of the obtained sample outcome against a sampling distribution of all possible sample outcomes. Our decision at the end of the test is also the same. If the obtained test statistic falls into the critical region (and thus is unlikely, given the assumption that the H_0 is true), we reject the H_0.

Of course, there are also some important differences in significance tests for sample proportions. These differences are best related in terms of the five-step model. In step 1, we assume that the variable is measured at the nominal level. In step 2, the symbols used to state the null hypothesis are different even though it is still a statement of "no difference."

In step 3, we will use only the standardized normal curve (the Z distribution) to find areas under the sampling distribution and locate the critical region. This will be appropriate as long as sample size is large. We will not consider small-sample tests of hypothesis for proportions in this text.

In step 4, computing the test statistic, the form of the formula remains the same. That is, the test statistic, Z (obtained), equals the sample statistic minus the mean of the sampling distribution, divided by the standard deviation of the sampling distribution. However, the symbols will change because we are basing the tests on sample proportions. The formula can be stated as

FORMULA 8.3

$$Z(\text{obtained}) = \frac{P_s - P_u}{\sqrt{P_u(1 - P_u)/N}}$$

Step 5, making a decision, is exactly the same as before. If the test statistic, Z(obtained), falls into the critical region, reject the H_0.

A TEST OF HYPOTHESIS USING SAMPLE PROPORTIONS

An example should clarify these procedures. A random sample of 122 households in a low-income neighborhood revealed that 53 (or a proportion of 0.43) of the households were headed by females. In the city as a whole, the proportion of female-headed households is 0.39. Are households in the lower-income neighborhood significantly different from the city as a whole in terms of this characteristic?

Step 1. Making Assumptions and Meeting Test Requirements.

Model: Random sampling
Level of measurement is nominal
Sampling distribution is normal in shape

Step 2. Stating the Null Hypothesis. The research question, as stated earlier, asks only if the sample proportion is *different from* the population proportion. Because we have not predicted a direction for the difference, a two-tailed test will be used.

$$H_0: P_u = 0.39$$
$$(H_1: P_u \neq 0.39)$$

Step 3. Selecting the Sampling Distribution and Establishing the Critical Region.

$$\text{Sampling distribution} = t \text{ distribution}$$
$$\alpha = 0.10, \text{ two-tailed test}$$
$$Z(\text{critical}) = \pm 1.65$$

Applying Statistics 8.2: Testing a Sample Proportion for Significance

Seventy-six percent of the respondents in a random sample drawn from the most affluent neighborhood in a community voted Republican in the most recent presidential election. For the community as a whole, 66% of the electorate voted Republican. Was the affluent neighborhood significantly more likely to have voted Republican?

Step 1. Making Assumptions and Meeting Test Requirements.

Model: Random sampling
    Level of measurement is nominal
    Sampling distribution is normal

This is a large sample, so we may assume a normal sampling distribution. The variable "percent Republican" is only nominal in level of measurement.

Step 2. Stating the Null Hypothesis (H_0). The null hypothesis says that the affluent neighborhood is not different from the community as a whole.

$$H_0: P_u = 0.66$$

The original question ("Was the affluent neighborhood *more* likely to vote Republican?") suggests a one-tailed research hypothesis:

$$(H_1: P_u > 0.66)$$

Step 3. Selecting the Sampling Distribution and Establishing the Critical Region.

$$\text{Sampling distribution} = Z \text{ distribution}$$
$$\alpha = 0.05$$
$$Z(\text{critical}) = +1.65$$

The research hypothesis says that we will be concerned only with outcomes in which the neighborhood is *more* likely to vote Republican or with

sample outcomes in the upper tail of the sampling distribution.

Step 4. Computing the Test Statistic. The information necessary for a test of the null hypothesis, expressed in the form of proportions, is

Neighborhood	Community
$P_s = 0.76$	$P_u = 0.66$
$N = 103$	

The test statistic, $Z(\text{obtained})$, would be

$$Z(\text{obtained}) = \frac{P_s - P_u}{\sqrt{P_u(1 - P_u)/N}}$$

$$Z(\text{obtained}) = \frac{0.76 - 0.66}{\sqrt{(0.66)(1 - 0.66)/103}}$$

$$Z(\text{obtained}) = \frac{0.10}{\sqrt{(0.2244)/103}}$$

$$Z(\text{obtained}) = \frac{0.100}{0.047}$$

$$Z(\text{obtained}) = 2.13$$

Step 5. Making a Decision and Interpreting Test Results. With alpha set at 0.05, one-tailed, the critical region begins at $Z(\text{critical}) = +1.65$. With an obtained Z score of 2.13, the null hypothesis is rejected. The difference between the affluent neighborhood and the community as a whole is statistically significant and in the predicted direction. Residents of the affluent neighborhood were significantly more likely to have voted Republican in the last presidential election.

ONE STEP AT A TIME	**Completing Step 4 of the Five-Step Model: Computing Z(obtained)**

To Compute the Test Statistic Using Formula 8.3

1: Start with the denominator of Formula 8.3 and substitute in the value for P_u. This value will be given in the statement of the problem.

2: Subtract the value of P_u from 1.

3: Multiply the value you found in step 2 by the value of P_u.

4: Divide the quantity you found in step 3 by N.

5: Take the square root of the value you found in step 4.

6: Subtract P_u from P_s.

7: Divide the quantity you found in step 6 by the quantity you found in step 5. This value is Z(obtained).

Step 4. Computing the Test Statistic.

$$Z(\text{obtained}) = \frac{P_s - P_u}{\sqrt{P_u(1 - P_u)/N}}$$

$$Z(\text{obtained}) = \frac{0.43 - 0.39}{\sqrt{(0.39)(0.61)/122}}$$

$$Z(\text{obtained}) = \frac{0.04}{0.044}$$

$$Z(\text{obtained}) = +0.91$$

Step 5. Making a Decision and Interpreting Test Results. The test statistic, Z(obtained), does not fall into the critical region. Therefore, we fail to reject the H_0. There is no statistically significant difference between the low-income community and the city as a whole in terms of the proportion of households headed by females. Figure 8.9 displays the sampling distribution, the critical region, and the Z(obtained). *(For practice in tests of significance using sample proportions, see problems 8.1c, 8.11 to 8.14, 8.15a to d, and 8.16.)*

FIGURE 8.9 **Sampling Distribution Showing Z(obtained) Versus Z(critical) ($\alpha = 0.10$, two-tailed test)**

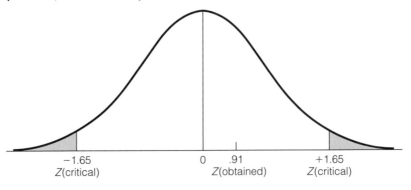

−1.65	0	.91	+1.65
Z(critical)		Z(obtained)	Z(critical)

ONE STEP AT A TIME	**Completing Step 5 of the Five-Step Model: Making a Decision and Interpreting Results**

1: Compare your Z(obtained) to your Z(critical). If Z(obtained) is *in* the critical region, *reject* the null hypothesis. If Z(obtained) is *not in* the critical region, *fail to reject* the null hypothesis.

2: Interpret the decision in terms of the original question. For example, our conclusion for the example

problem used in this section was "There is no significant difference between the low-income community and the city as a whole in the proportion of households that are headed by females."

SUMMARY

1. All the basic concepts and techniques for testing hypotheses were presented in this chapter. We saw how to test the null hypothesis of "no difference" for single-sample means and proportions. In both cases, the central question is whether the population represented by the sample has a certain characteristic.

2. All tests of a hypothesis involve finding the probability of the observed sample outcome, given that the null hypothesis is true. If the outcome has a low probability, we reject the null hypothesis. In the usual research situation, we will wish to reject the null hypothesis and thereby support the research hypothesis.

3. The five-step model will be our framework for decision making throughout the hypothesis-testing chapters. What we do during each step, however, will vary, depending on the specific test being conducted.

4. If we can predict a direction for the difference in stating the research hypothesis, a one-tailed test is called for. If no direction can be predicted, a two-tailed test is appropriate.

5. There are two kinds of errors in hypothesis testing. Type I, or alpha, error is rejecting a true null hypothesis; Type II, or beta, error is failing to reject a false null

hypothesis. The probabilities of committing these two types of error are inversely related and cannot be simultaneously minimized in the same test. In selecting an alpha level, we try to balance the probability of these two errors.

6. When testing sample means, the t distribution is used to find the critical region when the population standard deviation is unknown and sample size is small.

7. Sample proportions can also be tested for significance using the five-step model. Unlike tests using sample means, tests of sample proportions assume nominal level of measurement, use different symbols to state the null hypothesis, and use a different formula (8.3) to compute Z(obtained).

8. If you are still confused about the uses of inferential statistics described in this chapter, don't be alarmed or discouraged. A sizeable volume of complex material has been presented, and only rarely will someone fully comprehend the unique logic of hypothesis testing on first exposure. After all, it is not every day that you learn how to test a statement you don't believe (the null hypothesis) against a distribution that doesn't exist (the sampling distribution)!

SUMMARY OF FORMULAS

Single-sample means, large samples:

$$8.1 \qquad Z\text{(obtained)} = \frac{\overline{X} - \mu}{\sigma/\sqrt{N}}$$

Single-sample means when samples are small and population standard deviation is unknown:

$$8.2 \qquad t\text{(obtained)} = \frac{\overline{X} - \mu}{s/\sqrt{N-1}}$$

Single-sample proportions, large samples:

$$8.3 \qquad Z\text{(obtained)} = \frac{P_s - P_u}{\sqrt{P_u(1 - P_u)/N}}$$

GLOSSARY

Alpha level (α). The proportion of area under the sampling distribution that contains unlikely sample outcomes, given that the null hypothesis is true. Also, the probability of Type I error.

Critical region (region of rejection). The area under the sampling distribution that, in advance of the test itself, is defined as including unlikely sample outcomes, given that the null hypothesis is true.

Five-step model. A step-by-step guideline for conducting tests of hypotheses. A framework that organizes decisions and computations for all tests of significance.

Hypothesis testing. Statistical tests that estimate the probability of sample outcomes if assumptions about the population (the null hypothesis) are true.

Null hypothesis (H_0). A statement of "no difference." In the context of single-sample tests of significance, the population from which the sample was drawn is assumed to have a certain characteristic or value.

One-tailed test. A type of hypothesis test used when (1) the direction of the difference can be predicted or (2) concern focuses on outcomes in only one tail of the sampling distribution.

Research hypothesis (H_1). A statement that contradicts the null hypothesis. In the context of single-sample tests of significance, the research hypothesis says that the population from which the sample was drawn does not have a certain characteristic or value.

Significance testing. See **Hypothesis testing**.

Student's t distribution. A distribution used to find the critical region for tests of sample means when s is unknown and sample size is small.

t(critical). The t score that marks the beginning of the critical region of a t distribution.

t(obtained). The test statistic computed in step 4 of the five-step model. The sample outcome expressed as a t score.

Test statistic. The value computed in step 4 of the five-step model that converts the sample outcome into either a t score or a Z score.

Two-tailed test. A type of hypothesis test used when (1) the direction of the difference cannot be predicted or (2) concern focuses on outcomes in both tails of the sampling distribution.

Type I error (alpha error). The probability of rejecting a null hypothesis that is, in fact, true.

Type II error (beta error). The probability of failing to reject a null hypothesis that is, in fact, false.

Z(critical). The Z score that marks the beginnings of the critical region on a Z distribution.

Z(obtained). The test statistic computed in step 4 of the five-step model. The sample outcomes expressed as a Z score.

PROBLEMS

8.1 a. For each situation, find Z(critical).

Alpha	Form	Z(critical)
0.05	One-tailed	
0.10	Two-tailed	
0.06	Two-tailed	
0.01	One-tailed	
0.02	Two-tailed	

b. For each situation, find the critical t score.

Alpha	Form	N	t (critical)
0.10	Two-tailed	31	
0.02	Two-tailed	24	
0.01	Two-tailed	121	
0.01	One-tailed	31	
0.05	One-tailed	61	

c. Compute the appropriate test statistic (Z or t) for each situation:

1. $\mu = 2.40$ $\overline{X} = 2.20$
   $\sigma = 0.75$ $N = 200$

2. $\mu = 17.1$ $\overline{X} = 16.8$
   $s = 0.9$
   $N = 45$

3. $\mu = 10.2$ $\overline{X} = 9.4$
   $s = 1.7$
   $N = 150$

4. $P_u = .57$ $P_s = 0.60$
   $N = 117$

5. $P_u = 0.32$ $P_s = 0.30$
   $N = 322$

8.2 a. SOC The students at Littlewood Regional High School cut an average of 3.3 classes per month. A random sample of 117 seniors averages 3.8 cuts per month, with a standard deviation of 0.53. Are seniors significantly different from the student body as a whole? (*HINT: The wording of the research question suggests a two-tailed test. This means that the alternative, or research, hypothesis in step 2 will be stated as $H_1: \mu \neq 3.3$ and that the critical region will be split between the upper and lower tails of the sampling distribution. See Table 8.1 for values of Z(critical) for various alpha levels.*)

b. What if the research question were changed to "Do seniors cut a significantly *greater* number of classes"? How would the test conducted in problem 8.2a change? (*HINT: This wording implies a one-tailed test of significance. How would the research hypothesis change? For the alpha you used in problem 8.2a, what would the value of Z(critical) be?*)

8.3 SW **a.** Statewide, social workers average 10.2 years of experience. In a random sample, 203 social workers in greater metropolitan Shinbone average only 8.7 years, with a standard deviation of 0.52. Are social workers in Shinbone significantly less experienced? (*Note the wording of the research hypotheses. This situation may justify a one-tailed test of significance. If you chose a one-tailed test, what form would the research hypothesis take, and where would the critical region begin?*)

b. The same sample of social workers reports an average annual salary of $25,782, with a standard deviation of $622. Is this figure significantly higher than the statewide average of $24,509? (*The wording of the research hypotheses suggests a one-tailed test. What form would the research hypothesis take, and where would the critical region begin?*)

8.4 SOC Statewide, the average score on the college entrance exams (verbal test) is 453, with a standard deviation of 95. A random sample of 137 seniors at Littlewood Regional High School shows a mean score of 502. Is there a significant difference?

8.5 SOC A random sample of 423 Chinese Americans drawn from the population of a particular state has finished an average of 12.7 years of formal education, with a standard deviation of 1.7. Is this significantly different from the statewide average of 12.2 years?

8.6 SOC A sample of 105 sanitation workers for the city of Euonymus, Texas, earns an average of $24,375 per year. The average salary for all Euonymus city workers

is $24,230, with a standard deviation of $523. Are the sanitation workers overpaid? Conduct both one- and two-tailed tests.

8.7 GER **a.** Statewide, the population as a whole watches 6.2 hours of TV per day. A random sample of 1017 senior citizens in the state report watching an average of 5.9 hours per day, with a standard deviation of 0.7. Is the difference significant?

b. The same sample of senior citizens reports that they belong to an average of 2.1 voluntary organizations and clubs, with a standard deviation of 0.5. Statewide, the average is 1.7. Is the difference significant?

8.8 SOC A school system has assigned several hundred "chronic and severe underachievers" to an alternative educational experience. To assess the program, a random sample of 35 has been selected for comparison with all students in the system.

a. In terms of GPA, did the program work?

Systemwide GPA	Program GPA
$\mu = 2.47$	$\bar{X} = 2.55$
	$s = 0.70$
	$N = 35$

b. In terms of absenteeism (number of days missed per year), what can be said a about the success of the program?

Systemwide	Program
$\mu = 6.137$	$\bar{X} = 4.78$
	$s = 1.11$
	$N = 35$

c. In terms of standardized test scores in math and reading, was the program a Success?

Math Test—Systemwide	Math Test—Program
$\mu = 103$	$\bar{X} = 106$
	$s = 2.0$
	$N = 35$

Reading Test—Systemwide	Reading Test—Program
$\mu = 110$	$\bar{X} = 113$
	$s = 2.0$
	$N = 35$

(*HINT: Note the wording of the research questions. Is a one-tailed test justified? Is the program a success if the students in the program are no different from students systemwide? What if the program students were performing at lower levels? If a one-tailed test is used, what form should the research hypothesis take? Where will the critical region begin?*)

8.9 | SOC | A random sample of 26 local sociology graduates scored an average of 458 on the GRE advanced sociology test, with a standard deviation of 20. Is this significantly different from the national average ($\mu = 440$)?

8.10 | PA | Nationally, the per capita monthly fuel oil bill is $110. A random sample of 36 southeastern cities average $78, with a standard deviation of $4. Is the difference significant? Summarize your conclusions in a sentence or two.

8.11 | GER/CJ | A survey shows that 10% of the population is victimized by property crime each year. A random sample of 527 older citizens (65 years or more of age) shows a victimization rate of 14%. Are older people more likely to be victimized? Conduct both one- and two-tailed tests of significance.

8.12 | CJ | A random sample of 113 felons convicted of nonviolent crimes in a state prison system completed a program designed to improve their employment possibilities before being released on parole. Fifty-eight eventually became repeat offenders. Is this recidivism rate significantly different from the rate for all offenders in that state (57%)? Summarize your conclusions in a sentence or two. (*HINT: You must use the information given in the problem to compute a sample proportion. Remember to convert the population percentage to a proportion.*)

8.13 | PS | In a recent statewide election, 55% of the voters rejected a proposal to institute a state lottery. In a random sample of 150 urban precincts, 49% of the voters rejected the proposal. Is the difference significant? Summarize your conclusions in a sentence or two.

8.14 | CJ | Statewide, the police clear by arrest 35% of the robberies and 42% of the aggravated assaults reported to them. A researcher takes a random sample of all the robberies ($N = 207$) and aggravated assaults ($N = 178$) reported to a metropolitan police department in one year and finds that 83 of the robberies and 80 of the assaults were cleared by arrest. Are the local arrest rates significantly different from the statewide rate? Write a sentence or two interpreting your decision.

8.15 | SOC/SW | A researcher has compiled a file of information on a random sample of 317 families that have chronic, long-term patterns of child abuse. Reported here are some of the characteristics of the sample, along with values for the city as a whole. For each trait, test the null hypothesis of "no difference" and summarize your findings.

	Variable	City	Sample
a.	Mother's educational level (proportion completing high school)	$P_u = 0.65$	$P_s = 0.61$
b.	Family size (proportion of families with four or more children)	$P_u = 0.21$	$P_s = 0.26$
c.	Mother's work status (proportion of mothers with jobs outside the home)	$P_u = 0.51$	$P_s = 0.27$
d.	Relations with kin (proportion of families that have contact with kin at least once a week)	$P_u = 0.82$	$P_s = 0.43$
e.	Father's educational achievement (average years of formal schooling)	$\mu = 12.3$	$\bar{X} = 12.5$ $s = 1.7$
f.	Father's occupational stability (average years in present job)	$\mu = 5.2$	$\bar{X} = 3.7$ $s = 0.5$

8.16 [SW] You are the head of an agency seeking funding for a program to reduce unemployment among teenage males. Nationally, the unemployment rate for this group is 18%. A random sample of 323 teenage males in your area reveals an unemployment rate of 21.7%. Is the difference significant? Can you demonstrate a need for the program? Should you use a one-tailed test in this situation? Why? Explain the result of your test of significance as you would to a funding agency.

8.17 [PA] The city manager of Shinbone has received a complaint from the local union of firefighters to the effect that they are underpaid. Not having much time, the city manager gathers the records of a random sample of 27 firefighters and finds that their average salary is $38,073, with a standard deviation of $575. If she knows that the average salary nationally is $38,202, how can she respond to the complaint? Should she use a one-tailed test in this situation? Why? What would she say in a memo to the union that would respond to the complaint?

8.18 The following essay questions review the basic principles and concepts of inferential statistics. The order of the questions roughly follows the five-step model.

a. Hypothesis testing or significance testing can be conducted only with a random sample. Why?

b. Under what specific conditions can it be assumed that the sampling distribution is normal in shape?

c. Explain the role of the sampling distribution in a test of hypothesis.

d. The null hypothesis is an assumption about reality that makes it possible to test sample outcomes for their significance. Explain.

e. What is the critical region? How is the size of the critical region determined?

f. Describe a research situation in which a one-tailed test of hypothesis would be appropriate.

g. Thinking about the shape of the sampling distribution, why does use of the t distribution (as opposed to the Z distribution) make it more difficult to reject the null hypothesis?

h. What exactly can be concluded in the one-sample case when the test statistic falls into the critical region?

9

Hypothesis Testing II
The Two-Sample Case

LEARNING OBJECTIVES   By the end of this chapter, you will be able to

1. Identify and cite examples of situations in which the two-sample test of hypothesis is appropriate.
2. Explain the logic of hypothesis testing as applied to the two-sample case.
3. Explain what an independent random sample is.
4. Perform a test of hypothesis for two sample means or two sample proportions following the five-step model and correctly interpret the results.
5. List and explain each of the factors (especially sample size) that affect the probability of rejecting the null hypothesis. Explain the differences between statistical significance and importance.

In Chapter 8, we dealt with hypothesis testing in the one-sample case. In that situation, we were concerned with the significance of the difference between a sample statistic and a population parameter. In this chapter, we consider a new research situation where we will be concerned with the significance of the difference between two separate samples. For example, do men and women in the United States vary in their support for gun control? Obviously, we cannot ask every male and female for their opinions on this issue. Instead, we must draw random samples of both groups and use the information gathered from these samples to infer population patterns.

The question in hypothesis testing in the two-sample case is: Is the difference between the samples large enough to allow us to conclude (with a known probability of error) that the populations represented by the samples are different? Thus, if we find a large enough difference in support of gun control between random samples of men and women, we can argue that the difference between the samples did not occur by simple random chance but, rather, represents a real difference between men and women in the population.

In this chapter, we consider tests for the significance of the difference between sample means and sample proportions. In both tests, the five-step model will be the framework for our decision making. The hypothesis-testing process is very similar to that of the one-sample case, but we also need to consider some important differences.

The One-Sample Case Versus the Two-Sample Case

There are three important differences between the one-sample case considered in Chapter 8 and the two-sample case covered in this chapter. First, the one-sample case requires that the sample be selected following the principle of EPSEM (each case in the population must have an equal chance of being selected for the sample). The two-sample case requires that the samples be selected independently as well as randomly. Our procedure for selecting cases for one of the samples cannot affect the probability that any particular case will be selected for the other sample. In our example concerning gender differences in support of gun control, this would mean that the selection of a specific male for the sample would have no effect on the probability of selecting any particular female. This new requirement will be stated as **independent random sampling** in step 1.

The requirement of independent random sampling can be satisfied by drawing EPSEM samples from separate lists (for example, one for females and one for males). It is usually more convenient, however, to draw a single EPSEM sample from a single list of the population and then to subdivide the cases into separate groups (males and females, for example). As long as the original sample is selected randomly, any subsamples created by the researcher will meet the assumption of independent random samples.

The second important difference is in the form of the null hypothesis stated in step 2 of the five-step model. The null is still a statement of "no difference." Now, however, instead of saying that the population from which the sample is drawn has a certain characteristic, it will say that the two populations are the same. ("There is no significant difference between men and women in their support of gun control.") If the test statistic falls in the critical region, the null hypothesis of no difference between the populations can be rejected, and the argument that the populations are different will be supported.

A third important new element concerns the sampling distribution, or the distribution of all possible sample outcomes. In Chapter 8, the sample outcome was either a mean or a proportion. Now we are dealing with two samples (e.g., samples of men and women), and the sample outcome is the *difference between* the sample statistics. In terms of our example, the sampling distribution would include all possible differences in sample means for support of gun control between men and women. If the null hypothesis is true and men and women do *not* have different views about gun control, the difference between the population means would be zero, the mean of the sampling distribution would be zero, and the huge majority of differences between sample means would be zero or very close to zero. The greater the differences between the sample means, the further the sample outcome (the *difference* between the two sample means) will be from the mean of the sampling distribution (zero), and the more likely that the difference reflects a real difference between the populations represented by the samples.

Hypothesis Testing with Sample Means (Large Samples)

To illustrate the procedure for testing sample means, assume that a researcher has access to a nationally representative random sample and that the individuals in the sample have answered a survey that measures attitudes toward gun control. The sample is divided by sex, and sample statistics are computed for males and females. Assuming that the survey yields interval-ratio–level data, a test for the significance of the difference in sample means can be conducted.

As long as sample size is large (that is, as long as the combined number of cases in the two samples exceeds 100), the sampling distribution of the differences in sample means will be normal, and the normal curve (Appendix A) can be used to establish the critical regions. The test statistic, Z(obtained), will be computed by the usual formula: sample outcome (the difference between the sample means) minus the mean of the sampling distribution, divided by the standard deviation of the sampling distribution. The formula is presented as Formula 9.1. Note that numerical subscripts are used to identify the samples and populations. The "$\overline{X} - \overline{X}$" subscript attached to σ indicates that we are dealing with the sampling distribution of the *differences* in sample means.

FORMULA 9.1
$$Z(\text{obtained}) = \frac{(\overline{X}_1 - \overline{X}_2) - (\mu_1 - \mu_2)}{\sigma_{\overline{X} - \overline{X}}}$$

where $(\overline{X}_1 - \overline{X}_2)$ = the difference in the sample means

$(\mu_1 - \mu_2)$ = the difference in the population means

$\sigma_{\overline{X} - \overline{X}}$ = the standard deviation of the sampling distribution of the differences in sample means

Recall that tests of significance are always based on the assumption that the null hypothesis is true. If the means of the two populations are equal, then the term $(\mu_1 - \mu_2)$ will be zero and can be dropped from the equation. In effect, then, the formula we will actually use to compute the test statistic in step 4 will be

FORMULA 9.2
$$Z(\text{obtained}) = \frac{(\overline{X}_1 - \overline{X}_2)}{\sigma_{\overline{X} - \overline{X}}}$$

For large samples, the standard deviation of the sampling distribution of the difference in sample means is defined as

FORMULA 9.3
$$\sigma_{\overline{X} - \overline{X}} = \sqrt{\frac{\sigma_1^2}{N_1} + \frac{\sigma_2^2}{N_2}}$$

Since we will rarely, if ever, know the values of the population standard deviations (σ_1 and σ_2), we must use the sample standard deviations, corrected for bias, to estimate them. Formula 9.4 displays the equation used to estimate the standard deviation of the sampling distribution in this situation. This is called a **pooled estimate** since it combines information from both samples.

FORMULA 9.4
$$\sigma_{\overline{X} - \overline{X}} = \sqrt{\frac{s_1^2}{N_1 - 1} + \frac{s_2^2}{N_2 - 1}}$$

The sample outcomes for support of gun control are

Sample 1 (Men)	Sample 2 (Women)
$\bar{X}_1 = 6.2$	$\bar{X}_2 = 6.7$
$s_1 = 1.3$	$s_2 = 1.4$
$N_1 = 324$	$N_2 = 317$

We see from the sample statistics that men have a lower average score and are less supportive of gun control. The test of hypothesis will tell us if this difference is large enough to conclude that it did not occur by random chance alone but, rather, reflects an actual difference between the populations of men and women on this issue.

Step 1. Making Assumptions and Meeting Test Requirements. We now assume that the random samples are independent, but the rest of the model is the same as in the one-sample case.

Model: Independent random samples
Level of measurement is interval-ratio
Sampling distribution is normal

Step 2. Stating the Null Hypothesis. The null hypothesis states that the *populations* represented by the samples are not different on this variable. No direction for the difference has been predicted, so a two-tailed test is called for.

$$H_0: \mu_1 = \mu_2$$
$$H_1: \mu_1 \neq \mu_2$$

Step 3. Selecting the Sampling Distribution and Establishing the Critical Region. For large samples, the Z distribution can be used to find areas under the sampling distribution and establish the critical region. Alpha will be set at 0.05.

Sampling distribution $= Z$ distribution
Alpha $= 0.05$,
Z(critical) $= \pm 1.96$

Step 4. Computing the Test Statistic. The population standard deviations are unknown, so Formula 9.4 will be used to estimate the standard deviation of the sampling distribution. This value will then be substituted into Formula 9.2 and Z(obtained) will be computed.

$$\sigma_{\bar{X}-\bar{X}} = \sqrt{\frac{s_1^2}{N_1 - 1} + \frac{s_2^2}{N_2 - 1}}$$

$$\sigma_{\bar{X}-\bar{X}} = \sqrt{\frac{(1.3)^2}{324 - 1} + \frac{(1.4)^2}{317 - 1}}$$

$$\sigma_{\overline{X}-\overline{X}} = \sqrt{(0.0052) + (0.0062)}$$

$$\sigma_{\overline{X}-\overline{X}} = \sqrt{0.0114}$$

$$\sigma_{\overline{X}-\overline{X}} = 0.107$$

$$Z(\text{obtained}) = \frac{(\overline{X}_1 - \overline{X}_2)}{\sigma_{\overline{X}-\overline{X}}}$$

$$Z(\text{obtained}) = \frac{6.2 - 6.5}{0.107}$$

$$Z(\text{obtained}) = \frac{-0.300}{0.107}$$

$$Z(\text{obtained}) = -2.80$$

ONE STEP AT A TIME — **Completing Step 4 of the Five-Step Model: Computing Z(obtained)**

Use these procedures when samples are large. Solve Formula 9.4 first and then solve Formula 9.2.

To Solve Formula 9.4

1: Subtract 1 from N_1.

2: Square the value of the standard deviation for the first sample (s_1^2).

3: Divide the quantity you found in step 2 by the quantity you found in step 1.

4: Subtract 1 from N_2.

5: Square the value of the standard deviation for the second sample (s_2^2).

6: Divide the quantity you found in step 5 by the quantity you found in step 4.

7: Add the quantity you found in step 6 to the quantity you found in step 3.

8: Take the square root of the quantity you found in step 7.

To Solve Formula 9.2

1: Subtract $\overline{X}_2$ from $\overline{X}_1$.

2: Divide the value you found in step 1 of "To Solve Formula 9.2" by the quantity you found in step 8 of "To Solve Formula 9.4." This is Z(obtained).

ONE STEP AT A TIME — **Completing Step 5 of the Five-Step Model: Making a Decision and Interpreting Results**

1: Compare the Z(obtained) to your Z(critical). If Z(obtained) is *in* the critical region, *reject* the null hypothesis. If Z(obtained) is *not in* the critical region, *fail to reject* the null hypothesis.

2: Interpret the decision to reject or fail to reject the null hypothesis in terms of the original question. For example, our conclusion for the example problem was "There is a significant difference between men and women in their support for gun control."

Step 5. Making a Decision and Interpreting the Results of the Test.
Comparing the test statistic with the critical region:

$$Z\text{(obtained)} = -2.80$$

$$Z\text{(critical)} = \pm1.96$$

We see that the Z score falls into the critical region, which means that a difference as large as -0.30 ($6.2 - 6.5$) between the sample means is unlikely if the

---

Applying Statistics 9.1: A Test of Significance for Sample Means

A scale measuring satisfaction with family life has been administered to a sample of married respondents. On this scale, higher scores indicate greater satisfaction. The sample has been divided into respondents with no children and respondents with at least one child, and means and standard deviations have been computed for both groups. Is there a significant difference in satisfaction with family life between these two groups?

The sample information is:

Sample 1 (No Children)	Sample 2 (At Least One Child)
$\overline{X}_1 = 11.3$	$\overline{X}_2 = 10.8$
$s_1 = 0.6$	$s_2 = 0.5$
$N_1 = 78$	$N_2 = 93$

We can see from the sample results that respondents with no children are more satisfied. Is this difference significant?

Step 1. Making Assumptions and Meeting Test Requirements.

Model: Independent random samples
Level of measurement is interval-ratio
Sampling distribution is normal

Step 2. Stating the Null Hypothesis.

$$H_0: \mu_1 = \mu_2$$
$$H_1: \mu_1 \neq \mu_2$$

Step 3. Selecting the Sampling Distribution and Establishing the Critical Region.

Sampling distribution = Z distribution
Alpha = 0.05, two-tailed
$Z\text{(critical)} = \pm1.96$

Step 4. Computing the Test Statistic.

$$\sigma_{\overline{X}-\overline{X}} = \sqrt{\frac{s_1^2}{N_1 - 1} + \frac{s_2^2}{N_2 - 1}}$$

$$\sigma_{\overline{X}-\overline{X}} = \sqrt{\frac{(0.6)^2}{78 - 1} + \frac{(0.5)^2}{93 - 1}}$$

$$\sigma_{\overline{X}-\overline{X}} = \sqrt{0.007}$$

$$\sigma_{\overline{X}-\overline{X}} = 0.08$$

$$Z\text{(obtained)} = \frac{(\overline{X}_1 - \overline{X}_2)}{\sigma_{\overline{X}-\overline{X}}}$$

$$Z\text{(obtained)} = \frac{11.3 - 10.8}{0.08}$$

$$Z\text{(obtained)} = \frac{0.50}{0.08}$$

$$Z\text{(obtained)} = 6.25$$

Step 5. Making a Decision and Interpreting the Results of the Test.
Comparing the test statistic with the critical region,

$$Z\text{(obtained)} = 6.25$$
$$Z\text{(critical)} = \pm1.96$$

We reject the null hypothesis. Parents and childless couples are significantly different in their satisfaction with family life. Given the direction of the difference, we can also note that childless couples are significantly happier.

STATISTICS
IN EVERYDAY LIFE

The reading scores of American children have not improved significantly over recent years.* Tests are administered every other year to very large, representative samples of about 180,000 4th-graders, and scores on the test range from 0 to 500. The average score in 2009 was 221, which was significantly higher than the 2005 average (219) but exactly the same as the 2007 average (also 221). The tests were conducted with alpha set at 0.05. A different report† concludes that girls, on the average, have caught up with boys in math skills (that is, there is no significant difference between the average math scores for girls and boys) and remain well ahead of boys in verbal and reading skills. What sociological factors might explain these patterns?

*This report is available at http://nationsreportcard.gov/reading_2009/reading_2009_report/.
†This report is available at http://www.cep-dc.org/document/docWindow.cfm?fuseaction=document.viewDocument&documentid=304&documentFormatId=4644.

null hypothesis is true. The null hypothesis of no difference can be rejected, and the notion that men and women are different in terms of their support of gun control is supported. The decision to reject the null hypothesis has only a 0.05 probability (the alpha level) of being incorrect.

Note that the value for Z(obtained) is negative, indicating that men have significantly lower scores than women for support for gun control. The sign of the test statistics reflects our arbitrary decision to label men sample 1 and women sample 2. If we had reversed the labels and called women sample 1 and men sample 2, the sign of the Z(obtained) would have been positive, but its value (2.80) would have been exactly the same, as would our decision in step 5. *(For practice in testing the significance of the difference between sample means for large samples, see problems 9.1 to 9.6 and 9.15d to f.)*

Hypothesis Testing with Sample Means (Small Samples)

When the population standard deviation is unknown and sample size is small (combined N's of less than 100), the Z distribution cannot be used to find areas under the sampling distribution. Instead, we will use the t distribution to find the critical region and identify unlikely sample outcomes. To do this, we need to perform one additional calculation and make one additional assumption. The calculation is for degrees of freedom, which we need in order to use the t table (Appendix B). In the two-sample case, degrees of freedom are equal to $N_1 + N_2 - 2$.

The additional assumption is a more complex matter. When samples are small, we must assume that the population variances are equal in order to justify the assumption of a normal sampling distribution and to form a pooled estimate of the standard deviation of the sampling distribution. The assumption of equal variance in the population can be tested by a statistical technique known as the *analysis of variance*, or ANOVA (see Chapter 10). For our purposes here, however, we will simply assume equal population variances without formal testing. This assumption is safe as long as sample sizes are approximately equal.

THE FIVE-STEP MODEL AND THE t DISTRIBUTION

To illustrate this procedure, assume that a researcher believes that center-city families have significantly more children than suburban families. Random samples from both areas are gathered and the following sample statistics computed.

Sample 1 (Suburban)	Sample 2 (Center-City)
$\bar{X}_1 = 2.37$	$\bar{X}_2 = 2.78$
$s_1 = 0.63$	$s_2 = 0.95$
$N_1 = 42$	$N_2 = 37$

The sample data show a difference in the predicted direction. The significance of this observed difference can be tested with the five-step model.

Step 1. Making Assumptions and Meeting Test Requirements. Sample size is small, and the population standard deviation is unknown. Hence, we must assume equal population variances in the model.

> Model: Independent random samples
> Level of measurement is interval-ratio
> Population variances are equal
> Sampling distribution is normal

Step 2. Stating the Null Hypothesis. Since a direction has been predicted, a one-tailed test will be used. The research hypothesis is stated accordingly.

$$H_0: \mu_1 = \mu_2$$
$$H_1: \mu_1 < \mu_2$$

Step 3. Selecting the Sampling Distribution and Establishing the Critical Region. With small samples, the t distribution is used to establish the critical region. Alpha will be set at 0.05, and a one-tailed test will be used.

> Sampling distribution $= t$ distribution
> Alpha $= 0.05$, one-tailed
> Degrees of freedom $= N_1 + N_2 - 2 = 42 + 37 - 2 = 77$
> $t(\text{critical}) = -1.671$

Note that the critical region is placed in the lower tail of the sampling distribution in accordance with the direction specified in H_1.

Step 4. Computing the Test Statistic. With small samples, a different formula (Formula 9.5) is used for the pooled estimate of the standard deviation of the sampling distribution. This value is then substituted directly into the denominator of the formula for $t(\text{obtained})$ in Formula 9.6.

FORMULA 9.5

$$\sigma_{\bar{X}-\bar{X}} = \sqrt{\frac{N_1 s_1^2 + N_2 s_2^2}{N_1 + N_2 - 2}} \sqrt{\frac{N_1 + N_2}{N_1 N_2}}$$

$$\sigma_{\bar{X}-\bar{X}} = \sqrt{\frac{(42)(.63)^2 + (37)(.95)^2}{42 + 37 - 2}} \sqrt{\frac{42 + 37}{(42)(37)}}$$

$$\sigma_{\bar{X}-\bar{X}} = \sqrt{\frac{50.06}{77}} \sqrt{\frac{79}{1554}}$$

$$\sigma_{\bar{X}-\bar{X}} = (.81)(.23)$$

$$\sigma_{\bar{X}-\bar{X}} = 0.19$$

FORMULA 9.6

$$t(\text{obtained}) = \frac{(\bar{X}_1 - \bar{X}_2)}{\sigma_{\bar{X}-\bar{X}}}$$

$$t(\text{obtained}) = \frac{2.37 - 2.78}{0.19}$$

$$t(\text{obtained}) = \frac{-0.41}{0.19}$$

$$t(\text{obtained}) = -2.16$$

Step 5. Making a Decision and Interpreting the Results of the Test. The test statistic falls in the critical region:

$$t(\text{obtained}) = -2.16$$
$$t(\text{critical}) = -1.671$$

ONE STEP AT A TIME **Completing Step 4 of the Five-Step Model: Computing t(obtained)**

Solve Formula 9.5 first and then solve Formula 9.6 to compute the test statistic.

To Solve Formula 9.5

1: Add N_1 and N_2 and then subtract 2 from this total.

2: Square the standard deviation for the first sample (s_1^2) and multiply the result by N_1.

3: Square the standard deviation for the second sample (s_2^2) and multiply the result by N_2.

4: Add the quantities you found in steps 2 and 3.

5: Divide the quantity you found in step 4 by the quantity you found in step 1 and take the square root of the result.

6: Multiply N_1 by N_2.

7: Add N_1 and N_2.

8: Divide the quantity you found in step 7 by the quantity you found in step 6 and take the square root of the result.

9: Multiply the quantity you found in step 8 by the quantity you found in step 5.

To Solve Formula 9.6

1: Subtract $\bar{X}_2$ from $\bar{X}_1$.

2: Divide the difference between the sample means by the quantity you found in step 9 of "To Solve Formula 9.5." This is t(obtained).

FIGURE 9.1 The Sampling Distribution, with Critical Region and Test Statistic Displayed

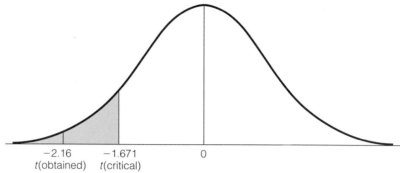

-2.16	-1.671	0
t(obtained)	t(critical)	

ONE STEP AT A TIME	**Completing Step 5 of the Five-Step Model: Making a Decision and Interpreting Results**

1: Compare the t(obtained) to your t(critical). If t(obtained) is *in* the critical region, *reject* the null hypothesis. If t(obtained) is *not in* the critical region, *fail to reject* the null hypothesis.

2: Interpret the decision to reject or fail to reject the null hypothesis in terms of the original question. For example, our conclusion for the example problem was "There is a significant difference between the size of center city and suburban families."

If the null hypothesis $\mu_1 = \mu_2$ were true, this would be a very unlikely outcome, so the null hypothesis can be rejected. There is a statistically significant difference (a difference so large that it is unlikely to be due to random chance) in the sizes of center-city and suburban families. Furthermore, center-city families are significantly larger in size. The test statistic and sampling distribution are depicted in Figure 9.1. *(For practice in testing the significance of the difference between sample means for small samples, see problems 9.7 and 9.8.)*

Hypothesis Testing with Sample Proportions (Large Samples)

Testing for the significance of the difference between two sample proportions is analogous to testing sample means. The null hypothesis states that there is no difference between the populations from which the samples are drawn. We compute a test statistic in step 4, which is then compared with the critical region. When sample sizes are large (combined N's of more than 100), the Z distribution may be used to find the critical region. In this text, we will not consider tests of significance for proportions based on small samples.

In order to find the value of the test statistics, several preliminary equations must be solved. Formula 9.7 uses the values of the two sample proportions (P_s) to give us an estimate of the population proportion (P_u), the proportion of cases in the population that have the trait under consideration assuming the null hypothesis is true.

FORMULA 9.7
$$P_u = \frac{N_1 P_{s1} + N_2 P_{s2}}{N_1 + N_2}$$

The value of P_u is then used to compute the standard deviation of the sampling distribution of the difference in sample proportions in Formula 9.8.

FORMULA 9.8
$$\sigma_{p-p} = \sqrt{P_u(1 - P_u)} \sqrt{\frac{N_1 + N_2}{N_1 N_2}}$$

This value is then substituted into the formula for computing the test statistic, presented as Formula 9.9.

FORMULA 9.9
$$Z(\text{obtained}) = \frac{(P_{s1} - P_{s2}) - (P_{u1} - P_{u2})}{\sigma_{p-p}}$$

where $(P_{s1} - P_{s2})$ = the difference between the sample proportions

$(P_{u1} - P_{u2})$ = the difference between the population proportions

σ_{p-p} = the standard deviation of the sampling distribution of the difference between sample proportions

As was the case with sample means, the second term in the numerator is assumed to be zero by the null hypothesis. Therefore, the formula reduces to

FORMULA 9.10
$$Z(\text{obtained}) = \frac{(P_{s1} - P_{s2})}{\sigma_{p-p}}$$

Remember to solve these equations in order, starting with Formula 9.7 (and skipping Formula 9.9).

ILLUSTRATING A TEST OF HYPOTHESIS BETWEEN TWO SAMPLE PROPORTIONS (LARGE SAMPLES)

An example will make these procedures clearer. Suppose we are researching social networks among senior citizens and wonder if blacks and whites differ in their number of memberships in clubs and other organizations. Random samples of black and white senior citizens have been selected and classified as high or low in terms of their number of memberships in voluntary associations.

Is there a statistically significant difference in the participation patterns of black and white elderly? The proportions of each group classified as "high" in participation and sample size for both groups are:

Sample 1 (Black Senior Citizens)	Sample 2 (White Senior Citizens)
$P_{s1} = 0.34$	$P_{s2} = 0.25$
$N_1 = 83$	$N_2 = 103$

Step 1. Making Assumptions and Meeting Test Requirements.

Model: Independent random samples
Level of measurement is nominal
Sampling distribution is normal

Step 2. Stating the Null Hypothesis. Since no direction has been predicted, this will be a two-tailed test.

$$H_0: P_{u1} = P_{u2}$$
$$H_1: P_{u1} \neq P_{u2}$$

Step 3. Selecting the Sampling Distribution and Establishing the Critical Region. Since sample size is large, the Z distribution will be used to establish the critical region. Setting alpha at 0.05, we have

Sampling distribution $= Z$ distribution
Alpha $= 0.05$, two-tailed
Z(critical) $= \pm 1.96$

Step 4. Computing the Test Statistic. Solve Formula 9.7 first, substitute the resultant value into Formula 9.8, and then solve for Z(obtained) with Formula 9.10.

$$P_u = \frac{N_1 P_{s1} + N_2 P_{s2}}{N_1 + N_2}$$

$$P_u = \frac{(83)(0.34) + (103)(0.25)}{83 + 103}$$

$$P_u = 0.29$$

$$\sigma_{p-p} = \sqrt{P_u(1 - P_u)} \sqrt{\frac{N_1 + N_2}{N_1 N_2}}$$

$$\sigma_{p-p} = \sqrt{(0.29)(0.71)} \sqrt{\frac{83 + 103}{(83)(103)}}$$

$$\sigma_{p-p} = (0.45)(0.15)$$

$$\sigma_{p-p} = 0.07$$

ONE STEP AT A TIME | **Completing Step 4 of the Five-Step Model: Computing Z(obtained)**

Solve Formulas 9.7, 9.8, and 9.10—in that order—to find the test statistic.

To Solve Formula 9.7

1: Add N_1 and N_2.

2: Multiply P_{s1} by N_1.

3: Multiply P_{s2} by N_2.

4: Add the quantity you found in step 3 to the quantity you found in step 2.

5: Divide the quantity you found in step 4 by the quantity you found in step 1. This is P_u.

To Solve Formula 9.8

1: Multiply P_u by $(1 - P_u)$.

2: Take the square root of the quantity you found in step 1.

3: Multiply N_1 by N_2.

4: Add N_1 and N_2. (NOTE: You already found this value in step 1 of "To Solve Formula 9.7.")

5: Divide the quantity you found in step 4 by the quantity you found in step 3.

6: Take the square root of the quantity you found in step 5.

7: Multiply the quantity you found in step 6 by the quantity you found in step 2.

To Solve Formula 9.10

1: Find the difference between the sample proportions.

2: Divide the quantity you found in step 1 by the quantity you found in step 7 of "To Solve Formula 9.8." This is Z(obtained).

$$Z(\text{obtained}) = \frac{(P_{s1} - P_{s2})}{\sigma_{p-p}}$$

$$Z(\text{obtained}) = \frac{0.34 - 0.25}{0.07}$$

$$Z(\text{obtained}) = 1.29$$

Step 5. Making a Decision and Interpreting the Results of the Test. Since the test statistic, Z(obtained) = 1.29, does not fall into the critical region as marked by the Z(critical) of ±1.96, we fail to reject the null hypothesis. The

---

ONE STEP AT A TIME | **Completing Step 5 of the Five-Step Model: Making a Decision and Interpreting Results**

1: Compare Z(obtained) to Z(critical). If Z(obtained) is *in* the critical region, *reject* the null hypothesis. If Z(obtained) is *not in* the critical region, *fail to reject* the null hypothesis.

2: Interpret the decision to reject or fail to reject the null hypothesis in terms of the original question. For example, our conclusion for the example problem was "There is no significant difference between the participation patterns of black and white senior citizens."

difference between the sample proportions is no greater than what would be expected if the null hypothesis were true and only random chance were operating. Black and white senior citizens are not significantly different in terms of participation patterns as measured in this test. *(For practice in testing the significance of the difference between sample proportions, see problems 9.10 to 9.14 and 9.15a to c.)*

Applying Statistics 9.2: Testing the Significance of the Difference Between Sample Proportions

Do attitudes toward premarital sex vary by gender? The proportion of each sex that feels that premarital sex is always wrong is

Females	Males
$P_{s1} = 0.35$	$P_{s2} = 0.32$
$N_1 = 450$	$N_2 = 417$

Females are more likely to say that premarital sex is always wrong. Is the difference significant? The table presents all the information we will need to conduct a test of the null hypothesis following the familiar five-step model with alpha set at 0.05, two-tailed test.

Step 1. Making Assumptions and Meeting Test Requirements.

> Model: Independent random samples
> Level of measurement is nominal
> Sampling distribution is normal

Step 2. Stating the Null Hypothesis.

$$H_0: P_{u1} = P_{u2}$$
$$H_1: P_{u1} \neq P_{u2}$$

Step 3. Selecting the Sampling Distribution and Establishing the Critical Region.

> Sampling distribution = Z distribution
> Alpha = 0.05, two-tailed
> Z(critical) = ±1.96

Step 4. Computing the Test Statistic. Remember to start with Formula 9.7, substitute the value for P_u into Formula 9.8, and then substitute that value into Formula 9.10 to solve for Z(obtained).

$$P_u = \frac{N_1 P_{s1} + N_2 P_{s2}}{N_1 + N_2}$$

$$P_u = \frac{(450)(0.35) + (417)(0.32)}{450 + 417}$$

$$P_u = \frac{290.94}{867}$$

$$P_u = 0.34$$

$$\sigma_{p-p} = \sqrt{P_u(1 - P_u)} \sqrt{\frac{N_1 + N_2}{N_1 N_2}}$$

$$\sigma_{p-p} = \sqrt{(0.34)(0.66)} \sqrt{\frac{450 + 417}{(450)(417)}}$$

$$\sigma_{p-p} = \sqrt{0.2244} \sqrt{0.0046}$$

$$\sigma_{p-p} = (0.47)(0.068)$$

$$\sigma_{p-p} = 0.032$$

$$Z(\text{obtained}) = \frac{(P_{s1} - P_{s2})}{\sigma_{p-p}}$$

$$Z(\text{obtained}) = \frac{0.35 - 0.32}{0.032}$$

$$Z(\text{obtained}) = \frac{0.030}{0.032}$$

$$Z(\text{obtained}) = 0.94$$

Step 5. Making a Decision and Interpreting the Results of the Test. With an obtained Z score of 0.94, we would fail to reject the null hypothesis. The proportion of females and males who feel that premarital sex is always wrong is not significantly different.

According to the Gallup polls, Republicans and Democrats have become increasingly polarized over the abortion issue. In 1975, one year after the landmark *Roe v. Wade* Supreme Court decision, 18% of Republicans and 19% of Democrats agreed that abortion should be legal "under any circumstances." In 2009, the percentage of Republicans approving abortion under any circumstances had fallen to 12% while the percentage of Democrats approving had risen to 31%. The 1 percentage point difference in 1975 might have been statistically significant but it was clearly not important. The 19% point difference in 2009 is statistically significant, important, and extremely consequential for the possibilities of mutual understanding and civil discourse in American politics. These results are based on random samples of about 1000 adult Americans.

How might we explain these trends? One possibility is that the Republican Party has lost many moderate, pro-choice members since the 1970s, leaving the party smaller and more ideologically homogenous. What information would you need to investigate this possibility?

Source: Saad, Lydia. 2010. "Republicans, Dems Abortion Views Grow More Polarized." Available at: http://www.gallup.com/poll/126374/Republicans-Dems-Abortion-Views-Grow-Polarized.aspx? version=print.

The Limitations of Hypothesis Testing: Significance Versus Importance

Given that we are usually interested in rejecting the null hypothesis, we should take a moment to consider systematically the factors that affect our decision in step 5. Generally speaking, the probability of rejecting the null hypothesis is a function of four independent factors:

1. The size of the observed difference

2. The alpha level

3. The use of one or two-tailed tests

4. The size of the sample

All of these factors except the first are under the direct control of the researcher. The relationship between alpha level and the probability of rejection is straightforward. The higher the alpha level, the larger the critical region, and the greater the probability of rejection. Thus, it is easier to reject the H_0 at the 0.05 level than at the 0.01 level, and easier still at the 0.10 level. The danger here, of course, is that higher alpha levels will lead to more frequent Type I errors, and we might find ourselves declaring small differences to be statistically significant. In similar fashion, using a one-tailed test will increase the probability of rejecting the null hypothesis (assuming that the proper direction has been predicted).

The final factor is sample size: With all other factors constant, the probability of rejecting H_0 increases with sample size. In other words, the larger the sample, the more likely we are to reject the null hypothesis and, with very large samples (say samples with thousands of cases), we may declare small, unimportant differences to be statistically significant.

This relationship may appear to be surprising, but the reasons for it can be appreciated with a brief consideration of the formulas used to compute test statistics in step 4. In all these formulas, for all tests of significance, sample size (N) is in the "denominator of the denominator." Algebraically, this is equivalent to being in the numerator of the formula and means that the value of the test statistic is directly proportional to N and that the two will increase together.

TABLE 9.1 Test Statistics for Single-Sample Means Computed from Samples of Various Sizes ($\overline{X} = 80$, $\mu = 79$, $s = 5$ throughout)

Sample Size	Test Statistic, Z(obtained)
100	1.99
200	2.82
500	4.47

To illustrate, consider Table 9.1, which shows the value of the test statistic for single-sample means from samples of various sizes. The value of the test statistic, Z(obtained), increases as N increases, even though none of the other terms in the formula change. This pattern of higher probabilities for rejecting H_0 with larger samples holds for all tests of significance.

On one hand, the relationship between sample size and the probability of rejecting the null should not alarm us unduly. Larger samples are, after all, better approximations of the populations they represent. Thus, decisions based on larger samples can be trusted more than decisions based on small samples.

On the other hand, this relationship clearly underlines what is perhaps the most important limitation of hypothesis testing. Simply because a difference is statistically significant does not guarantee that it is important in any other sense. Particularly with very large samples, relatively small differences may be statistically significant. Even with small samples, of course, differences that are otherwise trivial or uninteresting may be statistically significant. The crucial point is that statistical significance and theoretical or practical importance can be two very different things. Statistical significance is a necessary but not sufficient condition for theoretical or practical importance. A difference that is not statistically significant is almost certainly unimportant. However, significance by itself does not guarantee importance. Even when it is clear that the research results were not produced by random chance, the researcher must still assess their importance. Do they firmly support a theory or hypothesis? Are they clearly consistent with a prediction or analysis? Do they strongly indicate a line of action in solving some problem? These are the kinds of questions a researcher must ask when assessing the importance of the results of a statistical test.

Also, we should note that researchers have access to some very powerful ways of analyzing the importance (vs. the statistical significance) of research results. These statistics, which include bivariate measures of association and multivariate statistical techniques, are introduced in Parts III and IV of this text.

Interpreting Statistics: Are There Significant Differences in Income Between Men and Women?

In the United States, as in many other nations around the globe, concerted efforts have been made to equalize working conditions for men and women. How successful have these efforts been? Do significant differences in the earnings of men and women persist? Is there a "gender gap" in income?

Note that we could answer these questions with absolute certainty only if we knew the earnings of every single man and woman in the United States. The U.S. Census Bureau does publish information on income and gender but usually in terms of medians, not means. Because income information for the population is not readily available, we will investigate the relationship between gender and income by using statistics calculated on randomly selected samples of men and women. If the difference between the sample means of men and women are large enough, we can infer that there is a difference in average salaries between men and women in the population. The General Social Survey, or GSS, is given to randomly selected samples of Americans and will be used as a basis for this test.

Before conducting the test, we need to deal with two issues. First and more importantly, the GSS measures personal income with a series of categories rather than in actual dollars. In other words, respondents were asked to choose from a list of salary ranges (e.g., $10,000 to $12,499, $12,500 to $14,999, and so forth). Thus, income is measured at the ordinal level, and we need an interval-ratio dependent variable to compute sample means and test for the significance of the difference between them. To deal with this problem, we can convert the variable to a form that more closely approximates interval-ratio data by substituting the midpoints for each of the salary intervals. For example, instead of trying to work with the interval "12,500 to 14,999," we would use the midpoint ($13,750) as a score.

This technique makes income a more numerical variable, but it also creates possible inaccuracies. Estimates of the population parameters from this modified variable should be treated only as approximations. However, since we are concerned with the *difference* in income rather than actual income itself, we should still be able to come to some conclusions.

Second, we shouldn't compare the incomes of *all* men and *all* women because some difference could be the result of the fact that women are less likely to be in the paid labor market (i.e., they are more likely to be occupied as wives and mothers) and, thus, less likely to have an income. We will deal with this by restricting the comparison to respondents who work full-time.

Once these adjustments have been made, we can test for the significance of the difference in income. The sample information calculated from the full 2008 GSS database (not the smaller version used for SPSS exercises in this text) is:

Males	Females
$\bar{X}_1 = 55,161.86$	$\bar{X}_2 = 38,826.33$
$s_1 = 40,719.64$	$s_2 = 28,029.99$
$N_1 = 485$	$N_2 = 357$

It appears that there is a large gender gap in income and that males average over $16,000 a year more than females. Is this difference in sample means significant? Could it have occurred by random chance? We will use the large-sample test of significance for means. The null hypothesis is that males and females have the same average salary in the population ($\mu_1 = \mu_2$). We will skip the customary trip through the five-step model and simply report that the Z(obtained) calculated with Formula 9.2 (step 4) is 6.53, much greater than the customary Z(critical) score of ± 1.96 associated with an alpha level of 0.05.

The difference in sample means is so large that we must reject the null hypothesis and conclude (with a probability of error of 0.05) that the population means are different: Males earn significantly more than females.

Is the gender gap in income due to differences in levels of education? If females were significantly less educated than males, this would account for at least some of the difference in income. Using the 2008 General Social Survey again to get information on years of education for random samples of males and females, we have:

Males	Females
$\bar{X}_1 = 13.76$	$\bar{X}_2 = 14.04$
$s_1 = 3.03$	$s_2 = 2.75$
$N_1 = 564$	$N_2 = 438$

We can see that both males and females average more than 13 years of schooling and that there is little difference in the sample means. In fact, the test statistic computed in step 4 is a Z(obtained) of -1.49. The test statistic is not in the critical region, as marked by a Z(critical) score of ± 1.96 at the 0.05 level, so we must fail to reject the null hypothesis of no difference. Males and females have essentially equal levels of schooling in the population, and education cannot account for the differences in income.

These results suggest that females are getting a lower return in income for their education. Although females and males are essentially equal in preparation for work and careers (at least as measured by amount of schooling), a large and statistically significant gender gap in income persists. Why? To answer this question in detail would carry us far beyond the bounds of a statistics text. However, we can suggest that one important part of the answer lies in the tendency of men and women to pursue different kinds of careers and jobs. In particular, men tend to dominate the more lucrative, higher-prestige occupations (lawyer, doctor), while women are concentrated in jobs that have lower levels of remuneration (elementary school teacher, nurse). For example, in 2007, women comprised 92% of all registered nurses and 88% of all paralegals and legal assistants versus

STATISTICS

IN EVERYDAY LIFE

Is the gender gap closing? Comparing only full-time workers, the median income for women in 1955 was 64% of the median income for men. In 2007, the gap had closed to 78%. These figures are parameters (characteristics of the population) collected by the United States Census Bureau.

Note that it may not be possible to close this gap much further as long as there is a "glass ceiling" on women's chances for promotion to the highest (most lucrative) positions and relatively fewer women in the highest income brackets. For example, in 2008, comparing only people with full-time jobs, 14% of men earned more than $100,000 versus only 5.7% of women.*

What does the glass ceiling imply about the overall shape of the distribution of men's and women's income? Which distribution will be more positively skewed? What effect will that have on the mean? (See Chapter 4 for a review of the effect of skew on measures of central tendency.)

*Source: U.S Census Bureau, 2009. American Fact Finder. Data available at www.census.gov.

only 30% of all doctors and 33% of all lawyers.[1] Thus, the gender gap in income is sustained, in part, by the fact that women are more likely to be employed in occupations that command lower salaries. What sociological factors might account for these choices?

READING STATISTICS 5: Hypothesis Testing

Professional researchers present their results succinctly. This is partly because of space limitations in scientific journals and partly because professional researchers can assume a certain level of statistical literacy in their audiences. Thus, they omit many of the elements—such as the null hypothesis and the critical region—that we have been so careful to state.

Instead, researchers report only the sample values (for example, means or proportions), the value of the test statistic (for example, a Z or t score), the alpha level, the degrees of freedom (if applicable), and sample size. The results of the example problem used to illustrate the t test might be reported in the professional literature as "The difference between the sample means of 2.37 (suburban families) and 2.78 (center-city families) was tested and found to be significant ($t = -2.16$, $df = 77$, $p < 0.05$)." Note that the alpha level is reported as "$p < 0.05$." This is shorthand for "the probability of a difference of this magnitude occurring by chance alone, if the null hypothesis of no difference is true, is less than 0.05" and is a good illustration of how researchers can convey a great deal of information in just a few symbols. In a similar fashion, our somewhat long-winded statement "The test statistic falls in the critical region and, therefore, the null hypothesis is rejected" is rendered tersely and simply: "The difference . . . was . . . found to be significant."

When researchers need to report the results of many tests of significance, they will often use a summary table to report the sample information and whether the difference is significant at a certain alpha level. If you read the researcher's description and analysis of such tables, you should have little difficulty interpreting and understanding them. As a final note, these comments about how significance tests are reported in the literature apply to all of the tests of hypotheses covered in Part II of this text.

STATISTICS IN THE PROFESSIONAL LITERATURE

Movies aimed at teens and young adults are rich in their portrayals of gender and gender identity issues. Are these portrayals consistent with the behavior and concerns of their intended audiences? What effects—if any—do the portrayals have on their audiences?

Two researchers attempted to address some of these issues by, first, analyzing the content of the 20 top-grossing "teen" movies released in the U.S. between 1995 and 2005. Coders counted the number of times movie characters used aggressive tactics—including gossip, backstabbing, and public humiliation—to damage the status or self-esteem of others. They identified 337 such acts in the films and found that female characters were significantly more likely to be the perpetrators ($t = -3.15$, $p < 0.05$) and that aggression was significantly more likely to be rewarded in the film ($t = -2.29$, $p < 0.05$). These findings contrast with studies of "real life" that find that female friendships are more likely to be supportive and cooperative than friendships among males.

Do these portrayals have any effect on their audiences? Using different statistical techniques, the researchers find evidence that it does and raise the concern that movies are sending the message that aggression is an effective way of gaining and perpetuating status. Want to learn more? The citation is given below.

Behm-Morawitz, E., and Mastro, D. 2008. "Mean Girls? The Influence of Gender Portrayals in Teen Movies on Emerging Adults' Gender-Based Attitudes and Beliefs." *Journalism and Mass Communications Quarterly*, 85:131–146.

[1]U.S. Bureau of the Census, 2009. Statistical Abstract of the United States, 2009. p. 385. Available at www.census.gov

Applying Statistics 9.3: Testing the Significance of the Difference Between Sample Means

The World Values Survey (WVS) has been administered to random samples in a variety of nations periodically since 1981. Like the General Social Survey used for computer applications in this text, the WVS tests opinions on a variety of issues and concerns. Results of the surveys are available at http://www.worldvaluessurvey.org/.

One item on the WVS asks about support for abortion. Are there significant differences in support between males and females in Canada and the United States? Support for abortion was measured with a ten-point scale, with higher scores indicating more support. Results are for 2006.

Canada		United States	
Males	Females	Males	Females
$\overline{X}_1 = 4.6$	$\overline{X}_2 = 4.7$	$\overline{X}_1 = 4.6$	$\overline{X}_2 = 4.4$
$s_1 = 2.8$	$s_2 = 3.0$	$s_1 = 2.9$	$s_2 = 2.8$
$N_1 = 988$	$N_2 = 1{,}074$	$N_1 = 569$	$N_2 = 598$

The differences between the genders seem small. Are they significant? To conserve space, only the most essential steps of the five-step model will be reported. Significance level will be set at 0.05 and the tests will be two-tailed. The null hypothesis will be that there is no difference in support for abortion between *all* males and *all* females in the respective nations ($H_0: \mu_1 = \mu_2$).

We will use Formulas 9.2 and 9.4 to compute the test statistics (step 4).

For Canada:

$$\sigma_{\overline{X}-\overline{X}} = \sqrt{\frac{s_1^2}{N_1 - 1} + \frac{s_2^2}{N_2 - 1}}$$

$$\sigma_{\overline{X}-\overline{X}} = \sqrt{\frac{(2.8)^2}{988 - 1} + \frac{(3.0)^2}{1074 - 1}}$$

$$\sigma_{\overline{X}-\overline{X}} = \sqrt{\frac{7.84}{987} + \frac{9.0}{1073}}$$

$$\sigma_{\overline{X}-\overline{X}} = \sqrt{0.008 + 0.008}$$

$$\sigma_{\overline{X}-\overline{X}} = \sqrt{0.016}$$

$$\sigma_{\overline{X}-\overline{X}} = 0.13$$

$$Z(\text{obtained}) = \frac{\overline{X}_1 - \overline{X}_2}{\sigma_{\overline{X}-\overline{X}}}$$

$$Z(\text{obtained}) = \frac{4.6 - 4.7}{0.13} = \frac{-0.10}{0.13} = -0.77$$

For the United States:

$$\sigma_{\overline{X}-\overline{X}} = \sqrt{\frac{s_1^2}{N_1 - 1} + \frac{s_2^2}{N_2 - 1}}$$

$$\sigma_{\overline{X}-\overline{X}} = \sqrt{\frac{(2.9)^2}{568} + \frac{(2.8)^2}{597}}$$

$$\sigma_{\overline{X}-\overline{X}} = \sqrt{\frac{8.41}{568} + \frac{7.84}{597}}$$

$$\sigma_{\overline{X}-\overline{X}} = \sqrt{0.015 + 0.013}$$

$$\sigma_{\overline{X}-\overline{X}} = \sqrt{0.028} = 0.17$$

$$Z(\text{obtained}) = \frac{\overline{X}_1 - \overline{X}_2}{\sigma_{\overline{X}-\overline{X}}}$$

$$Z(\text{obtained}) = \frac{4.6 - 4.4}{0.17} = \frac{0.20}{0.17} = 1.18$$

For both tests, we fail to reject the null hypothesis. For both Canada and the United States, there is no statistically significant difference in support for abortion between men and women.

SUMMARY

1. A common research situation is to test for the significance of the difference between two populations. Sample statistics are calculated for random samples of each population, and then we test for the significance of the difference between the samples as a way of inferring differences between the specified populations.

2. When sample information is summarized in the form of sample means and N is large, the Z distribution is used to find the critical region. When N is small, the t distribution is used to establish the critical region. In the latter circumstance, we must also assume equal population variances before forming a pooled estimate of the standard deviation of the sampling distribution.

3. Differences in sample proportions may also be tested for significance. For large samples, the Z distribution is used to find the critical region.

4. In all tests of hypothesis, a number of factors affect the probability of rejecting the null hypothesis: the size of the difference, the alpha level, the use of one- versus two-tailed tests, and sample size. Statistical significance is not the same thing as theoretical or practical importance. Even after a difference is found to be statistically significant, the researcher must still demonstrate the relevance or importance of his or her findings. The statistics presented in Parts III and IV of this text will give us the tools we need to deal directly with issues beyond statistical significance.

SUMMARY OF FORMULAS

Test statistic for two sample means, large samples:

9.1 $\qquad Z\text{(obtained)} = \dfrac{(\overline{X}_1 - \overline{X}_2) - (\mu_1 - \mu_2)}{\sigma_{\overline{X}-\overline{X}}}$

Test statistic for two sample means, large samples (simplified formula):

9.2 $\qquad Z\text{(obtained)} = \dfrac{(\overline{X}_1 - \overline{X}_2)}{\sigma_{\overline{X}-\overline{X}}}$

Standard deviation of the sampling distribution of the difference in sample means, large samples:

9.3 $\qquad \sigma_{\overline{X}-\overline{X}} = \sqrt{\dfrac{\sigma_1^2}{N_1} + \dfrac{\sigma_2^2}{N_2}}$

Pooled estimate of the standard deviation of the sampling distribution of the difference in sample means, large samples:

9.4 $\qquad \sigma_{\overline{X}-\overline{X}} = \sqrt{\dfrac{s_1^2}{N_1 - 1} + \dfrac{s_2^2}{N_2 - 1}}$

Pooled estimate of the standard deviation of the sampling distribution of the difference in sample means, small samples:

9.5 $\qquad \sigma_{\overline{X}-\overline{X}} = \sqrt{\dfrac{N_1 s_1^2 + N_2 s_2^2}{N_1 + N_2 - 2}} \sqrt{\dfrac{N_1 + N_2}{N_1 N_2}}$

Test statistic for two sample means, small samples:

9.6 $\qquad t\text{(obtained)} = \dfrac{\overline{X}_1 - \overline{X}_2}{\sigma_{\overline{X}-\overline{X}}}$

Pooled estimate of population proportion, large samples:

9.7 $\qquad P_u = \dfrac{N_1 P_{s1} + N_2 P_{s2}}{N_1 + N_2}$

Standard deviation of the sampling distribution of the difference in sample proportions, large samples:

9.8 $\qquad \sigma_{p-p} = \sqrt{P_u(1 - P_u)}\sqrt{\dfrac{N_1 + N_2}{N_1 N_2}}$

Test statistic for two sample proportions, large samples:

9.9 $\qquad Z\text{(obtained)} = \dfrac{(P_{s1} - P_{s2}) - (P_{u1} - P_{u2})}{\sigma_{p-p}}$

Test statistic for two sample proportions, large samples (simplified formula):

9.10 $\qquad Z\text{(obtained)} = \dfrac{(P_{s1} - P_{s2})}{\sigma_{p-p}}$

GLOSSARY

Independent random samples. Random samples gathered in such a way that the selection of a particular case for one sample has no effect on the probability that any other particular case will be selected for the other samples.

Pooled estimate. An estimate of the standard deviation of the sampling distribution of the difference in sample means based on the standard deviations of both samples.

σ_{p-p}. Symbol for the standard deviation of the sampling distribution of the differences in sample proportions.

$\sigma_{\bar{X}-\bar{X}}$. Symbol for the standard deviation of the sampling distribution of the differences in sample means.

PROBLEMS

9.1 For each of the following, test for the significance of the difference in sample statistics using the five-step model. (*HINT: Remember to solve Formula 9.4 before attempting to solve Formula 9.2. Also, in Formula 9.4, perform the mathematical operations in the proper sequence. First square each sample standard deviation, then divide by the proper N, add the resultant values, and then find the square root of the sum.*)

a.

Sample 1	Sample 2
$\bar{X}_1 = 72.5$	$\bar{X}_2 = 76.0$
$s_1 = 14.3$	$s_2 = 10.2$
$N_1 = 136$	$N_2 = 257$

b.

Sample 1	Sample 2
$\bar{X}_1 = 107$	$\bar{X}_2 = 103$
$s_1 = 14$	$s_2 = 17$
$N_1 = 175$	$N_2 = 200$

9.2 [SOC] Gessner and Healey administered questionnaires to samples of undergraduates. Among other things, the questionnaires contained a scale that measured attitudes toward interpersonal violence (higher scores indicate greater approval of interpersonal violence). Test the results as reported here for sexual, racial, and social-class differences.

a.

Sample 1 (Males)	Sample 2 (Females)
$\bar{X}_1 = 2.99$	$\bar{X}_2 = 2.29$
$s_1 = 0.88$	$s_2 = 0.91$
$N_1 = 122$	$N_2 = 251$

b.

Sample 1 (Blacks)	Sample 2 (Whites)
$\bar{X}_1 = 2.76$	$\bar{X}_2 = 2.49$
$s_1 = 0.68$	$s_2 = 0.91$
$N_1 = 43$	$N_2 = 304$

c.

Sample 1 (White Collar)	Sample 2 (Blue Collar)
$\bar{X}_1 = 2.46$	$\bar{X}_2 = 2.67$
$s_1 = 0.91$	$s_2 = 0.87$
$N_1 = 249$	$N_2 = 97$

d. Summarize your results in terms of the significance and the direction of the differences. Which of these three factors seems to make the biggest difference in attitudes toward interpersonal violence?

9.3 [SOC] Do athletes in different sports vary in terms of their readiness for college? Reported here are College Board scores of random samples of college basketball and football players. Is there a significant difference? Write a sentence or two explaining the difference

a.

Sample 1 (Basketball Players)	Sample 2 (Football Players)
$\bar{X}_1 = 460$	$\bar{X}_2 = 442$
$s_1 = 92$	$s_2 = 57$
$N_1 = 102$	$N_2 = 117$

What about male and female college athletes?

b.

Sample 1 (Males)	Sample 2 (Females)
$\bar{X}_1 = 452$	$\bar{X}_2 = 480$
$s_1 = 88$	$s_2 = 75$
$N_1 = 107$	$N_2 = 105$

9.4 [PA] A number of years ago, the road and highway maintenance department of Sin City, Nevada, began recruiting minority group members through an affirmative action program. In terms of efficiency ratings, as compiled by their superiors, how do the affirmative action employees rate? The ratings of random samples of both groups were collected, and the results are reported here (higher ratings indicate greater efficiency).

Sample 1 (Affirmative Action)	Sample 2 (Regular)
$\bar{X}_1 = 15.2$	$\bar{X}_2 = 15.5$
$s_1 = 3.9$	$s_2 = 2.0$
$N_1 = 97$	$N_2 = 100$

Write a sentence or two of interpretation.

9.5 SOC Are middle-class families more likely than working-class families to maintain contact with kin? Write a paragraph summarizing the results of these tests.

a. A sample of middle-class families reported an average of 7.3 visits per year with close kin, while a sample of working-class families averaged 8.2 visits. Is the difference significant?

Visits	
Sample 1 (Middle Class)	Sample 2 (Working Class)
$\bar{X}_1 = 7.3$	$\bar{X}_2 = 8.2$
$s_1 = 0.3$	$s_2 = 0.5$
$N_1 = 89$	$N_2 = 55$

b. The middle-class families averaged 2.3 phone calls and 8.7 e-mail messages per month with close kin. The working-class families averaged 2.7 calls and 5.7 e-mail messages per month. Are these differences significant?

Phone Calls	
Sample 1 (Middle Class)	Sample 2 (Working Class)
$\bar{X}_1 = 2.3$	$\bar{X}_2 = 2.7$
$s_1 = 0.5$	$s_2 = 0.8$
$N_1 = 89$	$N_2 = 55$

E-Mail Messages	
Sample 1 (Middle Class)	Sample 2 (Working Class)
$\bar{X}_1 = 8.7$	$\bar{X}_2 = 5.7$
$s_1 = 0.3$	$s_2 = 1.1$
$N_1 = 89$	$N_2 = 55$

9.6 SOC Are college students who live in dormitories significantly more involved in campus life than students who commute to campus? The following data report the average number of hours per week students devote to extracurricular activities. Is the difference between these randomly selected samples of commuter and residential students significant?

Sample 1 (Residential)	Sample 2 (Commuter)
$\bar{X}_1 = 10.2$	$\bar{X}_2 = 12.4$
$s_1 = 1.9$	$s_2 = 2.0$
$N_1 = 173$	$N_2 = 158$

9.7 Are senior citizens who live in retirement communities more socially active than those who live in age-integrated communities? Write a sentence or two explaining the results of these tests. (*HINT: Remember to use the proper formulas for small sample sizes.*)

a. A random sample of senior citizens living in a retirement village reported that they had an average of 1.42 face-to-face interactions per day with their neighbors. A random sample of those living in age-integrated communities reported 1.58 interactions. Is the difference significant?

Sample 1 (Retirement Community)	Sample 2 (Age-Integrated Neighborhood)
$\bar{X}_1 = 1.42$	$\bar{X}_2 = 1.58$
$s_1 = 0.10$	$s_2 = 0.78$
$N_1 = 43$	$N_2 = 37$

b. Senior citizens living in the retirement village reported that they had an average of 7.43 telephone calls with friends and relatives each week, while those in the age-integrated communities reported an average of 5.50 calls. Is the difference significant?

Sample 1 (Retirement Community)	Sample 2 (Age-Integrated Neighborhood)
$\bar{X}_1 = 7.43$	$\bar{X}_2 = 5.50$
$s_1 = 0.75$	$s_2 = 0.25$
$N_1 = 43$	$N_2 = 37$

9.8 SW As the director of the local Boys Club, you have claimed for years that membership in your club reduces juvenile delinquency. Now a cynical member of your funding agency has demanded proof of your claim. Fortunately, your local sociology department is on your side and springs to your aid with student assistants, computers, and hand calculators at the ready. Random samples of members and nonmembers are gathered and interviewed with respect to their involvement in delinquent activities. Each respondent is asked to enumerate the number of delinquent acts he has engaged in over the past year. The results are in and reported here (the average number of admitted acts of delinquency). What can you tell the funding agency?

Sample 1 (Members)	Sample 2 (Nonmembers)
$\bar{X}_1 = 10.3$	$\bar{X}_2 = 12.3$
$s_1 = 2.7$	$s_2 = 4.2$
$N_1 = 40$	$N_2 = 55$

9.9 SOC A survey has been administered to random samples of respondents in each of five nations. For each nation, are men and women significantly different in terms of their reported levels of satisfaction? Respondents were asked: "How satisfied are you with your life as a whole?" Responses varied from 1 (very dissatisfied) to 10 (very satisfied). Conduct a test for the significance of the difference in mean scores for each nation.

France	
Males	Females
$\bar{X}_1 = 7.4$	$\bar{X}_2 = 7.7$
$s_1 = 0.20$	$s_2 = 0.25$
$N_1 = 1005$	$N_2 = 1234$

Nigeria	
Males	Females
$\bar{X}_1 = 6.7$	$\bar{X}_2 = 7.8$
$S_1 = 0.16$	$s_2 = 0.23$
$N_1 = 1825$	$N_2 = 1256$

China	
Males	Females
$\bar{X}_1 = 8.3$	$\bar{X}_2 = 7.1$
$s_1 = 0.21$	$s_2 = 0.11$
$N_1 = 1400$	$N_2 = 1200$

Mexico	
Males	Females
$\bar{X}_1 = 8.3$	$\bar{X}_2 = 9.1$
$s_1 = 0.29$	$s_2 = 0.30$
$N_1 = 1645$	$N_2 = 1432$

Japan	
Males	Females
$\bar{X}_1 = 8.8$	$\bar{X}_2 = 9.3$
$s_1 = 0.34$	$s_2 = 0.32$
$N_1 = 1621$	$N_2 = 1683$

9.10 For each problem, test the sample statistics for the significance of the difference.

a.

Sample 1	Sample 2
$P_{s2} = 0.20$	$P_{s1} = 0.17$
$N_2 = 114$	$N_1 = 101$

b.

Sample 1	Sample 2
$P_{s2} = 0.60$	$P_{s1} = 0.62$
$N_2 = 478$	$N_1 = 532$

9.11 CJ About half of the police officers in Shinbone, Kansas, have completed a special course in investigative procedures. Has the course increased their efficiency in clearing crimes by arrest? The proportions of cases cleared by arrest for samples of trained and untrained officers are reported here.

Sample 1 (Trained)	Sample 2 (Untrained)
$P_{s1} = 0.47$	$P_{s2} = 0.43$
$N_1 = 157$	$N_2 = 113$

9.12 SW A large counseling center needs to evaluate several experimental programs. Write a paragraph summarizing the results of these tests. Did the new programs work?

a. One program is designed for divorce counseling; the key feature of the program is its counselors, who are married couples working in teams. About half of all clients have been randomly assigned to this special program and half to the regular program, and the proportion of cases that eventually ended in divorce was recorded for both. The results for random samples of couples from both programs are reported here. In terms of preventing divorce, did the new program work?

Sample 1 (Special Program)	Sample 2 (Regular Program)
$P_{s1} = 0.53$	$P_{s2} = 0.59$
$N_1 = 78$	$N_2 = 82$

b. The agency is also experimenting with peer counseling for depressed children. About half of all clients were randomly assigned to peer counseling. After the program had run for a year, a random sample of children from the new program was compared with a random sample of children who did not receive peer counseling. In terms of the percentage of children who were judged to be "much improved," did the new program work?

Sample 1 (Peer Counseling)	Sample 2 (No Peer Counseling)
$P_{s1} = 0.10$	$P_{s2} = 0.15$
$N_1 = 52$	$N_2 = 56$

9.13 $\boxed{\text{SOC}}$ At St. Algebra College, the sociology and psychology departments have been feuding for years about the respective quality of their programs. In an attempt to resolve the dispute, you have gathered data about the graduate school experience of random samples of both groups of majors. The results are presented here: the proportion of majors who applied to graduate schools, the proportion of majors accepted into their preferred programs, and the proportion of these who completed their programs. As measured by these data, is there a significant difference in program quality?

a. Proportion of majors who applied to graduate school:

Sample 1 Sociology	Sample 2 Psychology
$P_{s1} = 0.53$	$P_{s2} = 0.40$
$N_1 = 150$	$N_2 = 175$

b. Proportion accepted by program of first choice:

Sample 1 Sociology	Sample 2 Psychology
$P_{s1} = 0.75$	$P_{s2} = 0.85$
$N_1 = 80$	$N_2 = 70$

c. Proportion completing the programs:

Sample 1 Sociology	Sample 2 Psychology
$P_{s1} = 0.75$	$P_{s2} = 0.69$
$N_1 = 60$	$N_2 = 60$

9.14 $\boxed{\text{CJ}}$ The mayor of Sin City, Nevada, started a "crimeline" program some years ago and wonders

if it's really working. The program publicizes unsolved violent crimes in the local media and offers cash rewards for information leading to arrests. Are "featured" crimes more likely to be cleared by arrest than other violent crimes? Results from random samples of both types of crimes are reported as follows:

Sample 1 Crimeline Crimes Cleared by Arrest	Sample 2 Noncrimeline Crimes Cleared by Arrest
$P_{s1} = 0.35$	$P_{s2} = 0.25$
$N_1 = 178$	$N_2 = 212$

9.15 $\boxed{\text{SOC}}$ Some results from a survey administered to a nationally representative sample are reported here in terms of differences by sex. Which of these differences, if any, are significant? Write a sentence or two of interpretation for each test.

a. Proportion favoring the legalization of marijuana:

Sample 1 (Males)	Sample 2 (Females)
$P_{s1} = 0.37$	$P_{s2} = 0.31$
$N_1 = 202$	$N_2 = 246$

b. Proportion strongly agreeing that "kids are life's greatest joy":

Sample 1 (Males)	Sample 2 (Females)
$P_{s1} = 0.47$	$P_{s2} = 0.58$
$N_1 = 251$	$N_2 = 351$

c. Proportion voting for President Bush in 2004:

Sample 1 (Males)	Sample 2 (Females)
$P_{s1} = 0.59$	$P_{s2} = 0.47$
$N_1 = 399$	$N_2 = 509$

d. Average hours spent with e-mail each week:

Sample 1 (Males)	Sample 2 (Females)
$\bar{X}_1 = 4.18$	$\bar{X}_2 = 3.38$
$s_1 = 7.21$	$s_2 = 5.92$
$N_1 = 431$	$N_2 = 535$

e. Average rate of church attendance:

Sample 1 (Males)	Sample 2 (Females)
$\bar{X}_1 = 3.19$	$\bar{X}_2 = 3.99$
$s_1 = 2.60$	$s_2 = 2.72$
$N_1 = 641$	$N_2 = 808$

f. Number of children:

Sample 1 (Males)	Sample 2 (Females)
$\bar{X}_1 = 1.49$	$\bar{X}_2 = 1.93$
$s_1 = 1.50$	$s_2 = 1.50$
$N_1 = 635$	$N_2 = 803$

YOU ARE THE RESEARCHER

Gender Gaps and Support for Traditional Gender Roles

There are two projects below. The first tests for significant differences in sample means between men and women on four variables of your own choosing. The second uses the **Compute** command to explore attitudes toward abortion or traditional gender roles. You are urged to complete both projects.

PROJECT 1: Exploring the Gender Gap

In this enlightened age, with its heavy stress on gender equality, how many important differences persist between the sexes? In this section, you will use SPSS to conduct tests for significant differences in sample means with sex as the independent variable. You will select four dependent variables and test to see if the sexes differ significantly in the areas measured by your variables.

Step 1: Choosing Dependent Variables

Select four variables from the 2008 GSS to serve as dependent variables. Choose *only* interval-ratio variables or ordinal variables with three or more scores or categories. As you select variables, you might keep in mind the issues at the forefront of the debate over gender equality: income, education, and other measures of equality. Or you might choose variables that relate to lifestyle choices and patterns of everyday life: religiosity, TV viewing habits, or political ideas.

List your four dependent variables in the table below.

Variable	SPSS Name	What exactly does this variable measure?
1		
2		
3		
4		

Step 2: Stating Hypotheses

For each dependent variable, state a hypothesis about the difference you expect to find between men and women. For example, you might hypothesize that men will be more liberal or women will be more educated. You can base your hypotheses on your own experiences or on the information about gender differences that you have acquired in your courses or from other sources.

Step 3: Getting the Output

SPSS for Windows includes several tests for the significance of the difference between means. In this demonstration, we'll use the **Independent-Samples T Test** to test for the significance of the difference between men and women. These tests are appropriate when we are working with large samples. If there are statistically significant differences between the sample means for men and women, we can conclude that there are differences between all U.S. men and U.S. women on this variable.

Start *SPSS for Windows* and load the 2008 GSS database. From the main menu bar, click **Analyze**, then **Compare Means**, and then **Independent-Samples T Test**. The **Independent-Samples T Test** dialog box will open with the usual list of variables on the left. Find and move the cursor over the names of the dependent variables you selected in Step 1. Click the top arrow in the middle of the window to move the variable names to the **Test Variable(s)** box.

Next, find and highlight *sex* and click the bottom arrow in the middle of the window to move *sex* to the **Grouping Variable** box. Two question marks will appear in the **Grouping Variable** box, and the **Define Groups** button will become active. SPSS needs to know which cases go in which groups and, in this case, the instructions we need to supply are straightforward. Males (indicated by a score of 1 on *sex*) go into group 1 and females (a score of 2) will go into group 2.

Click the **Define Groups** button, and the **Define Groups** window will appear. The cursor will be blinking in the box beside Group 1: SPSS is asking for the score that will determine which cases go into this group. Type a 1 in this box (for males) and then click the box next to Group 2 and type a 2 (for females). Click **Continue** to return to the **Independent-Samples T Test** window and click **OK** and your output will be produced.

Step 4: Reading the Output

To illustrate the appearance and interpretation of SPSS *t* test output, I will present a test for the significance of the gender difference in average age (note that age is not a good choice as a dependent variable; it works better as a cause than an effect). Here's what the output looks like. (*NOTE: Several columns of the output have been deleted to conserve space and improve clarity.*)

The first block of output ("Group Statistics") presents descriptive statistics. There were 681 males in the sample, and their average age was 48.21 with a standard deviation of 16.490. The 796 females averaged 48.12 years of age with a standard deviation of 17.957. We can see from this output that the sample means are different and that, on the average, females are slightly younger. Is the difference between the sample means significant?

The results of the test for significance are reported in the next block of output. *SPSS for Windows* does a separate test for each of two assumptions about the population variance, but we will look only at the "Equal Variances Assumed" reported in the top row. This is the model used in this chapter.

Skip over the first columns of the output block (which report the results of a test for equality of the population variances). In the top row, *SPSS for Windows* reports a *t* value (.095), the degrees of freedom ($df = 1475$), and a "Sig (2-tailed)" of .925. This last piece of information is an alpha level except that it is the *exact* probability of getting the observed difference in sample means if only chance is operating. Thus, there is no need to look up the test statistic in a *t* or *Z* table. This value is much greater than 0.05, our usual indicator of significance. We will fail to reject the null hypothesis and

Group Statistics

	RESPONDENTS SEX	N	Mean	Std. Deviation	Std. Error Mean
AGE OF RESPONDENT	MALE	681	48.21	16.490	.632
	FEMALE	796	48.12	17.957	.636

		Levene's Test for Equality of Variances				
		F	Sig	t	df	Sig (2-tailed)
AGE OF RESPONDENT	Equal variances Assumed	7.011	.008	.095	1475	.925
	Equal variances Not Assumed			.095	1467.61	.924

conclude that the difference is not statistically significant. There is no significant difference in average years of age between men and women in the population.

Step 5: Recording Your Results

Run *t* tests for your dependent variables and gender and record your results in the table below. Write the SPSS variable name in the first column and then record the descriptive statistics (mean, standard deviation, and *N*). Next, record the results of the test of significance, using the top row ("Equal Variance Assumed") of the INDEPENDENT SAMPLES output box. Record the *t* score, the degrees of freedom (*df*), and whether the difference is significant at the 0.05 level. If the value of "Sig (2-tailed)" is less than 0.05, reject the null hypothesis and write YES in this column. If the value of "Sig (2-tailed)" is more than 0.05, fail to reject the null hypothesis and write NO in the column.

Dependent Variable		Mean	*s*	*N*	*t* score	*df*	Sig (2-tailed) < 0.05?
	Men						
	Women						
	Men						
	Women						

(*continued*)

(*continued*)

Dependent Variable		Mean	s	N	t score	df	Sig (2-tailed) < 0.05?
	Men						
	Women						
	Men						
	Women						

Step 6: Interpreting Your Results

Summarize your findings. For each dependent variable, write (1) at least one sentence summarizing the test, in which you identify the variables being tested, the sample means for each group, N, the t score, and the significance level. In the professional research literature, you might find the results reported as: "For a sample of 1417 respondents, there was no significant difference between the average age of men (48.21) and the average age of women (48.12) ($t = 0.95$, $df = 1475$, $p > 0.05$)." (2) A sentence relating to your hypotheses. Were they supported? How?

PROJECT 2: Using the Compute Command to Explore Gender Differences

In this project, you will use the **Compute** command, which was introduced in Chapter 4, to construct a summary scale for either support for legal abortion or support for traditional gender roles. Do these attitudes vary significantly by gender? You will also choose a second independent variable, other than gender, to test for significant differences.

Step 1: Creating Summary Scales

To refresh your memory, I used the **Compute** command in Chapter 4 to create a summary scale (*abscale*) for attitudes toward abortion by adding the scores on the two constituent items (*abany* and *abnomore*). Remember that, once created, a computed variable is added to the active file and can be used like any of the variables actually recorded in the file. If you did not save the data file with *abscale* included, you can quickly recreate the variable by following the instructions in Chapter 4.

As an alternative, you can also create a scale to measure support for traditional gender roles. The GSS data set supplied with this text also includes two variables that measure gender attitudes. One of these (*fefam*) states, "It is much better for everyone involved if the man is the achiever outside the home and the woman takes care of the home and family." There are four possible responses to this item, ranging from "Strongly Agree" (1) to "Strongly Disagree" (4). Note that the lowest score (1) is most supportive of traditional gender roles.

The second item (*fechild*) states, "A working mother doesn't hurt the children." This item also has four possible responses and, again, the lowest score—strongly disagree (1)—is the most consistent with support of traditional gender roles. The scores of the two items vary in the same direction (for both, the lowest score indicates support for traditional gender roles), so we can create a summary scale by simply adding the two variables together. Follow the commands in Chapter 4 to create the scale, which we will call *fescale*. The computed variable will have a total of seven possible scores, with lower scores indicating more support for traditional gender roles and higher scores indicating less support.

Step 2: Stating Hypotheses

Choose either scale and state hypotheses about what differences you expect to find between men and women. Which gender will be more supportive (have a lower average score on *abscale*) of legal abortion? Why? Will men or women be more supportive (have a lower average score on *fescale*) of traditional gender roles?

Step 3: Getting and Interpreting the Output

Run the **Independent Samples T Test** as before with your chosen scale as the **Test Variable** and *sex* as the **Grouping Variable**. See the instructions for Project 1 above.

Step 4: Interpreting Your Results

Summarize your results as in Project 1, step 6. Was your hypothesis confirmed? How?

Step 5: Extending the Test by Selecting an Additional Independent Variable

What other independent variable besides gender might be related to attitudes toward abortion or traditional gender roles? Select another independent variable besides *sex* and conduct an additional *t* test with either *abscale* or *fescale* as the dependent variable. Remember that the *t* test requires that the independent variable have *only* two categories. For variables with more than two categories (*relig* or *racecen1*, for example), you can meet this requirement by using the **Define Groups** button in the **Grouping Variables** box to select specific categories of a variable. You could, for example, compare Protestants and Catholics on *relig* by choosing scores of 1 (Protestants) and 2 (Catholics).

Step 6: Stating Hypotheses

State a hypothesis about what differences you expect to find between the categories of your independent variable. Which category will be more supportive of legal abortion or more supportive of traditional gender roles? Why?

Step 7: Getting and Interpreting the Output

Run the **Independent Samples T Test** as before with the scale you selected as the **Test Variable** and your independent variable as the **Grouping Variable**.

Step 8: Interpreting Your Results

Summarize your results as in Project 1, step 6. Was your hypothesis confirmed? How?

10

Hypothesis Testing III
The Analysis of Variance

LEARNING OBJECTIVES   By the end of this chapter, you will be able to

1. Identify and cite examples of situations in which ANOVA is appropriate.
2. Explain the logic of hypothesis testing as applied to ANOVA.
3. Perform the ANOVA test, using the five-step model as a guide, and correctly interpret the results.
4. Define and explain the concepts of population variance, total sum of squares, sum of squares between, sum of squares within, mean square estimates, and post hoc tests.
5. Explain the difference between the statistical significance and the importance of relationships between variables.

In this chapter, we examine a very flexible and widely used test of significance called the **analysis of variance** (often abbreviated as **ANOVA**). This test is designed to be used with interval-ratio–level dependent variables and is a powerful tool for analyzing the most sophisticated and precise measurements you are likely to encounter.

It is perhaps easiest to think of ANOVA as an extension of the *t* test for the significance of the difference between two sample means, which was presented in Chapter 9. The *t* test can be used only in situations in which our independent variable has exactly two categories (e.g., Protestants and Catholics). The analysis of variance, on the other hand, is appropriate for independent variables with more than two categories (e.g., Protestants, Catholics, Jews, people with no religious affiliation, and so forth).

To illustrate, suppose we were interested in analyzing support for capital punishment. Why does opinion on this issue vary from person to person? Could there be a relationship between religion (the independent variable) and support for capital punishment (the dependent variable)? The death penalty has an obvious moral dimension and may well be affected by a person's religious background.

Suppose that we administered a scale that measures support for capital punishment at the interval-ratio level to a randomly selected sample that includes Protestants, Catholics, Jews, people with no religious affiliation ("Nones"), and people from other religions ("Others"). We have five categories of subjects, and we want to see if opinion varies significantly by the category (religion) into which a person is classified. We will also want to raise other issues: Which religion shows the most support for capital punishment? Are Protestants significantly more supportive than Catholics or Jews? How do people with no religious affiliation compare to other people? The analysis of variance provides a very useful statistical context in which the questions can be addressed.

The Logic of the Analysis of Variance

For ANOVA, the null hypothesis is that the populations from which the samples are drawn have the same score on the dependent variable. As applied to our problem, the null hypothesis could be phrased as "People from different religious denominations do not vary in their support for the death penalty" or, symbolically, as $\mu_1 = \mu_2 = \mu_3 = \cdots \mu_k$. (Note that this is an extended version of the null hypothesis for the two-sample t test.) As usual, the researcher will normally want to reject the null hypothesis and, in this case, show that support is related to religion.

If the null hypothesis of "no difference" in the populations is true, then any means calculated from randomly selected samples should be roughly equal in value. The average score for the Protestant sample should be about the same as the average score for the Catholics and the Jews, and so forth. Note that the averages are unlikely to be exactly the same value even if the null hypothesis really is true, since we will always encounter some error or chance fluctuations in the measurement process. We are *not* asking: "Are there differences between the samples (or, in our example, the religions)?" Rather, we are asking: "Are the differences between the samples large enough to reject the null hypothesis and justify the conclusion that the populations are different?"

Now consider what kinds of outcomes we might encounter if we actually administered a "Support of Capital Punishment Scale" and organized the scores by religion. Of the infinite variety of possibilities, let's focus on the two extreme outcomes presented in Tables 10.1 and 10.2. In the first set of hypothetical results (Table 10.1), we see that the means and standard deviations of the groups are quite similar. The average scores are about the same, and all five groups exhibit about the same dispersion. These results would be quite consistent with the null hypothesis of no difference between the populations on support for capital punishment. Neither the average score nor the dispersion of the scores varies in any important way by religion.

Now consider another set of fictitious results, as displayed in Table 10.2. Here we see substantial differences in average score, with Jews showing the lowest support and Protestants showing the highest. Also, the standard deviations are low and similar from category to category, indicating that there is not much variation within the religions. Table 10.2 shows marked differences

TABLE 10.1   Support for Capital Punishment by Religion (fictitious data)

	Protestant	Catholic	Jew	None	Other
Mean =	10.3	11.0	10.1	9.9	10.5
Standard deviation =	2.4	1.9	2.2	1.7	2.0

TABLE 10.2 **Support for Capital Punishment by Religion (fictitious data)**

	Protestant	Catholic	Jew	None	Other
Mean =	14.7	11.3	5.7	8.3	7.1
Standard deviation =	0.9	0.8	1.0	1.1	0.7

between religions combined with homogeneity (or low standard deviations) *within* religions. In other words, the religions are different from each other and there is not much variation within each religion. These results would contradict the null hypothesis and support the notion that support for the death penalty does vary by religion.

ANOVA proceeds by making the kinds of comparisons outlined above. The test compares the amount of variation between categories (for example, from Protestants to Catholics to Jews to "Nones" to "Others") with the amount of variation within categories (among Protestants, among Catholics, and so forth). The greater the difference *between* categories (as measured by the means) relative to the differences *within* categories (as measured by the standard deviations), the more likely that the null hypothesis of "no difference" is false and can be rejected. If support for capital punishment truly varies by religion, then the sample mean for each religion should be quite different from the others and dispersion within the categories should be relatively low.

The Computation of ANOVA

Even though we have been thinking of ANOVA as a test for the significance of the difference between sample means, the computational routine actually involves developing two separate estimates of the population variance, σ^2 (hence the name *analysis of variance*). Recall from Chapter 4 that the variance is the standard deviation squared. One estimate of the population variance is based on the amount of variation *within* each of the categories of the independent variable and the other is based on the amount of variation *between* categories.

Before constructing these estimates, we need to introduce some new concepts and statistics. The first new concept is the total variation of the scores, which is measured by a quantity called the **total sum of squares**, or **SST**:

FORMULA 10.1
$$\text{SST} = \sum X_i^2 - N\overline{X}^2$$

To solve this formula, first find the sum of the squared scores (in other words, square each score and then add up the squared scores). Next, square the mean of all scores, multiply that value by the total number of cases in the sample (N), and subtract that quantity from the sum of the squared scores.

Formula 10.1 may seem vaguely familiar. A similar expression— $\sum (X_i - \overline{X})^2$—appears in the formula for the standard deviation and variance (see Chapter 4). All three statistics incorporate information about the variation of the scores (or, in the case of SST, the squared scores) around the mean (or, in the case of SST, the square of the mean multiplied by N). In other words, all three statistics are measures of the variation, or dispersion, of the scores.

To construct the two separate estimates of the population variance, the total variation (SST) is divided into two components. One component reflects the pattern of variation *within* each of the categories and is called the **sum of squares within (SSW)**. In our example problem, SSW would measure the amount of variation in support for the death penalty within each of the religions.

The other component is based on the variation *between* categories and is called the **sum of squares between (SSB)**. Again using our example to illustrate, SSB measures how different people in each religion are from each other in their support for capital punishment. SSW and SSB are components of SST, as reflected in Formula 10.2:

FORMULA 10.2
$$\text{SST} = \text{SSB} + \text{SSW}$$

Let's start with the computation of SSB, our measure of the variation between categories. We use the category means as summary statistics to determine the size of the difference from category to category. In other words, we compare the average support for the death penalty for each religion with the average support for all religions to determine SSB. The formula for the sum of squares between (SSB) is

FORMULA 10.3
$$\text{SSB} = \sum N_k(\overline{X}_k - \overline{X})^2$$

where SSB = the sum of squares between the categories
N_k = the number of cases in a category
$\overline{X}_k$ = the mean of a category

To find SSB, subtract the overall mean of all scores $(\overline{X})$ from each category mean $(\overline{X}_k)$, square the difference, multiply by the number of cases in the category, and add the results across all the categories.

The second estimate of the population variance (SSW) is based on the amount of variation within the categories. Formula 10.2 shows that the total sum of squares (SST) is equal to the sum of SSW and SSB. This relationship means that we can find SSW by simple subtraction. Formula 10.4 rearranges the symbols in Formula 10.2:

FORMULA 10.4
$$\text{SSW} = \text{SST} - \text{SSB}$$

Let's pause for a second to remember what we are after here. If the null hypothesis is *true*, then there should not be much variation from category to category (see Table 10.1), and SSW and SSB should be roughly equal. If the null hypothesis is *not true*, there will be large differences between categories (see Table 10.2) relative to the differences within categories, and SSB should be much larger than SSW. SSB will increase as the difference *between* category means increases, especially when there is not much variation *within* the categories (SSW). The larger SSB is compared to SSW, the more likely we are to reject the null hypothesis.

The next step in the computational routine is to construct the estimates of the population variance. To do this, we divide each sum of squares by its respective degrees of freedom. To find the degrees of freedom associated with SSW, subtract the number of categories (k) from the number of cases (N). The degrees of freedom associated with SSB are the number of categories minus 1. In summary,

FORMULA 10.5
$$\text{dfw} = N - k$$

where dfw = degrees of freedom associated with SSW
N = total number of cases
k = number of categories

FORMULA 10.6
$$\text{dfb} = k - 1$$

where dfb = degrees of freedom associated with SSB
k = number of categories

The actual estimates of the population variance, called the **mean square estimates**, are calculated by dividing each sum of squares by its respective degrees of freedom:

FORMULA 10.7
$$\text{Mean square within} = \frac{\text{SSW}}{\text{dfw}}$$

FORMULA 10.8
$$\text{Mean square between} = \frac{\text{SSB}}{\text{dfb}}$$

The test statistic calculated in step 4 of the five-step model is called the **F ratio**, and its value is determined by the following formula:

FORMULA 10.9
$$F = \frac{\text{Mean square between}}{\text{Mean square within}}$$

As you can see, the value of the F ratio will be a function of the amount of variation between categories (based on SSB) compared to the amount of variation within the categories (based on SSW). The greater the variation between the categories relative to the variation within, the higher the value of the F ratio and the more likely we will reject the null hypothesis.

AN EXAMPLE OF COMPUTING THE ANALYSIS OF VARIANCE

Assume that we have administered our Support for Capital Punishment Scale to a sample of 20 individuals who are equally divided among the five religions. (Obviously, this sample is much too small for any serious research and is intended solely for purposes of illustration.) All scores are reported in

---

ONE STEP AT A TIME **Computing ANOVA**

To compute ANOVA, we will use Formulas 10.1, 10.3, and 10.4 to find SST, SSB, and SSW. Then we will calculate the degrees of freedom, mean square estimates of the population variance, and the obtained *F* ratio. It is strongly recommended that you use a computing table like Table 10.3 to organize the computations.

To Find SST Using Formula 10.1

1: Find $\sum X_i^2$ by squaring each score and adding all of the squared scores together.

2: Find $N\overline{X}^2$ by squaring the value of the mean and then multiplying by *N*.

3: Subtract the quantity you found in step 2 from the quantity you found in step 1.

To Find SSB Using Formula 10.3

1: Subtract the mean of all scores $(\overline{X})$ from the mean of each category $(\overline{X}_k)$ and then square each difference.

2: Multiply each of the squared differences you found in step 1 by the number of cases in the category (N_k).

3: Add the quantities you found in step 2.

To Find SSW Using Formula 10.4

1: Subtract the value of SSB from the value of SST.

To Calculate Degrees of Freedom

1: For dfw, use Formula 10.5. Subtract the number of categories (*k*) from the number of cases (*N*).

2: For dfb, use Formula 10.6. Subtract 1 from the number of categories (*k*).

To Construct the Two Mean Square Estimates of the Population Variance

1: To find the MSW estimate, use Formula 10.7. Divide SSW by dfw.

2: To find the MSB estimate, use Formula 10.8. Divide SSB by dfb.

To Find the Obtained *F* Ratio

1: Use Formula 10.9. Divide the MSB estimate by the MSW estimate .

Table 10.3 along with the other quantities needed to complete the computations. The scores (X_i) are listed for each of the five religions, and a column has been added for the squared scores (X_i^2). The sums of both X_i and X_i^2 are reported at the bottom of each column. The category means $(\overline{X}_k)$ show that the four Protestants averaged 12.5 on the Support for Capital Punishment Scale, the four Catholics averaged 21.0, and so forth. Finally, the overall mean (sometimes called the *grand mean*) is reported in the bottom row of the table. This shows that all 20 respondents averaged 16.6 on the scale.

To organize our computations, we'll follow the routine summarized in the One Step at a Time box. We begin by finding SST by means of Formula 10.1:

$$\text{SST} = \sum X_i^2 - N\overline{X}^2$$
$$\text{SST} = (666 + 1898 + 1078 + 1794 + 712) - (20)(16.6)^2$$
$$\text{SST} = 6148 - (20)(275.56)$$
$$\text{SST} = 6148 - 5511.2$$
$$\text{SST} = 636.80$$

The sum of squares between (SSB) is found by means of Formula 10.3:

$$\text{SSB} = \sum N_k(\overline{X}_k - \overline{X})^2$$
$$\text{SSB} = 4(12.5 - 16.6)^2 + 4(21.0 - 16.6)^2$$
$$+ 4(16.0 - 16.6)^2 + 4(20.5 - 16.6)^2 + 4(13.0 - 16.6)^2$$
$$\text{SSB} = 67.24 + 77.44 + 1.44 + 60.84 + 51.84$$
$$\text{SSB} = 258.80$$

Now SSW can be found by subtraction (Formula 10.4):

$$\text{SSW} = \text{SST} - \text{SSB}$$
$$\text{SSW} = 636.80 - 258.80$$
$$\text{SSW} = 378.00$$

To find the degrees of freedom for the two sums of squares, we use Formulas 10.5 and 10.6:

$$\text{dfw} = N - k = 20 - 5 = 15$$
$$\text{dfb} = k - 1 = 5 - 1 = 4$$

Finally, we are ready to construct the mean square estimates of the population variance. For the estimate based on SSW, we use Formula 10.7:

$$\text{Mean square within} = \frac{\text{SSW}}{\text{dfw}} = \frac{378.00}{15} = 25.20$$

For the between estimate, we use Formula 10.8:

$$\text{Mean square between} = \frac{\text{SSB}}{\text{dfb}} = \frac{258.80}{4} = 64.70$$

TABLE 10.3 Support for Capital Punishment by Religion for 16 Subjects (fictitious data)

Protestant		Catholic		Jewish		None		Other	
X_i	X_i^2	X_i	X_i^2	X_i	X_i^2	X_i	X_i^2	X_i	X_i^2
8	64	12	144	12	144	15	225	10	100
12	144	20	400	13	169	16	256	18	324
13	169	25	625	18	324	23	529	12	144
17	289	27	729	21	441	28	784	12	144
50	666	84	1898	64	1078	82	1794	52	712
$\overline{X}_k = 12.5$		$\overline{X}_k = 21.0$		$\overline{X}_k = 16.0$		$\overline{X}_k = 20.5$		$\overline{X}_k = 13.0$	
				$\overline{X} = 16.6$					

The test statistic, or F ratio, is found by means of Formula 10.9:

$$F = \frac{\text{Mean square between}}{\text{Mean square within}} = \frac{64.70}{25.20} = 2.57$$

This statistic must still be evaluated for its significance. *(Solve any of the end of chapter problems to practice computing these quantities and solving these formulas.)*

A Test of Significance for ANOVA

Now we will see how to test an F ratio for significance. We will also take a look at some of the assumptions underlying the ANOVA test. As usual, we will follow the five-step model as a convenient way of organizing the decision-making process.

Step 1. Making Assumptions and Meeting Test Requirements.

> Model: Independent random samples
> Level of measurement is interval-ratio
> Populations are normally distributed
> Population variances are equal

The model assumptions are quite strict and underscore the fact that ANOVA should be used only with dependent variables that have been carefully and precisely measured. However, as long as the categories are roughly equal in size, ANOVA can tolerate some violation of the model assumptions. In situations where you are uncertain or have samples of very different size, it is probably advisable to use an alternative test. (Chi square, described in Chapter 11, is one option.)

Step 2. Stating the Null Hypothesis.
For ANOVA, the null hypothesis always states that the means of the populations from which the samples were drawn are equal. For our example problem, we are concerned with five different populations, or categories, so our null hypothesis would be

$$H_0: \mu_1 = \mu_2 = \mu_3 = \mu_4 = \mu_5$$

where μ_1 represents the mean for Protestants, μ_2 the mean for Catholics, and so forth.

The alternative hypothesis states simply that at least one of the population means is different. The wording here is important. If we reject the null hypothesis, ANOVA does not identify which mean or means are significantly different. In the final section of the chapter, we briefly consider an advanced test that can help us identify which pairs of means are significantly different.

(H_1: At least one of the population means is different.)

Step 3. Selecting the Sampling Distribution and Establishing the Critical Region. The sampling distribution for ANOVA is the F distribution, which is summarized in Appendix D. Note that there are separate tables for alphas of 0.05 and 0.01. As with the t table, the value of the critical F score will vary by degrees of freedom. For ANOVA, there are two separate degrees of freedom, one for each estimate of the population variance. The numbers across the top of the table are the degrees of freedom associated with the between estimate (dfb), and the numbers down the side of the table are those associated with the within estimate (dfw). In our example, dfb is $(k - 1)$, or 4, and dfw is $(N - k)$, or 15 (see Formulas 10.5 and 10.6). So, if we set alpha at 0.05, our critical F score will be 3.06.

Summarizing these considerations:

$$\text{Sampling distribution} = F \text{ distribution}$$
$$\text{Alpha} = .05$$
$$\text{Degrees of freedom (within)} = (N - k) = 15$$
$$\text{Degrees of freedom (between)} = (k - 1) = 4$$
$$F(\text{critical}) = 3.06$$

Take a moment to inspect the two F tables and you will notice that all the values are greater than 1.00. This is because ANOVA is a one-tailed test and we are concerned only with outcomes in which there is more variance between categories than within categories. F values of less than 1.00 would indicate that the between estimate was lower in value than the within estimate, and, since we would always fail to reject the null in such cases, we simply ignore this class of outcomes.

Step 4. Computing the Test Statistic. This was done in the previous section, where we found an obtained F ratio of 2.57.

Step 5. Making a Decision and Interpreting the Results of the Test. Compare the test statistic with the critical value:

$$F(\text{critical}) = 3.06$$
$$F(\text{obtained}) = 2.57$$

Since the test statistic does not fall into the critical region, our decision would be to fail to reject the null. Support for capital punishment does not differ significantly by religion, and the variation we observed in the sample means is unimportant.

STATISTICS
IN EVERYDAY LIFE

Want to stay slim and trim? A study that followed over 34,000 women for 15 years is not particularly encouraging: On the average, the women in the sample put on pounds during the study. However, exercise did help to limit the weight gain. The women were divided into three groups, and those who exercised at the highest levels gained significantly fewer pounds than the women in the moderate and low exercise groups. This study used sophisticated statistics not covered in this text to analyze the data, but the statistical significance of the differences in mean weight gain for the three groups was established by an ANOVA technique.

Source: Lee, I., Djousse, L, Sesso, H., Wang, L., and Buring J. 2010. "Physical Activity and Weight Gain Prevention." *JAMA*, March 24–31, 1173–1179.

Applying Statistics 10.1: The Analysis of Variance

An experiment in teaching introductory biology was recently conducted at a large university. One section was taught by the traditional lecture–lab method, a second was taught by an all-lab/demonstration approach with no lectures, and a third was taught entirely by a series of recorded lectures and demonstrations that the students were free to view at any time and as often as they wanted. Students were randomly assigned to each of the three sections, and, at the end of the semester, random samples of final exam scores were collected from each section. Is there a significant difference in student performance by teaching method?

Final Exam Scores by Teaching Method

	Lecture		Demonstration		Recorded	
	X_i	X^2	X_i	X^2	X_i	X^2
	55	3025	56	3136	50	2500
	57	3249	60	3600	52	2704
	60	3600	62	3844	60	3600
	63	3969	67	4489	61	3721
	72	5184	70	4900	63	3969
	73	5329	71	5041	69	4761
	79	6241	82	6724	71	5041
	85	7225	88	7744	80	6400
	92	8464	95	9025	82	6724

$\sum X_i =$ 636 651 588

$\sum X^2 =$ 46,286 48,503 39,420

$\overline{X}_k =$ 70.67 72.33 65.33

$\overline{X} = 1875/27 = 69.44$

We can see by inspection that the "Recorded" group had the lowest average score and that the "Demonstration" group had the highest average score. The ANOVA test will tell us if these differences are large enough to justify the conclusion that they did not occur by chance alone. We can organize the computations following the steps described in One Step at a Time: Computing ANOVA (see page 247):

$SST = \sum X^2 - N\overline{X}^2$

$SST = (46{,}286 + 48{,}503 + 39{,}420) - 27(69.44)^2$

$SST = 134{,}209 - 130{,}191.67$

$SST = 4017.33$

$SSB = \sum N_k(\overline{X}_k - \overline{X})^2$

$SSB = (9)(70.67 - 69.44)^2 + (9)(72.33 - 69.44)^2$
$\quad\quad + (9)(65.33 - 69.44)^2$

$SSB = 13.62 + 75.17 + 152.03$

$SSB = 240.82$

$SSW = SST - SSB$

$SSW = 4017.33 - 240.82$

$SSW = 3776.51$

$dfw = N - k = 27 - 3 = 24$

$dfb = k - 1 = 3 - 1 = 2$

$$\text{Mean square within} = \frac{SSW}{dfw} = \frac{3776.51}{24} = 157.36$$

$$\text{Mean square between} = \frac{SSB}{dfb} = \frac{240.82}{2} = 120.41$$

$$F = \frac{\text{Mean square between}}{\text{Mean square within}}$$

$$F = \frac{120.41}{157.36}$$

$$F = 0.77$$

We can now conduct the test of significance.

Step 1. Making Assumptions and Meeting Test Requirements.

  Model: Independent random sample
  Level of measurement is interval-ratio
  Populations are normally distributed
  Population variances are equal

Step 2. Stating the Null Hypothesis.

$$H_0\colon \mu_1 = \mu_2 = \mu_3$$

(H_1: At least one of the population means is different.)

(*continued*)

Applying Statistics 10.1: (*continued*)

Step 3. Selecting the Sampling Distribution and Establishing the Critical Region.

Sampling distribution $= F$ distribution

Alpha $= .05$

Degrees of freedom (within) $= (N - k)$

$= (27 - 3) = 24$

Degrees of freedom (between) $= (k - 1)$

$= (3 - 1) = 2$

$F(\text{critical}) = 3.40$

Step 4. Computing the Test Statistic. We found an obtained F ratio of 0.77.

Step 5. Making a Decision and Interpreting the Results of the Test. Compare the test statistic with the critical value:

$F(\text{critical}) = 3.40$

$F(\text{obtained}) = 0.77$

We would clearly fail to reject the null hypothesis ("The population means are equal") and would conclude that the observed differences among the category means were the results of random chance. Student performance in this course does not vary significantly by teaching method.

An Additional Example of Computing and Testing the Analysis of Variance

In this section, we work through an additional example of the computation and interpretation of the ANOVA test. We first review matters of computation, find the obtained F ratio, and then test the statistic for its significance.

A researcher has been asked to evaluate the efficiency with which each of three social service agencies is administering a particular program. One area of concern is the speed of the agencies in processing paperwork and determining the eligibility of potential clients. The researcher has gathered information on the number of days required for processing a random sample of 10 cases in each agency. Is there a significant difference between the agencies? The data are reported in Table 10.4, which also includes some additional information we will need to complete our calculations.

To find SST by means of formula 10.1:

$$\text{SST} = \sum X^2 - N\overline{X}^2$$

$$\text{SST} = (524 + 1816 + 2462) - 30(11.67)^2$$

$$\text{SST} = 4802 - 30(136.19)$$

$$\text{SST} = 4802 - 4085.70$$

$$\text{SST} = 716.30$$

To find SSB by means of Formula 10.3:

$$\text{SSB} = \sum N_k(\overline{X}_k - \overline{X})^2$$

$$\text{SSB} = (10)(7.0 - 11.67)^2 + (10)(13.0 - 11.67)^2 + (10)(15.0 - 11.67)^2$$

TABLE 10.4 **Number of Days Required to Process Cases for Three Agencies (fictitious data)**

Client	Agency A X_i	Agency A X^2	Agency B X_i	Agency B X^2	Agency C X_i	Agency C X^2
1	5	25	12	144	9	81
2	7	49	10	100	8	64
3	8	64	19	361	12	144
4	10	100	20	400	15	225
5	4	16	12	144	20	400
6	9	81	11	121	21	441
7	6	36	13	169	20	400
8	9	81	14	196	19	361
9	6	36	10	100	15	225
10	6	36	9	81	11	121
$\sum X_i =$	70		130		150	
$\sum X^2 =$		524		1816		2462
$\overline{X}_k =$	7.0		13.0		15.0	

$$\overline{X} = 350/30 = 11.67$$

$$\text{SSB} = (10)(21.80) + (10)(1.77) + (10)(11.09)$$

$$\text{SSB} = 218.10 + 17.17 + 110.09$$

$$\text{SSB} = 346.70$$

Now we can find SSW using Formula 10.4:

$$\text{SSW} = \text{SST} - \text{SSB}$$

$$\text{SSW} = 716.30 - 346.70$$

$$\text{SSW} = 369.60$$

The degrees of freedom are found through Formulas 10.5 and 10.6:

$$\text{dfw} = N - k = 30 - 3 = 27$$

$$\text{dfb} = k - 1 = 3 - 1 = 2$$

The estimates of the population variances are found by means of Formulas 10.7 and 10.8:

$$\text{Mean square within} = \frac{\text{SSW}}{\text{dfw}} = \frac{369.60}{27} = 13.69$$

$$\text{Mean square between} = \frac{\text{SSB}}{\text{dfb}} = \frac{346.70}{3} = 173.35$$

The F ratio (Formula 10.9) is:

$$F\text{(obtained)} = \frac{\text{Mean square between}}{\text{Mean square within}}$$

$$F\text{(obtained)} = \frac{173.35}{13.69}$$

$$F\text{(obtained)} = 12.66$$

And we can now test this value for its significance.

Step 1. Making Assumptions and Meeting Test Requirements.

> Model: Independent random samples
> Level of measurement is interval-ratio
> Populations are normally distributed
> Population variances are equal

The researcher will always be in a position to judge the adequacy of the first two assumptions in the model. The second two assumptions are more problematical, but remember that ANOVA will tolerate some deviation from its assumptions as long as sample sizes are roughly equal.

Step 2. Stating the Null Hypothesis.

$$H_0: \mu_1 = \mu_2 = \mu_3$$

(H_1: At least one of the population means is different.)

Step 3. Selecting the Sampling Distribution and Establishing the Critical Region.

> Sampling distribution $= F$ distribution
> Alpha $= 0.05$
> Degrees of freedom (within) $= (N - k) = (30 - 3) = 27$
> Degrees of freedom (between) $= (k - 1) = (3 - 1) = 2$
> $F\text{(critical)} = 3.35$

Step 4. Computing the Test Statistic. We found an obtained F ratio of 12.66.

Step 5. Making a Decision and Interpreting the Results of the Test. Compare the test statistic with the critical value:

$$F\text{(critical)} = 3.35$$
$$F\text{(obtained)} = 12.66$$

The test statistic is in the critical region, and we would reject the null hypothesis of no difference. The differences between the three agencies are very unlikely to have occurred by chance alone. The agencies are significantly different in the speed with which they process paperwork and determine eligibility. *(For practice in conducting the ANOVA test, see problems 10.2 to 10.8. Begin with the lower-numbered problems since they have smaller data sets, fewer categories, and, therefore, the simplest calculations.)*

Applying Statistics 10.2: The Analysis of Variance

A random sample of 15 nations from three levels of development has been selected. "Low income" nations are largely agricultural and have the lowest quality of life. "High Income" nations are industrial and the most affluent and modern. "Middle income" nations are between these extremes. Are these general characteristics reflected in differences in life expectancy (the number of years the average citizen can expect to live at birth) among the three categories? The data for the 15 nations are for 2009.

Low Income		Middle Income		High Income	
Nation	Life Expectancy	Nation	Life Expectancy	Nation	Life Expectancy
Cambodia	61	China	73	Australia	81
Malawi	46	Indonesia	71	Canada	81
Nepal	64	Pakistan	66	Japan	83
Niger	53	South Korea	80	Russia	68
Sudan	58	Turkey	72	United Kingdom	79

Source: Population Reference Bureau. 2009 World Population Data Sheet. Available at http://www.prb.org/.

To find F(obtained) and conduct the ANOVA test, computations will be organized into table format:

Life Expectancy by Income Level

	X_i	X_i^2	X_i	X_i^2	X_i	X_i^2
	61	3721	73	5329	81	6561
	46	2116	71	5041	81	6561
	64	4096	66	4356	83	6889
	53	2809	80	6400	68	4624
	58	3364	72	5184	79	6241
$\sum X_i =$	282		362		392	
$\sum X^2 =$		16,106		26,310		30,876
$\overline{X}_k =$	56.4		72.4		78.4	

$$\overline{X} = 69.07$$

(continued)

Applying Statistics 10.2: (*continued*)

The ANOVA test will tell us if these differences are large enough to justify the conclusion that they did not occur by chance alone. Following the usual computational routine:

$$SST = \sum X^2 - N\overline{X}^2$$

$$SST = (16{,}106 + 26{,}310 + 30{,}876) - 15(69.07)^2$$

$$SST = 73{,}292 - 71{,}559.97$$

$$SST = 1732.03$$

$$SSB = \sum N_k(\overline{X}_k - \overline{X})^2$$

$$SSB = (5)(56.4 - 69.07)^2 + (5)(72.4 - 69.07)^2$$
$$\quad + (5)(78.4 - 69.07)^2$$

$$SSB = 802.64 + 55.44 + 435.24$$

$$SSB = 1293.32$$

$$SSW = SST - SSB$$

$$SSW = 1732.03 - 1293.32$$

$$SSW = 438.71$$

$$dfw = N - k = 15 - 3 = 12$$

$$dfb = k - 1 = 3 - 1 = 2$$

$$\text{Mean square within} = \frac{SSW}{dfw} = \frac{438.71}{12} = 36.56$$

$$\text{Mean square between} = \frac{SSB}{dfb} = \frac{1293.32}{2} = 646.66$$

$$F = \frac{MSB}{MSW} = \frac{646.66}{36.56} = 17.69$$

We can now conduct the test of significance.

Step 1. Making Assumptions and Meeting Test Requirements.

Model: Independent random samples
Level of measurement is interval-ratio
Populations are normally distributed
Population variances are equal

Step 2. Stating the Null Hypothesis.

$$H_0: \mu_1 = \mu_2 = \mu_3$$

(H_1: At least one of the population means is different.)

Step 3. Selecting the Sampling Distribution and Establishing the Critical Region.

Sampling distribution = F distribution
Alpha = 0.05

$$dfw = (N - k) = (15 - 3) = 12$$

$$dfb = (k - 1) = (3 - 1) = 2$$

$$F(\text{critical}) = 3.88$$

Step 4. Computing the Test Statistic. We found an obtained F ratio of 17.69.

Step 5. Making a Decision and Interpreting the Results of the Test. Compare the test statistic with the critical value

$$F(\text{critical}) = 3.88$$
$$F(\text{obtained}) = 17.69$$

The null hypothesis ("The population means are equal") can be rejected. The differences in life expectancy between nations at different income levels are statistically significant.

The Limitations of the Test

ANOVA is appropriate whenever you want to test differences between the means of an interval-ratio–level dependent variable across three or more categories of an independent variable. This application is called a **one-way analysis of variance**, since it involves the effect of a single variable (for example, religion) on another (for example, support for capital punishment). This is the simplest application of ANOVA, and you should be aware that the technique has numerous more advanced and complex forms. For example, you may encounter research projects in which the effects of two separate variables (for example, religion and gender) on some third variable were observed.

One important limitation of ANOVA is that it requires an interval-ratio dependent variable and roughly equal numbers of cases in each category of the independent variable. The former condition may be difficult to meet with complete confidence for many variables of interest to the social sciences. The latter condition may create problems when the research hypothesis calls for comparisons between groups that are, by their nature, unequal in numbers (for example, white versus black Americans) and may call for some unusual sampling schemes in the data-gathering phase of a research project. Neither of these limitations should be particularly crippling, since ANOVA can tolerate some deviation from its model assumptions, but you should be aware of these limitations in planning your own research as well as in judging the adequacy of research conducted by others.

A second limitation of ANOVA, introduced in Chapter 9, actually applies to all forms of significance testing. Tests of significance are designed to detect nonrandom differences or differences so large that they are very unlikely to be produced by random chance alone. The problem is that differences that are statistically significant are not necessarily important in any other sense. Parts III and IV of this text provide some statistical techniques that can assess the importance of results directly.

A final limitation of ANOVA relates to the research hypothesis. This hypothesis is not specific; it simply asserts that at least one of the population means is different from the others. Obviously, when we reject the null hypothesis, we would like to know *which* differences between the sample means are significant. We can sometimes make this determination by simple inspection. In our problem involving social service agencies, for example, it is pretty clear from Table 10.4 that Agency A is the source of most of the differences. This informal, "eyeball" method can be misleading, however, and you should exercise caution in drawing conclusions. In the next section, we provide an extended example of the interpretation of ANOVA and introduce a technique (called **post hoc**, or "after the fact," analysis) that permits us to identify significant differences between the sample means reliably. The computational routines for post hoc analysis are beyond the scope of this text, but the tests are commonly available in computerized statistical packages such as SPSS.

Interpreting Statistics: Does Sexual Activity Vary by Marital Status?

Does marital status make a difference in people's level of sexual activity? Do married individuals have sex more often? Do unmarried individuals report having more sexual partners than married individuals? We will use the data from the 2008 General Social Survey to answer these questions scientifically. The respondents to this survey are a representative sample (see Chapter 6) of the population of the United States.

This analysis involves three variables. The independent variable is marital status, and there will be two separate dependent variables: the number of different

TABLE 10.5 Descriptive Statistics for Sexual Activity Variables

	Number of Sex Partners, Last Five Years	Number of Times Respondent Had Sex During Last Year
Mean =	2.98	51.65
Standard deviation =	8.72	61.69
N =	1,708	1,686

sexual partners in the past five years and the number of times the respondent had sex during the last 12 months. Marital status is a nominal-level variable with five categories. The two dependent variables are basically ordinal in level of measurement rather than the interval-ratio level required by ANOVA. For example, respondents were asked to describe their level of sexual activity by selecting from a number of broad categories such as "2 or 3 times a month" (see Appendix G for a complete listing of the scores of the two variables that measure sexual activity). We encountered a similar problem in Chapter 9 when we assessed the gender gap in income. Then, we solved the problem by substituting midpoints for each of the intervals of the variables, and we will do the same here. For example, on number of sex partners, we will substitute 7.5 for the category "5 to 10," and, for the sake of simplification, a value of 100 will be substituted for the "More than 100 partners" category. The variable measuring frequency of sexual activity has been transformed in a similar way: A value of 12 has been substituted for "about once a month," 52 for "about once a week," and so forth.

The summary statistics for the two sexual activity variables are presented in Table 10.5. The sample averaged almost three sex partners (2.98) over the last five years and had sex at a rate of about once a week over the past year (51.65). Note that the standard deviation is high relative to the mean for both variables. This is caused by extreme positive skews: Most cases are grouped in the low range of the variable (e.g., they had only one or two sexual partners), but a few respondents have very high scores.

The means and standard deviations for number of partners over the last five years for each marital status are reported in Table 10.6. The F ratio for these differences is 26.80, which is significant at less than the 0.001 level. We would conclude that marital status does make a significant difference in the number of

TABLE 10.6 Number of Sexual Partners in the Last Five Years by Marital Status

	Marital Status					ANOVA	
	Married	Widowed	Divorced	Separated	Never Married	F ratio	Alpha
Mean =	1.48	0.57	2.94	3.69	6.29	26.80	<0.001
Standard Deviation =	4.64	0.90	7.01	8.34	13.91		
N =	1050	197	397	78	578		

TABLE 10.7   Frequency of Sex by Marital Status

	Marital Status					ANOVA	
	Married	Widowed	Divorced	Separated	Never Married	F ratio	Alpha
Mean =	59.45	8.78	40.41	29.80	59.15	25.85	<0.001
Standard Deviation =	59.55	29.97	58.14	44.97	68.70		
N =	1050	198	401	78	606		

sex partners. Comparing the category means, the results are what one would expect. Widowed respondents had the lowest number of partners, on the average, and married individuals had the second lowest, averaging 1.48 partners. The divorced and separated averaged more sex partners than the married and widowed, and individuals who were never married had the highest average number.

Table 10.7 reports the means and standard deviations for the number of times the respondent reported having sex in the last 12 months by marital status. Once again, widowed individuals had by far the lowest rate. Separated respondents had the second-lowest rate (although much higher than the widowed), followed by the divorced. The never married and married respondents report nearly equal rates of sexual activity, on the average, of almost 60 times a year. The F ratio of 25.85 is significant at less than 0.001, so we can conclude that the rate of sexual activity is significantly affected by marital status.

POST HOC ANALYSIS

Tables 10.6 and 10.7 indicate significant relationships, but which differences in the tables are most important and contribute the most to the significant F ratios? We can see by inspection that the widowed group is very different from the others, but what other differences might be important? A post hoc analysis of the differences in sample means can provide objective answers to these questions.

We will not present the computational routines for post hoc tests, but the interpretation of the results of these tests is straightforward. The tests essentially compare the means of all possible pairs of categories (i.e., married with divorced, widowed with separated, and so forth) and tell us exactly which combinations of means contribute the most to a significant F ratio. Comparing sample means in this way increases the probability of making an alpha error (falsely rejecting a true null hypothesis—see Chapter 8); to correct for this, post hoc tests use more stringent criteria to identify significant differences.

Table 10.8 presents the results of post hoc tests for both measures of sexual activity. The categories are listed at the left-hand side of the table and again across the top, along with the category means. The entries in the table are the differences in the means for the category listed at the left of the table and the category listed across the top. A positive difference indicates that the category listed to the left had a higher score, and a negative difference means that the category

TABLE 10.8   A Post Hoc Test for Differences in Sexual Activity by Marital Status

Marital Status	Mean Differences in Number of Partners Over Last Five Years				
	Married (1.48)	Widowed (0.57)	Divorced (2.94)	Separated (3.69)	Never Married (6.29)
Married (1.48)		0.91	−1.46*	−2.21*	−4.81*
Widowed (0.57)			−2.37*	−3.12*	−5.72*
Divorced (2.94)				−0.75	−3.35*
Separated (3.69)					−2.60*

Marital Status	Mean Difference in Frequency of Sex				
	Married (59.45)	Widowed (8.78)	Divorced (40.41)	Separated (29.80)	Never Married (59.15)
Married (59.45)		50.67*	19.04*	29.65*	0.30
Widowed (8.78)			−31.63*	−21.02*	−50.37*
Divorced (40.41)				10.61	−18.74*
Separated (29.80)					−29.35*

*Mean difference is significant at the 0.05 level, Modified Least Significant Differences Test.

listed across the top had the higher score. For example, the average number of partners over the last five years was 1.48 for married respondents (category listed on the left) and 0.61 for widowed respondents (category listed across the top). The difference between the two category means is 1.48 − 0.57, or 0.91, the value reported at the intersection of the two categories. This value is positive, which indicates that married respondents (category to the left) had a higher score than widowed respondents (category across the top). An asterisk (*) next to the value means that the difference is significant at the 0.05 level.

Looking first at the top portion of the table (differences for number of sexual partners over the last five years), we can see that the "never married" respondents had a significantly higher number of sex partners than all other categories. Divorced respondents had significantly more partners than every category except separated respondents, and the widowed had significantly fewer partners than every other category except married.

The differences for frequency of sexual activity (reported in the bottom portion of the table) show similar patterns. The widowed had sex significantly less often than every other group. However, the test also finds some less obvious differences: Divorced respondents had sex significantly less often than married respondents and the "never married" had sex significantly more often than divorced and separated respondents. The fact that the widowed group was the "most different" was obvious from Tables 10.6 and 10.7, but these additional significant differences might well have escaped our attention had a post hoc test not been conducted.

In conclusion, the analysis of variance test shows that marital status makes a significant difference in levels of sexual activity for adult Americans. Married individuals had sex significantly more often with significantly fewer partners. People who "never married" have significantly more sex partners, and widowed people rank the lowest on both measures of sexual activity.

STATISTICS
IN EVERYDAY LIFE

How can you tell when someone is lying? What behavioral cues tell you that someone is honest and truthful? This is an especially important question when it involves possible crimes: How can police, judges, and juries ascertain the veracity of accusations, alibis, and testimony?

There is a lot of folk wisdom about detecting lies (e.g., "Liars won't make eye contact"), but social science research also addresses the issue. One study* used college students to compose accounts of traumatic events, some real and some fictitious, and then asked others to judge them. Using analysis of variance to identify significant differences in the structure of the stories, the researchers found that, on the average, fabricated accounts were seen as less plausible on their face, partly because they included significantly fewer specifics (times, places, names, and so forth), and were less rich in detail. It's always a challenge to tell when someone is lying, but research like this can improve accuracy and, conceivably, lead to more accurate and just findings in our criminal justice system.

*Porter, S., Peace, K., and Emmet, K. 2007. "You Protest Too Much, Methinks: Investigating the Features of Truthful and Fabricated Reports of Traumatic Experiences. *Canadian Journal of Behavioural Science*, 39:79–91.

READING STATISTICS 6: Stress, Minority Group Status, and Children

We can use a recent research article to illustrate how to read ANOVA results in the professional literature. Researchers Choi, Meininger, and Roberts administered a battery of standardized tests that measured stress and mental health to over 300 middle school students in the Houston area. In addition, they measured self-esteem and family cohesion, resources the students could use to combat stress and mental health problems. Since the researchers were concerned with a variety of dependent variables, they reported their results in a summary table, a version of which is presented here. Note that the table lists the independent variable (racial or ethnic group) in the columns and five of the dependent variables used in the study in the rows. The means for each test are noted in the body of the table and you should inspect these statistics to look for patterns in the differences between groups.

The right-hand columns of the table list the *F* ratios—*F*(obtained) in our terminology—for each test and the exact probability (*p*) that the differences occurred by random chance alone. The *p* values are another way to represent what we call the alpha level, and values in this column that are less than 0.05 are statistically significant (we would say that the "null hypothesis has been rejected").

In two of the five tests reported here, the differences are significant at the 0.05 level. For the test measuring stress, European American children had the lowest average score; for the test of self-esteem, they had higher scores than two of the other three groups. For two of the other tests (family cohesion and suicidal ideation), the differences in sample means are clearly not significant. In the fifth test (depression), the differences approach significance and, again, European American children show the most favorable score.

What do these tests show? Looking over the full array of their evidence—not just the partial results reported here—the authors conclude that minority group adolescents are more vulnerable to stress and to some kinds of mental health problems. They also have access to resources (e.g., cohesive families) that can be used to deal with these problems but,

(*continued*)

Stress, Resources, and Mental Distress by Group

	Group					
	European Americans	African Americans	Hispanic Americans	Asian Americans		
Measure	Mean	Mean	Mean	Mean	*F*	*p*
General Social Stress	24.93	28.15	34.84	32.26	9.42	0.000
Self-Esteem	32.29	35.29	29.78	26.28	6.30	0.000
Family Cohesion	6.75	6.65	6.44	5.72	1.17	0.321
Depression	40.52	42.91	45.26	43.25	2.21	0.087
Suicidal Ideation	4.07	4.25	4.47	3.80	0.85	0.470

Based on Table 2, p. 273.

overall, minority group children require more attention from health care providers. Want to learn more? The citation is given on top of the page.

Choi, Heeseung, Meininger, Janet, and Roberts, Robert. 2006. "Ethnic Differences in Adolescents' Mental Distress, Social Stress, and Resources." *Adolescence*, 41:263–283.

SUMMARY

1. One-way analysis of variance is a powerful test of significance that is commonly used when comparisons across more than two categories or samples are of interest. It is perhaps easiest to conceptualize ANOVA as an extension of the test for the difference in sample means.

2. ANOVA compares the amount of variation within the categories to the amount of variation between categories. If the null hypothesis of no difference is false, there should be relatively great variation between categories and relatively little variation within categories. The greater the differences from category to category relative to the differences within the categories, the more likely we will be able to reject the null hypothesis.

3. The computational routine for even simple applications of ANOVA can quickly become quite complex. The basic process is to construct separate estimates of the population variance based on the variation within the categories and the variation between the categories. The test statistic is the *F* ratio, which is based on a comparison of these two estimates. The basic computational routine is summarized at the end of Section 10.4, and this is probably an appropriate time to mention the widespread availability of statistical packages such as SPSS, the purpose of which is to perform complex

calculations such as these accurately and quickly. If you haven't yet learned how to use such programs, ANOVA may provide you with the necessary incentive.

4. The ANOVA test can be organized into the familiar five-step model for testing the significance of sample outcomes. Although the model assumptions (step 1) require high-quality data, the test can tolerate some deviation as long as sample sizes are roughly equal. The null hypothesis takes the familiar form of stating that there is no difference of any importance among the population values, while the alternative hypothesis asserts that at least one population mean is different. The sampling distribution is the *F* distribution, and the test is always one-tailed. The decision to reject or to fail to reject the null is based on a comparison of the obtained *F* ratio with the critical *F* ratio as determined for a given alpha level and degrees of freedom. The decision to reject the null hypothesis indicates only that one or more of the population means is different from the others. We can often determine which of the sample means accounts for the difference by inspecting the sample data, but this informal method should be used with caution, and post hoc tests are more reliable indicators of significant differences.

SUMMARY OF FORMULAS

Total sum of squares:

10.1 $\quad$ SST $= \sum X_i^2 - N\overline{X}^2$

The two components of the total sum of squares:

10.2 $\quad$ SST = SSB + SSW

Sum of squares between:

10.3 $\quad$ SSB $= \sum N_k(\overline{X}_k - \overline{X})^2$

Sum of squares within:

10.4 $\quad$ SSW = SST − SSB

Degrees of freedom for SSW:

10.5 $\quad$ dfw $= N - k$

Degrees of freedom for SSB:

10.6 $\quad$ dfb $= k - 1$

Mean square within:

10.7 $\quad$ Mean square within $= \dfrac{\text{SSW}}{\text{dfw}}$

Mean square between:

10.8 $\quad$ Mean square between $= \dfrac{\text{SSB}}{\text{dfb}}$

F ratio:

10.9 $\quad$ $F = \dfrac{\text{Mean square between}}{\text{Mean square within}}$

GLOSSARY

Analysis of variance. A test of significance appropriate for situations in which we are concerned with the differences among more than two sample means.

ANOVA. See **Analysis of variance**.

F ratio. The test statistic computed in step 4 of the ANOVA test.

Mean square estimate. An estimate of the variance calculated by dividing the sum of squares within (SSW) or the sum of squares between (SSB) by the proper degrees of freedom.

One-way analysis of variance. Applications of ANOVA in which the effect of a single independent variable on a dependent variable is observed.

Post hoc test. A technique for determining which pairs of means are significantly different.

Sum of squares between (SSB). The sum of the squared deviations of the sample means from the overall mean, weighted by sample size.

Sum of squares within (SSW). The sum of the squared deviations of scores from the category means.

Total sum of squares (SST). The sum of the squared deviations of the scores from the overall mean.

PROBLEMS

NOTE: The number of cases in these problems is very low—a fraction of the sample size necessary for any serious research—in order to simplify computations.

10.1 Conduct an ANOVA test for each of the following sets of scores. (*HINT: Follow the computational shortcut outlined in Section 10.4, and keep track of all sums and means by constructing computational tables like Table 10.3 or 10.4.*)

a.

Category		
A	B	C
5	10	12
7	12	16
8	14	18
9	15	20

b.

Category		
A	B	C
1	2	3
10	12	10
9	2	7
20	3	14
8	1	1

c.

Category			
A	B	C	D
13	45	23	10
15	40	78	20
10	47	80	25
11	50	34	27
10	45	30	20

10.2 SOC **a.** The Society for Ethical Culture provides non-theistic services in a congregational setting. Until three years ago, the membership of the Persipolis, New York, society grew steadily every year, but then it leveled off. The leaders have therefore decided to launch a membership drive, but before they begin, they would like to find out what types of people are most likely to join and attend services each week. A random sample of 15 existing members has been asked how many times they attended services over the past five weeks and for certain other information. Which independent variables have significant relationships with attendance?

b. Attendance by education:

Less Than High School	High School	College
0	1	0
1	3	3
2	3	4
3	4	4
4	5	4

c. Attendance by length of residence in present community:

Less Than 2 Years	2–5 Years	More Than 5 Years
0	0	1
1	2	3
3	3	3
4	4	4
4	5	4

d. Attendance by extent of television watching:

Little or None	Moderate	High
0	3	4
0	3	4
1	3	4
1	3	4
2	4	5

e. Attendance by number of children:

None	One Child	More Than One Child
0	2	0
1	3	3
1	4	4
3	4	4
3	4	5

10.3 SOC In a local community, a random sample of 18 couples has been assessed on a scale that measures the extent to which power and decision making are shared (lower scores) or monopolized by one party (higher scores) and on marital happiness (lower scores indicate lower levels of unhappiness). The couples were also classified by type of relationship: traditional (only the husband works outside the home), dual-career (both parties work), and cohabitational (parties living together but not legally married, regardless of work patterns). Does decision making or happiness vary significantly by type of relationship?

a.

	Decision Making	
Traditional	Dual-Career	Cohabitational
7	8	2
8	5	1
2	4	3
5	4	4
7	5	1
6	5	2

b.

	Happiness	
Traditional	Dual-Career	Cohabitational
10	12	12
14	12	14
20	12	15
22	14	17
23	15	18
24	20	22

10.4 CJ Two separate after-school programs for at-risk teenagers have been implemented in the city of Redland. One is a big-brother/big-sister mentoring program, in which adult volunteers are paired with troubled teenagers. The second involves extracurricular team sports competitions. In terms of the percentage reduction in crimes reported to the police over a one-year period, were the programs successful? The results are for random samples of 18 neighborhoods drawn from the entire city.

Mentoring Program	Competition Program	No Program
−10	−21	+30
−20	−15	−10
+10	−80	+14
+20	−10	+80
+70	−50	+50
+10	−10	−20

10.5 Are sexually active teenagers any better informed about AIDS and other potential health problems related to sex than teenagers who are sexually inactive? A 15-item test of general knowledge about sex and health was administered to random samples of teens who are sexually inactive, teens who are sexually active but with only a single partner ("going steady"), and teens who are sexually active with more than one partner. Is there any significant difference in the test scores?

Inactive	Active— One Partner	Active— More Than One Partner
10	11	12
12	11	12
8	6	10
10	5	4
8	15	3
5	10	15

10.6 SOC Does the rate of voter turnout vary significantly by the type of election? A random sample of voting precincts displays the following pattern of voter turnout by election type. Assess the results for significance.

Local Only	State	National
33	35	42
78	56	40
32	35	52
28	40	66
10	45	78
12	42	62
61	65	57
28	62	75
29	25	72
45	47	51
44	52	69
41	55	59

10.7 GER Do older citizens lose interest in politics and current affairs? A brief quiz on recent headline stories was administered to random samples of respondents from each of four different age groups. Is there a significant difference? The following data represent numbers of correct responses.

High School (15–18)	Young Adult (21–30)	Middle-Aged (30–55)	Retired (65+)
0	0	2	5
1	0	3	6
1	2	3	6
2	2	4	6
2	4	4	7
2	4	5	7
3	4	6	8
5	6	7	10
5	7	7	10
7	7	8	10
7	7	8	10
9	10	10	10

10.8 SOC A small random sample of respondents has been selected from the General Social Survey database. Each respondent has been classified as either a city dweller, a suburbanite, or a rural dweller. Are there statistically significant differences by place of residence for any of the variables listed here? (See Appendix G for definitions and scores of the dependent variables.)

a. Occupational Prestige (*prestg80*)

Urban	Suburban	Rural
32	40	30
45	48	40
42	50	40
47	55	45
48	55	45
50	60	50
51	65	52
55	70	55
60	75	55
65	75	60

b. Number of Children (*childs*)

Urban	Suburban	Rural
1	0	1
1	1	4
0	0	2
2	0	3
1	2	3
0	2	2

(*continued next page*)

(*continued*)

Number of Children (*childs*)

Urban	Suburban	Rural
2	3	5
2	2	0
1	2	4
0	1	6

c.

Family Income (*income06*)

Urban	Suburban	Rural
5	6	5
7	8	5
8	11	11
11	12	10
8	12	9
9	11	6
8	11	10
3	9	7
9	10	9
10	12	8

d.

Church Attendance (*attend*)

Urban	Suburban	Rural
0	0	1
7	0	5
0	2	4
4	5	4
5	8	0
8	5	4
7	8	8
5	7	8
7	2	8
4	6	5

e.

Hours of TV Watching per Day (*tvhours*)

Urban	Suburban	Rural
5	5	3
3	7	7
12	10	5
2	2	0
0	3	1
2	0	8
3	1	5
4	3	10
5	4	3
9	1	1

10.9 SOC Does support for suicide ("death with dignity") vary by social class? Is this relationship different in different nations? Small samples in three nations were asked if it is ever justified for a person with an incurable disease to take his or her own life. Respondents answered in terms of a 10-point scale, on which 10 was "always justified" (the strongest support for "death with dignity") and 1 was "never justified" (the lowest level of support). Results are reported here.

Mexico

Lower Class	Working Class	Middle Class	Upper Class
5	2	1	2
2	2	1	4
4	1	3	5
5	1	4	7
4	6	1	8
2	5	2	10
3	7	1	10
1	2	5	9
1	3	1	8
3	1	1	8

Canada

Lower Class	Working Class	Middle Class	Upper Class
7	5	1	5
7	6	3	7
6	7	4	8
4	8	5	9
7	8	7	10
8	9	8	10
9	5	8	8
9	6	9	5
6	7	9	8
5	8	5	9

United States

Lower Class	Working Class	Middle Class	Upper Class
4	4	4	1
5	5	6	5
6	1	7	8
1	4	5	9
3	3	8	9
3	3	9	9
3	4	9	8
5	2	8	6
3	1	7	9
6	1	2	9

Why Are Some People Liberal (or Conservative)?

This project investigates political ideology using *polviews*—a seven-point scale that measures how liberal or conservative a person is—as the dependent variable. This variable is ordinal in level of measurement but will be treated as interval-ratio for this test.

Before you begin the project, read Demonstrations 1 and 2, in which I show you how to conduct an ANOVA test with SPSS and introduce you to a new SPSS command that you will need to complete the project. The new SPSS command, called **Recode**, enables us to collapse scores on a variable. This operation is extremely useful because it allows us to change a variable to fit a particular task.

Demonstration 1: Recoding Variables

Here, I will demonstrate how we can take a variable like *age*, which has a wide range of scores (respondents in the 2008 GSS ranged in age from 18 to 89) and transform it into a variable with just a few scores, which we can use as an independent variable in ANOVA. We will create a new version of *age* that has three categories. When we are finished, we will have two versions of the same variable in the data set: the original interval-ratio version with age measured in years and a new ordinal-level version with collapsed categories. If we wish, the new version of age can be added to the permanent data file and used in the future.

The decision to use three categories for *age* is arbitrary; we could easily have decided on four, five, or even six categories for the new, recoded independent variable. If we find that we are unhappy with the three-category version of the variable, we can always recode the original version of *age* into a different set of categories.

We will collapse the values of *age* into three broad categories with roughly equal numbers of cases in each category. How can we define these new categories? Begin by running the **Frequencies** command for *age* to inspect the distribution of the scores. I used the cumulative percent column of the frequency distribution to find ages that divided the variable into thirds. I found that 33% of the 2008 GSS sample was younger than 38 and that 66.6% were younger than 55. I decided to use these ages as the cutting points for the new categories, as summarized below:

Ages	Percent of Sample
18–38	32.8%
39–55	33.8%
56–89	33.4%

To recode *age* into these categories, follow these steps:

1. In the **SPSS Data Editor** window, click **Transform** from the menu bar and then click **Recode**. A window will open that gives us two choices: "**into same variable**" or "**into different variable**." If we choose **into same variable**, the new version of the variable will *replace* the old and the variable *age* (measured in years) would disappear. We definitely do *not* want this to happen, so we will choose (click on) **into different variable**. This option will allow us to keep both the old and new versions of the variable.

2. The **Recode into Different Variable** window will open. A box containing an alphabetical list of variables will appear on the left. Use the cursor to highlight *age* and

then click on the arrow button to move the variable to the **Input Variable → Output Variable** box. The input variable is the old version of age, and the output variable is the new, recoded version we will soon create.

3. In the **Output Variable** box on the right, click in the **Name** box and type a name for the new (output) variable. I suggest *ager* ("**age r**ecoded") for the new variable, but you can assign any name as long as it does not duplicate the name of some other variable in the data set and is no longer than eight characters. Click the **Change** button and the expression *age → ager* will appear in the **Input Variable → Output Variable** box.

4. Click on the **Old and New Values** button in the middle of the screen, and a new dialog box will open. Read down the left-hand column until you find the **Range** button. Click on the button, and the cursor will move to the small box immediately below. In these boxes we will specify the low and high points of each interval of the new variable *ager*.

5. Type 18 (the youngest age in the sample) into the left-hand **Range** dialog box and then click on the right-hand box and type 38. In the **New Value** box in the upper-right-hand corner of the screen, click the **Value** button. Type 1 in the **Value** dialog box and then click the **Add** button directly below. The expression 18–38 → 1 will appear in the **Old → New** dialog box.

6. Return to the **Range** dialog boxes. Type 39 in the left-hand box and 55 in the right-hand box and then click the **Value** button in the **New Values** box. Type 2 in the **Value** dialog box and then click the **Add** button. The expression 39–55 → 2 appears in the **Old → New** dialog box.

7. Finish by returning to the **Range** dialog box and entering the values 56 and 89 in the correct boxes. Click the **Value** button in the **New Values** box. Type 3 in the **Value** dialog box and then click the **Add** button. The expression 56–89 → 3 appears in the **Old → New** dialog box

8. Click the **Continue** button at the bottom of the screen, and you will return to the **Recode into Different Variable** dialog box. Click **OK**, and SPSS will execute the transformation.

You now have a data set with an additional variable named *ager* (or whatever name you gave the recoded variable). You can find the new variable in the last column to the right in the data window. You can make the new variable a permanent part of the data set by saving the data file at the end of the session. If you do not wish to save the new, expanded data file, click **No** when you are asked if you want to save the data file. If you are using the student version of *SPSS for Windows*, remember that you are limited to a maximum of 50 variables and you will not be able to save the new variable.

Demonstration 2: Using ANOVA to Analyze the Effect of Age on Number of Sex Partners (*partnrs5*)

To demonstrate how to run ANOVA with SPSS, I will conduct a test with recoded *age* as the independent variable and scores on *partnrs5*, a measure of sexual activity that asks how many different sexual partners the respondent has had over the past five years, as the dependent variable. Note the coding scheme for *partnrs5* (see Appendix G or click **Utilities → Variables** on the SPSS menu bar). The first five scores (zero partners to four partners) are actual numbers but the higher scores represent broad categories (e.g., "5" means the respondent had between five and 10 different partners). This variable is a combination of interval-ratio and ordinal scoring and we will have to treat the means with some caution.

SPSS provides several different ways of conducting the analysis of variance test. The procedure summarized below is the most accessible of these, but it still incorporates options and capabilities that we have not covered in this chapter. If you wish to explore these possibilities, please use the online **Help** facility.

To use the ANOVA procedure, click **Analyze, Compare Means**, and then **One-way Anova**. The **One-way Anova** window appears. Find *partnrs5* in the variable list on the left and click the arrow to move the variable name into the **Dependent List** box. Note that you can request more than one dependent variable at a time. Next, find the name of the recoded *age* variable (perhaps called *ager*?) and click the arrow to move the variable name into the **Factor** box.

Click **Options** and then click the box next to **Descriptive** in the **Statistics** box to request means and standard deviations along with the analysis of variance. Click **Continue** and then click **OK**, and the following output will be produced.

Descriptives: HOW MANY SEX PARTNERS R HAD IN LAST 5 YEARS

	N	Mean	Std. Deviation	Std. Error
1.00	428	2.40	1.823	.088
2.00	415	1.46	1.361	.067
3.00	396	.90	.895	.045
Total	1239	1.60	1.551	.044

NOTE: This table has been edited and will not look exactly like the raw SPSS output.

ANOVA: HOW MANY SEX PARTNERS R HAD IN LAST 5 YEARS

	Sum of Squares	df	Mean Square	F	Sig.
Between Groups	476.097	2	238.049	117.582	.000
Within Groups	2502.326	1236	2.025		
Total	2978.423	1238			

The output box labeled **Descriptives** presents the category means and shows, not surprisingly, that the youngest respondents (category 1 or people 18 to 38 years old) had the highest average number of different partners. The oldest respondents (category 3) had the lowest number of sex partners and the overall or grand mean was 1.60 (see the Total row).

The output box labeled **ANOVA** includes the various degrees of freedom, all of the sums of squares, the Mean Square estimates, the F ratio (117.582), and, at the far right, the exact probability ("Sig.") of getting these results if the null hypothesis is true. This is reported as .000, much lower than our usual alpha level of 0.05. The differences in *partnrs5* for the various age groups are statistically significant.

Your turn.

PROJECT: Political Ideology (*polviews*)

In this project, you will investigate political ideology using scores on *polviews*—a seven-point scale that measures how liberal or conservative a person is—as the dependent variable. This variable is ordinal in level of measurement but will be treated as interval-ratio for this test. As you analyze and interpret your results, remember that *polviews* is coded so that higher scores indicate greater conservatism (see Appendix G or click **Utilities → Variables** on the SPSS menu bar).

Step 1: Choosing Independent Variables

Select three variables from the 2008 GSS to serve as independent variables. What factors might help to explain why some people are more liberal and some are more conservative? Choose *only* independent variables with three to six scores or categories. Among other possibilities, you might consider education (use *degree*), religious denomination, age (use the recoded version) or social class. Use the recode command to collapse independent variables with more than six categories into three or four categories. In general, variables that measure characteristics or traits (like gender or race) will work better than those that measure attitude or opinion (like *cappun* or *gunlaw*), which are more likely to be manifestations of political ideology, not causes.

List your independent variables in the table below:

Variable	SPSS Name	What exactly does this variable measure?
1		
2		
3		

Step 2: Stating Hypotheses

For each independent variable, state a hypothesis about its relationship with *polviews*. For example, you might hypothesize that people with greater education will be more liberal or that the more religious will be more conservative. You can base your hypotheses on your own experiences or on information you have acquired in your courses.

Step 3: Getting and Reading the Output

We discussed these tasks in the demonstration above.

Step 4: Recording Results

Record the results of your ANOVA tests in the table below, using as many rows for each independent variable as necessary. Write the SPSS variable name in the first column and then write the names of the categories of that independent variable in the next column. Record the descriptive statistics (mean, standard deviation, and N) in the designated columns. Write the value of the F ratio and, in the far right-hand column, indicate whether or not the results are significant at the 0.05 level. If the value in the "Sig." column of the ANOVA output is less than 0.05, write YES in this column. If the value in the "Sig." column of the ANOVA output is more than 0.05, write NO in this column. Finally, for each independent variable, record the grand or overall means, standard deviations, and sample sizes in the row labeled "Totals = ."

Independent Variables:	Categories	Mean	Std. Dev.	N	F Ratio	Sig. at 0.05 Level?
1. _____	1.				_____	_____
	2.					
	3.					
	4.					
	5.					
	6.					
	Totals =					
2. _____	1.				_____	_____
	2.					
	3.					
	4.					
	5.					
	6.					
	Totals =					
3. _____	1.				_____	_____
	2.					
	3.					
	4.					
	5.					
	6.					
	Totals =					

Step 6: Interpreting your Results

Summarize your findings. For each test, write:

1. At least one sentence summarizing the test in which you identify the variables being tested, the sample means for each group, N, the F ratio, and the significance level. In the professional research literature, you might find the results reported as: "For a sample of 1417 respondents, there was no significant difference between the average age of Southerners (46.50), Northerners (44.20), Midwesterners (47.80), and Westerners (47.20) ($F = 1.77$, $p > 0.05$)."
2. A sentence relating to your hypotheses. Were they supported? How?

11

Hypothesis Testing IV
Chi Square

LEARNING OBJECTIVES   By the end of this chapter, you will be able to

1. Identify and cite examples of situations in which the chi square test is appropriate.
2. Explain the structure of a bivariate table and the concept of independence as applied to expected and observed frequencies in a bivariate table.
3. Explain the logic of hypothesis testing as applied to a bivariate table.
4. Perform the chi square test using the five-step model and correctly interpret the results.
5. Explain the limitations of the chi square test and, especially, the difference between statistical significance and importance.

The **chi square (χ^2) test** has probably been the most frequently used test of hypothesis in the social sciences, a popularity that is due largely to the fact that the assumptions and requirements in step 1 of the five-step model are easy to satisfy. The test can be conducted with variables measured at the nominal level (the lowest level of measurement) and is **nonparametric**, which means that it requires no assumption at all about the shape of the population or sampling distribution.

Why is it an advantage to have easy-to-satisfy assumptions and requirements? The decision to reject the null hypothesis (step 5) is not specific: It means only that one statement in the model (step 1) *or* the null hypothesis (step 2) is wrong. Usually, of course, we single out the null hypothesis for rejection. The more certain we are of the model, the greater our confidence that the null hypothesis is the faulty assumption. A "weak" or easily satisfied model means that our decision to reject the null hypothesis can be made with even greater certainty.

Chi square has also been popular for its flexibility: It can be used with variables that have many categories or scores. For example, in Chapter 9, we tested the significance of the difference in the proportions of black and white citizens who were "highly participatory" in voluntary associations. What if the researcher wished to expand the test to include Hispanic and Asian Americans? The two-sample test would no longer be applicable, but chi square handles the more complex variable easily. Also, unlike the ANOVA test presented in Chapter 10, the chi square test can be conducted with variables at any level of measurement.

The Bivariate Table

Chi square is computed from **bivariate tables**, so called because they display the scores of cases on two different variables at the same time. Bivariate tables are used to ascertain if there is a significant relationship between the variables and for other purposes that we will investigate in later chapters. In fact, these tables are very commonly used in research, and a detailed examination of them is in order.

First of all, bivariate tables have (of course) two dimensions. We refer to the horizontal (across) dimension in terms of **rows** and the vertical dimension (up and down) in terms of **columns**. Each column or row represents a score on a variable, and the intersections of the row and columns (**cells**) represent the combined scores on both variables.

Let's use an example to clarify. Suppose that a researcher is interested in senior citizens and, specifically, if their participation in voluntary groups, community-service organizations, and so forth is affected by their marital status. To simplify the analysis, the researcher has confined the sample to people who are presently married or not married (including people who are single and divorced) and has measured involvement in voluntary associations as a simple dichotomy: people have been classified as either high or low.

By convention, the independent variable (the variable that is taken to be the cause) is placed in the columns and the dependent variable in the rows. In the example at hand, marital status is the causal variable (the question was "Is membership *affected by* marital status?"), and each column will represent a score on this variable. Each row, on the other hand, will represent a score on level of involvement (high or low). Table 11.1 displays the outline of the bivariate table for a sample of 100 senior citizens.

Note some further details of the table. First, subtotals have been added to each column and row. These are called the row or column **marginals**, and in this case they tell us that 50 members of the sample are married and 50 are not married (the column marginals), and 50 are rated as high in participation and 50 are rated low (the row marginals). Second, the total number of cases in the sample ($N = 100$) is reported at the intersection of the row and column marginals. Finally, take careful note of the labeling of the table. Each row and column is identified, and the table has a descriptive title that includes the names of the variables, with the dependent variable listed first. Clear, complete labels and concise titles must be included in *all* tables, graphs, and charts.

TABLE 11.1 **Rates of Participation in Voluntary Associations by Marital Status for 100 Senior Citizens**

	Marital Status		
Participation Rates	Married	Unmarried	
High			50
Low			50
	50	50	100

As you have noticed, Table 11.1 lacks one piece of crucial information: the numbers in the body of the table. To finish the table, we need to classify the marital status and level of participation of each member of the sample, keep count of how often each combination of scores occurs, and record these numbers in the appropriate cell of the table. Since each variable has two scores, four combinations of scores are possible, each corresponding to a cell in the table. For example, married people with high levels of participation would be counted in the upper-left-hand cell, unmarried people with low levels of participation would be counted in the lower-right-hand cell, and so forth. When we are finished counting, each cell will display the number of times each *combination* of scores occurred.

Finally, note how we could expand the bivariate table to accommodate variables with more scores. If we wished to include people with other marital statuses (widowed, separated, and so forth), we would simply add columns. More elaborate dependent variables could also be easily accommodated. If we had measured participation rates with three categories (e.g., high, moderate, and low) rather than two, we would simply add a row to the table.

The Logic of Chi Square

The chi square test has several different uses, and most of this chapter deals with an application called the *chi square test for independence*. We have encountered the term *independence* in connection with the requirements for the two-sample case (Chapter 9) and for the ANOVA test (Chapter 10). In the context of chi square, **independence** refers to the relationship between the variables, not between the samples. Two variables are independent if the classification of a case into a particular category of one variable has no effect on the probability that the case will fall into any particular category of the second variable. For example, marital status and participation in voluntary associations would be independent of each other if the classification of a case as married or not married has no effect on the classification of the case as high or low on participation. In other words, the variables are independent if level of participation and marital status were completely unrelated to each other.

Consider Table 11.1 again. If the variables are independent, the cell frequencies will be determined solely by random chance and we would find that, just as an honest coin will show heads about 50% of the time, about half of the married respondents will rank high on participation and half will rank low. The same pattern would hold for the 50 unmarried respondents and, therefore, each of the four cells would have about 25 cases in it, as illustrated in Table 11.2. This pattern of cell frequencies indicates that marital status has no effect on the probability that a person would be either high or low in participation. The probability of being classified as high or low would be 0.50 for both marital statuses, and the variables would therefore be independent.

The null hypothesis for chi square is that the variables are independent. Under the assumption that the null hypothesis is true, the cell frequencies we would expect to find if only random chance were operating are computed. These frequencies, called **expected frequencies** (symbolized f_e), are then compared, cell by cell, with the frequencies actually observed in the table (**observed frequencies**,

TABLE 11.2 **The Cell Frequencies That Would Be Expected If Rates of Participation and Marital Status Were Independent**

	Marital Status		
Participation Rates	Married	Unmarried	
High	25	25	50
Low	25	25	50
	50	50	100

symbolized f_o). If the null hypothesis is true and the variables are independent, then there should be little difference between the expected and observed frequencies. If the null hypothesis is false, however, there should be large differences between the two. The greater the differences between expected (f_e) and observed (f_o) frequencies, the less likely that the variables are independent and the more likely that we will be able to reject the null hypothesis.

The Computation of Chi Square

As with all tests of hypothesis, with chi square we compute a test statistic, χ^2**(obtained)**, from the sample data and then place that value on the sampling distribution of all possible sample outcomes. Specifically, the χ^2(obtained) will be compared with the value of χ^2**(critical)**, which will be determined by consulting a chi square table (Appendix C) for a particular alpha level and degrees of freedom. Prior to conducting the formal test of hypothesis, let us take a moment to consider the calculation of chi square, as defined by Formula 11.1:

FORMULA 11.1
$$\chi^2(\text{obtained}) = \sum \frac{(f_o - f_e)^2}{f_e}$$

where f_o = the cell frequencies observed in the bivariate table

f_e = the cell frequencies that would be expected if the variables were independent

We must work cell by cell to solve this formula. To compute chi square, subtract the expected frequency from the observed frequency for each cell, square the result, divide by the expected frequency for that cell, and then sum the resultant values for all cells.

This formula requires an expected frequency for each cell in the table. In Table 11.2, the marginals are the same value for all rows and columns, and the expected frequencies are obvious by intuition: $f_e = 25$ for all four cells. In the more usual case, the expected frequencies will not be obvious, marginals will be unequal, and we must use Formula 11.2 to find the expected frequency for each cell:

FORMULA 11.2
$$f_e = \frac{\text{Row marginal} \times \text{Column marginal}}{N}$$

TABLE 11.3 Employment of 100 Social Work Majors by Accreditation Status of Undergraduate Program

	Accreditation Status		
Employment Status	Accredited	Not Accredited	Totals
Working as a social worker	30	10	40
Not working as a social worker	25	35	60
Totals	55	45	100

That is, the expected frequency for any cell is equal to the total number of cases in the row (the row marginal) times the total number of cases in the column (the column marginal) divided by the total number of cases in the table (N).

An example using Table 11.3 should clarify these procedures. A random sample of 100 social work majors have been classified in terms of whether the Council on Social Work Education has accredited their undergraduate programs (the column, or independent, variable) and whether they were hired in social work positions within three months of graduation (the row, or dependent, variable).

Beginning with the upper-left-hand cell (graduates of accredited programs who are working as social workers), the expected frequency for this cell, using Formula 11.2, is (40 × 55)/100, or 22. For the other cell in this row (graduates of nonaccredited programs who are working as social workers), the expected frequency is (40 × 45)/100, or 18. For the two cells in the bottom row, the expected frequencies are (60 × 55)/100, or 33, and (60 × 45)/100, or 27, respectively. The expected frequencies for all four cells are displayed in Table 11.4.

Note that the row and column marginals as well as the total number of cases in Table 11.4 are exactly the same as those in Table 11.3. The row and column marginals for the expected frequencies must *always* equal those of the observed frequencies, a relationship that provides a convenient way of checking your arithmetic to this point.

The value for chi square for these data can now be found by solving Formula 11.1. It will be helpful to use a computing table, such as Table 11.5, to organize the several steps required to compute chi square. The table lists the observed frequencies (f_o) in column 1 in order from the upper-left-hand cell to the lower-right-hand cell, moving left to right across the table and top to

TABLE 11.4 Expected Frequencies for Table 11.3

	Accreditation Status		
Employment Status	Accredited	Not Accredited	Totals
Working as a social worker	22	18	40
Not working as a social worker	33	27	60
Totals	55	45	100

TABLE 11.5 Computational Table for Table 11.3

(1)	(2)	(3)	(4)	(5)
f_o	f_e	$f_o - f_e$	$(f_o - f_e)^2$	$(f_o - f_e)^2/f_e$
30	22	8	64	2.91
10	18	−8	64	3.56
25	33	−8	64	1.94
35	27	8	64	2.37
$N = 100$	$N = 100$	0		$\chi^2(\text{obtained}) = 10.78$

bottom. Column 2 lists the expected frequencies (f_e) in exactly the same order. Double-check to make sure you have listed the cell frequencies in the same order for both of these columns.

The next step is to subtract the expected frequency from the observed frequency for each cell and list these values in column 3. To complete column 4, square the value in column 3 and then, in column 5, divide the column 4 value by the expected frequency for that cell. Finally, add up column 5. The sum of this column is $\chi^2(\text{obtained})$:

$$\chi^2(\text{obtained}) = 10.78$$

ONE STEP AT A TIME **Computing Chi Square**

Begin by preparing a computing table similar to Table 11.5. List the observed frequencies (f_o) in column 1. The total for column 1 is the number of cases (N).

To Find the Expected Frequencies (f_e) Using Formula 11.2

1: Start with the upper-left-hand cell and multiply the row marginal by the column marginal for that cell.

2: Divide the quantity you found in step 1 by N. The result is the expected frequency (f_e) for that cell. Record this value in the second column of the computing table. Make sure you place the value of f_e in the same row as the observed frequency for that cell.

3: Repeat steps 1 and 2 for each cell in the table. Double-check to make sure that you are using the correct row and column marginals. Record each f_e in the second column of the computational table.

4: Find the total of the expected frequencies column. This total *must* equal the total of the observed

frequencies column (which is the same as N). If the two totals do not match (within rounding error), recompute the expected frequencies.

To Find Chi Square Using Formula 11.1

1: For each cell, subtract the expected frequency (f_e) from the observed frequency (f_o), and list these values in the third column of the computational table $(f_o - f_e)$. Find the total for this column. If this total is not zero, you have made a mistake and need to check your computations.

2: Square each of the values in the third column of the table, and record the result in the fourth column, labeled $(f_o - f_e)^2$.

3: Divide each value in column 4 by the expected frequency for that cell, and record the result in the fifth column, labeled $(f_o - f_e)^2/f_e$.

4: Find the total for the fifth column. This value is $\chi^2(\text{obtained})$.

Note that the totals for columns 1 and 2 (f_o and f_e) are exactly the same. This will always be the case, and if the totals do not match, you have made a computational error, probably in the calculation of the expected frequencies. Also note that the sum of column 3 will always be zero, another convenient way to check your math to this point.

This sample value for chi square must still be tested for its significance. *(For practice in computing chi square, see any problem at the end of this chapter, starting with the smallest tables.)*

The Chi Square Test for Independence

We are now ready to conduct the chi square test for independence. Recall that if the variables are independent, the score of a case on one variable will have no relationship with its score on the other variable. As always, the five-step model for significance testing will provide the framework for organizing our decision making. The data presented in Table 11.3 will serve as our example.

Step 1. Making Assumptions and Meeting Test Requirements. Note that, since the test is nonparametric, we make no assumptions at all about the shape of the sampling distribution.

> Model: Independent random samples
> Level of measurement is nominal

Step 2. Stating the Null Hypothesis. The null hypothesis states that the two variables are independent. If the null hypothesis is true, the differences between the observed and expected frequencies will be small. As usual, the research hypothesis directly contradicts the null hypothesis. Thus, if we reject H_0, the research hypothesis will be supported.

> H_0: The two variables are independent.
> (H_1: The two variables are dependent.)

Step 3. Selecting the Sampling Distribution and Establishing the Critical Region. The sampling distribution of sample chi squares, unlike the Z and t distributions, is positively skewed, with higher values of sample chi squares in the upper tail of the distribution (to the right). Thus, with the chi square test, the critical region is established in the upper tail of the sampling distribution.

Values for χ^2(critical) are given in Appendix C. This table is similar to the t table, with alpha levels arrayed across the top and degrees of freedom down the side. A major difference, however, is that degrees of freedom (df) for chi square are found by the following formula:

FORMULA 11.3
$$df = (r - 1)(c - 1)$$

A table with two rows and two columns (a 2×2 table) has one degree of freedom regardless of the number of cases in the sample.[1] A table with two rows and three columns would have $(2 - 1)(3 - 1)$, or 2 degrees of freedom. Our sample problem involves a 2×2 table with df $= 1$, so if we set alpha at 0.05, the critical chi square score would be 3.841. Summarizing these decisions, we have

$$\text{Sampling distribution} = \chi^2 \text{ distribution}$$
$$\text{Alpha} = 0.05$$
$$\text{Degrees of freedom} = 1$$
$$\chi^2(\text{critical}) = 3.841$$

Step 4. Computing the Test Statistic. The mechanics of these computations were introduced in the previous section. As you recall, we had

$$\chi^2(\text{obtained}) = \sum \frac{(f_o - f_e)^2}{f_e}$$

$$\chi^2(\text{obtained}) = 10.78$$

Step 5. Making a Decision and Interpreting the Results of the Test. Comparing the test statistic with the critical region,

$$\chi^2(\text{obtained}) = 10.78$$
$$\chi^2(\text{critical}) = 3.841$$

we see that the test statistic falls into the critical region and, therefore, we reject the null hypothesis of independence. The pattern of cell frequencies observed in Table 11.3 is unlikely to have occurred by chance alone. The variables are dependent. Specifically, based on these sample data, the probability of securing employment in the field of social work is dependent on the accreditation status of the program. *(For practice in conducting and interpreting the chi square test for independence, see problems 11.2 to 11.15.)*

Let us stress exactly what the chi square test does and does not tell us. A significant chi square means that the variables are (probably) dependent on each other in the population: Accreditation status makes a difference in whether or not a person is working as a social worker. What chi square does *not* tell us is the exact

---

[1]Degrees of freedom are the number of values in a distribution that are free to vary for any particular statistic. A 2×2 table has one degree of freedom because, for a given set of marginals, once one cell frequency is determined, all other cell frequencies are fixed (that is, they are no longer free to vary). In Table 11.3, for example, if any cell frequency is known, all others are determined. If the upper-left-hand cell is known to be 30, the remaining cell in that row must be 10, since there are 40 cases total in the row and $40 - 30 = 10$. Once the frequencies of the cells in the top row are established, cell frequencies for the bottom row are determined by subtraction from the column marginals. Incidentally, this relationship can be used to good advantage when computing expected frequencies. For example, in a 2×2 table, only one expected frequency needs to be computed. The f_e's for all other cells can then be found by subtraction.

Applying Statistics 11.1: The Chi Square Test

Do men and women vary in their opinions about cohabitation? A random sample of 47 males and females has been rated as high or low in their support for "living together." The results are:

Support for Cohabitation	Group		
	Males	Females	Totals
High	15	5	20
Low	10	17	27
Totals	25	22	47

The frequencies we would expect to find if the null hypothesis (H_0: The variables are independent) were true are:

Support for Cohabitation	Group		
	Males	Females	Totals
High	10.64	9.36	20.00
Low	14.36	12.64	27.00
Totals	25.00	22.00	47.00

Expected frequencies are found on a cell-by-cell basis by means of the formula

$$f_e = \frac{\text{Row marginal} \times \text{Column marginal}}{N}$$

and the calculation of chi square will be organized into a computational table.

(1)	(2)	(3)	(4)	(5)
f_o	f_e	$f_o - f_e$	$(f_o - f_e)^2$	$\frac{(f_o - f_e)^2}{f_e}$
15	10.64	4.36	19.01	1.79
5	9.36	−4.36	19.01	2.03
10	14.36	−4.36	19.01	1.32
17	12.64	4.36	19.01	1.50
$N = 47$	$N = 47.00$	0.00	χ^2(obtained) = 6.64	

$$\chi^2\text{(obtained)} = 6.64$$

Step 1. Making Assumptions and Meeting Test Requirements.

Model: Independent random samples
Level of measurement is nominal

Step 2. Stating the Null Hypothesis.

H_0: The two variables are independent.
(H_1: The two variables are dependent.)

Step 3. Selecting the Sampling Distribution and Establishing the Critical Region.

Sampling distribution = χ^2 distribution
Alpha = 0.05
Degrees of freedom = 1
χ^2(critical) = 3.841

Step 4. Computing the Test Statistic.

$$\chi^2\text{(obtained)} = \sum \frac{(f_o - f_e)^2}{f_e}$$

$$\chi^2\text{(obtained)} = 6.64$$

Step 5. Making a Decision and Interpreting the Results of the Test. With an obtained χ^2 of 6.64, we would reject the null hypothesis of independence. For this sample there is a statistically significant relationship between gender and support for cohabitation.

To complete the analysis, it would be useful to know exactly how the two variables are related. We can determine this by computing and analyzing column percents:

Support for Cohabitation	Group		
	Males	Females	Totals
High	60.00%	22.73%	42.55%
Low	40.00%	77.27%	57.45%
Totals	100.00%	100.00%	100.00%

The column percents show that 60% of males in this sample are highly supportive versus only 23% of females. We have already concluded that the relationship is significant, and now we know the pattern of the relationship: Males are more supportive.

ONE STEP AT A TIME	Computing Column Percentages

1: Start with the upper-left-hand cell. Divide the cell frequency (the number of cases in the cell) by the total number of cases in that column (or the column marginal). Multiply the result by 100 to convert to a percentage.

2: Move down one cell and repeat step 1. Continue moving down the column, cell by cell, until you have converted all cell frequencies to percentages.

3: Move to the next column. Start with the cell in the top row and repeat step 1 (making sure that you use the correct column total in the denominator of the fraction).

4: Continue moving down the second column until you have converted all cell frequencies to percentages.

5: Continue these operations, moving from column to column one at a time, until you have converted all cell frequencies to percentages.

nature of the relationship. In our example, it does not tell us which type of graduate is more likely to be working as a social worker. To make this determination, we must perform some additional calculations. We can figure out how the independent variable (accreditation status) is affecting the dependent variable (employment as a social worker) by computing "column percentages" or by calculating percentages within each column of the bivariate table. This procedure is analogous to calculating percentages for frequency distributions (see Chapter 2).

To calculate column percentages, divide each cell frequency by the total number of cases in the column (the column marginal) and multiply the result by 100. For Table 11.3, starting in the upper-left-hand cell, we see that there are 30 cases in this cell and 55 cases in the column. So 30 of the 55 graduates of accredited programs are working as social workers. The column percentage for this cell is therefore $(30/55) \times 100 = 54.55\%$. For the lower-left-hand cell, the column percent is $(25/55) \times 100 = 45.45\%$. For the two cells in the right-hand column (graduates of nonaccredited programs), the column percents are $(10/45) \times 100 = 22.22$ and $(35/45) \times 100 = 77.78$. Table 11.6 displays all column percents for Table 11.3.

Column percents make the relationship between the variables more obvious, and we can see easily from Table 11.6 that it's the students from accredited programs who are more likely to be working as social workers. Nearly 55% of these students are working as social workers versus about 22% of the students from nonaccredited programs. We already know that this relationship is significant (unlikely to be caused by random chance), and now, with the aid of

TABLE 11.6 Column Percents for Table 11.3

Employment Status	Accreditation Status		Totals
	Accredited	Not Accredited	
Working as a social worker	54.55%	22.22%	40.00%
Not working as a social worker	45.45%	77.78%	60.00%
Totals	100.00% (55)	100.00% (45)	100.00%

There seems to be a gender gap in American politics: Women tend to be more sympathetic to Democratic presidential candidates while men tend to prefer the Republican. In the 2008 election, Democratic candidate Obama captured 57% of women voters but only 52% of male voters while Republican McCain attracted 48% of the male voters and only 43% of the female voters. The relationship between gender and presidential vote is statistically significant at well below the 0.05 level. What additional information would you need to confirm or disprove the hypothesis that there is a gender gap in politics?

Presidential Choice by Gender

Candidate	Women	Men
B. Obama (Dem.)	57%	52%
J. McCain (Rep.)	43%	48%
	100%	100%

Source: U. S. Bureau of the Census. http://www.census .gov/prod/2009pubs/10statab/election.pdf.

column percents, we know how the two variables are related. According to these results, graduates from accredited programs have a decided advantage in securing social work jobs.

Let's highlight two points in summary:

1. Chi square is a test of statistical significance. It tests the null hypothesis that the variables are independent in the population. If we reject the null hypothesis, we are concluding, with a known probability of error (determined by the alpha level), that the variables are dependent on each other in the population. In the terms of our example, this means that accreditation status makes a difference in the likelihood of finding work as a social worker. By itself, however, chi square does not tell us the nature of the relationship.

2. Column percentages allow us to examine the bivariate relationship in more detail. The column percentages show us exactly how the independent variable affects the dependent variable. In this case, they reveal that graduates of accredited programs are more likely to find work as social workers. We explore column percentages more extensively when we discuss bivariate association in Chapters 12–14.

The Chi Square Test: An Additional Example

To this point, we have confined our attention to 2×2 tables, that is, tables with two rows and two columns. For purposes of illustration, we will work through the computational routines and decision-making process for a larger table. As you will see, larger tables require more computations (because they have more cells), but in all other essentials they are dealt with in the same way as the 2×2 table.

A researcher is concerned with the possible effects of marital status on the academic progress of college students. Do married students, with their extra

TABLE 11.7   **Grade-Point Average (GPA) by Marital Status for 453 College Students**

	Marital Status		
GPA	Married	Unmarried	Totals
Good	70	90	160
Average	60	110	170
Poor	45	78	123
Totals	175	278	453

burden of family responsibilities, suffer academically as compared to unmarried students? Is academic performance dependent on marital status? A random sample of 453 students is gathered, and each student is classified as either married or unmarried and—using grade-point average (GPA) as a measure—as a good, average, or poor student. Results are presented in Table 11.7.

For the top-left-hand cell (married students with good GPAs) the expected frequency would be $(160 \times 175)/453$, or 61.8. For the other cell in this row, expected frequency is $(160 \times 278)/453$, or 98.2. In similar fashion, all expected frequencies are computed (being very careful to use the correct row and column marginals) and displayed in Table 11.8.

The next step is to solve the formula for χ^2(obtained), being very careful to be certain that we are using the proper f_o's and f_e's for each cell. Once again, we will use a computational table (Table 11.9) to organize the calculations and then test the obtained chi square for its statistical significance. Remember that obtained chi square is equal to the total of column 5.

The value of the obtained chi square (2.79) can now be tested for its significance.

Step 1. Making Assumptions and Meeting Test Requirements.

Model: Independent random samples
Level of measurement is nominal

Step 2. Stating the Null Hypothesis.

H_0: The two variables are independent.
(H_1: The two variables are dependent.)

TABLE 11.8   **Expected Frequencies for Table 11.7**

	Marital Status		
GPA	Married	Unmarried	Totals
Good	61.8	98.2	160
Average	65.7	104.3	170
Poor	47.5	75.5	123
Totals	175.0	278.0	453

TABLE 11.9 Computational Table for Table 11.7

(1) f_o	(2) f_e	(3) $f_o - f_e$	(4) $(f_o - f_e)^2$	(5) $(f_o - f_e)^2/f_e$
70	61.8	8.2	67.24	1.09
90	98.2	−8.2	67.24	0.69
60	65.7	−5.7	32.49	0.49
110	104.3	5.7	32.49	0.31
45	47.5	−2.5	6.25	0.13
78	75.5	2.5	6.25	0.08
$N = 453$	$N = 453.0$	0.0		χ^2(obtained) = 2.79

Step 3. Selecting the Sampling Distribution and Establishing the Critical Region.

$$\text{Sampling distribution} = \chi^2 \text{ distribution}$$
$$\text{Alpha} = 0.05$$
$$\text{Degrees of freedom} = (r - 1)(c - 1) = (3 - 1)(2 - 1) = 2$$
$$\chi^2(\text{critical}) = 5.991$$

Step 4. Computing the Test Statistic.

$$\chi^2(\text{obtained}) = \sum \frac{(f_o - f_e)^2}{f_e}$$

$$\chi^2(\text{obtained}) = 2.79$$

Step 5. Making a Decision and Interpreting the Results of the Test.
The test statistic, χ^2(obtained) = 2.79, does not fall into the critical region, which, for alpha = 0.05, df = 2, begins at χ^2(critical) of 5.991. Therefore, we fail to reject the null hypothesis. The observed frequencies are not significantly different from the frequencies we would expect to find if the variables were independent and only random chance were operating. Based on these sample results, we can conclude that the academic performance of college students is not dependent on their marital status. Since we failed to reject the null hypothesis, we will not examine column percentages as we did for Table 11.3.

Chi Square and the Goodness-of-Fit Test[2]

In addition to the test for independence between two variables, chi square can be used to test the significance of the distribution of scores for a single variable. This application is called the **goodness-of-fit test**; it is used to compare an

---

[2]This section is optional.

actual or observed distribution of scores against a set of expected frequencies. The greater the difference between the observed and expected scores, the more likely that the observed pattern did not occur by random chance alone. If the observed and expected frequencies are similar, it is said that there is a "good fit" (hence the name of this application), and we would conclude that the two distributions were not significantly different.

A major difference in this new application is that the null hypothesis is used to determine the expected frequencies, not Formula 11.2. An example should clarify this process. Suppose you were gambling on coin tosses and suspected that a particular coin was biased toward heads. Over a series of tosses, what percentage of heads would you expect to observe from an unbiased coin? Your null hypothesis would be that the coin was unbiased and that half of all tosses should be heads and half tails. Notice what we have just done: We figured out the expected frequencies (half of the flips should be heads) from the null hypothesis (the coin is unbiased) rather than by computing them.

Let's consider another example of a situation in which the goodness-of-fit test is appropriate. Is there a seasonal rhythm to the crime rate? If we gathered crime statistics on a monthly basis for a given jurisdiction, what would we expect to find? The problem involves only one variable (crime rate), and we are concerned with how that variable is distributed over a set of categories (by month). If the crime rate does *not* vary by month, we would expect that about 1/12 of all crimes committed in a year would be committed in each and every month. Our null hypothesis would be that the crime rate does not vary across time, and our expected frequencies would be calculated by dividing the actual (observed) number of crimes by the number of months. (To conserve time and space, we will ignore the slight complexities that would be introduced by taking account of the varying number of days per month.)

In a particular jurisdiction, a total of 2172 crimes were committed last year. The actual distribution of crimes by month is displayed in Table 11.10. The expected distribution of crimes per month would be found by dividing the total number of crimes by 12: 2172/12 = 181. Note that the expected frequency

TABLE 11.10 Number of Crimes per Month

Month	Number of Crimes
January	190
February	152
March	121
April	110
May	147
June	199
July	250
August	247
September	201
October	150
November	193
December	212
	2172

will be the same for every month. With these values determined, we can calculate chi square by using Formula 11.1.

$$\chi^2(\text{obtained}) = \sum \frac{(f_o - f_e)^2}{f_e}$$

$$\chi^2(\text{obtained}) = \frac{(190 - 181)^2}{181} + \frac{(152 - 181)^2}{181} + \frac{(121 - 181)^2}{181}$$

$$+ \frac{(110 - 181)^2}{181} + \frac{(147 - 181)^2}{181} + \frac{(199 - 181)^2}{181}$$

$$+ \frac{(250 - 181)^2}{181} + \frac{(247 - 181)^2}{181} + \frac{(201 - 181)^2}{181}$$

$$+ \frac{(150 - 181)^2}{181} + \frac{(193 - 181)^2}{181} + \frac{(212 - 181)^2}{181}$$

$$\chi^2(\text{obtained}) = \frac{81}{181} + \frac{841}{181} + \frac{3600}{181}$$

$$+ \frac{5041}{181} + \frac{1156}{181} + \frac{324}{181}$$

$$+ \frac{4761}{181} + \frac{4356}{181} + \frac{400}{181}$$

$$+ \frac{961}{181} + \frac{144}{181} + \frac{961}{181}$$

$$\chi^2(\text{obtained}) = 0.45 + 4.65 + 19.89 + 27.85 + 6.39 + 1.79 + 26.30$$
$$+ 24.07 + 2.21 + 5.31 + 0.80 + 5.31$$

$$\chi^2(\text{obtained}) = 125.02$$

The χ^2 of 125.02 can now be tested for significance in the usual manner.

Step 1. Making Assumptions and Meeting Test Requirements.

Model: Random sampling
Level of measurement is nominal

In this application of significance testing, we are assuming that the observed frequencies represent a random sample of all possible frequency distributions.

Step 2. Stating the Null Hypothesis.

H_0: There is no difference in the crime rate by month.
(H_1: There is a difference in the crime rate by month)

Step 3. Selecting the Sampling Distribution and Establishing the Critical Region. In the goodness-of-fit test, degrees of freedom are equal to the number of categories minus 1, or df $= k - 1$. In the problem under

consideration, there are 12 months or categories and, therefore, 11 degrees of freedom.

$$\text{Sampling distribution} = \chi^2 \text{ distribution}$$
$$\text{Alpha} = 0.05$$
$$\text{Degrees of freedom} = k - 1 = 12 - 1 = 11$$
$$\chi^2(\text{critical}) = 19.675$$

Step 4. Computing the Test Statistic.

$$\chi^2(\text{obtained}) = 125.02$$

Step 5. Making a Decision and Interpreting the Results of the Test. The test statistic (125.02) clearly falls into the critical region (which begins at 19.675), so we reject the null hypothesis. These data suggest that, at least for this jurisdiction, crime rate does vary by month in a nonrandom fashion. *(For practice in conducting and interpreting the chi square goodness-of-fit test, see problems 11.16 and 11.17.)*

The Limitations of the Chi Square Test

Like any other test, chi square has limits, and you should be aware of several potential difficulties. First, chi square becomes difficult to interpret when the variables have many categories. For example, two variables with five categories each would generate a 5 × 5 table with 25 cells—far too many combinations of scores to be easily absorbed or understood. As a very rough rule of thumb, the chi square test is easiest to interpret and understand when both variables have four or fewer scores.

Two further limitations of the test are related to sample size. When sample size is small, we can no longer assume that the sampling distribution of all possible sample outcomes is accurately described by the chi square distribution. For chi square, a small sample is defined as one where a high percentage of the cells have expected frequencies (f_e) of 5 or less. Various rules of thumb have been developed to help the researcher decide what constitutes a "high percentage of cells." Probably the safest course is to take corrective action whenever *any* of the cells have expected frequencies of 5 or less.

For 2 × 2 tables, the value of $\chi^2(\text{obtained})$ can be adjusted by applying Yates' correction for continuity, the formula for which is

FORMULA 11.4

$$\chi_c^2 = \sum \frac{(|f_o - f_e| - 0.5)^2}{f_e}$$

where χ_c^2 = corrected chi square

$|f_o - f_e|$ = the absolute values of the difference between observed and expected frequency for each cell

The correction factor is applied by reducing the absolute value[3] of the term $(f_o - f_e)$ by 0.5 before squaring the difference and dividing by the expected frequency for the cell.

For tables larger than 2×2, there is no correction formula for computing χ^2(obtained) for small samples. It may be possible to combine some of the categories of the variables and thereby increase cell sizes. Obviously, however, this course of action should be taken only when it is sensible to do so. In other words, distinctions that have clear theoretical justifications should not be erased merely to conform to the requirements of a statistical test. When you feel that categories cannot be combined to build up cell frequencies and the percentage of cells with expected frequencies of 5 or less is small, it is probably justifiable to continue with the uncorrected chi square test as long as the results are regarded with a suitable amount of caution.

A second potential problem related to sample size occurs with large samples. I pointed out in Chapter 9 that all tests of hypothesis are sensitive to sample size. That is, the probability of rejecting the null hypothesis increases as the number of cases increases, regardless of any other factor. Chi square is especially sensitive to sample size: The value of χ^2(obtained) will increase at the same rate as sample size. If N is doubled, for example, the value of χ^2(obtained) will double. Thus, larger samples may lead to a decision to reject the null when the actual relationship is trivial. *(For an illustration of this principle, see problem 11.14.)*

You should be aware of this relationship between sample size and the value of chi square because it once again raises the distinction between statistical significance and theoretical importance. On one hand, tests of significance play a crucial role in research. When we are working with random samples, we must know if our research results could have been produced by mere random chance.

On the other hand, like any other statistical technique, tests of hypothesis are limited in the range of questions they can answer. Specifically, these tests will tell us whether our results are statistically significant or not. They will not necessarily tell us if the results are important in any other sense. To deal more directly with questions of importance, we must use an additional set of statistical techniques called *measures of association*. We previewed these techniques in this chapter when we used column percents, and measures of association are the subject of Part III of this text.

Interpreting Statistics: Family Values and Social Class

When sociologists examine patterns of family life in the United States, they commonly find that parents of different social classes raise their children in distinctly different ways. These differences include disciplinary techniques, expectations regarding success in school, and a host of other variables. In this installment of Interpreting Statistics, we examine differences in the goals

---

[3]Absolute values ignore plus and minus signs.

parents of different social classes might have for their children. Specifically, we look at an item from the 2008 General Social Science survey that asks parents: "If you had to choose, which thing on this list would you pick as the most important for a child to learn to prepare him or her for life?" Respondents were offered a list of traits from which to choose; we focus on two of these: "To obey" and "To think for himself or herself." Respondents could rank each trait from 1 ("most important") to 5 ("least important"), but, to simplify the analysis, I have collapsed the scale to three scores:

1 = Most important
2 = Second or third most important
3 = Fourth or fifth most important

Why would parents of different social classes vary in their socialization values? One possible causal factor is the nature of the work associated with each of the social classes. For people in the higher social classes, work tends to stress individual initiative and creativity and be less supervised. Lower-class and working-class occupations, in contrast, are more likely to be highly routinized, regimented, and closely supervised.

Thus, in a reflection of their work situations, we can hypothesize that parents in the higher social classes will be more likely to stress independent thinking, while parents in the lower social classes will be more likely to endorse obedience as the most important socialization goal. In the language of chi square, we can hypothesize that socialization goals will be dependent on social class.

Tables 11.11 and 11.12 show the results of two chi square tests, one for the relationship between social class and "think for self" and one for the relationship between social class and "obey." These results are based on the entire 2008 GSS, not the smaller version supplied with this text. Before analyzing the results, remember that the General Social Survey is administered to random samples of adult Americans, selected in accordance with the rules of EPSEM (see Chapter 6). Thus, we can assume that the conclusions of this analysis will apply to U.S. society in general.

Both tables have 6 degrees of freedom ($4 - 1 \times 3 - 1 = 3 \times 2 = 6$), so, at the 0.05 alpha level, the critical chi square value is 12.59. The relationship

TABLE 11.11   Socialization Goal ("Think for Self") by Social Class

Importance of "Thinking for Self"	Social Class				
	Lower Class	Working Class	Middle Class	Upper Class	Totals
Highest	38	190	199	15	442
Moderate	40	161	141	12	354
Lowest	14	116	63	5	198
Totals =	92	467	403	32	994

Chi square = 16.49

between social class and "think for self" is statistically significant and the relationship with "obey" is very close to significance at the 0.05 level. We can conclude that our hypothesis is partly supported.

Next, we should examine the column percentages to see if they conform to our hypotheses. Recall that we hypothesized that obedience would be more strongly endorsed by lower- and working-class respondents and that independent thinking would be more strongly endorsed by the middle and upper classes. Do the patterns in Tables 11.13 and 11.14 conform to these hypotheses?

TABLE 11.12 Socialization Goal ("Obey") by Social Class

| Importance of "Obey" | Social Class | | | | |
	Lower Class	Working Class	Middle Class	Upper Class	Totals
Highest	21	78	52	6	157
Moderate	26	143	106	6	281
Lowest	45	246	245	20	556
Totals =	92	467	403	32	994

Chi square = 11.57

TABLE 11.13 Socialization Goal ("Think for Self") by Social Class (percentages)

| Importance of "Thinking for Self" | Social Class | | | |
	Lower Class	Working Class	Middle Class	Upper Class
Highest	41.3%	40.7%	49.4%	46.9%
Moderate	43.5%	34.5%	35.0%	37.5%
Lowest	15.2%	24.8%	15.6%	15.6%
Totals =	100.0%	100.0%	100.0%	100.0%

TABLE 11.14 Socialization Goal ("Obey") by Social Class (percentages)

| Importance of "Obey" | Social Class | | | |
	Lower Class	Working Class	Middle Class	Upper Class
Highest	22.8%	16.7%	12.9%	18.8%
Moderate	28.3%	30.6%	26.3%	18.8%
Lowest	48.9%	52.7%	60.8%	62.5%
Totals =	100.0%	100.0%	100.0%	100.0%

Starting with the top row of Table 11.13, we can see that independent thinking had a lot of support in all classes, and there is little difference between the column percentages. Working- and lower-class respondents were the least likely to rate the goal as the highest, middle-class respondents are the most likely, and upper-class respondents are a little less supportive. Overall, this pattern is consistent with our hypothesis but the differences from class to class are not particularly striking.

Turning to Table 11.14, and again focusing on the top row, we see that a minority of the respondents from each class support obedience as a socialization goal. Consistent with our hypothesis, lower-class respondents were the most

---

Applying Statistics 11.2: Using Chi Square to Test for Cross-National Differences in Socialization Values

The World Values Survey, which is administered periodically to randomly selected samples of citizens from around the globe, allows us to test theories in a variety of settings. We have examined the relationship between socialization goals and social class for a random sample of adults in the United States. In this section, we test the significance of the relationship between support for obedience as a socialization goal and gender using randomly selected samples of residents of Canada and Great Britain. Do males place more emphasis on this trait than females? Does the relationship between these two variables change from nation to nation?

Respondents to the World Values Survey were given a list of possible socialization goals and asked to select the goals they thought were most important. Those who included obedience in their list were coded as "Yes" in the tables that follow. Both samples were tested in 2006.

And chi square is computed with the following table:

(1)	(2)	(3)	(4)	(5)
f_o	f_e	$f_o - f_e$	$(f_o - f_e)^2$	$(f_o - f_e)^2/f_e$
220	230.29	−10.29	105.86	0.46
260	249.71	10.29	105.86	0.42
278	267.71	10.29	105.86	0.40
280	290.29	−10.29	105.86	0.36
1038	1038.00	0.00		$\chi^2 = 1.64$

This is a 2×2 table, so there is one degree of freedom (df = $1 \times 1 = 1$), and the critical chi square is 3.841. The relationship is not significant, but we will still look at the pattern of the column percentages:

Great Britain (frequencies)

Obedience Mentioned?	Gender		Totals
	Male	Female	
Yes	220	260	480
No	278	280	558
Totals	498	540	1,038

Great Britain (percentages)

Obedience Mentioned?	Gender		Totals
	Male	Female	
Yes	44.2%	48.1%	46.2%
No	55.8%	51.9%	53.8%
Totals	100.0%	100.0%	100.0%

(continued)

Applying Statistics 11.2: (continued)

Not only is the relationship not significant, but there is very little difference between British males and females in their support for obedience.

Are the variables related in the Canadian sample?

Canada (frequencies)

Obedience Mentioned?	Gender		Totals
	Male	Female	
Yes	304	340	644
No	729	781	1510
Totals	1033	1121	2154

And we will use a computing table to get the value of chi square:

(1)	(2)	(3)	(4)	(5)
f_o	f_e	$f_o - f_e$	$(f_o - f_e)^2$	$(f_o - f_e)^2/ f_e$
304	308.85	−4.85	23.52	0.08
340	335.15	4.85	23.52	0.07
729	724.15	4.85	23.52	0.03
781	785.85	−4.85	23.52	0.03
2154	2154.00	0.00		$\chi^2 = 0.21$

With one degree of freedom, the critical chi square is 3.841. As was the case for the British sample, there is no significant relationship between gender and support for this socialization goal in the Canadian sample.

Looking at the column percentages, there is virtually no difference between males and females in their support for obedience:

Canada (percentages)

Obedience Mentioned?	Gender		Totals
	Male	Female	
Yes	29.4%	30.3%	29.9%
No	70.6%	69.7%	70.1%
Totals	100.0%	100.0%	100.0%

In summary, there is no relationship between gender and support for obedience as a socialization goal in either nation. Note, however, that there is a difference in support for obedience between the two nations. About 46% of the British mentioned obedience but only 30% of the Canadians did likewise.

likely to rate obedience highly (but remember that the relationship between the variables is not significant).

What can we conclude? These results show only partial support for our hypothesis. Only one of the relationships is statistically significant and our expectations about the pattern of column percentages were only partly confirmed. Perhaps the problem with our hypothesis is that it is more appropriate for an industrial society, in which a large portion of the workforce is engaged in manual labor, blue-collar, manufacturing jobs. The United States is now a postindustrial society, in which good jobs increasingly demand high levels of education and well-developed critical thinking abilities. If people of all class backgrounds have come to recognize this economic reality, we would expect to find less difference in socialization goals by social class than in the past.

READING STATISTICS 7: A Profile of Children Who Attack Parents

Researchers Walsh and Krienert used FBI records to build a statistical portrait of juveniles who attack their parents. They point out that there is not much research on this topic and that little is known about the dynamics and characteristics of child–parent violence. They focus on gender differences, and some of their findings for aggravated parental assault are presented in Table 11.15. To conserve space, only column percentages—not frequencies—are presented. Significant relationships are identified with *($p < 0.05$) or **($p < 0.001$).

The authors conclude that female offenders are significantly younger than males and significantly more likely to attack their mothers or stepmothers. There was no significant difference in the age of the victim but female offenders were significantly more likely than male offenders to use a knife in the assault.

The authors note that this is only a first step in developing a research knowledge base on this crime. Want to learn more? The citation is given below.

Walsh, Jeffrey, and Krienert, Jessie. 2007. "Child-Parent Violence: An Empirical Analysis of Offender, Victim, and Event Characteristics in a National Sample of Reported Incidents." *Journal of Family Violence*, 22:563–574.

TABLE 11.15   Aggravated Parental Assault by Gender (percentages)

Variable:	Female Offenders	Male Offenders
OFFENDER AGE:**		
13 and under	18.4%	15.4%
14–17	61.9%	53.7%
18–21	19.6%	30.9%
VICTIM SEX:**		
Male	21.2%	45.4%
Female	73.1%	54.6%
VICTIM AGE:		
30 and under	2.8%	2.2%
31–40	41.1%	37.4%
41–50	42.8%	44.7%
51 and older	13.3%	15.6%
KNIFE USED?**		
No	57.7%	65.1%
Yes	42.3%	34.9%

SUMMARY

1. The chi square test for independence is appropriate for situations in which the variables of interest have been organized into table format. The null hypothesis is that the variables are independent, or that the classification of a case into a particular category on one variable has no effect on the probability that the case will be classified into any particular category of the second variable.

2. Since chi square is nonparametric and requires only nominally measured variables, its model assumptions are easily satisfied. Furthermore, since it is computed from bivariate tables, in which the number of rows and columns can be easily expanded, the chi square test can be used in many situations in which other tests are inapplicable.

3. In the chi square test, we first find the frequencies that would appear in the cells if the variables were independent (f_e) and then compare those frequencies, cell by cell, with the frequencies actually observed in the cells (f_o). If the null is true, expected and observed frequencies should be quite close in value. The greater the difference between the observed and expected frequencies, the greater the possibility of rejecting the null.

4. The chi square test has several important limitations. It is often difficult to interpret when tables have many (more than four or five) dimensions. Also, as sample size (N) decreases, the chi square test becomes less trustworthy, and corrective action may be required. Finally, with very large samples, we may declare relatively trivial relationships to be statistically significant. As is the case with all tests of hypothesis, statistical significance is not the same thing as "importance" in any other sense. As a general rule, statistical significance is a necessary but not sufficient condition for theoretical or practical importance.

SUMMARY OF FORMULAS

Chi square (obtained):

11.1 $\chi^2(\text{obtained}) = \sum \dfrac{(f_o - f_e)^2}{f_e}$

Expected frequencies:

11.2 $f_e = \dfrac{\text{Row marginal} \times \text{Column marginal}}{N}$

Degrees of freedom, bivariate tables:

11.3 $\text{df} = (r - 1)(c - 1)$

Yates' correction for continuity:

11.4 $\chi_c^2 = \sum \dfrac{(|f_o - f_e| - 0.5)^2}{f_e}$

GLOSSARY

Bivariate table. A table that displays the joint frequency distributions of two variables.

Cells. The cross-classification categories of the variables in a bivariate table.

χ^2(critical). The score on the sampling distribution of all possible sample chi squares that marks the beginning of the critical region.

χ^2(obtained). The test statistic as computed from sample results.

Chi square (χ^2) test. A nonparametric test of hypothesis for variables that have been organized into a bivariate table.

Column. The vertical dimension of a bivariate table. By convention, each column represents a score on the independent variable.

Expected frequency (f_e). The cell frequencies that would be expected in a bivariate table if the variables were independent.

Goodness-of-fit test. An additional use for chi square that tests the significance of the distribution of a single variable.

Independence. The null hypothesis in the chi square test. Two variables are independent if, for all cases, the classification of a case on one variable has no effect on the probability that the case will be classified in any particular category of the second variable.

Marginals. The row and column subtotals in a bivariate table.

Nonparametric. A "distribution-free" test. These tests do not assume a normal sampling distribution.

Observed frequency (f_o). The cell frequencies actually observed in a bivariate table.

Row. The horizontal dimension of a bivariate table, conventionally representing a score on the dependent variable.

PROBLEMS

11.1 For each of the following tables, calculate the obtained chi square. (*HINT: Calculate the expected frequencies for each cell with Formula 11.2. Double-check to make sure you are using the correct row and column marginals for each cell. It may be helpful to record the expected frequencies in table format as well—see Tables 11.2, 11.4, and 11.7. Next, use a computational table to organize the calculation for Formula 11.1—see Tables 11.5 and 11.8. For each cell, subtract expected frequency from observed frequency and record the result in column 3. Square the value in column 3 and record the result in column 4, and then divide the value in column 4 by the expected frequency for that cell and record the result in column 5. Remember that the sum of column 5 in the computational table is the*) *obtained chi square. As you proceed, double-check to make sure you are using the correct values for each cell.*)

a.

20	25	45
25	20	45
45	45	90

c.

25	15	40
30	30	60
55	45	100

b.

10	15	25
20	30	50
30	45	75

d.

20	45	65
15	20	35
35	65	100

11.2 PS Many analysts have noted a "gender gap" in elections for the U.S. presidency, with women more likely to vote for the Democratic candidate. A sample of university faculty has been asked about their political party preference. Do their responses indicate a significant relationship between gender and party preference?

| | Gender | | |
Party Preference	Male	Female	Totals
Democrat	10	15	25
Republican	15	10	25
Totals	25	25	50

a. Is there a statistically significant relationship between gender and party preference?
b. Compute column percents for the table to determine the pattern of the relationship. Which gender is more likely to prefer the Democrats?

11.3 |SW| A local politician is concerned that a program for the homeless in her city is discriminating against blacks and other minorities. The following data were taken from a random sample of black and white homeless people.

| | Race | | |
Received Services?	Black	White	Totals
Yes	6	7	13
No	4	9	13
Totals	10	16	26

a. Is there a statistically significant relationship between race and whether or not the person has received services from the program?
b. Compute column percents for the table to determine the pattern of the relationship. Which group was more likely to get services?

11.4 |SOC| A sample of 25 cities have been classified as high or low on their homicide rates and on the number of handguns sold within the city limits. Is there a relationship between these two variables? Do cities with higher homicide rates have significantly higher gun sales? Explain your results in a sentence or two.

| | Homicide Rate | | |
Volume of Gun Sales	Low	High	Totals
High	8	5	13
Low	4	8	12
Totals	12	13	25

11.5 |PA| Is there a relationship between salary levels and unionization for public employees? The following data represent this relationship for fire departments in a random sample of 100 cities of roughly the same size. Salary data have been dichotomized at the median. Summarize your findings.

| | Status | | |
Salary	Union	Nonunion	Totals
High	21	29	50
Low	14	36	50
Totals	35	65	100

a. Is there a statistically significant relationship between these variables?
b. Compute column percents for the table to determine the pattern of the relationship. Which group was more likely to get high salaries?

11.6 |CJ| A local judge has been allowing some individuals convicted of "driving under the influence" to work in a hospital emergency room as an alternative to fines, suspensions, and other penalties. A random sample of offenders has been drawn. Do participants in this program have lower rates of recidivism for this offense?

| | Status | | |
Recidivist?	Participants	Nonparticipants	Totals
Yes	60	123	183
No	55	108	163
Totals	115	231	346

a. Is there a statistically significant relationship between these variables?
b. Compute column percents for the table to determine the pattern of the relationship. Which group is more likely to be rearrested for driving under the influence?

11.7 |SOC| The state Department of Education has rated a sample of local school systems for compliance with state-mandated guidelines for quality. Is the quality of a school system significantly related to the affluence of the community as measured by per capita income?

| | Per Capita Income | | |
Quality	Low	High	Totals
Low	16	8	24
High	9	17	26
Totals	25	25	50

a. Is there a statistically significant relationship between these variables?

b. Compute column percents for the table to determine the pattern of the relationship. Are high- or low-income communities more likely to have high-quality schools?

11.8 SOC A program of pet therapy has been running at a local nursing home. Are the participants in the program more alert and responsive than nonparticipants? The results, drawn from a random sample of residents, are reported here.

	Status		
Alertness	Participants	Non-participants	Totals
High	23	15	38
Low	11	18	29
Totals	34	33	67

a. Is there a statistically significant relationship between participation and alertness?

b. ompute column percents for the table to determine the pattern of the relationship. Which group was more likely to be alert?

11.9 SOC Is there a relationship between length of marriage and satisfaction with marriage? The necessary information has been collected from a random sample of 100 respondents drawn from a local community. Write a sentence or two explaining your decision.

	Length of Marriage (in years)			
Satisfaction	Less Than 5	5–10	More Than 10	Totals
Low	10	20	20	50
High	20	20	10	50
Totals	30	40	30	100

a. Is there a statistically significant relationship between these variables?

b. Compute column percents for the table to determine the pattern of the relationship. Which group is more likely to be highly satisfied?

11.10 PS Is there a relationship between college class standing and political orientation? Are juniors and seniors significantly different from freshmen and sophomores on this variable? The following table reports the relationship between these two variables for a random sample of 267 college students.

	College Class Standing		
Political	Freshmen and Orientation	Juniors and Seniors Sophomores	Totals
Liberal	40	43	83
Moderate	50	50	100
Conservative	44	40	84
Totals	134	133	267

a. Is there a statistically significant relationship between these variables?

b. Compute column percents for the table to determine the pattern of the relationship. Which group is more likely to be conservative?

11.11 SOC At a large urban college, about half of the students live off campus in various arrangements, and the other half live in dormitories on campus. Is academic performance dependent on living arrangements? The results based on a random sample of 300 students are presented here.

	Residential Status			
GPA	Off Campus with Roommates	Off Campus with Parent	On Campus	Totals
Low	22	20	48	90
Moderate	36	40	54	130
High	32	10	38	80
Totals	90	70	140	300

a. Is there a statistically significant relationship between these variables?

b. Compute column percents for the table to determine the pattern of the relationship. Which group is more likely to have a high GPA?

11.12 SOC An urban planning commissioner has built up a database describing a sample of the neighborhoods in her city and has developed a scale by which each area can be rated for the "quality of life" (this includes measures of pollution, noise, open space, services available, and so on). She has also asked samples of residents of these areas about their level of satisfaction with their neighborhoods. Is there significant agreement between the commissioner's objective ratings of quality and the respondents' self-reports of satisfaction?

Satisfaction	Quality of Life			
	Low	Moderate	High	Totals
Low	21	15	6	42
Moderate	12	25	21	58
High	8	17	32	57
Totals	41	57	59	157

a. Is there a statistically significant relationship between these variables?

b. Compute column percents for the table to determine the pattern of the relationship? Which group is most likely to say that their satisfaction is high?

11.13 SOC Does support for the legalization of marijuana vary by region of the country? The table displays the relationship between the two variables for a random sample of 1020 adult citizens. Is the relationship significant?

Legalize?	Region				
	North	Midwest	South	West	Totals
Yes	60	65	42	78	245
No	245	200	180	150	775
Totals	305	265	222	228	1020

a. Is there a statistically significant relationship between these variables?

b. Compute column percents for the table to determine the pattern of the relationship. Which region is most likely to favor the legalization of marijuana?

11.14 SOC A researcher is concerned with the relationship between attitudes toward violence and violent behavior. If attitudes "cause" behavior (a very debatable proposition), then people who have positive attitudes toward violence should have high rates of violent behavior. A pretest was conducted on 70 respondents; among other things, the respondents were asked, "Have you been involved in a violent incident of any kind over the past six months?" The researcher established the following relationship:

Involvement	Attitude Toward Violence		
	Favorable	Unfavorable	Totals
Yes	16	19	35
No	14	21	35
Totals	30	40	70

The chi square calculated on these data is 0.23, which is not significant at the 0.05 level (confirm this conclusion with your own calculations). Undeterred by this result, the researcher proceeded with the project and gathered a random sample of 7000. In terms of percentage distributions, the results for the full sample were exactly the same as for the pretest:

Involvement	Attitude Toward Violence		
	Favorable	Unfavorable	Totals
Yes	1600	1900	3500
No	1400	2100	3500
Totals	3000	4000	7000

However, the chi square obtained is a very healthy 23.4 (confirm with your own calculations). Why is the full-sample chi square significant when the pretest was not? What happened? Do you think that the second result is important?

11.15 SOC Some results from a survey administered to a nationally representative sample are presented here. For each table, conduct the chi square test of significance and compute column percents. Write a sentence or two of interpretation for each test.

a. Support for the legal right to an abortion by age:

Support?	Age			
	Younger than 30	30–49	50 and Older	Totals
Yes	154	360	213	727
No	179	441	429	1049
Totals	333	801	642	1776

b. Support for the death penalty by age:

Support?	Age			
	Younger than 30	30–49	50 and Older	Totals
Favor	361	867	675	1903
Oppose	144	297	252	693
Totals	505	1164	927	2596

c. Fear of walking alone at night by age:

Fear?	Age			
	Younger than 30	30–49	50 and Older	Totals
Yes	147	325	300	772
No	202	507	368	1077
Totals	349	832	668	1849

d. Support for legalizing marijuana by age:

	Age			
Legalize?	Younger than 30	30–49	50 and Older	Totals
Should	128	254	142	524
Should not	224	534	504	1262
Totals	352	788	646	1786

e. Support for suicide when a person has an incurable disease by age:

	Age			
Support?	Younger than 30	30–49	50 and Older	Totals
Yes	225	537	367	1129
No	107	270	266	643
Totals	332	807	633	1772

11.16 SOC (*This problem is optional.*) The director of athletics at the local high school wonders if the sports program is getting a proportional amount of support from each of the four classes. If there are roughly equal numbers of students in each of the classes, what does the following breakdown of attendance figures from a random sample of students in attendance at a recent basketball game suggest?

Class	Frequencies
Freshman	200
Sophomore	150
Junior	120
Senior	110
Totals	580

11.17 SOC (*This problem is optional.*) A small western town has roughly equal numbers of Hispanic, Asian, and Anglo-American residents. Are the three groups equally represented at town meetings? The attendance figures for a random sample drawn from those attending a meeting were

Group	Frequencies
Hispanic	74
Asian	55
Anglo	53
Totals	182

Is there a statistically significant pattern here?

YOU ARE THE RESEARCHER

Understanding Political Beliefs

Two projects are presented below, and I urge you to apply your understanding of the chi square test to both. In the first, you will examine the sources of people's beliefs about some of the most hotly debated topics in U.S. society: capital punishment, assisted suicide, gay marriage, and immigration. In the second, you will compare various independent variables to see which has the most significant relationship with your chosen dependent variable.

Using Crosstabs

In these projects you will use a new procedure called **Crosstabs** to produce bivariate tables, chi square, and column percentages. This procedure is very commonly used in social science research at all levels, and you will see many references to **Crosstabs** in chapters to come.

To begin the procedure, click **Analyze**, then click **Descriptive Statistics** and **Crosstabs**. The **Crosstabs** dialog box appears with the variables listed in a box on the left. Highlight the name(s) of your dependent variable(s) and click the arrow to move the variable name into the **Rows** box. Next, find the name of your independent variable(s) and move it into the **Columns** box. SPSS will process all combinations of variables in the row and column boxes at one time.

Click the **Statistics** button at the bottom of the window and click the box next to Chi-square. Return to the **Crosstabs** window and click the **Cells** button and select "column" in the **Percentages** box. This will generate column percents for the table. Return to the **Crosstabs** window and click **Continue** and **OK** to produce your output.

Crosstabs Demonstration

I will demonstrate this command by examining the relationship between gender (*sex*) and support for legal abortion "for any reason" (*abany*). Gender is my independent variable and I listed it in the column box and I placed *abany* in the row box. Here is the output:

ABORTION IF WOMAN WANTS FOR ANY REASON* RESPONDENTS SEX Crosstabulation

			RESPONDENTS SEX		Total
			MALE	FEMALE	
ABORTION IF WOMAN WANTS FOR ANY REASON	YES	Count	191	206	397
		% within RESPONDENTS SEX	44.8%	39.4%	41.8%
	NO	Count	235	317	552
		% within RESPONDENTS SEX	55.2%	60.6%	58.2%
Total		Count	426	523	949
		% within RESPONDENTS SEX	100.0%	100.0%	100.0%

Chi-Square Tests

	Value	df	Asymp. Sig. (2-sided)	Exact Sig. (2-sided)	Exact Sig. (1-sided)
Pearson Chi-Square	2.863[b]	1	.091		
Continuity Correction[a]	2.644	1	.104		
Likelihood Ratio	2.861	1	.091		
Fisher's Exact Test				.098	.052
Linear-by-Linear Association	2.860	1	.091		
N of Valid Cases	949				

[a]Computed only for a 2 × 2 table.
[b]0 cells (.0%) have expected count less than 5. The minimum expected count is 178.21.

Read the crosstab table cell by cell. Each cell displays the number of cases in the cell and the column percent for that cell. For example, starting with the upper-left-hand cell, there were 191 respondents who were male and who said "yes" to *abany*, and these were 44.8% of all men in the sample. In contrast, 206 of the females (39.4%) also supported legal abortion "for any reason." The column percentages are quite similar, which would indicate that *sex* and *abany* will not have a significant relationship.

The results of the chi square test are reported in the output block that follows the table. The value of chi square (obtained) is 2.863 and there was 1 degree of freedom. The exact significance of the chi square, reported in the column labeled "Asymp. Sig (2-sided)," is .091. This is above the standard indicator of a significant result (alpha = 0.05), so we may conclude, as we saw with the column percentages, that there is no statistically significant relationship between these variables. Support for legal abortion is not dependent on gender.

PROJECT 1: Explaining Beliefs

In this project, you will analyze beliefs about capital punishment (*cappun*), assisted suicide (*suicide1*), gay marriage (*marhomo*), and immigration (*letin1*). You will select an independent variable, use SPSS to generate chi squares and column percentages, and analyze and interpret your results.

Step 1: Choose an Independent Variable

Select an independent variable that seems likely to be an important cause of people's attitudes about these issues. *Be sure to select an independent variable that has only 2–5 categories!* If you select an independent variable with more than 5 scores, use the recode command to reduce the number of categories. You might consider gender, level of education (use *degree*), religion, or age (the recoded version—see Chapter 10) as possible independent variables, but there are many others. Record the variable name and state exactly what the variable measures in the box below.

SPSS Name	What exactly does this variable measure?

Step 2: Stating Hypotheses

State hypotheses about the relationships you expect to find between your independent variable and each of the four dependent variables. State these hypotheses in terms of which category of the independent variable you expect to be associated with which category of the dependent variable (for example, "I expect that men will be more supportive of the legal right to an abortion for any reason").

Step 3: Running Crosstabs

Click **Analyze** → **Descriptives** → **Crosstabs** and place the four dependent variables (*cappun, suicide1, letin1,* and *marhomo*) in the **Rows:** box and the independent variable you selected in the **Columns:** box. Click the **Statistics** button to get chi square and the **Cells** button for column percentages.

Step 4: Recording Results

Your output will consist of four tables and it will be helpful to summarize your results in the following table. Remember that the significance of the relationship is found in the column labeled "Asymp. Sig (2-sided)" of the second box in the output.

Dependent Variable	Chi square	Degrees of Freedom	Significance
cappun			
*letdie*1			
*letin*1			
marhomo			

Step 5: Analyzing and Interpreting Results

Write a short summary of results of each test in which you:

1. Identify the variables being tested, the value and significance of chi square, *N*, and the pattern (if any) of the column percentages. In the professional research literature, you might find the results reported as: "For a sample of 949 respondents, there was no significant relationship between gender and support for abortion (chi square = 2.863, df = 1, $p > 0.05$). About 45% of the men supported the legal right to an abortion 'for any reason' vs. about 39% of the women."
2. Explain if your hypotheses were supported.

PROJECT 2 Exploring the Impact of Various Independent Variables

In this project, you will examine the relative ability of a variety of independent variables to explain or account for a single dependent variable. You will again use the **Crosstabs** procedure in SPSS to generate chi squares and column percentages and use the value of alpha to judge which independent variable has the most important relationship with your dependent variable.

Step 1: Choosing Variables

Select a dependent variable. You may use any of the four from Project 1 or select a new dependent variable from the 2008 GSS. Be sure that your dependent variable has no more than five values or scores. Use the recode command as necessary to reduce the number of categories. Good choices for dependent variables include any measure of attitudes or opinions. *Do not* select characteristics like race, sex, or religion as dependent variables.

Select three independent variables that seem likely to be important causes of the dependent variable you selected. Your independent variable should have no more than five or six categories. You might consider gender, level of education, religiosity, or age (the recoded version—see Chapter 10) as possibilities, but there are many others.

Record the variable names and state exactly what each variable measures in the table below.

SPSS Name	What exactly does this variable measure?
Dependent variable	
Independent variables	

Step 2: Stating Hypotheses

State hypotheses about the relationships you expect to find between your independent variables and the dependent variable. State these hypotheses in terms of which category of the independent variable you expect to be associated with which category of the dependent variable (for example, "I expect that men will be more supportive of the legal right to an abortion for any reason").

Step 3: Running Crosstabs

Click **Analyze → Descriptives → Crosstabs** and place your dependent variable in the **Rows:** box and all three of your independent variables in the **Columns:** box. Click the **Statistics** button to get chi square and the **Cells** button for column percentages.

Step 4: Recording Results

Your output will consist of three tables, and it will be helpful to summarize your results in the following table. Remember that the significance of the relationship is found in the column labeled "Asymp. Sig (2-sided)" of the second box in the output.

Independent Variables	Chi square	Degrees of Freedom	Significance

Step 5: Analyzing and Interpreting Results

Write a short summary of results of each test in which you:

1. Identify the variables being tested, the value and significance of chi square, N, and the pattern (if any) of the column percentages.
2. Explain whether your hypotheses were supported.
3. Explain which independent variable had the most significant relationship (lowest value in the "Asymp. Sig 2-tailed" column) with your dependent variable.

PART II CUMULATIVE EXERCISES

1. Conduct the appropriate test of significance for each research situation. Problems are stated in no particular order and include research situations from each chapter in Part II.

a. Is there a gender gap in use of the Internet? Random samples of men and women have been questioned about the average number of minutes they spend each week on the Internet for any purpose. Is the difference significant?

Women	Men
$\overline{X}_1 = 55$	$\overline{X}_2 = 60$
$s_1 = 2.5$	$s_2 = 2.0$
$N_1 = 520$	$N_2 = 515$

b. For high school students, is there a relationship between social class and involvement in activities such as clubs and sports? Data have been gathered for a random sample of students. Is the relationship significant?

	Social Class		
Involvement	Middle	Lower	Totals
High	11	19	30
Moderate	19	21	40
Low	8	12	20
Totals	38	52	90

c. The General Social Survey asks respondents how many total sex partners they have had over their lifetimes. For a subsample of 23 respondents, does the number vary significantly by educational level?

Less Than High School	High School	At Least Some College
1	1	2
1	2	1
2	3	5
1	8	9
9	10	11
2	5	2
9	4	1
	3	1

d. A random sample of the U.S. population was asked how many times they had moved since they were 18 years of age. The results are presented here. On the average, how many times do adult Americans move?

$$\overline{X} = 3.5$$
$$s = 0.5$$
$$N = 1450$$

e. On the average, school districts in a state receive budget support from the state government of $1623 per student. A random sample of 107 rural schools reports they received an average of $1605 per pupil. Is the difference significant?

$$\mu = 1623 \qquad \overline{X} = 1605$$
$$s = 701$$
$$N = 107$$

2. Following are a number of research questions that can be answered by the techniques presented in Chapters 7–11. For each question, select the most appropriate test, compute the necessary statistics, and state your conclusions.

In order to complete some problems, you must first calculate the sample statistics (for example, means or proportions) that are used in the test of hypothesis. Use alpha = 0.05 throughout. The questions are presented in random order. There is at least one problem for each chapter in this Part, but not for every statistical procedure covered in the chapters.

In selecting tests and procedures, you need to consider the question, the number of samples or categories being compared, and the level of measurement of the variables. The database for this problem is based on the General Social Survey (GSS). The actual questions asked and the complete response codes are presented in Appendix G.

Abbreviated codes, along with SPSS variable names, are listed under the "Survey Items" head. The scores of some variables have been simplified or collapsed for this exercise. These problems are based on a small sample, and you may have to violate some assumptions about sample size in order to complete this exercise.

a. Is there a statistically significant difference in average hours of TV watching by income level? By race?

b. Is there a statistically significant relationship between age and happiness?

c. Estimate the average number of hours spent watching TV for the entire population.

d. If Americans currently average 2.3 children, is this sample representative of the population?

e. Are the educational levels of Catholics and Protestants significantly different?

f. Does average hours of TV watching vary by level of happiness?

g. Based on the sample data, estimate the proportion of black Americans in the population.

SURVEY ITEMS

1. How many children have you ever had (*childs*)? Scores are actual numbers.

2. Respondent's educational level (*degree*)
   0. Less than HS
   1. HS
   2. At least some college

3. Race (*race*)
   1. White
   2. Black

4. Age (*age*)
   1. Younger than 35
   2. 35 and older

5. Number of hours of TV watched per day (*tvhours*). Values are actual number of hours.

6. What is your religious preference? (*relig*)
   1. Protestant
   2. Catholic

7. Respondent's income (*income06*)
   1. $24,999 or less
   2. $25,000 or more

8. Respondent's overall level of happiness (*happy*)
   1. Very happy
   2. Pretty happy
   3. Not too happy

Case	No. of Children	Educational Level	Race	Age	TV Hours	Religious Preference	Income	Happiness
1	3	1	1	1	3	1	1	2
2	2	0	1	1	1	1	2	3
3	4	2	1	2	3	1	1	1
4	0	2	1	1	2	1	1	1
5	5	1	1	1	2	1	2	2
6	1	1	1	2	3	1	2	1
7	9	0	1	1	6	1	1	1
8	6	1	1	2	4	1	1	2
9	4	2	1	1	2	2	2	2
10	2	1	1	1	1	1	2	3
11	2	0	1	2	4	1	1	3
12	4	1	2	1	5	2	1	2
13	0	1	1	2	2	2	1	2
14	2	1	1	2	2	1	1	2
15	3	1	2	2	4	1	1	1
16	2	0	1	2	2	1	2	1
17	2	1	1	2	2	1	2	3
18	0	2	1	2	2	1	2	1
19	3	0	1	2	5	2	1	1
20	2	1	2	1	10	1	1	3
21	2	1	1	2	4	1	2	1
22	1	0	1	2	5	1	2	1
23	0	2	1	1	2	2	1	1
24	0	1	1	2	0	2	2	1
25	2	2	1	1	1	2	2	1

Part III

Bivariate Measures of Association

This section covers the computation and interpretation of a class of statistics known as *measures of association*. These statistics are extremely useful in scientific research and are commonly reported in the professional literature. They provide, in a single number, an indication of the strength and—if appropriate—the direction of a bivariate association.

It is important to remember the difference between statistical significance, covered in Part II, and association, the topic of this part. Tests for statistical significance answer certain questions: Were the differences or relationships observed in the sample caused by mere random chance? What is the probability that the sample results reflect patterns in the population(s) from which the sample(s) were selected? Measures of association address a different set of questions: How strong is the relationship between the variables? What is the direction or pattern of the relationship?

Thus, measures of association provide information that complements that supplied by tests of significance. Association and significance are two different things. It is most satisfying to find results those that are *both* statistically significant and strong, but it is common to find mixed or ambiguous results: relationships that are statistically significant but weak, not statistically significant but strong, and so forth.

Chapter 12 introduces the basic ideas behind the analysis of association in terms of bivariate tables and column percentages and presents measures of association for use with nominal-level variables, Chapter 13 covers measures for ordinal-level variables, and Chapter 14 presents Pearson's *r*, the most important measure of association and the only one designed for interval-ratio–level variables.

12

Association Between Variables Measured at the Nominal Level

LEARNING OBJECTIVES   By the end of this chapter, you will be able to

1. Explain how we can use measures of association to describe and analyze the importance of relationships (vs. their statistical significance).
2. Define *association* in the context of bivariate tables and in terms of changing conditional distributions.
3. List and explain the three characteristics of a bivariate relationship: existence, strength, and pattern or direction.
4. Investigate a bivariate association by properly calculating percentages for a bivariate table and interpreting the results.
5. Compute and interpret measures of association for variables measured at the nominal level.

As we have seen over the past several chapters, tests of statistical significance are extremely important in social science research. When we work with random samples rather than populations, these tests are indispensable for dealing with the possibility that our research results are the products of mere random chance. However, tests of significance are often merely the first step in the analysis of research results. These tests do have limitations, and statistical significance is not necessarily the same thing as relevance or importance. Furthermore, all tests of significance are affected by sample size: Tests performed on large samples may result in decisions to reject the null hypothesis when, in fact, the observed differences are quite minor.

Beginning with this chapter, we will be working with **measures of association**. Whereas tests of significance detect nonrandom relationships, measures of association provide information about the strength and direction of relationships, information that is highly relevant for assessing the importance of relationships and testing the power and validity of our theories. The theories that guide scientific research are almost always stated in cause-and-effect terms (for example: "Variable *X* causes variable *Y*"). As an example, recall our discussion of the contact hypothesis in Chapter 1. In that theory, the causal (or independent) variable was equal status contacts between groups, and the effect (or dependent) variable was prejudice. The theory asserts that involvement in equal-status-contact situations *causes* prejudice to decline. Measures of association help us trace causal relationships among variables, and they are our most important and powerful statistical tools for documenting, measuring, and analyzing cause-and-effect relationships.

As useful as they are, measures of association, like any class of statistics, do have their limitations. Most importantly, these statistics cannot *prove* that two variables are causally related. Even if there is a strong (and significant) statistical association between two variables, we cannot necessarily conclude that one variable is a cause of the other. We will explore causation in more detail in Part IV, but for now you should keep in mind that causation and association are two different things. We can use a statistical association between variables as evidence for a causal relationship, but association by itself is not proof that a causal relationship exists.

Another important use for measures of association is prediction. If two variables are associated, we can predict the score of a case on one variable from the score of that case on the other variable. For example, if equal status contacts and prejudice are associated, we can predict that people who have experienced many such contacts will be less prejudiced than those who have had few or no contacts. Note that prediction and causation can be two separate matters. If variables are associated, we can predict from one to the other even if the variables are not causally related.

Bivariate Association: Introduction and Basic Concepts

In this chapter you will be introduced to the concept of **association** between variables in the context of bivariate tables. You will also learn how to use percentages and several different measures of association to analyze a bivariate relationship. By the end of this chapter, you will have an array of statistical tools you can use to analyze the strength and pattern of an association between two variables.

ASSOCIATION AND BIVARIATE TABLES

Most generally, two variables are said to be associated if the distribution of one of them changes under the various categories or scores of the other. For example, suppose that an industrial sociologist was concerned with the relationship between job satisfaction and productivity for assembly-line workers. If these two variables are associated, then scores on productivity will change under the different conditions of satisfaction. Highly satisfied workers will have different scores on productivity than workers who are low on satisfaction, and levels of productivity will vary by levels of satisfaction.

We can clarify this relationship with a bivariate table. As you recall (see Chapter 11), bivariate tables display the scores of cases on two different variables. By convention, the **independent**, or *X*, **variable** (that is, the variable taken as causal) is arrayed in the columns, and the **dependent**, or *Y*, **variable** in the rows.[1] That is, each column of the table (the vertical dimension) represents a score or category of the independent variable (*X*), and each row (the horizontal dimension) represents a score or category of the dependent variable (*Y*).

Table 12.1 displays a relationship between productivity and job satisfaction for a fictitious sample of 173 factory workers. We focus on the columns to

---

[1] In the material that follows, we will often, for the sake of brevity, refer to the independent variable as *X* and the dependent variable as *Y*.

TABLE 12.1 Productivity by Job Satisfaction (frequencies)

Productivity (Y)	Job Satisfaction (X)			Totals
	Low	Moderate	High	
Low	30	21	7	58
Moderate	20	25	18	63
High	10	15	27	52
Totals	60	61	52	173

detect the presence of an association between variables displayed in table format. Each column shows the pattern of scores on the dependent variable for each score on the independent variable. For example, the left-hand column indicates that 30 of the 60 workers who were low on job satisfaction were low on productivity, 20 were moderately productive, and 10 were high on productivity. The middle column shows that 21 of the 61 moderately satisfied workers were low on productivity, 25 were moderately productive, and 15 were high on productivity. Of the 52 workers who are highly satisfied (the right-hand column), 7 were low on productivity, 18 were moderate, and 27 were high.

As we read the table from column to column we can observe the effects of the independent variable on the dependent variable (provided, of course, that the table is constructed with the independent variable in the columns). These "within-column" frequency distributions are called the **conditional distributions of Y**, since they display the distribution of scores on the dependent variable for each condition (or score) of the independent variable.

Table 12.1 indicates that productivity and satisfaction are associated: The distribution of scores on Y (productivity) changes across the various conditions of X (satisfaction). For example, half of the workers who were low on satisfaction were also low on productivity (30 out of 60). On the other hand, over half of the workers who were high on satisfaction were high on productivity (27 out of 52).

Although it is a test of significance, chi square provides another way to detect an association between variables organized into table format. Any nonzero value for obtained chi square indicates that the variables are associated. For example, the obtained chi square for Table 12.1 is 24.2, a value that affirms our previous conclusion, based on the conditional distributions of Y, that an association of some sort exists between job satisfaction and productivity.

Often, the researcher will have already computed chi square before considering bivariate association. In such cases, we can use chi square instead of the conditional distributions of Y to ascertain whether or not there is an association. If the obtained chi square is zero, the two variables are independent and not associated. Any value other than zero indicates some association between the variables. Remember, however, that statistical significance and association are two different things. It is perfectly possible for two variables to be associated (as indicated by a nonzero chi square) but still independent (if we fail to reject the null hypothesis).

In this section, we have defined, in a general way, the concept of association between two variables. We have also shown two different ways to detect the presence of an association. In the next section, we extend the analysis and, in a systematic way, show how additional very useful information about the relationship between two variables can be developed.

THREE CHARACTERISTICS OF BIVARIATE ASSOCIATIONS

Bivariate associations possess three different characteristics, each of which must be analyzed for a full investigation of the relationship. Investigating these characteristics may be thought of as a process of finding answers to three questions:

1. Does an association exist?

2. If an association does exist, how strong is it?

3. What is the pattern and/or the direction of the association?

We will consider each of these questions separately.

Does an Association Exist?   We have already seen that we can detect an association by observing the conditional distributions of Y in a table or by using chi square. In Table 12.1, we know that the two variables are associated to some extent because the conditional distributions of productivity (Y) are different across the various categories of satisfaction (X) and because the chi square statistic is a nonzero value.

Comparisons from column to column in Table 12.1 are relatively easy to make because the column totals are roughly equal. This will not usually be the case, and it is helpful to compute percentages to control for varying column totals. These **column percentages**, introduced in Chapter 11, are computed within each column separately and make the pattern of association more visible.

The general procedure for detecting association with bivariate tables is to compute percentages within the columns (vertically, or down each column) and then compare column to column across the table (horizontally, or across the rows). Table 12.2 presents column percentages calculated from the data in Table 12.1. Note that this table reports the row and column marginals, but in parentheses. Besides controlling for any differences in column totals, tables in percentage form are usually easier to read because changes in the conditional distributions of Y are easier to detect.

In Table 12.2, we can see that the largest cell changes position from column to column. For workers who are low on satisfaction, the single largest cell is in the top row (low on productivity). For the middle column (moderate on satisfaction), the largest cell is in the middle row (moderate on productivity), and, for the right-hand column (high on satisfaction), it is in the bottom row (high on

TABLE 12.2   Productivity by Job Satisfaction (percentages)

Productivity (Y)	Job Satisfaction (X)			Totals
	Low	Moderate	High	
Low	50.00%	34.43%	13.46%	33.53% (58)
Moderate	33.33%	40.98%	34.62%	36.42% (63)
High	16.67%	24.59%	51.92%	30.06% (52)
Totals	100.00%	100.00%	100.00%	100.01%
	(60)	(61)	(52)	(173)

productivity). Even a cursory glance at Table 12.2 reinforces our conclusion that an association does exist between these two variables.

If two variables are not associated, then the conditional distributions of *Y* will not change across the columns. The distribution of *Y* would be the same for each condition of *X*. Table 12.3 illustrates a "perfect nonassociation" between height and productivity. Table 12.3 is only one of many patterns that indicate "no association." The important point is that the conditional distributions of *Y* are the same. Levels of productivity do not change at all for the various heights; therefore, no association exists between these variables. Also, the obtained chi square computed from this table would have a value of zero, again indicating no association.[2]

How Strong Is the Association? Once we know there is an association between two variables, we need to know how strong it is. This is essentially a matter of determining the amount of change in the conditional distributions of *Y*. At one extreme, of course, is the case of "no association," where the conditional distributions of *Y* do not change at all (see Table 12.3). At the other extreme is a perfect association, the strongest possible relationship. A perfect association exists between two variables if each value of the dependent variable is associated with one and only one value of the independent variable.[3] In a bivariate table, the variables would have a perfect relationship if all cases in each column are located in a single cell and there is no variation in *Y* for a given value of *X* (see Table 12.4).

TABLE 12.3 Productivity by Height (an illustration of no association)

Productivity (*Y*)	Height (*X*)		
	Short	Medium	Tall
Low	33.33%	33.33%	33.33%
Moderate	33.33%	33.33%	33.33%
High	33.33%	33.33%	33.33%
Totals	100.00%	100.00%	100.00%

TABLE 12.4 Productivity by Height (an illustration of perfect association)

Productivity (*Y*)	Height (*X*)		
	Short	Medium	Tall
Low	0%	0%	100%
Moderate	0%	100%	0%
High	100%	0%	0%
Totals	100%	100%	100%

[2]See Chapter 11 for instructions on computing column percentages.
[3]Each measure of association to be introduced in the following chapters incorporates its own definition of a "perfect association," and these definitions vary somewhat, depending on the specific logic and mathematics of the statistic. That is, for different measures computed from the same table, some measures will possibly indicate perfect relationships when others will not. We will note these variations in the mathematical definitions of a perfect association at the appropriate times.

A perfect relationship would be taken as very strong evidence of a causal relationship between the variables, at least for the sample at hand. In fact, the results presented in Table 12.4 would indicate that, for this sample, height is the sole cause of productivity. Also, in the case of a perfect relationship, predictions from one variable to the other could be made without error. If we know that a particular worker is short, for example, we could be sure that he or she is highly productive.

Of course, the huge majority of relationships will fall somewhere between the two extremes of no association and perfect association, and we need to develop some way of describing these intermediate relationships consistently and meaningfully. For example, Tables 12.1 and 12.2 show an association between productivity and job satisfaction. How could this relationship be described in terms of strength? How close is the relationship to perfect? How far away from no association?

To answer these questions, we rely on measures of association for precise, objective indicators of the strength of a relationship. Virtually all of these statistics are designed so that they have a lower limit of 0.00 and an upper limit of 1.00 (± 1.00 for ordinal and interval-ratio measures of association). A measure that equals 0.00 indicates no association between the variables (the conditional distributions of Y do not vary), and a measure of 1.00 (± 1.00 in the case of ordinal and interval-ratio measures) indicates a perfect relationship. The exact meaning of values between 0.00 and 1.00 varies from measure to measure, but, for all measures, the closer the value is to 1.00, the stronger the relationship (the greater the change in the conditional distributions of Y).

We begin to consider the many measures of association later in this chapter. At this point, we consider a more informal way of assessing the strength of a relationship based on comparing column percentages across the rows, called the **maximum difference**. This technique should be used as a "quick and easy" method for assessing the strength of a relationship: easy to apply but limited in its usefulness. To find the maximum difference, compute the column percentages as usual and then skim the table across each of the rows to find the largest difference—in any row—between column percentages. For example, the largest difference in column percentages in Table 12.2 is in the top row between the "Low" column and the "High" column: 50.00% – 13.46% = 36.54%. The maximum difference in the middle row is between "moderates" and "highs" (40.98% – 33.33% = 7.65%), and in the bottom row it is between "highs" and "lows" (51.92% – 16.67% = 35.25%). Both of these values are less than the maximum difference in the top row.

Once you have found the maximum difference in the table, the scale presented in Table 12.5 can be used to describe the strength of the relationship. For instance, we could describe the relationship between productivity and job satisfaction in Table 12.2 as strong.

Be aware that the relationships between the size of the maximum difference and the descriptive terms (weak, moderate, and strong) in Table 12.5 are arbitrary and approximate. We will get more precise and useful information when we compute and analyze a measure of association. Also, maximum differences are easiest to find and most useful for smaller tables. In large tables, with many (say, more than three) columns and rows, it can be cumbersome to find the high and low percentages, and it is advisable to consider only measures of association as indicators of the strength for these tables. Finally, note that the maximum

TABLE 12.5 The Relationship between the Maximum Difference and the Strength of the Relationship

Maximum Difference	Strength
If the maximum difference is:	The strength of the relationship is:
between 0 and 10 percentage points	Weak
between 10 and 30 percentage points	Moderate
more than 30 percentage points	Strong

TABLE 12.6 Library Use by Education (an illustration of a positive relationship)

Library Use	Education		
	Low	Moderate	High
Low	60%	20%	10%
Moderate	30%	60%	30%
High	10%	20%	60%
Total	100%	100%	100%

TABLE 12.7 Amount of Television Viewing by Education (an illustration of a negative relationship)

Television Viewing	Education		
	Low	Moderate	High
Low	10%	20%	60%
Moderate	30%	60%	30%
High	60%	20%	10%
Total	100%	100%	100%

difference is based on only two values (the high and low column percentages within any row). Like the range (see Chapter 4), this statistic can give a misleading impression of the overall strength of the relationship.

As a final caution, remember that chi square does not measure the strength of relationships. Even very large values for chi square do not necessarily mean that the relationship is strong. Significance and association are two separate matters and chi square, by itself, is not a measure of association. While a nonzero value indicates some association between the variables, the magnitude of chi square bears no particular relationship to the strength of the association. (*For practice in computing percentages and judging the existence and strength of an association, see any of the problems at the end of this chapter.*)

What Is the Pattern and/or the Direction of the Association? To investigate the pattern of the association, we need to ascertain which values or categories of one variable are associated with which values or categories of the other. We have already remarked on the pattern of the relationship between productivity and satisfaction. Table 12.2 indicates that low scores on satisfaction are associated with low scores on productivity, moderate satisfaction with moderate productivity, and high satisfaction with high productivity.

When one or both variables in a bivariate table are measured at the nominal level, we can discuss the pattern of the relationship only.[4] However, when both

---

[4]The scores of variables measured at the nominal level have no numerical order (by definition). Therefore, associations including nominal-level variables, while they may have a pattern, cannot have a direction.

variables are at least ordinal in level of measurement, the association between the variables may also be described in terms of direction. The direction of the association can be either positive or negative. An association is positive if the variables vary in the same direction. That is, in a **positive association**, high scores on one variable are associated with high scores on the other variable, and low scores on one variable are associated with low scores on the other. In a positive association, as one variable increases in value, the other also increases; and as one variable decreases, the other also decreases. Table 12.6 displays, with fictitious data, a positive relationship between education and use of public libraries. As education increases (as you move from left to right across the table), library use also increases (the percentage of "high" users increases). The association between job satisfaction and productivity, as displayed in Tables 12.1 and 12.2, is also a positive association.

In a **negative association**, the variables vary in opposite directions. High scores on one variable are associated with low scores on the other, and increases in one variable are accompanied by decreases in the other. Table 12.7 displays a negative relationship, again with fictitious data, between education and television viewing. The amount of television viewing decreases as education increases. In other words, as you move from left to right across the top of the table (as education increases), the percentage of heavy viewers decreases.

Measures of association for ordinal and interval-ratio variables are designed so that they will be positive for positive associations and negative for negative associations. Thus, a measure of association preceded by a plus sign indicates a positive relationship between the two variables, with the value +1.00 indicating a perfect positive relationship. A negative sign indicates a negative relationship, with −1.00 indicating a perfect negative relationship. *(For practice in determining the pattern of an association, see any of the end-of-chapter problems. For practice in determining the direction of a relationship, see problems 12.1, 12.6, 12.7, 12.8, 12.9, 12.10, and 12.12.)*

STATISTICS IN EVERYDAY LIFE	Even after professional football was integrated in the middle of the 20th century, racial discrimination continued on a number of levels. Most obviously, perhaps, was the perfect relationship between race and the quarterback position. This central leadership role was reserved for whites, a practice that reflected the widespread stereotype that blacks lacked the intelligence and decision-making ability required to play quarterback. Today, racial discrimination in professional sports is much weaker than in the past and minority-group athletes face little overt discrimination in their competition for any position, including quarterback. In fact, African Americans are vastly overrepresented in the National Football League (NFL): They are about 13% of the general population and about 67% of all NFL players. This makes the NFL one of the most integrated areas of U.S. society, but African Americans are still only 17% of quarterbacks,* a value that may reflect lingering—if more subtle—prejudices and stereotypes. What data would you need to collect to account for these patterns?

*Source: Lapchik, R., Kamke, C. and McMechan, D. 2010. *The 2009 Racial and Gender Report Card: National Football League.* Available at http://www.tidesport.org/racialgenderreportcard.html.

The Importance of Percentages and Some Errors of Interpretation

Percentages are among the humblest of statistics, but they can provide important information about the relationship between two variables. Of course, this information will be clear and accurate only when bivariate tables are properly constructed, percentages are computed within columns, and comparisons are made from column to column. Errors and misunderstandings can occur when there is confusion about which variable is the cause (or independent variable) and which is the effect (or dependent variable), or when the researcher asks the wrong questions about the relationship.

To illustrate these problems, let's review the process by which we analyze bivariate tables. When we compare column percentages across the table, we are asking: "Does Y (the dependent variable) vary by X (the independent variable)?" We conclude that there is evidence for a causal relationship if the values of Y change under the different values of X.

To illustrate further, consider Table 12.8, which shows the relationship between race or ethnic group (limited to black, white, and Hispanic Americans for the sake of simplicity) and support for affirmative action. The data is taken from the 2008 General Social Survey, a representative national sample. Race or ethnicity must be the independent or causal variable in this relationship, since it may shape people's attitudes and opinions. The reverse cannot be true: A person's opinion cannot cause their race or ethnicity. Group membership is the column variable in the table and the percentages are computed in the proper direction. A quick inspection shows that support for affirmative action varies by group, indicating that there may be a causal relationship between these variables. The maximum difference between the columns is about 32 percentage points, indicating that the relationship is moderate to strong.

What if we had misunderstood this causal relationship? If we had computed percentages within each row, for example, race would become the dependent variable. We would be asking "Does race vary by support for affirmative action?" Table 12.9 shows the results of asking this incorrect question.

A casual glance at the top row of the table seems to indicate a causal relationship since 55.1% of the supporters of affirmative action are white, only 39.5% are black, and a miniscule 5.4% are Hispanic. If we looked *only* at the top row of the table (as people sometimes do), we would conclude that whites are more supportive of affirmative action than other groups. But the second row shows that whites

TABLE 12.8 Support for Affirmative Action by Racial Group (frequencies and percentages)

Support Affirmative Action?	Racial or Ethnic Group			
	White	Black	Hispanic	
Yes	81 (*11.6%*)	58 (*43.3%*)	8 (*19.0%*)	147 (16.9%)
No	615 (*88.4%*)	76 (*56.7%*)	34 (*81.0%*)	725 (83.1%)
	696 (*100.0%*)	134 (*100.0%*)	42 (*100.0%*)	872 (100.0%)

TABLE 12.9   Row Percentages

Support Affirmative Action?	Racial or Ethnic Group			
	White	Black	Hispanic	
Yes	55.1%	39.5%	5.4%	100.0%
No	84.8%	10.5%	4.7%	100.0%

are also the huge majority (84.8%) of those who *oppose* the policy. How can this be? The row percentages in this table simply reflect the fact that whites vastly outnumber the other groups: There are 5 times as many whites as blacks in the sample and 16 times as many whites as Hispanics. Computing percentages within the rows would make sense only if racial or ethnic group could vary by attitude or opinion, and Table 12.9 could easily lead to false conclusions about this relationship.

Professional researchers sometimes compute percentages in the wrong direction or ask a question about the relationship incorrectly, and you should always check bivariate tables to make sure that the analysis agrees with the patterns in the table.

Applying Statistics 12.1: Association and Bivariate Tables

Why are many Americans attracted to movies that emphasize graphic displays of violence? One idea is that the fans of "slash" movies feel threatened by violence in their daily lives and use these movies as a means of coping with their fears. In the safety of the theater, violence can be vicariously experienced, and feelings and fears can be expressed privately. Also, violent movies almost always provide a role model of one character who does deal with violence successfully (usually, of course, with more violence).

Is fear of violence associated with frequent attendance at high-violence movies? The following table reports the joint frequency distributions of "fear" and "attendance" in percentages for a fictitious sample.

Attendance at Violent Movies by Fear of Violence

Attendance	Fear		
	Low	Moderate	High
Rare	50%	20%	30%
Occasional	30%	60%	30%
Frequent	20%	20%	40%
Totals	100%	100%	100%

The conditional distributions of attendance (*Y*) change across the values of fear (*X*), so these variables are associated. The clustering of cases in the diagonal from upper left to lower right suggests a substantial relationship in the predicted direction. People who are low on fear attend violent movies infrequently, and people who are high on fear attend frequently. Since the maximum difference in column percentages in the table is 30 (in both the top and middle rows), the relationship can be characterized as moderate to strong

These results do suggest an important relationship between fear and attendance. Notice, however, that they also pose an interesting causal problem. The table supports the idea that fearful and threatened people attend violent movies as a coping mechanism (*X* causes *Y*) but is also consistent with the reverse causal argument: Attendance at violent movies increases fears for one's personal safety (*Y* causes *X*). The results support *both* causal arguments and remind us that association is not the same thing as causation.

Measures of Association

Column percentages provide useful and detailed information about the bivariate associations and should always be computed and analyzed. However, they can be awkward and cumbersome to use, particularly for larger tables. Measures of association, on the other hand, summarize the overall strength (and direction for ordinal-level variables) of a bivariate association in a single number, a more compact and convenient format for interpretation and discussion.

There are many measures of association, but we will focus on a few of the most widely used. We will cover these statistics by the level of measurement for which they are most appropriate. In this chapter, we consider measures appropriate for nominal variables; in the next chapter, we will cover measures for ordinal-level variables. Finally, in Chapter 14, we will consider Pearson's r, a measure of association or correlation for interval-ratio–level variables. To analyze relationships for variables that are at different levels (for example, one nominal-level variable and one ordinal-level variable), we generally use the measures of association appropriate for the lower of the two levels of measurement.

CHI SQUARE–BASED MEASURES OF ASSOCIATION

Over the years, social science researchers have relied heavily on measures of association based on the value of chi square. When the value of chi square is already known, these measures are easy to calculate. To illustrate, let us reconsider Table 11.3, which displayed, with fictitious data, a relationship between accreditation and employment for social work majors. For the sake of convenience, this table is reproduced here as Table 12.10.

We saw in Chapter 11 that this relationship is statistically significant ($\chi^2 = 10.78$, which is significant at $\alpha = 0.05$), but the question now concerns the *strength* of the association. A brief glance at Table 12.10 shows that the conditional distributions of employment status do change, so the variables are associated. To emphasize this point, it is always helpful to calculate column percentages, as in Table 12.11.

So far, we know that the relationship between these two variables is statistically significant and that there is an association of some kind between accreditation and employment. To assess the strength of the association, we will compute a **phi** (ϕ). This statistic is a frequently used chi square–based measure of association appropriate for 2×2 tables (that is, tables with two rows and two columns).

TABLE 12.10 Employment of 100 Social Work Majors by Accreditation Status of Undergraduate Program (fictitious data)

Employment Status	Accreditation Status		Totals
	Accredited	Not Accredited	
Working as a social worker	30	10	40
Not working as a social worker	25	35	60
Totals	55	45	100

TABLE 12.11   Employment by Accreditation Status (percentages)

	Accreditation Status		
Employment Status	Accredited	Not Accredited	Totals
Working as a social worker	54.5%	22.2%	40.0%
Not working as a social worker	45.5%	77.8%	60.0%
Totals	100.00%	100.00%	100.00%

Before considering the computation of phi, it will be helpful to establish some general guidelines for interpreting the value of measures of association for nominally measured variables, similar to the guidelines we used for interpreting the maximum difference in column percentages. For phi and the other measures introduced in this chapter, the general relationship between the value of the statistic and the strength of the relationship is presented in Table 12.12.

As was the case for Table 12.5, these relationships between the numerical values and the descriptive terms are arbitrary and meant as general guidelines only. Measures of association generally have mathematical definitions that yield interpretations more meaningful and exact than these.

Calculating Phi (ϕ). One of the attractions of phi is that it is easy to calculate. Simply divide the value of the obtained chi square by N and take the square root of the result. Expressed in symbols, the formula for phi is

FORMULA 12.1

$$\phi = \sqrt{\frac{\chi^2}{N}}$$

For the data displayed in Table 12.8, the chi square was 10.78. Therefore, phi is

$$\phi = \sqrt{\frac{\chi^2}{N}}$$

$$\phi = \sqrt{\frac{10.78}{100}}$$

$$\phi = 0.33$$

TABLE 12.12   The Relationship Between the Value of Nominal-Level Measures of Association and the Strength of the Relationship

Value	Strength
If the value is	*The strength of the relationship is*
between 0.00 and 0.10	Weak
between 0.11 and 0.30	Moderate
greater than 0.30	Strong

For a 2 $\times$2 table, phi ranges in value from 0 (no association) to 1.00 (perfect association). The closer to 1.00, the stronger the relationship; the closer to 0.00, the weaker the relationship. For Table 12.10, we already knew that the relationship was statistically significant at the 0.05 level. Phi, as a measure of association, adds information about the strength of the relationship: There is a moderate-to-strong relationship between these two variables. As for the pattern of the association, the column percentages in Table 12.11 shows that graduates of accredited programs were more often employed as social workers.

Calculating Cramer's V. For tables larger than 2 $\times$ 2 (specifically, for tables with more than two columns and more than two rows), the upper limit of phi can exceed 1.00. This makes phi difficult to interpret, and a more general form of the statistic called **Cramer's V** must be used for larger tables. The formula for Cramer's V is

FORMULA 12.2

$$V = \sqrt{\frac{\chi^2}{(N)(\min r - 1, c - 1)}}$$

where $(\min r - 1, c - 1)$ = the minimum value of $r - 1$ (number of rows minus 1) or $c - 1$ (number of columns minus 1)

In words: To calculate V, find the lesser of the number of rows minus 1 ($r - 1$) or the number of columns minus 1 ($c - 1$), multiply this value by N, divide the result into the value of chi square, and then find the square root. Cramer's V has an upper limit of 1.00 for a table of any size and will be the same value as phi if the table has either two rows or two columns. Like phi, Cramer's V can be interpreted as an index that measures the strength of the association between two variables.

To illustrate the computation of V, suppose you had gathered the data displayed in Table 12.13, which shows the relationship between membership in student organizations and academic achievement for a sample of college students. The obtained chi square for this table is 31.5, which is significant at the 0.05 level. Cramer's V is

$$V = \sqrt{\frac{\chi^2}{(N)(\min r - 1, c - 1)}}$$
$$V = \sqrt{\frac{31.50}{(75)(2)}}$$
$$V = \sqrt{\frac{31.50}{150}}$$
$$V = \sqrt{0.21}$$
$$V = 0.46$$

Since Table 12.13 has the same number of rows and columns, we may use either $(r - 1)$ or $(c - 1)$ in the denominator. In either case, the value of the denominator is N multiplied by $(3 - 1)$, or 2. The computed value of V of 0.46 means there is a strong association between club membership and academic achievement. Column percentages are presented in Table 12.14 to help identify

TABLE 12.13   Academic Achievement by Club Membership

Academic Achievement	Membership			Totals
	Fraternity or Sorority	Other Organization	No Memberships	
Low	4	4	17	25
Moderate	15	6	4	25
High	4	16	5	25
Totals	23	26	26	75

TABLE 12.14   Academic Achievement by Club Membership (percentages)

Academic Achievement	Membership			Totals
	Fraternity or Sorority	Other Organization	No Memberships	
Low	17.4%	15.4%	65.4%	33.3%
Moderate	65.2%	23.1%	15.4%	33.3%
High	17.4%	61.5%	19.2%	33.3%
Totals	100.0%	100.0%	100.0%	100.00%

the pattern of this relationship. Fraternity and sorority members tend to be moderate, members of other organizations tend to be high, and nonmembers tend to be low in academic achievement.

ONE STEP AT A TIME   **Computing and Interpreting Phi and Cramer's *V***

To Calculate Phi, Solve Formula 12.1

1: Divide the value of chi square by *N*.

2: Find the square root of the quantity you found in step 1. The resulting value is phi.

3: Consult Table 12.12 to help interpret the value of phi.

To Calculate Cramer's *V*, Solve Formula 12.2

1: Find the number of rows (*r*) and number of columns (*c*) in the table. Subtract 1 from the lesser of these two numbers to find (min *r* − 1, *c* − 1).

2: Multiply the value you found in step 1 by *N*.

3: Divide the value of chi square by the value you found in step 2.

4: Take the square root of the quantity you found in step 3. The resulting value is *V*.

5: Consult Table 12.12 to help interpret the value of *V*.

The Limitations of Phi and _V._ One limitation of phi and Cramer's _V_ is that they are only general indicators of the strength of the relationship. Of course, the closer these measures are to 0.00, the weaker the relationship and the closer to 1.00, the stronger the relationship. Values between 0.00 and 1.00 can be described as weak, moderate, or strong, according to the general convention introduced in Table 12.12, but they have no direct or meaningful interpretation. On the other hand, phi and _V_ are easy to calculate (once the value of chi square has been obtained) and are commonly used indicators of the importance of an association.[5] _(For practice in computing phi and Cramer's V, see any of the problems at the end of this chapter. Remember that for tables that have either two rows or two columns, phi and Cramer's V will have the same value.)_

A measure of Association Based on Proportional Reduction in Error

In recent years, a group of measures based on a logic known as **proportional reduction in error (PRE)** have been developed to complement the older chi square–based measures of association.

THE LOGIC OF PROPORTIONAL REDUCTION IN ERROR

The logic of these measures is based on two different predictions about the scores of cases on the dependent variable. In one prediction, we ignore information about the independent variable and, in the second prediction, we take the independent variable into account. A PRE measure of association like lambda tells us how much knowledge of the independent variable improves our predictions of the dependent variable. An example will make the logic clearer.

For nominal-level variables, we first predict the category into which each case will fall on the dependent variable (_Y_) while ignoring the independent variable (_X_). Since we would be predicting blindly in this case, we would make many errors of prediction. That is, we would often predict the value of a case on the dependent variable incorrectly.

The second prediction allows us to take the independent variable into account. If the two variables are associated, the additional information supplied by the independent variable will reduce our errors of prediction—that is, we should misclassify fewer cases. The stronger the association between the variables, the greater the reduction in errors. In the case of a perfect association, we would make no errors at all when predicting the score on _Y_ from the score on _X_. When there is no association between the variables, on the other hand, knowledge of the independent variable will not improve the accuracy of our predictions. We would make just as many errors of prediction with knowledge of the independent variable as we did without knowledge of the independent variable.

---

[5]Two other chi square based measures of association, T^2 and C (the contingency coefficient) are sometimes reported in the literature. Both of these measures have serious limitations. T^2 has an upper limit of 1.00 only for tables with an equal number of rows and columns, and the upper limit of C varies, depending on the dimensions of the table. These characteristics make these measures more difficult to interpret and thus less useful than phi or Cramer's _V_.

To illustrate further, suppose you were placed in the rather unusual position of having to predict whether each of the next 100 people you meet will be shorter or taller than 5 feet 9 inches in height under the condition that you would have no knowledge of these people at all. With absolutely no information about these people, your predictions will be wrong quite often (you will frequently misclassify a tall person as short, and vice versa).

Now assume that you must go through this ordeal twice, but that on the second round you know the sex of the person whose height you must predict. Since height is associated with sex and females are, on the average, shorter than males, the optimal strategy would be to predict that all females are short and all males are tall. Of course, you will still make errors on this second round; but, if the variables are associated, the number of errors on the second round will be less than the number of errors on the first. That is, using information about the independent variable will reduce the number of errors (if, of course, the two variables are related). How can these unusual thoughts be translated into a useful statistic?

LAMBDA

One hundred individuals have been categorized by gender and height, and the data are displayed in Table 12.15. It is clear, even without percentages, that the two variables are associated. To measure the strength of this association, a PRE measure called **lambda** (symbolized by the Greek letter λ) will be calculated. Following the logic introduced in the previous section, we must find two quantities: first, the number of prediction errors made while ignoring the independent variable (gender) and, second, the number of prediction errors made while taking gender into account. These two sums will then be compared to derive the statistic.

First, the information given by the independent variable (gender) can be ignored by working only with the row marginals. Two different predictions can be made about height (the dependent variable) by using these marginals. We can predict either that all subjects are tall or that all subjects are short.[6] The first prediction (all subjects are tall) will result in 48 errors. That is, for this prediction, all 100 cases would be placed in the first row. Only 52 of the cases actually belong in this row, so this prediction would result in $(100 - 52)$, or 48, errors. If we had predicted that all subjects were short, on the other hand, we would have made 52 errors $(100 - 48 = 52)$. We will take the *lesser* of these two numbers and refer to this quantity as E_1, for the number of errors made while ignoring the independent variable. So $E_1 = 48$.

TABLE 12.15   Height by Gender

Height	Gender		
	Male	Female	Totals
Tall	44	8	52
Short	6	42	48
Totals	50	50	100

[6]Other predictions are, of course, possible, but these are the only two permitted by lambda.

Next, we predict the score on Y (height) again, but this time we take X (gender) into account by moving from column to column. Since each column is a category of X, we thus take X into account in making our predictions. For each column, we predict that all cases will fall in the cell with the largest frequency. For the left-hand column (males), we predict that all 50 cases will be tall and make six errors ($50 - 44 = 6$). For the second column (females), our prediction is that all females are short, and eight errors will be made. By moving from column to column, we have taken X into account and have made a total of 14 errors of prediction, a quantity we will label E_2 ($E_2 = 6 + 8 = 14$).

If the variables are associated, we will make fewer errors under the second procedure than under the first. In other words, E_2 will be smaller than E_1. In this case, we made fewer errors of prediction while taking gender into account ($E_2 = 14$) than while ignoring gender ($E_1 = 48$), so gender and height are clearly associated. Our errors were reduced from 48 to only 14. To find the *proportional* reduction in error, use Formula 12.3:

FORMULA 12.3
$$\lambda = \frac{E_1 - E_2}{E_1}$$

For the sample problem, the value of lambda would be

$$\lambda = \frac{E_1 - E_2}{E_1}$$

$$\lambda = \frac{48 - 14}{48}$$

$$\lambda = \frac{34}{48}$$

$$\lambda = 0.71$$

The value of lambda ranges from 0.00 to 1.00. Of course, a value of 0.00 means that the variables are not associated at all (E_1 is the same as E_2), and a value of 1.00 means that the association is perfect (E_2 is zero and scores on the dependent variable can be predicted without error from the independent variable). Unlike phi or V, however, the numerical value of lambda between the extremes of 0.00 and 1.00 has a precise meaning: It tells us how much the independent variable (X) helps us to predict (or, more loosely, understand) the dependent variable (Y). When multiplied by 100, the value of lambda indicates the strength of the association in terms of the percentage reduction in error. Thus, the lambda calculated for Table 12.15 could be interpreted by saying that knowledge of gender improves our ability to predict height by 71%. That is, we are 71% better off knowing gender when attempting to predict height.

An Additional Example for Calculating and Interpreting Lambda. In this section, we work through another example in order to state the computational routine for lambda in more general terms. Suppose a researcher was concerned with the relationship between religious denomination and attitude toward capital punishment and had collected the data presented in Table 12.16 from a sample of 130 respondents.

TABLE 12.16   **Attitude Toward Capital Punishment by Religious Denomination (fictitious data)**

Attitude	Religion				
	Catholic	Protestant	Other	None	Totals
Favors	10	9	5	14	38
Neutral	14	12	10	6	42
Opposed	11	4	25	10	50
Totals	35	25	40	30	130

Find E_1, the number of errors made while ignoring X (religion, in this case). Subtract the largest row total from N. For Table 13.7, E_1 will be

$$E_1 = N - \text{(Largest row total)}$$
$$E_1 = 130 - 50$$
$$E_1 = 80$$

To find E_2, begin with the left-hand column (Catholics) and subtract the largest cell frequency from the column total. Repeat these steps for each column and then add the subtotals:

$$\text{For Catholics: } 35 - 14 = 21$$
$$\text{For Protestants: } 25 - 12 = 13$$
$$\text{For "Others": } 40 - 25 = 15$$
$$\text{For "None": } 30 - 14 = \underline{16}$$
$$E_2 = 65$$

Substitute the values of E_1 and E_2 into Formula 12.3:

$$\lambda = \frac{80 - 65}{80}$$
$$\lambda = \frac{15}{80}$$
$$\lambda = 0.19$$

ONE STEP AT A TIME   **Computing and Interpreting Lambda**

To Calculate Lambda, Solve Formula 12.3

1: To find E_1, subtract the largest row subtotal (marginal) from N.

2: Starting with the far left-hand column, subtract the largest cell frequency in the column from the column total. Repeat this step for all columns in the table.

3: Add up all the values you found in step 2. The result is E_2.

4: Subtract E_2 from E_1.

5: Divide the quantity you found in step 4 by E_1. The result is lambda.

To Interpret Lambda

1: Multiply the value of lambda by 100. This percentage tells us the extent to which our predictions of the dependent variable are improved by taking the independent variable into account. Also, lambda may be interpreted using the descriptive terms in Table 12.10.

The Limitations of Lambda. Lambda has two characteristics that should be stressed. First, lambda is asymmetric. This means that the value of the statistic will vary, depending on which variable is taken as independent. For example, in Table 12.16, lambda would be 0.14 if attitude toward capital punishment had been taken as the independent variable. Thus, you should exercise some caution in the designation of an independent variable. If you consistently follow the convention of arraying the independent variable in the columns and compute lambda as outlined earlier, the asymmetry of the statistic should not be confusing.

Second, when one of the row totals is much larger than the others, lambda can be misleading. It can be 0.00 even when other measures of association are greater than 0.00 and the conditional distributions for the table indicate that there is an association between the variables. This anomaly is a function of the way lambda is calculated and suggests that great caution should be exercised in the interpretation of lambda when the row marginals are very unequal. In fact, in the case of

---

Applying Statistics 12.2: Measures of Association

A random sample of students at a large urban university has been classified as either "traditional" (18–23 years of age and unmarried) or "nontraditional" (24 or older or married). Subjects have also been classified as "vocational" if their primary motivation for college attendance is career or job oriented, or "academic" if their motivation is to pursue knowledge for its own sake. Are these two variables associated?

	Type		
Motivation	Traditional	Non-traditional	Totals
Vocational	25	60	85
Academic	75	15	90
Totals	100	75	175

Since this is a 2 ×2 table, we can compute phi as a measure of association. The chi square for the table is 51.89, so phi is

$$\phi = \sqrt{\frac{\chi^2}{N}}$$

$$\phi = \sqrt{\frac{51.89}{175}}$$

$$\phi = \sqrt{0.30}$$

$$\phi = 0.55$$

which indicates a strong relationship between the two variables.

A lambda can also be computed as an additional measure of association:

$$E_1 = 175 - 90 = 85$$

$$E_2 = (100 - 75) + (75 - 60)$$

$$= 25 + 15 = 40$$

$$\lambda = \frac{E_1 - E_2}{E_1}$$

$$\lambda = \frac{85 - 40}{85}$$

$$\lambda = \frac{45}{85}$$

$$\lambda = 0.53$$

A lambda of 0.53 indicates that we would make 53% fewer errors in predicting motivation (Y) from student type (X), as opposed to predicting motivation while ignoring student type. The association is strong, and, by inspection of the table, we can see that traditional students are more likely to have academic motivations (75%) and that nontraditional types are more likely to be vocational in motivation (80%).

very unequal row marginals, a chi square–based measure of association would be the preferred measure of association. *(For practice in computing lambda, see any of the problems at the end of this chapter or Chapter 11. As with phi and Cramer's V, it's probably a good idea to start with small samples and 2 × 2 tables.)*

STATISTICS
IN EVERYDAY LIFE

How would you feel if you learned that a Muslim clergyman with strong anti-U.S. views had been invited to speak on your campus? Do you suppose that your gender might affect your feelings? According to the 2008 General Social Survey, administered to a representative sample of Americans, there is a relationship between these two variables, as illustrated in this table:

Support for Allowing Anti-U.S. Muslim Clergyman to Speak by Gender: Frequencies and (*Percentages*)

	GENDER		
ALLOW?	Males	Females	Total
Yes	237 (*54.0%*)	197 (*36.9%*)	434 (*44.6%*)
No	202 (*46.0%*)	337 (*63.1%*)	539 (*55.4%*)
	439 (*100.0%*)	534 (*100.0%*)	973 (*100.0%*)

Chi square = 28.50 ($p < 0.05$), maximum difference = 17.1, phi = 0.17, lambda = 0.08

This relationship is statistically significant and (disregarding lambda) moderate in strength. A slight majority of men would allow the speaker but a strong majority of women are opposed. What might account for this difference between the genders?

READING STATISTICS 8: Bivariate Tables

The conventions for constructing and interpreting bivariate tables presented in this text are commonly but not universally followed in the professional literature. Tables will usually be constructed with the independent variable in the columns, the dependent variable in the rows, and percentages calculated in the columns. However, you should be careful to check the format of every table you attempt to read to see if these conventions have been observed. If the table is presented with the independent variable in the rows, for example, you will have to reorient your analysis (or redraw the table) to account for this. Above all, you should convince yourself that the percentages have been calculated in the correct direction. Even skilled professionals occasionally calculate percentages incorrectly and misinterpret the data.

Once you have assured yourself that the table is properly presented, you can apply the analytical techniques developed in this chapter. By comparing the conditional distributions of the dependent variable, you can ascertain for yourself if the variables are associated and check the strength, the pattern, and (for tables with ordinal-level variables) the direction of the association. You may then compare your conclusions with those of the researchers. Finally, note the measures of association (if any) and compare the interpretation presented by the researchers with the guidelines presented in this chapter. What does the measure of association say about the strength of the relationship?

(continued)

READING STATISTICS 8: (*continued*)

STATISTICS IN THE PROFESSIONAL LITERATURE

A team of researchers, noting that smoking rates in the U.S. have not declined in a decade, sought to identify some of the factors that helped people kick this habit. They used data already collected from a large sample of men in Finland and compared "sustained quitters" (who had refrained from smoking for $3\frac{1}{2}$ years) with "relapsers." Some of the findings, including background and behavioral variables are presented in Table 1.

The researchers noted that quitters were significantly more likely to be married and had smoked significantly fewer cigarettes per day. In other results, quitters were significantly less likely to report symptoms of anxiety, depression, and insomnia and were less likely to consume alcohol. For more information and other conclusion, see the original article.

Source: Augustson, E., Wanke, K., Rogers, S., Bergen, A., Chatterjee, N, Snyder, K., Albanes, D., Taylor, P., and Caporaso, N. 2008. "Predictors of Sustained Smoking Cessation: A Prospective Analysis of Chronic Smokers from the Alpha-Tocopherol Beta-Carotene Cancer Prevention Study," Table 1, p. 551, *American Journal of Public Health*, March 2008; 98:549–555. Used by permission.

TABLE 1 **Demographic and Behavioral Differences Between Sustained Quitters and Relapsers**

Demographics	Sustained Quitters	Relapsers	Significance
Age mean	57.3	56.6	<0.001
Marital Status			<0.001
Married	88.0%	82.6%	
Never married	4.3%	5.1%	
Divorced	5.1%	9.4%	
Widowed	2.6%	2.9%	
Education	17.5%	19.3%	0.37

Smoking Characteristics			
Always inhale (% yes)	44.2%	50.2%	0.001
No. of cigs per day			<0 .001
1–9	17.7%	10.8%	
10–19	35.8%	31.0%	
≥20	46.5%	58.2%	
Age started (mean)	20.3	20.0	0.13
Years smoked (mean)	33.4	33.9	0.21
Pack years (mean)	29.2	33.0	<0.001

SUMMARY

1. Analyzing the association between variables provides information that is complementary to tests of significance. The latter are designed to detect nonrandom relationships, whereas measures of association are designed to quantify the importance or strength of a relationship.

2. Relationships between variables have three characteristics: the existence of an association, the strength of the association, and the direction or pattern of the association. These three characteristics can be investigated by calculating percentages for a bivariate table in the direction of the independent variable (vertically) and then comparing in the opposite direction (horizontally). It is often useful (as well as quick and easy) to assess the strength of a relationship by finding the maximum difference in column percentages in any row of the table.

3. Tables 12.1 and 12.2 can be analyzed in terms of these three characteristics. Clearly, a relationship does exist between job satisfaction and productivity, since the conditional distributions of the dependent variable (productivity) are different for the three different conditions of the independent variable (job satisfaction). Even without a measure of association, we can see that the association is substantial in that the change in Y (productivity) across the three categories of X (satisfaction) is marked. The maximum difference of 36.54% confirms that the relationship is substantial (moderate

to strong). Furthermore, the relationship is positive in direction. Productivity increases as job satisfaction rises, and workers who report high job satisfaction tend also to be high on productivity. Workers with little job satisfaction tend to be low on productivity.

4. Given the nature and strength of the relationship, it could be predicted with fair accuracy that highly satisfied workers tend to be highly productive ("Happy workers are busy workers"). These results might be taken as evidence of a causal relationship between these two variables, but they cannot, by themselves, prove that a causal relationship exists: Association is not the same thing as causation. In fact, although we have presumed that job satisfaction is the independent variable, we could have argued the reverse causal sequence ("Busy workers are happy workers"). The results presented in Tables 12.1 and 12.2 are consistent with both causal arguments.

5. Phi, Cramer's V, and lambda are measures of association and each is appropriate for a specific situation. Phi is used for nominal-level variables in a 2 × 2 table and Cramer's V is used for tables larger than 2 × 2. Lambda is a proportional reduction in error (PRE) measure appropriate for nominal-level variables. These statistics express information about the strength of the relationship *only*. An inspection of the column percentages will reveal patterns in the bivariate relationship as well as other useful information.

SUMMARY OF FORMULAS

Phi

12.1 $\qquad \phi = \sqrt{\dfrac{\chi^2}{N}}$

Cramer's V

12.2 $\qquad V = \sqrt{\dfrac{\chi^2}{(N)(\min r - 1, c - 1)}}$

Lambda

12.3 $\qquad \lambda = \dfrac{E_1 - E_2}{E_1}$

GLOSSARY

Association. The relationship between two (or more) variables. Two variables are said to be associated if the distribution of one variable changes for the various categories or scores of the other variable.

Column percentages. Percentages computed with each column of a bivariate table.

Conditional distribution of Y. The distribution of scores on the dependent variable for a specific score or category of the independent variable when the variables have been organized into table format.

Cramer's V. A chi square–based measure of association for nominally measured variables that have been organized into a bivariate table with any number of rows and columns.

Dependent variable. In a bivariate relationship, the variable that is taken as the effect.

Independent variable. In a bivariate relationship, the variable that is taken as the cause.

Lambda. A proportional reduction in error (PRE) measure for nominally measured variables that have been organized into a bivariate table.

Maximum difference. A way to assess the strength of an association between variables that have been organized into a bivariate table. The maximum difference is the largest difference between column percentages for any row of the table.

Measures of association. Statistics that quantify the strength of the association between variables.

Negative association. A bivariate relationship where the variables vary in opposite directions. As one variable increases, the other decreases, and high scores on one variable are associated with low scores on the other.

Phi (ϕ). A chi square–based measure of association for nominally measured variables that have been organized into a bivariate table with two rows and two columns (a 2 × 2 table).

Positive association. A bivariate relationship where the variables vary in the same direction. As one variable increases, the other also increases, and high scores on one variable are associated with high scores on the other.

Proportional reduction in error (PRE). The logic that underlies the definition and computation of lambda. The statistic compares the number of errors made when predicting the dependent variable while ignoring the independent variable (E_1) with the number of errors made while taking the independent variable into account (E_2).

X. Symbol used for any independent variable.

Y. Symbol used for any dependent variable.

PROBLEMS

12.1 PA Various supervisors in the city government of Shinbone, Kansas, have been rated on the extent to which they practice authoritarian styles of leadership and decision making. The efficiency of each department has also been rated, and the results are summarized below. Use column percentages, the maximum difference, and measures of association to describe the strength and pattern of this association.

	Authoritarianism		
Efficiency	Low	High	Totals
Low	10	12	22
High	17	5	22
Totals	27	17	44

12.2 SOC The administration of a local college campus has proposed an increase in the mandatory student fee in order to finance an upgrading of the intercollegiate football program. A survey sampling the opinions of the faculty regarding this proposal has been completed. Is there any association between support for raising fees and the gender, discipline, or tenured status of the faculty? Use column percentages, the maximum difference, and measures of association to describe the strength and pattern of these associations.

a. Support for raising fees by gender:

	Gender		
Support	Males	Females	Totals
For	12	8	20
Against	15	12	27
Totals	27	20	47

b. Support for raising fees by discipline:

	Discipline		
Support	Liberal Arts	Science & Business	Totals
For	6	13	19
Against	14	14	28
Totals	20	27	47

c. Support for raising fees by tenured status:

	Status		
Support	Tenured	Nontenured	Totals
For	15	4	19
Against	18	10	28
Totals	33	14	47

12.3 PS How consistent are people in their voting habits? Do people vote for the same party from election to election? Below are the results of a poll in which people were asked if they had voted Democrat or Republican in each of the last two presidential elections. Use column percentages, the maximum difference, and measures of association to describe the strength and pattern of the association.

2008 Election	2004 Election		
	Democrat	Republican	Totals
Democrat	117	23	140
Republican	17	178	195
Totals	134	201	335

12.4 SOC A needs assessment survey has been distributed in a large corporation. Employees have been asked if they thought that a new policy of casual dress on Fridays would boost company morale. Use column percentages, the maximum difference, and measures of association to describe the strength and direction of the association between the type of employee and the positive or negative attitude toward casual Fridays. Write a few sentences describing the relationship.

Casual Fridays?	Employee Type		
	Supervisor	Worker	Totals
Yes	121	426	547
No	175	251	426
Totals	296	677	973

12.5 SW As the state director of mental health programs, you note that some local mental health facilities have very high rates of staff turnover. You believe that part of this problem is a result of the fact that some of the local directors have very little training in administration and poorly developed leadership skills. Before implementing a program to address this problem, you collect some data to make sure that your beliefs are supported by the facts. Is there a relationship between staff turnover and the administrative experience of the directors? Use column percentages, the maximum difference, and measures of association to describe the strength and direction of the association. Write a few sentences describing the relationship.

Turnover	Director Experienced?		
	No	Yes	Totals
Low	4	9	13
Moderate	9	8	17
High	15	5	20
Totals	28	22	50

12.6 CJ About half the neighborhoods in a large city have instituted programs to increase citizen involvement in crime prevention. Do these areas experience less crime? Write a few sentences describing the relationship in terms of pattern and strength of the association. Use column percentages, the maximum difference, and measures of association to describe the strength and direction of the association. Write a few sentences describing the relationship.

Crime Rate	Program		
	No	Yes	Totals
Low	28	16	44
Moderate	33	27	60
High	52	45	97
Totals	113	88	201

12.7 GER A survey of senior citizens who live in either a housing development specifically designed for retirees or an age-integrated neighborhood has been conducted. Is type of living arrangement related to sense of social isolation?

Sense of Isolation	Living Arrangement		
	Housing Development	Integrated Neighborhood	Totals
Low	80	30	110
High	20	120	140
Totals	100	150	250

12.8 SOC A researcher has conducted a survey on sexual attitudes for a sample of 317 teenagers. The respondents were asked whether they considered premarital sex to be "always wrong" or "OK under certain circumstances." The tables below summarize the relationship between responses to this item and several other variables. For each table, assess the strength and pattern of the relationship and write a paragraph interpreting these results.

a. Attitudes toward premarital sex by gender:

Premarital Sex	Gender Female	Male	Totals
Always wrong	90	105	195
Not always wrong	65	57	122
Totals	155	162	317

b. Attitudes toward premarital sex by courtship status:

Premarital Sex	Ever "Gone Steady" No	Yes	Totals
Always wrong	148	47	195
Not always wrong	42	80	122
Totals	190	127	317

c. Attitudes toward premarital sex by social class:

Premarital Sex	Social Class Blue Collar	White Collar	Totals
Always wrong	72	123	195
Not always wrong	47	75	122
Totals	119	198	317

12.9 SOC Below are the five dependent variables cross-tabulated against gender as an independent variable used in problem 11.15. Use column percentages, the maximum difference, and an appropriate measure of association to conduct the analysis. Summarize the results of your analysis in a paragraph that describes the strength and pattern of each relationship.

a. Support for the legal right to an abortion by gender:

Right to Abortion?	Gender Male	Female	Totals
Yes	310	418	728
No	432	618	1050
Totals	742	1036	1778

b. Support for capital punishment by gender:

Capital Punishment?	Gender Male	Female	Totals
Favor	908	998	1906
Oppose	246	447	693
Totals	1154	1445	2599

c. Approve of suicide for people with incurable disease by gender:

Right to Suicide?	Gender Male	Female	Totals
Yes	524	608	1132
No	246	398	644
Totals	770	1006	1776

d. Support for sex education in public schools by gender:

Sex Education?	Gender Male	Female	Totals
Favor	685	900	1585
Oppose	102	134	236
Totals	787	1034	1821

e. Support for traditional gender roles by gender:

Women Should Take Care of Running Their Homes and Leave Running the Country to Men	Gender Male	Female	Totals
Agree	116	164	280
Disagree	669	865	1534
Totals	785	1029	1814

Understanding Political Beliefs, Part II

At the end of Chapter 11, you investigated possible causes of people's beliefs on four controversial issues. Now you will extend your analysis by using **Crosstabs** to get the statistics presented in this chapter. There will be two projects and, once again, we will begin with a brief demonstration using the relationship between *abany* and *sex*.

Demonstration

With the 2008 GSS loaded, click **Analyze**, **Descriptive Statistics**, and **Crosstabs** and name *abany* as the row (dependent) variable and *sex* as the column (independent) variable. Click the **Cells** button and request column percentages by clicking the box next to **Column** in the **Percentages** box. Also, click the **Statistics** button and request chi square, phi, Cramer's *V*, and lambda by clicking the appropriate boxes. Click **Continue** and **OK**, and your task will be executed.

As you will see, SPSS generates a good deal of output for this procedure and I have condensed the information into a single table, based on how this test might be presented in the professional research literature. Note that the cells of the bivariate table include both frequencies and percentages and that all the statistical information is presented in a single, short line beneath the table.

Support for Legal Abortion for Any Reason (*abany*) by Sex

		RESPONDENTS SEX		
		MALE	FEMALE	Total
Should a woman be able to get a legal abortion if She wants it for any reason?	YES	191	206	397
		44.8%	39.4%	41.8%
	NO	235	317	552
		55.2%	60.6%	58.2%
	Total	426	523	949
		100.0%	100.0%	100.0%

Chi square = 2.863 (df = 1, $p > 0.05$); phi = 0.06; lambda = 0.00.

The three questions about bivariate association introduced in this chapter can provide a useful framework for reading and analyzing these results:

1. ***Is there an association?*** The column percentages in the bivariate table change, so there is an association between support for abortion and gender. This conclusion is verified by the fact that chi square is a nonzero value.
2. ***How strong is the association?*** We can assess strength in several different ways. First, the maximum difference is 44.8 – 39.4 or 5.4 percentage points. (This calculation is based on the top row but, since this is a 2 × 2 table, using the bottom row would result in exactly the same value). According to Table 12.5, this means that the relationship between the variables is weak.

    Second, we can (and should) use a measure of association to assess the strength of the relationship. We have a choice of two different measures: lambda and (since this is a 2 × 2 table) phi. Looking in the "Directional Measures" output box, we see several different values for lambda. Recall that lambda is an asymmetric measure and that it changes value depending on which variable is seen as dependent. In this case, the dependent variable is "ABORTION IF WOMAN WANTS FOR ANY REASON" and the associated lambda is reported as .000. This indicates that there is no association between the variables; but we have already seen that the variables *are* associated, if only weakly. Remember that lambda can be zero when the variables are associated but the row totals in the table are very unequal. Lambda might be a little misleading, but it's still telling us that the relationship is weak.

Turning to phi (under "Symmetric Measures"), we see a value of .06. This indicates that the relationship is weak (see Table 12.12), a conclusion that is consistent with the value of the maximum difference and the fact that the relationship is not significant at the 0.05 level.

3. *What is the pattern of the relationship?* Since sex is a nominal-level variable, the relationship cannot have a direction (that is, it cannot be positive or negative). We can, however, discuss the pattern of the relationship: how values of the variables seem to go together. Although the maximum difference is small (5.4%), males are more supportive of abortion than females.

In summary, using chi square, the column percentages, and phi, we can say that the relationship between support for legal abortion and gender is weak and not statistically significant. If we were searching for an important cause of attitudes about abortion, we would discard this independent variable and seek another. Your turn.

PROJECT 1: Explaining Beliefs

In this project, you will once again analyze beliefs about capital punishment (*cappun*), assisted suicide (*letdie1*), gay marriage (*marhomo*), and immigration (*letin1*). You will select an independent variable *other than the one you used in Chapter 11* and use SPSS to generate chi square, column percentages, phi or Cramer's *V*, and lambda. You will use all of these statistics to help analyze and interpret your results.

Step 1: Choose an Independent Variable

Select an independent variable that seems likely to be an important cause of people's attitudes about the death penalty, assisted suicide, gay marriage, and immigration. Be sure to select an independent variable that has *only* 2–5 categories and use the recode command if necessary. You might consider gender, level of education, religion, or age (the recoded version—see Chapter 10) as possibilities, but there are many others. Record the variable name and state exactly what the variable measures in the table below:

SPSS Name	What exactly does this variable measure?

Step 2: Stating Hypotheses

State hypotheses about the relationships you expect to find between your independent variable and each of the four dependent variables. State these hypotheses in terms of which category of the independent variable you expect to be associated with which category of the dependent variable (for example, "I expect that men will be more supportive of the legal right to an abortion for any reason").

Step 3: Running Crosstabs

Click **Analyze** → **Descriptives** → **Crosstabs** and place the four dependent variables (*cappun*, *letdie1*, *letin1*, and *marhomo*) in the **Rows:** box and the independent variable you selected in the **Columns:** box. Click the **Statistics** button to get chi square, phi and Cramer's *V*, and lambda and the **Cells** button for column percentages.

Step 4: Recording Results

These commands will generate a lot of output and it will be helpful to summarize your results in the following table.

Dependent Variable	Chi Square Significant at < 0.05?	Maximum Difference	Phi or Cramer's *V*	Lambda
cappun				
letdie1				
letin1				
marhomo				

Step 5: Analyzing and Interpreting Results

Write a short summary of results for each dependent variable. The summary needs to identify the variables being tested, the results of the chi square test, and the strength and pattern of the relationship. It is probably best to characterize the relationship in general terms and then cite the statistical values in parentheses. For example, we might summarize our test of the relationship between gender and support for abortion by saying: "The relationship between gender and support for abortion was not significant and weak (chi square = 1.853, df = 1, $p > 0.05$, phi = 0.06). Men were slightly more supportive of the right to a legal abortion for any reason than women." You should also note whether or not your hypotheses were supported.

PROJECT 2: Exploring the Impact of Various Independent Variables

In this project, you will examine the relative ability of a variety of independent variables to explain or account for a single dependent variable. You will again use the **Crosstabs** procedure in SPSS to generate statistics and use the alpha levels and measures of association to judge which independent variable has the most important relationship with your dependent variable.

Step 1: Choosing Variables

Select a dependent variable. You may use any of the four from Project 1 in this chapter or select a new dependent variable from the 2008 GSS. Be sure that your dependent variable has no more than five values or scores. Good choices for dependent variables include any measure of attitudes or opinions. Do not select characteristics like race, sex, or religion as dependent variables.

Select three independent variables that seem likely to be important causes of the dependent variable you selected. Your independent variable should have no more than five or six categories. You might consider gender, level of education, religion, or age (the recoded version—see Chapter 10) as possibilities, but there are many others.

Record the variable names and state exactly what each variable measures in the table below.

SPSS Name	What exactly does this variable measure?
Dependent Variable	
Independent Variables	

Step 2: Stating Hypotheses

State hypotheses about the relationships you expect to find between your independent variables and the dependent variable. State these hypotheses in terms of which category of the independent variable you expect to be associated with which category of the dependent variable (for example, "I expect that men will be more supportive of the legal right to an abortion for any reason").

Step 3: Running Crosstabs

Click **Analyze** → **Descriptives** → **Crosstabs** and place your dependent variable in the **Rows:** box and all three of your independent variables in the **Columns:** box. Click the **Statistics** button to get chi square, phi and Cramer's *V*, and lambda. Click the **Cells** button for column percentages.

Step 4: Recording Results

Your output will consist of three tables, and it will be helpful to summarize your results in the following table. Remember that the significance of the relationship is found in the column labeled "Asymp. Sig (2-sided) of the second box in the output.

Dependent Variable	Chi Square Significant at Less than 0.05?	Maximum Difference	Phi or Cramer's *V*	Lambda

Step 5: Analyzing and Interpreting Results

Write a short summary of results of each test using the same format as in Project 1. Remember to explain whether or not your hypotheses were supported. Finally, assess which of the independent variables had the most important relationship with your dependent variable. Use the alpha (or the "Asymp. Sig 2-tailed") level and the value of the measures of association to make this judgment.

13

Association Between Variables Measured at the Ordinal Level

By the end of this chapter, you will be able to

1. Calculate and interpret gamma and Spearman's rho.
2. Explain the logic of proportional reduction in error in terms of gamma.
3. Use gamma and Spearman's rho to analyze and describe a bivariate relationship in terms of the three questions introduced in Chapter 12.
4. Test gamma and Spearman's rho for significance.

There are two common types of ordinal-level variables. Some have many possible scores and resemble interval-ratio–level variables. We will call these *continuous ordinal variables*. A scale that measured an attitude with many different items would produce this type of variable.

The second type, which we will call a *collapsed ordinal variable*, has only a few (no more than five or six) scores and can be created either by collecting data in collapsed form or by collapsing a continuous scale. For example, we would produce collapsed ordinal variables by measuring social class as upper, middle, or lower or by reducing the scores on an attitude scale into just a few categories (such as high, moderate, and low).

A number of measures of association have been invented for use with collapsed ordinal-level variables. Rather than attempt to cover them all, we will concentrate on **gamma (G)**. Other measures suitable for collapsed ordinal-level data (Somer's *d* and Kendall's tau-*b*) are covered at the Web site for this text. For "continuous" ordinal variables, a statistic called **Spearman's rho (r_s)** is commonly used, and we cover this measure of association toward the end of this chapter.

This chapter will enable you to expand your understanding of how bivariate associations can be described and analyzed, but it is important to remember that we are still addressing the three questions raised in Chapter 12: Are the variables associated? How strong is the association? What is the direction of the association?

Proportional Reduction in Error (PRE) for Collapsed Ordinal-Level Variables: Gamma

As we saw in Chapter 12, for nominal-level variables, the logic of PRE is based on two different predictions of the scores of cases on the dependent variable (Y): one that ignores the independent variable (X) and a second that takes the independent variable into account. The value of lambda shows the extent to which taking the independent variable into account improves the accuracy when predicting the score of the dependent variable. The PRE logic for ordinal-level variables is similar, and gamma, like lambda, measures the proportional reduction in error gained by predicting one variable while taking the other into account. The major difference lies in the way predictions are made.

For gamma, we predict the *order* of pairs of cases rather than a score on the dependent variable. That is, we predict whether one case will have a higher or lower score than the other. First, we predict the order of a pair of cases on one variable while ignoring their order on the other variable. Second, we repeat the prediction while taking the order on the other variable into account.

To illustrate, assume that a researcher is concerned about the causes of "burnout" (that is, demoralization and loss of commitment) among elementary school teachers and wonders if there is a relationship between burnout and years of service. One way to state the research question is: Do teachers with more years of service have higher levels of burnout? Another way to ask the same question is: Do teachers who *rank higher* on years of service also *rank higher* on burnout? If we knew that teacher A had more years of service than teacher B, could we predict that teacher A is also more "burned out" than teacher B? That is, would knowledge of the order of this pair of cases on one variable help us predict their order on the other?

If the two variables are associated, we will reduce our errors when our prediction about one variable is based on knowledge of the other. Furthermore, the stronger the association, the fewer the errors we will make. When there is no association between the variables, gamma will be 0.00, and knowledge of the order of a pair of cases on one variable will not improve our ability to predict their order on the other. A gamma of ±1.00 denotes a perfect relationship: The order of all pairs of cases on one variable would be predictable without error from their order on the other variable.

With nominal-level variables, we analyzed the pattern of the relationship between the variables. That is, we looked to see which value on one variable (e.g., "male") was associated with which value on the other variable (e.g., "tall" on height). Recall that ordinally measured variables have scores that can be rank ordered from high to low or from more to less (see Chapter 1). This means that relationships between ordinal-level variables can have a direction as well as a pattern. In terms of the logic of gamma, variables have a *positive* relationship if cases tend to be ranked in the same order on both variables. If Case A ranks above Case B on one variable, Case A would also be ranked above Case B on the second variable. The relationship suggested earlier between years of service and burnout is positive.

In a *negative* relationship, the order of the cases would be reversed between the two variables. If Case A ranked above Case B on one variable, it would rank below Case B on the second variable. If there is a negative relationship between prejudice and education and Case A was more educated than Case B (that is, ranked above Case B on education), then Case A would be less prejudiced (that is, would rank below Case B on prejudice).

THE COMPUTATION OF GAMMA

Table 13.1 summarizes the relationship between "length of service" and "burnout" for a fictitious sample of 100 teachers. We need two sums to compute

TABLE 13.1 Burnout by Length of Service (fictitious data)

Burnout	Length of Service			Totals
	Low	Moderate	High	
Low	20	6	4	30
Moderate	10	15	5	30
High	8	11	21	40
Totals	38	32	30	100

gamma: the number of pairs of cases that are ranked the same on both variables (labeled N_s) and the number of pairs of cases ranked differently on the variables (N_d). We find these sums by working with the cell frequencies.

To find the number of pairs of cases ranked the same (N_s), begin with the cell containing the cases that were ranked the lowest on both variables. In Table 13.1, this would be the upper-left-hand cell. (*NOTE: Not all tables are constructed with values increasing from left to right across the columns and from top to bottom down the rows. Check to make sure that you have located the proper cell.*) The 20 cases in the upper-left-hand cell all rank low on both burnout and length of service, and we will refer to them as "low-lows," or "LLs."

Now form a pair of cases by selecting one case from this cell and one from any other cell—for example, the middle cell in the table. All 15 cases in this cell are moderate on both variables and can be labeled moderate-moderates, or MMs. Any pair of cases formed between these two cells will be ranked the same on both variables. That is, all LLs are lower than all MMs on both variables (on *X*, low is less than moderate; on *Y*, low is less than moderate). The total number of pairs of cases is given by multiplying the cell frequencies. So the contribution of these two cells to the total N_s is (20)(15), or 300.

Gamma ignores all pairs of cases that are tied on either variable. For example, any pair of cases formed between the LLs and any other cell in the top row (low on burnout) or the left-hand column (low on length of service) will be tied on one variable. Also, any pair of cases formed within any cell will be tied on both *X* and *Y*. Gamma ignores all pairs of cases formed within the same row, column, or cell. This means that, in computing N_s, we will work with only the pairs of cases that can be formed between each cell and the cells below and to the right of it.

In summary: To find the total number of pairs of cases ranked the same on both variables (N_s), multiply the frequency in each cell by the total of all frequencies below and to the right of that cell. Repeat this procedure for each cell and add the resultant products. The total of these products is N_s. This procedure is displayed in Figure 13.1 for each cell in Table 13.1. Note that none of the cells in the bottom row or the right-hand column of Table 13.1 can contribute to N_s because they have no cells below and to the right of them. Figure 13.1 shows the direction of multiplication for each of the four cells that, in a 3 × 3 table, can contribute to N_s. Computing N_s for Table 13.1, we find that a total of 1831 pairs of cases are ranked the same on both variables.

	Contribution to N_s	
For LLs, 20(15 + 5 + 11 + 21)	=	1040
For MLs, 6(21 + 5)	=	156
For HLs, 4(0)	=	0
For LMs, 10(11 + 21)	=	320
For MMs, 15(21)	=	315
For HMs, 5(0)	=	0
For LHs, 8(0)	=	0
For MHs, 11(0)	=	0
For HHs, 21(0)	=	0
	$N_s =$	1831

Our next step is to find the number of pairs of cases ranked differently (N_d) on both variables. To do this, multiply the frequency in each cell by the total of all frequencies below and to the left of that cell. Note that the pattern for computing N_d is the reverse of the pattern for N_s. This time, we begin with the upper-right-hand cell (high-lows, or HLs) and multiply the number of cases in the cell by the total frequency of cases below and to the left. The four cases in the upper-right-hand cell of Table 13.1 are low on Y and high on X. If a pair of cases is formed with any case from this cell and any cell below and to the left, the cases will be ranked differently on the two variables. For example, if a pair is formed between any HL case and any case from the middle cell (MMs), the HL case would be less than the MM case on Y ("low" is less than "moderate") but more than the MM case on X ("high" is greater than "moderate"). The computation of N_d is detailed below and shown graphically in Figure 13.2. In the computations, we have omitted cells that cannot contribute to N_d because they have no cells below and to the left of them.

FIGURE 13.1 Computing N_s in a 3 × 3 Table

	Contribution to N_d
For LLs, 4(10 + 15 + 8 + 11)	= 176
For MLs, 6(10 + 8)	= 108
For HLs, 5(8 + 11)	= 95
For MMs, 15(8)	= 120
	N_d = 499

Table 13.1 has 499 pairs of cases ranked in different order and 1831 pairs of cases ranked in the same order.

The formula for computing gamma is

FORMULA 13.1
$$G = \frac{N_s - N_d}{N_s + N_d}$$

where N_s = the number of pairs of cases ranked in the same order on both variables

N_d = the number of pairs of cases ranked in different order on the variables

For Table 13.1, the value of gamma would be

$$G = \frac{N_s - N_d}{N_s + N_d}$$

$$G = \frac{1831 - 499}{1831 + 499}$$

$$G = \frac{1332}{2330}$$

$$G = 0.57$$

FIGURE 13.2 Computing N_d in a 3 × 3 Table

For HL's

	L	M	H
L			▓
M	▓	▓	
H	▓	▓	

For ML's

	L	M	H
L		▓	
M	▓		
H	▓		

For HM's

	L	M	H
L			
M			▓
H	▓	▓	

For MM's

	L	M	H
L			
M		▓	
H	▓		

THE INTERPRETATION OF GAMMA

A gamma of 0.57 means that we would make 57% fewer errors if we predicted the order of pairs of cases on one variable from the order of pairs of cases on the other (as opposed to predicting order while ignoring the other variable). Length of service is associated with degree of burnout, and the relationship is positive. Knowing the rankings of two teachers on length of service (Case A is higher on length of service than Case B) will help us predict their ranking on burnout (we would predict that Case A will also be higher than Case B on burnout).

Table 13.2 provides some additional assistance for interpreting gamma in a format similar to Tables 12.5 and 12.12. As before, the descriptive terms are arbitrary and intended as general guidelines only. Note, in particular, that strength of a relationship is independent of direction. That is, a gamma of -0.35 is exactly as strong as a gamma of $+0.35$ but is opposite in direction.

To use the computational routine for gamma presented earlier, you must arrange the table in the manner of Table 13.1, with the column variable increasing in value as you move from left to right and the row variable increasing from top to bottom. Be careful to construct your tables according to this

---

ONE STEP AT A TIME   **Computing and Interpreting Gamma**

To Compute Gamma

1: Make sure the table is arranged with the column variable increasing from left to right and the row variable increasing from top to bottom.

2: To compute N_s, start with the upper-left-hand cell. Multiply the number of cases in this cell by the total number of cases in all cells below and to the right. Repeat this process for each cell in the table. Add up these subtotals to find N_s.

3: To compute N_d, start with the upper-right-hand-cell. Multiply the number of cases in this cell by the total number of cases in all cells below and to the left. Repeat this process for each cell in the table. Add up these subtotals to find N_d.

4: Subtract the value of N_d from N_s.

5: Add the value of N_d to N_s.

6: Divide the value you found in step 4 by the value you found in step 5. This value is gamma.

Interpreting the Strength of the Relationship

1: Always begin with the column percentages: The bigger the change in column percentages, the stronger the relationship.

2: There are two ways to use gamma to interpret strength. You can use either or both:

a. Use Table 13.2 to describe strength in general terms.

b. Use the logic of proportional reduction in error. Multiply gamma by 100. This value represents the percentage improvement in our predictions of the dependent variable by taking the independent variable into account.

Interpreting the Direction of the Relationship

1: Always begin by looking at the pattern of the column percentages. If the cases tend to fall in a diagonal from upper left to lower right, the relationship is positive. If the cases tend to fall in a diagonal from lower left to upper right, the relationship is negative.

2: The sign of the gamma also tells the direction of the relationship. However, be very careful when interpreting direction with ordinal-level variables. Remember that coding schemes for these variables are arbitrary and that a positive gamma may mean that the actual relationship is negative, and vice versa.

TABLE 13.2 **Relationship Between the Value of Gamma and the Strength of the Relationship**

Value	Strength
If the value is	*The strength of the relationship is*
between 0.00 and 0.30	Weak
between 0.31 and 0.60	Moderate
greater than 0.60	Strong

format; if you are working with data already in table format, you may have to rearrange the table or rethink the direction of patterns. Gamma is a symmetrical measure of association; that is, the value of gamma will be the same regardless of which variable is taken as independent. *(To practice computing and interpreting gamma, see problems 13.1 to 13.10 and 13.15. Begin with some of the smaller, 2 × 2 tables until you are comfortable with these procedures.)*

DETERMINING THE DIRECTION OF RELATIONSHIPS

Nominal measures of association like phi and lambda measure only the strength of a bivariate association. Ordinal measures of association, like gamma, are more sophisticated and add information about the overall direction of the relationship (positive or negative). In one way, it is easy to determine

---

Applying Statistics 13.1: Gamma

A group of 40 nations have been rated as high or low on religiosity (based on the percentage of a random sample of citizens who described themselves as "very religious") and as high or low in their support for single mothers (based on the percentage of a random sample of citizens who said they would approve of a woman's choosing to be a single parent). Are more religious nations less approving of single mothers?

	Religiosity		
Approval	Low	High	Totals
Low	4	9	13
High	11	16	27
Totals	15	25	40

The number of pairs of cases ranked in the same order on both variables (N_s) would be

$$N_s = 4(16) = 64$$

The number of pairs of cases ranked in different order on both variables (N_d) would be

$$N_d = 9(11) = 99$$

Gamma is

$$G = \frac{N_s - N_d}{N_s + N_d} = \frac{64 - 99}{64 + 99} = \frac{-35}{163} = -0.22$$

A gamma of -0.22 means that we would make 22% fewer errors predicting the order of pairs of cases on the dependent variable (approval of single mothers) when taking the independent variable (religiosity) into account. There is a moderate-to-weak, negative association between these two variables. As religiosity increases, approval decreases (that is, more religious nations are less approving of single mothers).

Applying Statistics 13.2: Gamma

For local political elections, do voters turn out at higher rates when the candidates for office spend more money on media advertising? Each of 177 localities has been rated as high or low on voter turnout. Also, the total advertising budgets for all candidates in each locality have been classified as high, moderate, or low. Are these variables associated?

Voter Turnout	Expenditure			Totals
	Low	Moderate	High	
Low	35	32	17	84
High	23	27	43	93
Totals	58	59	60	177

We will use gamma to summarize the strength and direction of the association. The number of pairs of cases ranked in the same order on both variables (N_s) would be

$$N_s = 35(27 + 43) + 32(43) = 2450 + 1376 = 3826$$

The number of pairs of cases ranked in different order on both variables (N_d) would be

$$N_d = 17(27 + 23) + 32(23) = 850 + 736 = 1586$$

Gamma is

$$G = \frac{N_s - N_d}{N_s + N_d}$$

$$G = \frac{3826 - 1586}{3826 + 1586}$$

$$G = \frac{2240}{5412}$$

$$G = 0.41$$

A gamma of 0.41 means that we would make 41% fewer errors by using candidates' advertising expenditures to predict turnout. This is a moderate, positive association.

---

direction: If gamma is a plus value, the relationship is positive. A minus sign for gamma indicates a negative relationship. Often, however, direction is confusing when working with ordinal-level variables, so it will be helpful if we focus on the matter specifically.

With gamma, a positive relationship means that the *scores* of cases tend to be ranked in the same order on both variables and that the variables change in the same direction. That is, as scores on one variable increase (or decrease), scores on the other variable also increase (or decrease). Cases tend to have scores in the same range on both variables (i.e., low scores go with low scores, moderate with moderate, and so forth). Table 13.3 illustrates the general shape of a positive relationship. Note that the cases tend to fall along a diagonal from upper left to lower right (assuming, of course, that tables have been constructed with the column variable increasing from left to right and the row variable from top to bottom).

Tables 13.4 and 13.5 present an example of a positive relationship with actual data. The sample consists of 186 preindustrial societies from around the globe. Each has been rated in terms of its degree of stratification or inequality and the type of political institution it has. In the societies that are the lowest in stratification, people share resources equally and the degree of inequality increases from left to right across the columns of the table. In "stateless" societies, there is no formal political institution and the political institution becomes more elaborate and stronger as you move down the rows from top to bottom. The gamma for this table is 0.86, so the relationship is strong and

TABLE 13.3   A Generalized Positive Relationship

	Variable X		
Variable Y	Low	Moderate	High
Low	X		
Moderate		X	
High			X

TABLE 13.4   State Structure by Degree of Stratification* (frequencies)

	Degree of Stratification		
Type of State	Low	Medium	High
Stateless	77	5	0
Semi-state	28	15	4
State	12	19	26
Totals	117	39	30

*Data are from the Human Relation Area File, standard cross-cultural sample.

positive. By inspection, we can see that most cases fall in the diagonal from upper left to lower right.

The percentages in Table 13.5 make it even clearer that societies with little inequality tend to be stateless and that the political institution becomes more elaborate as inequality increases. The great majority of the least stratified societies had no political institution, and none of the highly stratified societies were stateless.

In negative relationships, low scores on one variable are associated with high scores on the other and high scores with low scores. This means that the cases will tend to fall along a diagonal from lower left to upper right. Table 13.6 illustrates a generalized negative relationship. The cases with higher scores on variable X tend to have lower scores on variable Y, and score on Y decreases as score on X increases.

Tables 13.7 and 13.8 present an example of a negative relationship, with data taken from a public-opinion poll. The independent variable is church attendance, and the dependent variable is approval of cohabitation ("Is it all right for a couple to live together without intending to get married?"). Note that rates of attendance increase from left to right, and approval of cohabitation increases from top to bottom of the table.

Once again, the percentages in Table 13.8 make the pattern more obvious. The great majority of people who were low on attendance ("never") were high on approval of cohabitation, and most people who were high on attendance were low on approval. As attendance increases, approval of cohabitation tends to

TABLE 13.5   State Structure by Degree of Stratification (percentages)

	Degree of Stratification		
Type of State	Low	Medium	High
Stateless	65.8%	12.8%	0.0%
Semi-state	23.9%	38.5%	13.3%
State	10.3%	48.7%	86.7%
Totals	100.0%	100.0%	100.0%

TABLE 13.6   A Generalized Negative Relationship

	Variable X		
Variable Y	Low	Moderate	High
Low			X
Moderate		X	
High	X		

TABLE 13.7   Approval of Cohabitation by Church Attendance (frequencies)

	Attendance		
Approval	Never	Monthly or Yearly	Weekly
Low	37	186	195
Moderate	25	126	46
High	156	324	52
Totals	218	636	293

TABLE 13.8   Approval of Cohabitation by Church Attendance (percentages)

	Attendance		
Approval	Never	Monthly or Yearly	Weekly
Low	17.0	29.2	66.6
Moderate	11.5	19.8	15.7
High	71.6	50.9	17.8
Totals	100.1%	99.9%	100.1%

decrease. The gamma for this table is -0.57, indicating a moderate-to-strong, negative relationship between these variables.

You should be aware of an additional complication. The coding for ordinal-level variables is arbitrary, and a higher score may mean "more" or "less" of the variable being measured. For example, if we measured social class as upper, middle, or lower, we could assign scores to the categories in either of two ways:

A	B
(1) Upper	(3) Upper
(2) Middle	(2) Middle
(3) Lower	(1) Lower

While coding scheme B might seem preferable (because higher scores go with higher class position), *both* schemes are perfectly legitimate, and the direction of a relationship will change depending on which scheme is selected. Using scheme B, we would find a positive gamma between social class and (for example) education: As education increased, so would class.

With scheme A, however, the same relationship would appear to be negative because the numerical scores (1, 2, 3) are coded in reverse order: The highest social class is assigned the lowest score, and so forth. If you didn't check the coding scheme, you might conclude that class decreases as education increases, when, actually, the opposite is true.

Unfortunately, this source of confusion cannot be avoided when working with ordinal-level variables. Coding schemes will always be arbitrary for these variables, so you need to exercise additional caution when interpreting the direction of ordinal-level variables.

INTERPRETING ASSOCIATION WITH BIVARIATE TABLES

Up to this point, we have discussed bivariate association using tables, column percentages, and both nominal and ordinal measures of association. In this section, we summarize this material by using these techniques to address the relationship between level of development and the extent of political and civil rights for a sample of 89 nations.

The independent variable has three categories. High-income nations generally are industrial or postindustrial and have high levels of affluence, comfort, education, and health. Low-income nations, also called "Least Developed Countries" (LDCs), have less productive economies and low levels of education and health. Middle-income nations lie between the two extremes.

On the dependent variable (political and civil rights), nations that score high are strong democracies with free and fair elections, multiple political parties competing for office, universal suffrage, and institutionalized guarantees of free speech and the right of dissent. Citizens of nations that score low on this variable have few or no political rights, few or no meaningful choices between rival political parties, and little or no ability to express their grievances and concerns. These nations may be governed by a single political party or by a single person. In either case, the citizens have little or no control over the leadership of the nation or the policies of the state. Nations in the intermediate category have some—but not all—of the characteristics of a democratic political system.

Why should these variables be related? A full consideration of this question would carry us far beyond the bounds of a statistics text, of course, but we can suggest some relevant connections. A modern, industrial economy requires an educated, literate workforce and industrialization is always accompanied by mandatory schooling, free or low-cost public education, and an upgraded national school system. An educated population is more critical, sophisticated, and demanding and more likely to organize in pursuit of its self-interests. A literate, organized working and middle class will insist on full participation in affairs of state and demand the fullest possible civil and political rights. Thus, we can state a hypothesis: The higher the income level of a nation (the more industrialized its economy), the higher the level of political and civil rights. These variables should be substantially and positively related. Table 13.9 presents the results.

A gamma of 0.81 shows that there is a strong, positive relationship between income level and level of political and civil rights. The majority (65.2%) of low-income nations score low on the dependent variable and none score high. The great majority (81.5%) of high-income nations score high on political and civil rights. The two high-income nations that scored low (Saudi Arabia and the United Arab Emirates) base their affluence on oil production—not industrial or postindustrial economies—and lack the highly educated and literate working and middle classes of other high-income nations. A slight majority of middle-income nations have moderate scores on political and civil rights, but they are also more variable than nations at the other income levels.

TABLE 13.9 Political and Civil Rights by Income Level (frequencies and percentages)

| RIGHTS | INCOME LEVEL | | |
	Low	Middle	High
Low	15 (65.2%)	12 (30.8%)	2 (7.4%)
Moderate	8 (34.8%)	20 (51.3%)	3 (11.1%)
High	0 (0.0%)	7 (17.9%)	22 (81.5%)
Totals	23 (100.0%)	39 (100.0%)	27 (100.0%)

Chi square = 50.70, df = 4, $p < 0.000$, gamma = 0.81.

In the spring of 2010, the Pew Research Center published a poll showing that the American public was very unhappy with the federal government. Only 25% of the respondents said that the government could be trusted "all or most of the time," and high percentages expressed anger (21%) and frustration (56%) with the government. These were among the highest levels of discontent recorded over the past half century.

These emotions were not equally shared across the population. As shown in the table below, negative feelings were particularly likely to be expressed by older Americans. The relationship between negative feelings and age is statistically significant, moderate to weak in strength, and positive in direction (gamma = 0.31). This means that anger tended to increase as age increased or that older respondents were more likely to express anger than younger respondents. Note, in particular, that the age groups have very similar percentages of "frustrated" members but differ substantially in the percentage of angry respondents, with older people almost twice as likely to express this emotion.

Feelings Toward the Federal Government by Age

Feeling	Age	
	Younger (18–49)	Older (50 and older)
Content	27.1%	14.9%
Frustrated	56.4%	57.4%
Angry	16.5%	27.7%
	100.0%	100.0%
	(1064)	(1344)

$\chi^2 = 76.00$, df = 2, $p < 0.001$, gamma = 0.31.

Source: Pew Research Center. 2010. The People and Their Government. Available at http://people-press.org/reports/pdf/606.pdf.

Overall, the cases tend to fall in a diagonal from upper left to lower right: Low scores on income are associated with low scores on "rights" and almost all high-income nations score high on "rights." The expectations stated in the hypothesis are confirmed. Also, we should note that the relationship is statistically significant (although these 89 nations are not a random sample).

Spearman's RHO (r_s)

In the previous section, we considered ordinal variables that have a limited number of categories (possible values) and are presented in tables. However, many ordinal-level variables have a broad range and many distinct scores. These variables may be collapsed into a few broad categories (such as high, moderate, and low), organized into a bivariate table, and analyzed with gamma. Collapsing scores in this manner may be desirable in many instances, but it may also obscure or lose some important distinctions between cases.

For example, suppose a researcher wished to test the claim that jogging is beneficial, not only physically but also psychologically. Do joggers have an

TABLE 13.10 **The Scores of 10 Subjects on Involvement in Jogging and a Measure of Self-Esteem**

Jogger	Involvement in Jogging (X)	Self-Esteem (Y)
Wendy	18	15
Debbie	17	18
Phyllis	15	12
Stacey	12	16
Evelyn	10	6
Tricia	9	10
Christy	8	8
Patsy	8	7
Marsha	5	5
Lynn	1	2

enhanced sense of self-esteem? To deal with this issue, 10 female joggers are evaluated on two scales, the first measuring involvement in jogging and the other measuring self-esteem. Scores are reported in Table 13.10.

These variables could be collapsed by splitting the scores into two values (high and low). There are two potential difficulties with collapsing scores in this way. First, the scores seem continuous, and there are no obvious or natural division points in the distribution that would allow us to distinguish between high scores and low ones. Second, and more important, grouping these cases into broader categories will lose information. That is, if both Wendy and Debbie were placed in the category "high" on involvement, the fact that they had different scores would be lost. If differences like this are important and meaningful, then we should choose a measure of association that permits the retention of as much detail and precision in the scores as possible.

THE COMPUTATION OF RHO

Spearman's rho (r_s) is a measure of association for ordinal-level variables that have a broad range of many different scores and few ties between cases on either variable. Scores on ordinal-level variables cannot, of course, be manipulated mathematically except for judgments of "greater than" or "less than." To compute Spearman's rho, first the cases are ranked from high to low on each variable and then the ranks (not the scores) are manipulated to produce the final measure. Table 13.11 displays the original scores and the rankings of the cases on both variables.

To rank the cases, first find the highest score on each variable and assign it rank 1. Wendy has the high score on X (18) and is thus ranked number 1. Debbie, on the other hand, is highest on Y and is ranked first on that variable. All other cases are then ranked in descending order of scores. If any cases have the same score on a variable, assign them the average of the ranks they would have used up had they not been tied. Christy and Patsy have identical scores of 8 on involvement. Had they not been tied, they would have used up ranks 7 and 8. The average of these two ranks is 7.5, and this average of used ranks is assigned to all tied cases. (For example, if Marsha had also had a score of 8, three ranks—7, 8, and 9—would have been used, and all three tied cases would have been ranked eighth.)

The formula for Spearman's rho is

FORMULA 13.2
$$r_s = 1 - \frac{6\Sigma D^2}{N(N^2 - 1)}$$

where ΣD^2 = the sum of the differences in ranks, the quantity squared

To compute ΣD^2, the rank of each case on Y is subtracted from its rank on X (D is the difference between rank on Y and rank on X). A column has been provided in Table 13.11 to record the differences case by case. Note that the sum of this column (ΣD) is 0. That is, the negative differences in rank are equal to the positive differences, as will always be the case, and you should find the total of this column as a check on your computations to this point. If ΣD is not equal to 0, you have made a mistake either in ranking the cases or in subtracting the differences.

ONE STEP AT A TIME   **Computing and Interpreting Spearman's Rho**

To Compute Spearman's Rho

1: Set up a computing table like Table 13.11 to organize the computations. In the far left-hand column, list the cases in order, with the case with the highest score on the independent variable (X) stated first.

2: In the next column, list the scores on X.

3: In the third column, list the rank of each case on X, beginning with rank 1 for the highest score. If any cases have the same score, assign them the average of the ranks they would have used up had they not been tied.

4: In the fourth column, list the score of each case on Y, and then, in the fifth column, rank the cases on Y from high to low. Assign the rank of 1 to the case with the highest score on Y, and assign any tied cases the average of the ranks they would have used up had they not been tied.

5: For each case, subtract the rank on Y from the rank on X and write the difference (D) in the sixth column. Add up this column. If the sum is not zero, you have made a mistake and need to recompute.

6: Square the value of each D and record the result in the seventh column. Add this column up to find ΣD^2, and substitute this value into the numerator of Formula 13.2.

7: Multiply ΣD^2 by 6.

8: Square N and subtract 1 from the result.

9: Multiply the quantity you found in step 8 by N.

10: Divide the quantity you found in step 7 by the quantity you found in step 9.

11: Subtract the quantity you found in step 10 from 1. The result is r_s.

Interpreting the Strength of the Relationship

1: Use either or both of the following to interpret strength

a. Use Table 13.2 to describe strength in general terms.

b. Square the value of rho and multiply by 100. This value represents the percentage improvement in our predictions of the dependent variable by taking the independent variable into account.

Interpreting the Direction of the Relationship

1: Use the sign of r_s. Be careful when interpreting direction with ordinal-level variables. Remember that coding schemes for these variables are arbitrary and that a positive r_s may mean that the actual relationship is negative, and vice versa.

TABLE 13.11 Computing Spearman's RHO

	Involvement (X)	Rank	Self-Image (Y)	Rank	D	D²
Wendy	18	1	15	3	−2	4
Debbie	17	2	18	1	1	1
Phyllis	15	3	12	4	−1	1
Stacey	12	4	16	2	2	4
Evelyn	10	5	6	8	−3	9
Tricia	9	6	10	5	1	1
Christy	8	7.5	8	6	1.5	2.25
Patsy	8	7.5	7	7	0.5	0.25
Marsha	5	9	5	9	0	0
Lynn	1	10	2	10	0	0
					$\Sigma D = 0$	$\Sigma D^2 = 22.50$

STATISTICS
IN EVERYDAY LIFE

Which nations offer the highest quality of life? Where is the most desirable place to live? A number of organizations have developed ranking systems—for cities as well as nations—to help people answer these questions. One such ranking is published by an organization called International Living (IL) (http://www.internationalliving.com), which seeks to find places around the globe that offer very comfortable lifestyles at bargain prices. Their annual ranking rates nations on nine variables, including cost of living, health care, and safety. For 2010, its top ten nations (with several tied scores) are:

Rank	Nation
1	France
2	Austria
	Switzerland
	Germany
5	New Zealand
6	Luxembourg
	United States
	Belgium
9	Canada
	Italy

It may seem at first glance that the rankings simply reflect levels of income, industrialization, and affluence. However, IL tries to capture a variety of other factors in their rankings. For example, a Spearman's rho calculated between the rankings of all 194 nations included in the IL study and gross domestic product per capita (a measure of income and affluence) indicates a strong—but not perfect—association of 0.69. Thus, the IL rankings reflect features of social life other than affluence.

Applying Statistics 13.3: Spearman's Rho

Five cities have been rated on an index that measures the quality of life. Also, the percentage of the population that has moved into each city over the past year has been determined. Have cities with higher quality-of-life scores attracted more new residents? The following table summarizes the scores, ranks, and differences in ranks for each of the five cities.

City	Quality of Life	Rank	% New Residents	Rank	D	D^2
A	30	1	17	1	0	0
B	25	2	14	3	−1	1
C	20	3	15	2	1	1
D	10	4	3	5	−1	1
E	2	5	5	4	1	1
					$\Sigma D = 0$	$\Sigma D^2 = 4$

Spearman's rho for these variables is

$$r_s = 1 - \frac{6\Sigma D^2}{N(N^2 - 1)}$$

$$r_s = 1 - \frac{(6)(4)}{5(25 - 1)}$$

$$r_s = 1 - \left(\frac{24}{120}\right)$$

$$r_s = 1 - 0.20$$

$$r_2 = 0.80$$

These variables have a strong, positive association. The higher the quality-of-life score, the greater the percentage of new residents. The value of r_s^2 is 0.64 ($0.80^2 = 0.64$), which indicates that we will make 64% fewer errors when predicting rank on one variable from rank on the other, as opposed to ignoring rank on the other variable.

In the column headed D^2, each difference is squared to eliminate negative signs. The sum of this column is ΣD^2, and this quantity is entered directly into the formula. For our sample problem:

$$r_s = 1 - \frac{6\Sigma D^2}{N(N^2 - 1)}$$

$$r_s = 1 - \frac{6(22.5)}{10(100 - 1)}$$

$$r_s = 1 - \frac{135}{990}$$

$$r_s = 1 - 0.14$$

$$r_s = 0.86$$

THE INTERPRETATION OF RHO

Spearman's rho is an index of the strength of association between the variables; it ranges from 0 (no association) to ±1.00 (perfect association). A perfect positive association ($r_s = +1.00$) would exist if there were no disagreements in ranks between the two variables (if cases were ranked in exactly the same order on both variables). A perfect negative relationship ($r_s = -1.00$) would exist if the ranks were in perfect disagreement (if the case ranked highest on one variable were lowest on the other, and so forth). A Spearman's rho of 0.86 indicates a strong, positive relationship between these two variables. The respondents who were highly involved in jogging also ranked high on self-image. These results are supportive of claims regarding the psychological benefits of jogging.

Spearman's rho is an index of the relative strength of a relationship, and values between 0 and ±1.00 have no direct interpretation. However, if the value of rho is squared, a PRE interpretation is possible. Rho squared (r_s^2) represents the proportional reduction in errors of prediction when predicting rank on one variable from rank on the other variable, as compared to predicting rank while ignoring the other variable. In our example, r_s was 0.86 and r_s^2 would be 0.74. Thus, our errors of prediction would be reduced by 74% if we used the rank on jogging to predict rank on self-image. *(For practice in computing and interpreting Spearman's rho, see problems 13.11 to 13.14. Problem 13.11 has the smallest number of cases and is probably a good choice for a first attempt at these procedures.)*

Testing the Null Hypothesis of "No Association" with Gamma and Spearman's Rho

If we are working with random samples, we will need to ascertain if the sample findings can be generalized to the population. In Part II of this text, we considered various ways that information taken from samples—for example, the difference between two sample means—could be generalized to the populations from which the samples were drawn. Measures of association like gamma and Spearman's rho can also be tested for significance.

For nominal-level variables, the statistical significance of a relationship is usually judged with the chi square statistic. We could also use this test for tables displaying the relationship between ordinal-level variables. However, chi square tests the probability that the observed cell frequencies occurred by chance alone and is not a direct test of the significance of the measure of association (gamma or Spearman's rho) itself.

When testing gamma and Spearman's rho for statistical significance, the null hypothesis will state that there is no association between the variables in the population and that, therefore, the population value for the measure is 0.00. Population values will be denoted by the Greek letters gamma (γ) and rho (ρ_s). For both measures, the test procedures will be organized around the familiar five-step model (see Chapter 8).

TESTING GAMMA FOR SIGNIFICANCE

To illustrate the test of significance for gamma, we will use Table 13.1, where gamma was 0.57.

Step 1. Making Assumptions and Meeting Test Requirements. When sample size is greater than 10, we can assume that the sampling distribution of all possible sample gammas is normal in shape.

Model: Random sampling
Level of measurement is ordinal
Sampling distribution is normal

Step 2. Stating the Null Hypothesis.

$$H_0: \gamma = 0.0$$
$$(H_1: \gamma \pm 0.0)$$

Step 3. Selecting the Sampling Distribution and Establishing the Critical Region. For samples of 10 or more, the Z distribution (Appendix A) can be used to find areas under the sampling distribution:

Sampling distribution $= Z$ distribution
Alpha $= 0.05$
Z(critical) $= \pm 1.96$

Step 4. Computing the Test Statistic.

FORMULA 13.3

$$Z(\text{obtained}) = G\sqrt{\frac{N_s + N_d}{N(1 - G^2)}}$$

$$Z(\text{obtained}) = G\sqrt{\frac{N_s + N_d}{N(1 - G^2)}} = 0.57\sqrt{\frac{1831 - 499}{100(1 - 0.33)}} = 0.57\sqrt{\frac{2330}{100(0.67)}}$$

$$Z(\text{obtained}) = 0.57\sqrt{34.78} = 3.36$$

Step 5. Making a Decision and Interpreting the Results of the Test. Comparing the Z(obtained) with the Z(critical), we get

$$Z(\text{obtained}) = 3.36$$
$$Z(\text{critical}) = \pm 1.96$$

The null hypothesis can be rejected. The sample gamma is unlikely to have occurred by chance alone, and we may conclude that these variables are related in the population from which the sample was drawn. *(For practice in conducting and interpreting the test of significance for gamma, see problems 13.2, 13.4, 13.7, 13.10, and 13.15.)*

TESTING SPEARMAN'S RHO FOR SIGNIFICANCE

For Spearman's rho, the null hypothesis states that the population value (ρ_s) is actually 0 and, therefore, the value of the sample Spearman's rho (r_s) is the result of mere random chance. When the number of cases in the sample is 10 or more, the sampling distribution of Spearman's rho approximates the *t* distribution, and we will use this distribution to conduct the test. To illustrate, we will use the Spearman's rho computed for the data presented in Table 13.11.

Step 1. Making Assumptions and Meeting Test Requirements.

Model: Random sampling
Level of measurement is ordinal
Sampling distribution is normal

Step 2. Stating the Null Hypothesis.

$$H_0: \rho_s = 0.0$$

$$(H_1: \rho_s \pm 0.0)$$

Step 3. Selecting the Sampling Distribution and Establishing the Critical Region.

Sampling distribution = t distribution
Alpha = 0.05
Degrees of freedom = $N - 2 = 8$
t(critical) = ± 2.306

Step 4. Computing the Test Statistic.

FORMULA 13.4

$$t(\text{obtained}) = r_s \sqrt{\frac{N - 2}{1 - r_s^2}}$$

$$t(\text{obtained}) = r_s \sqrt{\frac{N - 2}{1 - r_s^2}} = 0.86 \sqrt{\frac{8}{1 - 0.74}} = 0.86 \sqrt{\frac{8}{0.26}}$$

$$t(\text{obtained}) = 0.86 \sqrt{30.77} = (0.86)(5.55) = 4.77$$

Step 5. Making a Decision and Interpreting the Results of the Test.

Comparing the test statistic with the critical region:

$$t(\text{obtained}) = 4.77$$
$$t(\text{critical}) = \pm 2.306$$

The null hypothesis can be rejected. We may conclude, with a 0.05 chance of making an error, that the variables are related in the population from which the samples were drawn. *(For practice in conducting and interpreting the test of significance for Spearman's rho, see problems 13.11 to 13.14.)*

READING STATISTICS 9: Bivariate Tables and Associated Statistics

The statistics associated with any bivariate table will usually be reported directly below the table itself. This information would include the name of the measure of association, its value, and (if relevant) its sign. Also, if the research involves a random sample, the results of the chi square test or a test for the significance of the measure itself will be reported. So the information might look something like this:

$$\chi^2 = 13.23 \ (2df, \ p < 0.05), \ \lambda = 0.47$$

Note that the alpha level is reported in the "p" format, which we have discussed previously.

Besides simply reporting the value of these statistics, the researcher will interpret them in the text of the article. These statistics might be characterized by the statement, "The association between the variables is statistically significant and strong." Again we see that researchers will avoid our rather wordy (but more careful) style of stating results. Where we might say, "A lambda of 0.47 indicates that we will reduce our errors of prediction by 47% when predicting the dependent variable from the independent variable, as opposed to predicting the dependent while ignoring the independent," the researcher will simply report the value of lambda and characterize that value in a word or two (e.g., "The association is strong"). The researcher assumes that his or her audience is statistically literate and can make the more detailed interpretations.

STATISTICS IN THE PROFESSIONAL LITERATURE

While gender inequality seems to be a nearly universal characteristic of human society, the extent of male advantage is highly variable. Why are some societies more unequal than others? Sociologists Stephen Sanderson, Alex Heckert, and Joshua Dubrow sought some answers to this question by examining the correlates of gender inequality in a sample of preindustrial societies. They tested a number of hypotheses but found the most support for what they called "materialist" theories, which argued that "the greater the extent to which women are involved in economic production, the higher their status tends to be" (p. 1426). Table 1 reports their findings for the hypotheses derived from this theory. The entries in the table are gammas, and the asterisks indicate the statistical significance of the relationship based on a chi square test.

To read this table, note that the three variables in the columns measure the dependent variable, or the status of women. The four row variables measure the independent variable, or the importance of women for various economic activities. The entries in the table are gammas. The first three row variables are self-explanatory, and the higher the score of the society on these variables, the greater the economic importance of women. The variable in the bottom row measures the extent to which the society depends on the plow as an instrument of production, and its meaning may not be so obvious. Since it requires a good deal of upper-body strength, plow agriculture is

TABLE 1

Measures Related to Materialist Theories (Independent Variables)	Measures of Women's Status (Dependent Variables)		
	Domestic Authority	Control of Sexuality	Female Solidarity
Female contribution to gathering	0.23	0.38*	0.45***
Female contribution to subsistence	0.05	0.13	−0.01
Female contribution to agriculture	0.46***	0.12	0.57***
Use of the plow	−0.51**	−0.63**	−0.58***

N's range from 67 to 93.

*$p < 0.05$; **$p < 0.01$; ***$p < 0.001$.

(*continued*)

generally "men's work," and the greater the extent to which the society uses plows, the lower the economic importance of women.

Is the materialist theory supported? The relationships between the first three measures of women's economic importance and women's status (the three column variables) are positive in direction (with one exception) and weak to moderate in strength. The positive relationships indicate that women's status increases as their economic importance increases, and the strength and significance of the relationships provide moderate (but not overwhelming) support for the theory being tested. Looking at the bottom row, we see somewhat stronger support for the materialist theory. All three relationships are statistically significant at less than the 0.05 level, moderate to strong, and negative—the direction predicted by the materialist

theory ("The greater the reliance on the plow, the lower the status of women").

Although I have not reproduced them here, the tests of competing theories showed weaker and less significant relationships, and the authors conclude that the materialist perspective is most consistent with the empirical relationships they were able to examine. Although these results cannot be considered proof (remember that correlation is not the same thing as causation), they are very consistent with the predictions of materialist theory.

Source: Sanderson, Stephen, Heckert, Alex, and Dubrow, Joshua. 2005. "Militarist, Marxian, and Non-Marxian Theories of Gender Inequality: A Cross-Cultural Test." *Social Forces*, 83:1425–1441.

SUMMARY

1. A measure of association for variables with collapsed ordinal scales (gamma) was covered along with a measure (Spearman's rho) appropriate for "continuous" ordinal variables. Both measures summarize the overall strength and direction of the association between the variables.

2. Gamma is a PRE-based measure that shows the improvement in our ability to predict the order of pairs of cases on one variable from the order of pairs of cases on the other variable, as opposed to ignoring the order of the pairs of cases on the other variable.

3. Spearman's rho is computed from the ranks of the scores of the cases on two "continuous" ordinal variables and, when squared, can be interpreted by the logic of PRE.

4. Both gamma and Spearman's rho should be tested for their statistical significance when computed for a random sample drawn from a defined population. The null hypothesis is that the variables are not related in the population, and the test can be organized by using the familiar five-step model.

SUMMARY OF FORMULAS

Gamma	13.1	$G = \dfrac{N_s - N_d}{N_s + N_d}$
Spearman's rho	13.2	$r_s = 1 - \dfrac{6\sum D^2}{N(N^2 - 1)}$
Z(obtained) for gamma	13.3	$Z(\text{obtained}) = G\sqrt{\dfrac{N_s + N_d}{N(1 - G^2)}}$
t(obtained) for Spearman's rho	13.4	$t(\text{obtained}) = r_s\sqrt{\dfrac{N - 2}{1 - r_s^2}}$

GLOSSARY

Gamma (*G*). A measure of association appropriate for variables measured with "collapsed" ordinal scales that have been organized into table format; *G* is the symbol for any sample gamma, γ is the symbol for any population gamma.

N_d. The number of pairs of cases ranked in different order on two variables.

N_s. The number of pairs of cases ranked in the same order on two variables.

Spearman's rho (r_s). A measure of association appropriate for ordinally measured variables that are "continuous" in form; r_s is the symbol for any sample Spearman's rho; ρ_s is the symbol for any population Spearman's rho.

PROBLEMS

For problems 13.1 to 13.10 and 13.15, calculate percentages for the bivariate tables as described in Chapter 12. Use the percentages to help analyze the strength and direction of the association.

13.1 [SOC] A small sample of non-English-speaking immigrants to the United States has been interviewed about their level of assimilation. Is the pattern of adjustment affected by length of residence in the United States? For each table compute gamma and summarize the relationship in terms of strength and direction. (*HINT: In 2 × 2 tables, only two cells can contribute to N_s or N_d. To compute N_s, multiply the number of cases in the upper-left-hand cell by the number of cases in the lower-right-hand cell. For N_d, multiply the number of cases in the upper-right-hand cell by the number of cases in the lower-left-hand cell.*)

a. Facility in English:

	Length of Residence		
English Facility	Less Than Five Years (Low)	More Than Five Years (High)	Totals
Low	20	10	30
High	5	15	20
Totals	25	25	50

b. Total family income:

	Length of Residence		
Income	Less Than Five Years (Low)	More Than Five Years (High)	Totals
Below national average (1)	18	8	26
Above national average (2)	7	17	24
Totals	25	25	50

c. Extent of contact with country of origin:

	Length of Residence		
Contact	Less Than Five Years (Low)	More Than Five Years (High)	Totals
Rare (1)	5	20	25
Frequent (2)	20	5	25
Totals	25	25	50

13.2 Compute gamma for the table presented in problem 12.1. If you computed a nominal measure of association for this table previously, compare the measures of association. Are they similar in value? Do they characterize the strength for the association in the same way? What information about the relationship does gamma provide that is not available from nominal measures of association?

13.3 [CJ] A random sample of 150 cities has been classified as small, medium, or large by population and as high or low on crime rate. Is there a relationship between city size and crime rate?

	City Size			
Crime Rate	Small	Medium	Large	Totals
Low	21	17	8	46
High	29	33	42	104
Totals	50	50	50	150

a. Describe the strength and direction of the relationship.
b. Is the relationship significant?

13.4 [SOC] Some research has shown that families vary by how they socialize their children to sports, games, and other leisure-time activities. In middle-class families, such activities are carefully monitored by parents and are, in general, dominated by adults (for example, Little League baseball). In working-class families, children more often organize and initiate such activities themselves, and parents are much less involved (for example, sandlot or playground baseball games). Are the following data consistent with these findings? Summarize your conclusions in a few sentences.

As a Child, Did You Play Mostly Organized or Sandlot Sports?	Social Class		
	Middle Class	Working Class	Totals
Organized	155	123	278
Sandlot	101	138	239
Totals	256	261	517

13.5 Is there a relationship between age and support for traditional gender roles? Is the relationship between the variables different for different nations? The World Values Survey has been administered to random samples drawn from Canada, the United States, and Mexico. Respondents were asked if they agree or disagree that a university education is more important for boys than for girls. Compute column percentages and gamma for each table. Is there a relationship? Describe the strength and direction of the relationship. Which age group is most supportive of traditional gender roles (or: agrees with the statement). How does the relationship change from nation to nation?

a. Canada

"University Is More Important for Boys"	Age Group		
	15–29	30–49	50 and older
Agree	17	24	64
Disagree	384	804	841
	401	828	905

b. United States

"University Is More Important for Boys"	Age Group		
	15–29	30–49	50 and older
Agree	29	29	29
Disagree	238	418	478
	267	447	507

c. Mexico

"University Is More Important for Boys"	Age Group		
	15–29	30–49	50 and older
Agree	109	154	114
Disagree	385	489	259
	494	643	373

13.6 PA All applicants to the Allegory, New Mexico, Police Academy are given an aptitude test, but the test has never been evaluated to see if test scores are in any way related to job performance. The following table reports aptitude test scores and job performance ratings for a random sample of 78 Allegory policemen.

Efficiency Ratings	Test Scores			
	Low	Moderate	High	Totals
Low	12	6	7	25
Moderate	9	11	9	29
High	5	9	10	24
Totals	26	26	26	78

a. Are these two variables associated? Describe the strength and direction of the relationship in a sentence or two.

b. Is gamma statistically significant?

c. Should the aptitude test continue to be administered? Why or why not?

13.7 SW A sample of children has been observed and rated for symptoms of depression. Their parents have been rated for authoritarianism. Is there any relationship between these variables? Write a few sentences stating your conclusions.

Symptoms of Depression	Authoritarianism			
	Low	Moderate	High	Totals
Few	7	8	9	24
Some	15	10	18	43
Many	8	12	3	23
Totals	30	30	30	90

13.8 SOC Are prejudice and level of education related? State your conclusion in a few sentences.

| Prejudice | Level of Education | | | | |
	Elementary School	High School	Some College	College Graduate	Totals
Low	48	50	61	42	201
High	45	43	33	27	148
Totals	93	93	94	69	349

13.9 | SOC | In a recent survey, a random sample of respondents was asked to indicate how happy they were with their situations in life. Are their responses related to income level?

| Happiness | Income | | | |
	Low	Moderate	High	Totals
Not happy	101	82	36	219
Pretty happy	40	227	100	367
Very happy	216	198	203	617
Totals	357	507	339	1203

a. Describe the strength and direction of the relationship.

b. Is the relationship significant?

13.10 | SOC | In a random sampling, 11 neighborhoods in Leonia, Illinois, have been rated by an urban sociologist on a "quality-of-life" scale (which includes measures of affluence, availability of medical care, and recreational facilities) and a social cohesion scale. The results are presented here in scores. Higher scores indicate higher "quality of life" and greater social cohesion.

Neighborhood	Quality of Life	Social Cohesion
Beaconsdale	17	8.8
Brentwood	40	3.9
Chesapeake Shores	47	4.0
College Park	90	3.1
Denbigh Plantation	35	7.5
Kingswood	52	3.5
North End	23	6.3
Phoebus	67	1.7
Riverview	65	9.2
Queens Lake	63	3.0
Windsor Forest	100	5.3

a. Are the two variables associated? What is the strength and direction of the association?

Summarize the relationship in a sentence or two. (*HINT: Don't forget to square the value of Spearman's rho for a PRE interpretation.*)

b. Conduct a test of significance for this relationship. Summarize your findings.

13.11 | SW | Several years ago, a job-training program began, and a team of social workers screened the candidates for suitability for employment. Now the screening process is being evaluated, and the actual work performance of a sample of hired candidates has been rated. Did the screening process work? Is there a relationship between the original scores and performance evaluation on the job?

Case	Original Score	Performance Evaluation
A	17	78
B	17	85
C	15	82
D	13	92
E	13	75
F	13	72
G	11	70
H	10	75
I	10	92
J	10	70
K	9	32
L	8	55
M	7	21
N	5	45
O	2	25

13.12 | SOC | Following are the scores of a sample of 15 nations on a measure of ethnic diversity (the higher the number, the greater the diversity) and a measure of economic inequality (the higher the score, the greater the inequality). Are these variables related? Are ethnically diverse nations more economically unequal?

Nation	Diversity	Inequality
India	91	29.7
South Africa	87	58.4
Kenya	83	57.5
Canada	75	31.5
Malaysia	72	48.4

(*continued on next page*)

(*continued*)

Nation	Diversity	Inequality
Kazakhstan	69	32.7
Egypt	65	32.0
United States	63	41.0
Sri Lanka	57	30.1
Mexico	50	50.3
Spain	44	32.5
Australia	31	33.7
Finland	16	25.6
Ireland	4	35.9
Poland	3	27.2

13.13 Twenty ethnic, racial, or national groups were rated by a random sample of white and black students on a Social Distance Scale. Lower scores represent less social distance and less prejudice. How similar are these rankings? Is the relationship statistically significant?

	Average Social Distance Scale Score	
Group	White Students	Black Students
1 White Americans	1.2	2.6
2 English	1.4	2.9
3 Canadians	1.5	3.6
4 Irish	1.6	3.6
5 Germans	1.8	3.9
6 Italians	1.9	3.3
7 Norwegians	2.0	3.8
8 American Indians	2.1	2.7
9 Spanish	2.2	3.0
10 Jews	2.3	3.3
11 Poles	2.4	4.2
12 Black Americans	2.4	1.3
13 Japanese	2.8	3.5
14 Mexicans	2.9	3.4
15 Koreans	3.4	3.7
16 Russians	3.7	5.1
17 Arabs	3.9	3.9
18 Vietnamese	3.9	4.1
19 Turks	4.2	4.4
20 Iranians	5.3	5.4

13.14 In problem 12.9, we looked at the relationships between five dependent variables and sex. In this exercise, we use income as an independent variable and assess its relationship with this same set of variables. For each table, calculate percentages and gamma. Describe the strength, direction, and statistical significance of each relationship in a few sentences. *Be careful in interpreting direction.*

a. Support for the legal right to an abortion by income:

Right to an Abortion?	Income			Totals
	Low	Moderate	High	
Yes	220	218	226	664
No	366	299	250	915
Totals	586	517	476	1579

b. Support for capital punishment by income:

Capital Punishment?	Income			Totals
	Low	Moderate	High	
Favor	567	574	552	1693
Oppose	270	183	160	613
Totals	837	757	712	2306

c. Approval of suicide for people with an incurable disease by income:

Right to Suicide?	Income			Totals
	Low	Moderate	High	
Approve	343	341	338	1022
Oppose	227	194	147	568
Totals	570	535	485	1590

d. Support for sex education in public schools by income

Sex Education?	Income			Totals
	Low	Moderate	High	
For	492	478	451	1421
Against	85	68	53	206
Totals	577	546	504	1627

e. Support for traditional gender roles by income:

Women Should Take Care of Running Their Homes and Leave Running the Country to Men	Income			Totals
	Low	Moderate	High	
Agree	130	71	39	240
Disagree	448	479	461	1388
Totals	578	550	500	1628

Exploring Sexual Attitudes and Behavior

Two projects are presented to help you apply the skills developed in this chapter. Both focus on sex, a subject of great fascination to many people. The first project uses bivariate tables and gamma to explore the possible causes of attitudes and opinions about premarital sex. The second uses Spearman's rho to investigate sexual behavior, specifically the number of different sexual partners people have had over the past five years.

PROJECT 1: Who Approves of Premarital Sex?

What type of person is most likely to approve of sex before marriage? In this exercise, you will take attitudes toward premarital sex (*premarsx*) as the dependent variable. The wording and coding scheme for the variable are presented in the table below. Remember that the coding scheme for ordinal-level variables is arbitrary. In this case, higher scores indicate greater approval of premarital sex and lower scores mean greater disapproval. Keep the coding scheme in mind as you analyze the direction of the relationships you find.

There's been a lot of discussion about the way morals and attitudes about sex are changing in this country. If a man and a woman have sex relations before marriage, do you think this is	
1	Always wrong
2	Almost always wrong
3	Sometimes wrong
4	Not wrong at all

Step 1: Choosing Independent Variables

Select four variables from the 2008 GSS that you think might be important causes of attitudes toward premarital sex and list the variables in the table below. Your independent variables *cannot* be nominal in level of measurement and *cannot* have more than four to five categories or scores. Some of the ordinal-level variables in which you might be interested (such as *attend*, a measure of church attendance) have more than four to five categories and should be recoded, as should interval-ratio variables such as *age* or *income06*. See Chapter 10 for instructions on recoding. Select independent variables that seem likely to be an important cause of people's attitudes about sex. Be sure to note the coding scheme for each variable.

SPSS Variable Name	What exactly does this variable measure?
_____	_____
_____	_____
_____	_____
_____	_____

Step 2: Stating Hypotheses

State hypotheses about the relationships you expect to find between your independent variables and *premarsx*. State these hypotheses in terms of the direction of the relationship you expect to find. For example, you might hypothesize that approval of premarital sex will decline as age increases (a negative relationship).

Step 3: Running Crosstabs

Click **Analyze → Descriptives → Crosstabs** and place the dependent variable (*premarsx*) in the **Rows:** box and all of the independent variables you selected in the **Columns:** box. Click the **Statistics** button to get chi square and gamma and the **Cells** button to get column percentages.

Step 4: Recording Results

These commands will generate a lot of output, and it will be helpful to summarize your results in the following table.

Independent Variable	Chi Square Significant at < 0.05? (Write *Yes* or *No*)	Gamma (Be sure to note the sign or direction as well as numerical value)
_____	_____	_____
_____	_____	_____
_____	_____	_____
_____	_____	_____

Step 5: Analyzing and Interpreting Results

For each independent variable, analyze and interpret the significance, strength, and direction of the relationship. For example, you might say the following: "There was a moderate negative relationship between age and approval of premarital sex (chi square = 4.26; df = 2; $p < 0.05$, gamma = 0.35). Older respondents tended to be more opposed." Be sure to *explain* the direction of the relationship; don't just characterize it as negative or positive. *Be careful when interpreting direction with ordinal variables.*

Independent Variable	Interpretation
1	
2	
3	
4	

Step 6: Testing Hypotheses

Write a few sentences of overall summary for this test. Were your hypotheses supported? Which independent variable had the strongest relationship with approval of premarital sex? Why do you think this is so?

PROJECT 2: Sexual Behavior

In this exercise, your dependent variable will be *partnrs5*, which measures how many different sex partners a person has had over the past five years. The wording and coding scheme for the variable are presented in the table below. For this variable, higher scores mean a greater number of partners, so interpreting the direction of relationships should be relatively straightforward.

How many sex partners have you had over the past five years?	
0	No partners
1	1 partner
2	2 partners
3	3 partners
4	4 partners
5	5–10 partners
6	11–20 partners
7	21–100 partners
8	More than 100 partners

Note that, like the *sexfreq* variable we used in chapter 10, *partnrs5* is interval-ratio for the first five categories and then becomes ordinal for higher scores. In other words, the variable has a true zero point (a score of 0 means no sex partners at all) and increases by equal, defined units from one partner through four partners. Higher scores, however, represent broad categories, not exact numbers. This mixture of levels of measurement makes Spearman's rho an appropriate statistic to use to measure the strength and direction of relationships

To access Spearman's rho, click **Analyze → Correlate → Bivariate**, and the **Bivariate Correlations** window will open. The variables are listed in the window on the left. Below that window is a box labeled **Correlation Coefficients**. Click **Spearman** from the options to get Spearman's rho. Next, select *partnrs5* and your independent variables (see below) from the list on the left; click the arrow to move them to the **Variables:** window on the right. SPSS will compute Spearman's rho for all pairs of variables included in the **Variables:** window.

Let's take a brief look at the output produced by this procedure. I looked at some relationships that are *not* particularly sensible so as not to usurp any of the more interesting ideas you might have for this (or other) projects. I take *concong* (the respondent's degree of confidence in the U.S. Congress) as the dependent variable and *attend* (frequency of church attendance) and *papres80* (the relative prestige of the respondent's father's occupation) as independent variables.

The output from the **Correlate** procedure is a table showing the bivariate correlations of all possible combinations of variables, including the relationship of the variable with itself. The table is called a correlation matrix and will look like the table below. We will deal with the correlation matrix in more detail in Chapter 14. (Note that this table has been edited to fit this space and will not look exactly like the SPSS output.)

Correlations

		CONFIDENCE IN US CONGRESS	HOW OFTEN R ATTENDS RELIGIOUS SERVICES	FATHER'S OCCUPATIONAL PRESTIGE SCORE (1980)
CONFIDENCE IN US CONGRESS	Correlation Coefficient	1.000	-.119(**)	-.016
	Sig. (2-tailed)	.	.000	.653
	N	994	990	798
HOW OFTEN R ATTENDS RELIGIOUS SERVICES	Correlation Coefficient	-.119(**)	1.000	-.046
	Sig. (2-tailed)	.000	.	.113
	N	990	1478	1170
FATHER'S OCCUPATIONAL PRES'TIGE SCORE (1980)	Correlation Coefficient	-.016	-.046	1.000
	Sig. (2-tailed)	.653	.113	.
	N	798	1170	1176

**Correlation is significant at the 0.01 level (2-tailed).

Since we have three variables, the table has nine cells, and there are three pieces of information in each cell: the value of Spearman's rho, the statistical significance of the r_s, and the number of cases. The cell on the upper left shows the relationship between *concong* and itself, the next cell to the right shows the relationship between *concong* and *attend*, and so forth. The output shows a weak negative relationship between *concong* and *attend* and a very weak negative relationship between *concong* and *papres80*. Now it's your turn.

Step 1: Choosing Independent Variables for *partnrs5*

Select four variables from the 2008 GSS that you think might be important causes of this dimension of sexual behavior. Your independent variables *cannot* be nominal in

FIGURE 14.2 Positive (a), Negative (b), and Zero (c) Relationships

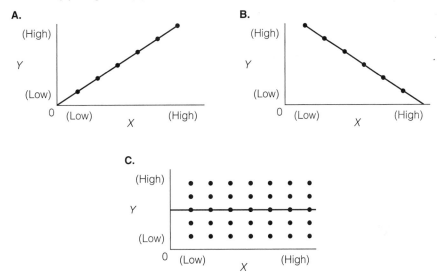

housework (*Y*) also increases. Husbands in families with more children tend to do more housework. If the relationship had been negative, the regression line would have sloped in the opposite direction to indicate that high scores on one variable were associated with low scores on the other.

To summarize these points, Figure 14.2 shows a perfect positive and a perfect negative relationship and a "zero relationship," or "nonrelationship," between two variables.

LINEARITY

One key assumption underlying the statistical techniques to be introduced later in this chapter is that the two variables have an essentially **linear relationship**. In other words, the observation points or dots in the scattergram must form a pattern that can be approximated with a straight line. Significant departures from linearity would require the use of statistical techniques beyond the scope of this text. Examples of some common curvilinear relationships are presented in Figure 14.3. If the scattergram shows that the variables have a nonlinear relationship, the techniques described in this chapter should be used with great caution or not at all. Checking for the linearity of the relationship is perhaps the most important reason for examining the scattergram before proceeding with the statistical analysis. If the relationship is nonlinear, you might need to treat the variables as if they were ordinal rather than interval-ratio in level of measurement. (*For practice in constructing and interpreting scattergrams, see problems 14.1 to 14.4.*)

USING SCATTERGRAMS FOR PREDICTION

A final use of the scattergram is to predict scores of cases on one variable from their score on the other. To illustrate, suppose that, based on the relationship between number of children and husband's housework displayed in Figure 14.1,

FIGURE 14.3 Some Nonlinear Relationships

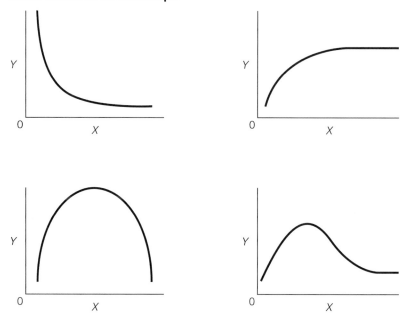

we wish to predict the number of hours of housework a husband with a family of six children would do each week. The sample has no families with six children, but if we extend the axes and regression line in Figure 14.1 to incorporate this score, a prediction is possible. Figure 14.4 reproduces the scattergram and illustrates how the prediction would be made.

The predicted score on Y—which is symbolized as Y' ("Y-prime") to distinguish predictions of Y from actual Y scores—is found by first locating the relevant score on X ($X = 6$ in this case) and then drawing a straight line from

FIGURE 14.4 Predicting Husband's Housework

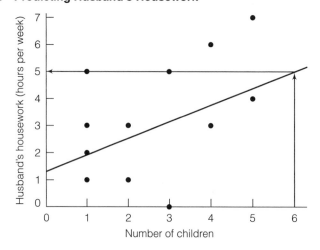

that point to the regression line. From the regression line, another straight line parallel to the X axis is drawn across to the Y axis. The predicted Y score (Y') is found at the point where the line crosses the Y axis. In our example, we would predict that, in a dual–wage-earner family with six children, the husband would devote about five hours per week to housework.

THE REGRESSION LINE

Of course, this prediction technique is crude, and the value of Y' can change, depending on how accurately the freehand regression line is drawn. One way to eliminate this source of error would be to find the straight line that most accurately describes the relationship between the two variables. How can we find this line?

Recall that our criterion for the freehand regression line was that it touch all the dots or come as close to doing so as possible. Also, recall that the dots above each value of X can be thought of as conditional distributions of Y, the dependent variable. Within each conditional distribution of Y, the mean is the point around which the variation of the scores is at a minimum. In Chapter 3, we noted that the mean of any distribution of scores is the point around which the variation of the scores, as measured by squared deviations, is minimized:

$$\sum (X_i - \overline{X})^2 = \text{minimum}$$

Thus, if the regression line is drawn so that it touches each **conditional mean of** Y, it would be the straight line that comes as close as possible to all the scores.

Conditional means are found by summing all Y values for each value of X and then dividing by the number of cases. For example, four families had one child ($X = 1$), and the husbands of these four families devoted 1, 2, 3, and 5 hours of housework. The conditional mean of Y for $X = 1$ is 2.75 (11/4 = 2.75). Husbands in families with one child worked an average of 2.75 hours per week doing housekeeping chores. Conditional means of Y are computed in the same way for each value of X. They are displayed in Table 14.2 and plotted in Figure 14.5.

TABLE 14.2 Conditional Means of Y (husband's housework) for Various Values of X (number of children)

Number of Children (X)	Husband's Housework (Y)	Conditional Mean of Y
1	1,2,3,5	2.75
2	3,1	2.00
3	5,0	2.50
4	6,3	4.50
5	7,4	5.50

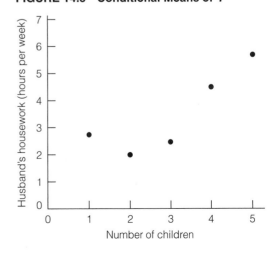

FIGURE 14.5 Conditional Means of Y

Let us quickly remind ourselves of the reason for these calculations. We are seeking the single best-fitting regression line for summarizing the relationship between X and Y, and we have seen that a line drawn through the conditional means of Y will minimize the spread of the observation points. It will come as close to all the scores as possible and will therefore be the single best-fitting regression line.

Now, a line drawn through the points on Figure 14.5 (the conditional means of Y) will be the best-fitting line we are seeking, but you can see from the scattergram that the line will not be straight. In fact, only rarely (when there is a perfect relationship between X and Y) will conditional means fall in a perfectly straight line. Since we still must meet the condition of linearity, let us revise our criterion and define the regression line as the unique straight line that touches all conditional means of Y or comes as close to doing so as possible. Formula 14.1 defines the "least-squares" regression line, or the single straight regression line that best fits the pattern of the data points.

FORMULA 14.1

$$Y = a + bX$$

where Y = score on the dependent variable
a = the Y intercept, or the point where the regression line crosses the Y axis
b = the slope of the regression line, or the amount of change produced in Y by a unit change in X
X = score on the independent variable

The formula introduces two new concepts. First, the **Y intercept (a)** is the point at which the regression line crosses the vertical, or Y, axis. Second, the **slope (b)** of the least-squares regression line is the amount of change produced in the dependent variable (Y) by a unit change in the independent variable (X). Think of the slope of the regression line as a measure of the effect of the X variable on the Y variable. If the variables have a strong association, then changes in the value of X will be accompanied by substantial changes in the value of Y, and the slope (b) will have a high value. The weaker the effect of X on Y (the weaker the association between the variables), the lower the value of the slope (b). If the

STATISTICS

IN EVERYDAY LIFE

Regression analysis has many practical applications and is used routinely in many professions—city planning, criminal justice, business, demography, and even military science—to model and predict trends. For example, medical researchers are using regression analysis to study the relationship between weather conditions at mass events (e.g., the Indianapolis 500 or rock concerts) and the need for emergency services.* This knowledge may be extremely useful for planners, who might be able to accurately forecast the need for medical services.

Of course, these applications use techniques that are much more sophisticated than the bivariate relationships explored in this chapter. However, the basic regression equation ($Y = a + bX$) is the foundation for even the most advanced and powerful of these models.

*Baird, M., O'Conner, R., Williamson, A., Sojka, B., and Brady, W. 2010. "The Impact of Warm Weather on Mass Event Medical Need: A Review of the Literature." *American Journal of Emergency Medicine*, 28:224–229.

two variables are unrelated, the least-squares regression line would be parallel to the X axis, and b would be 0.00 (the line would have no slope).

With the least-squares formula (Formula 14.1), we can predict values of Y in a much less arbitrary and impressionistic way than through mere eyeballing. This will be so, remember, because the least-squares regression line is the single straight line that best fits the data and comes as close as possible to all of the conditional means of Y. Before seeing how predictions of Y can be made, however, we must first calculate a and b. (*For practice in using the regression line to predict scores on Y from scores on X, see problems 14.1 to 14.3 and 14.5.*)

Computing and Interpreting the Regression Coefficients (*a* and *b*)

In this section, we cover how to compute and interpret the coefficients in the equation for the regression line: the slope (b) and the Y intercept (a). Since the value of b is needed to compute a, we begin with the computation of the slope.

COMPUTING THE SLOPE (*b*)

The formula for the slope is

FORMULA 14.2
$$b = \frac{\Sigma(X - \overline{X})(Y - \overline{Y})}{\Sigma(X - \overline{X})^2}$$

The numerator of this formula is called the *covariation* of X and Y. It is a measure of how X and Y vary together, and its value will reflect both the direction and the strength of the relationship. The denominator is the sum of the squared deviations around the mean of X.

The calculations necessary for computing the slope should be organized into a computational table, as in Table 14.3, which has a column for each of the quantities needed to solve the formula. The data are from the dual–wage-earner family sample (see Table 14.1).

The first column of the table lists the original X scores for each case and the second column shows the deviations of these scores around their mean. The third and fourth columns repeat this information for the Y scores and the deviations of the Y scores. Column 5 shows the covariation of the X and Y scores. The entries in this column are found by multiplying the deviation of the X score (column 2) by the deviation of the Y score (column 4) for each case. Finally, the entries in column 6 are found by squaring the value in column 2 for each case.

Table 14.3 gives us all the quantities we need to solve Formula 14.2. Substitute the total of column 5 in Table 14.3 in the numerator and the total of column 6 in the denominator:

$$b = \frac{\Sigma(X - \overline{X})(Y - \overline{Y})}{\Sigma(X - \overline{X})^2}$$

$$b = \frac{18.32}{26.68}$$

$$b = 0.69$$

TABLE 14.3 Computation of the Slope (*b*)

1 X	2 $X - \overline{X}$	3 Y	4 $Y - Y'$	5 $(X - \overline{X})(Y - \overline{Y})$	6 $(X - \overline{X})^2$
1	−1.67	1	−2.33	3.89	2.79
1	−1.67	2	−1.33	2.22	2.79
1	−1.67	3	−0.33	0.55	2.79
1	−1.67	5	1.67	−2.79	2.79
2	−0.67	3	−0.33	0.22	0.45
2	−0.67	1	−2.33	1.56	0.45
3	0.33	5	1.67	0.55	0.11
3	0.33	0	−3.33	−1.10	0.11
4	1.33	6	2.67	3.55	1.77
4	1.33	3	−0.33	−0.44	1.77
5	2.33	7	3.67	8.55	5.43
5	2.33	4	0.67	1.56	5.43
32	−0.04	40	0.04	18.32	26.68

$$\overline{X} = \frac{32}{12} = 2.67$$

$$\overline{Y} = \frac{40}{12} = 3.33$$

A slope of 0.69 indicates that, for each unit change in X, there is an increase of 0.69 units in Y. For our example, the addition of each child (an increase of one unit in X) results in an increase of 0.69 hours of housework being done by the husband (an increase of 0.69 units—or hours—in Y).

COMPUTING THE *Y* INTERCEPT (*a*)

Once the slope has been calculated, finding the intercept (*a*) is relatively easy. We computed the means of X and Y while calculating the slope, and we enter these figures into Formula 14.3.

FORMULA 14.3

$$a = \overline{Y} - b\overline{X}$$

For our sample problem, the value of *a* would be

$$a = \overline{Y} - b\overline{X}$$
$$a = 3.33 - (0.69)(2.67)$$
$$a = 3.33 - 1.84$$
$$a = 1.49$$

Thus, the least-squares regression line will cross the Y axis at the point where Y equals 1.49.

Now that we have values for the slope and the Y intercept, we can state the full least-squares regression line for our sample data:

$$Y = a + bX$$
$$Y = 1.49 + (0.69)X$$

PREDICTING SCORES ON *Y* WITH THE LEAST-SQUARES REGRESSION LINE

The regression formula can be used to estimate, or predict, scores on *Y* for any value of *X*. Previously, we used the freehand regression line to predict a score on *Y* (husband's housework) for a family with six children ($X = 6$). Our prediction was that, in families of six children, husbands would contribute about five hours per week to housekeeping chores. By using the least-squares regression line, we can see how close our impressionistic, eyeball prediction was.

$$Y = a + bX$$
$$Y = 1.49 + (0.69)(6)$$
$$Y = 1.49 + 4.14$$
$$Y = 5.63$$

Based on the least-squares regression line, we would predict that in a dual–wage-earner family with six children, the husband would devote 5.63 hours a week to housework. What would our prediction of the husband's housework be for a family of seven children ($X = 7$)?

Note that our predictions of *Y* scores are "educated guesses." We will be unlikely to predict values of *Y* exactly except in the (relatively rare) case where the bivariate relationship is perfect and perfectly linear. Note also, however, that the accuracy of our predictions will increase as relationships become stronger. This is because the dots are more clustered around the least-squares regression line in stronger relationships. *(The slope and Y intercept may be computed for any problem at the end of this chapter, but see problems 14.1 to 14.5 in particular. These problems have smaller data sets and will provide good practice until you are comfortable with these calculations.)*

ONE STEP AT A TIME | **Computing the Slope (*b*)**

To Compute the Slope (*b*), Solve Formula 14.2

1: Use a computing table like Table 14.3 to help organize the computations. List the scores of the cases on the independent variable (*X*) in column 1.

2: Compute the mean of *X* by dividing the total of column 1 by the number of cases.

3: Subtract the mean of *X* from each *X* score and list the results in column 2. *NOTE: The sum of column 2 must be zero (except for rounding error). If this sum is not zero, you have made a mistake in computations.*

4: List the score of each case on *Y* in column 3. Compute the mean of *Y* by dividing the total of column 3 by the number of cases.

5: Subtract the mean of *Y* from each *Y* score and list the results in column 4. *NOTE: The sum of column 4 must be zero (except for rounding error). If this sum is not zero, you have made a mistake in computations.*

6: For each case, multiply the value in column 2 by the value in column 4. Place the result in column 5. Find the sum of this column.

7: Square each value in column 2 and place the result in column 6. Find the sum of this column.

8: Divide the sum of column 5 by the sum of column 6. The result is the slope.

ONE STEP AT A TIME   **Computing the Y Intercept (a)**

To Compute the Y Intercept (a), Solve Formula 14.3

1: Multiply the slope (b) by the mean of X.

2: Subtract the value you found in step 1 from the mean of Y. This value is a, or the Y intercept.

ONE STEP AT A TIME   **Using the Regression Line to Predict Scores on Y**

1: Choose a value for X. Multiply this value by the value of the slope (b).

2: Add the value you found in step 1 to the value of a, the Y intercept. The resulting value is the predicted score on Y.

The Correlation Coefficient (Pearson's r)

The slope of the least-squares regression line (b) is a measure of the effect of X on Y, and it will increase in value as the relationship increases in strength. However, b does not vary between zero and 1 and is therefore awkward to use as a measure of association. Instead, researchers rely heavily (almost exclusively) on a statistic called **Pearson's r**, or the correlation coefficient, to measure association between interval-ratio variables. Like the ordinal measures of association discussed in Chapter 13, Pearson's r varies from 0.00 to ±1.00, with 0.00 indicating no association and +1.00 and −1.00 indicating perfect positive and perfect negative relationships, respectively. The formula for Pearson's r is:

FORMULA 14.4

$$r = \frac{\Sigma(X - \overline{X})(Y - \overline{Y})}{\sqrt{[\Sigma(X - \overline{X})^2][\Sigma(Y - \overline{Y})^2]}}$$

Note that the numerator of this formula is the covariation of X and Y, as was the case with Formula 14.2. A computing table like Table 14.3, with a column added for the sum of $(Y - \overline{Y})^2$, is strongly recommended as a way of organizing the quantities needed to solve this equation (see Table 14.4).

For our sample problem involving dual–wage-earner families, the quantities displayed in Table 14.4 can be substituted directly into Formula 14.4:

$$r = \frac{\Sigma(X - \overline{X})(Y - \overline{Y})}{\sqrt{[\Sigma(X - \overline{X})^2][\Sigma(Y - \overline{Y})^2]}}$$

$$r = \frac{18.32}{\sqrt{(26.68)(50.68)}}$$

$$r = \frac{18.32}{\sqrt{1352.14}}$$

$$r = \frac{18.32}{36.77}$$

$$r = 0.50$$

TABLE 14.4   Computation of the Pearson's *r*

1 X	2 $X - \overline{X}$	3 Y	4 $Y - \overline{Y}$	5 $(X - \overline{X})(Y - \overline{Y})$	6 $(X - \overline{X})^2$	7 $(Y - \overline{Y})^2$
1	−1.67	1	−2.33	3.89	2.79	5.43
1	−1.67	2	−1.33	2.22	2.79	1.77
1	−1.67	3	−0.33	0.55	2.79	0.11
1	−1.67	5	1.67	−2.79	2.79	2.79
2	−0.67	3	−0.33	0.22	0.45	0.11
2	−0.67	1	−2.33	1.56	0.45	5.43
3	0.33	5	1.67	0.55	0.11	2.79
3	0.33	0	−3.33	−1.10	0.11	11.09
4	1.33	6	2.67	3.55	1.77	7.13
4	1.33	3	−0.33	−0.44	1.77	0.11
5	2.33	7	3.67	8.55	5.43	13.47
5	2.33	4	0.67	1.56	5.43	0.45
32	−0.04	40	0.04	18.32	26.68	50.68

An *r* value of 0.50 indicates a moderately strong, positive linear relationship between the variables. As the number of children in the family increases, the hourly contribution of husbands to housekeeping duties also increases. *(Every problem at the end of this chapter requires the computation of Pearson's r. It is probably a good idea to practice with smaller data sets and easier computations first—see problem 14.1 in particular.)*

INTERPRETING THE CORRELATION COEFFICIENT: r^2

Pearson's *r* is an index of the strength of the linear relationship between two variables. While a value of 0.00 indicates no linear relationship and a value of

---

ONE STEP AT A TIME	**Computing Pearson's *r***

To Begin

1: Add a column to the computing table you used to compute the slope (*b*). Square the value of $(Y - \overline{Y})$ and record the result in this column (column 7).

2: Find the sum of column 7.

To Find the Value of Pearson's *r* by Solving Formula 14.4

1: Multiply the sum of column 6 by the sum of column 7.

2: Take the square root of the value you found in step 1.

3: Divide the quantity you found in step 2 into the sum of column 5, or the sum of the cross products. The result is Pearson's *r*.

To Interpret the Strength of Pearson's *r*, You Can Either

Use Table 13.2 to describe strength in general terms.

or

Square the value of *r* and multiply by 100. This value represents the percentage of variation in *Y* that is explained by *X*.

To Interpret the Direction of the Relationship

Look at the sign of *r*. If *r* has a plus sign (or if there is no sign), the relationship is positive and the variables change in the same direction. If *r* has a minus sign, the relationship is negative and the variables change in opposite directions.

±1.00 indicates a perfect linear relationship, values between these extremes have no direct interpretation. We can, of course, describe relationships in terms of how closely they approach the extremes (for example, coefficients approaching 0.00 can be described as "weak" and those approaching ±1.00 as "strong"), but this description is somewhat subjective. Also, we can use the guidelines stated in Table 13.2 for gamma to attach descriptive words to the specific values of Pearson's *r*. In other words, values between 0.00 and 0.30 would be described as weak, values between 0.30 and 0.60 would be moderate, and values greater than 0.60 would be strong. Remember, of course, that these labels are arbitrary guidelines and will not be appropriate or useful in all possible research situations.

Fortunately, there is a less arbitrary, more direct interpretation of *r* by calculating an additional statistic called the **coefficient of determination**. This statistic, which is simply the square of Pearson's *r* (r^2), can be interpreted with a type of logic akin to proportional reduction in error (PRE). As you recall, the logic of PRE measures of association is to predict the value of the dependent variable under two different conditions. First, *Y* is predicted while ignoring the information supplied by *X*; second, the independent variable is taken into account. With r^2, both the method of prediction and the construction of the final statistic require the introduction of some new concepts.

When working with variables measured at the interval-ratio level, the predictions of the *Y* scores under the first condition (while ignoring *X*) will be the mean of the *Y*. Given no information on *X*, this prediction strategy will be optimal because we know that the mean of any distribution is closer to all the scores than any other point in the distribution. Recall the principle of minimized variation introduced in Chapter 3 and expressed as

$$\sum(Y - \overline{Y})^2 = \text{minimum}$$

The scores of any variable vary less around the mean than around any other point. If we predict the mean of *Y* for every case, we will make fewer errors of prediction than if we predict any other value for *Y*.

Of course, we will still make many errors in predicting *Y* even if we faithfully follow this strategy. The amount of error is represented in Figure 14.6, which displays the relationship between number of children and husband's housework, with the mean of *Y* noted. The vertical lines from the actual scores to the predicted score represent the amount of error we would make when predicting *Y* while ignoring *X*.

We can define the extent of our prediction error under the first condition (while ignoring *X*) by subtracting the mean of *Y* from each actual *Y* score and squaring and summing these deviations. The resultant figure, which can be noted as $\sum(Y - \overline{Y})^2$, is called the **total variation** in *Y*. We now have a visual representation (Figure 14.6) and a method for calculating the error we incur by predicting *Y* without knowledge of *X*. As we shall see, we do not actually need to calculate the total variation to find the value of the coefficient of determination, r^2.

Our next step will be to determine the extent to which knowledge of *X* improves our ability to predict *Y*. If the two variables have a linear relationship, then predicting scores on *Y* from the least-squares regression equation will

FIGURE 14.6   Predicting *Y* Without *X* (dual-career families)

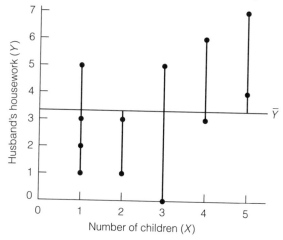

incorporate knowledge of *X* and reduce our errors of prediction. So, under the second condition, our predicted *Y* score for each value of *X* will be

$$Y' = a + bX$$

Figure 14.7 displays the data from the dual-career families with the regression line drawn in. The vertical lines from each data point to the regression line represent the amount of error in predicting *Y* that remains even after *X* has been taken into account.

As was the case under the first condition, we can precisely define the reduction in error that results from taking *X* into account. Specifically, two different sums can be found and then compared with the total variation of *Y* to construct a statistic that will indicate the improvement in prediction.

The first sum, called the **explained variation**, represents the improvement in our ability to predict *Y* when taking *X* into account. This sum is found by

FIGURE 14.7   Predicting *Y* with *X* (dual-career families)

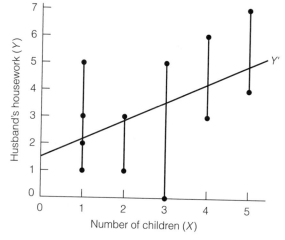

subtracting $\overline{Y}$ (our predicted Y score without X) from the score predicted by the regression equation (Y', or the Y score predicted with knowledge of X) for each case and then squaring and summing these differences. These operations can be summarized as $\Sigma(Y' - \overline{Y})^2$ and the resultant figure could then be compared with the total variation in Y to ascertain the extent to which our knowledge of X improves our ability to predict Y. Specifically, it can be shown mathematically that

FORMULA 14.5

$$r^2 = \frac{\Sigma(Y' - \overline{Y})}{\Sigma(Y - \overline{Y})} = \frac{\text{Explained variation}}{\text{Total variation}}$$

Thus, the coefficient of determination, or r^2, is the proportion of the total variation in Y attributable to, or explained by, X. Like other PRE measures, r^2 indicates precisely the extent to which X helps us predict, understand, or explain Y.

Earlier, we referred to the improvement in predicting Y with X as the explained variation. The use of this term suggests that some of the variation in Y will be "unexplained," or not attributable to the influence of X. In fact, the vertical lines in Figure 14.7 represent the **unexplained variation**, or the difference between our best prediction of Y with X and the actual scores. The unexplained variation is thus the scattering of the actual scores around the regression line; it can be found by subtracting the predicted Y scores from the actual Y scores for each case and then squaring and summing the differences. These operations can be summarized as $\Sigma(Y - Y')^2$, and the resultant sum would measure the amount of error in predicting Y that remains even after X has been taken into account. The proportion of the total variation in Y unexplained by X can be found by subtracting the value of r^2 from 1.00. Unexplained variation is usually attributed to the influence of some combination of other variables, measurement error, and random chance.

As you may have recognized by this time, the explained and unexplained variations bear a reciprocal relationship with one another. As one increases in value, the other decreases. Also, the stronger the linear relationship between X and Y, the greater the value of the explained variation and the lower the unexplained variation. In the case of a perfect relationship ($r = \pm 1.00$), the unexplained variation would be 0 and r^2 would be 1.00. This would indicate that X explains, or accounts for, all the variation in Y and that we could predict Y from X without error. On the other hand, when X and Y are not linearly related ($r = 0.00$), the explained variation would be 0 and r^2 would be 0.00. In such a case, we would conclude that X explains none of the variation in Y and does not improve our ability to predict Y.

Relationships intermediate between these two extremes can be interpreted in terms of how much X increases our ability to predict, or explain, Y. For the dual-career families, we calculated an r of 0.50. Squaring this value yields a coefficient of determination of 0.25 ($r^2 = 0.25$), which indicates that number of children (X) explains 25% of the total variation in husband's housework (Y). When predicting the number of hours per week that husbands in such families would devote to housework, we will make 25% fewer errors by basing the predictions on number of children and predicting from the regression line, as opposed to ignoring the X variable and predicting the mean of Y for every case.

Also, 75% of the variation in Y is unexplained by X and presumably due to some combination of the influence of other variables, measurement error, and random chance. *(For practice in the interpretation of r^2, see any of the problems at the end of this chapter.)*

STATISTICS
IN EVERYDAY LIFE

Does popularity pay? According to a recent study,* having many friends in high school is correlated with economic success later in life. Using a regression-based statistical model, researchers found that each additional friend in high school correlates with a 2% gain in wages 35 years later.[†] Statistically speaking, popularity seemed to have an effect that was roughly equal to about half the gain from an extra year of schooling—in other words, two additional high school friends would statistically equal the benefit of one additional year of college.

Remembering that correlation is not the same thing as causation, how else could this relationship be explained? What additional information would you need to demonstrate that the relationship between popularity and income was causal?

*Conti, G., Galeotti, A., Mueller, G., and Pudney, S. 2009. "Popularity." Available at http://www.iser .essex.ac.uk/publications/working-papers/iser/2009-03.pdf.

[†]Although it is based on more advanced statistics, this statement could be interpreted as a characterization of the slope of the bivariate relationship between popularity and economic success.

Applying Statistics 14.1: Computing the Regression Coefficients and Pearson's r

For five cities, information has been collected on number of civil disturbances (riots, strikes, and so forth) over the past year and on unemployment rate. Are these variables associated? The data are presented in the following table. Columns have been added for all sums necessary for the computation of the slope (b) and Pearson's r.

City	Unemployment Rate (X)	$(X - \overline{X})$	Civil Disturbances (Y)	$(Y - \overline{Y})$	$(X - \overline{X})(Y - \overline{Y})$	$(X - \overline{X})^2$	$(Y - \overline{Y})^2$
A	22	6.8	25	14.4	97.92	46.24	207.36
B	20	4.8	13	2.4	11.52	23.04	5.76
C	10	−5.2	10	−0.6	3.12	27.04	0.36
D	15	−0.2	5	−5.6	1.12	0.04	31.36
E	9	−6.2	0	−10.6	65.72	38.44	112.36
	$\overline{76}$	$\overline{0.0}$	$\overline{53}$	$\overline{0.0}$	$\overline{179.40}$	$\overline{134.80}$	$\overline{357.20}$

$$\overline{X} = 15.2$$
$$\overline{Y} = 10.6$$

(continued)

Applying Statistics 14.1: (*continued*)

The slope (*b*) is

$$b = \frac{\Sigma(X - \overline{X})(Y - \overline{Y})}{\Sigma(X - \overline{X})^2}$$

$$b = \frac{179.4}{134.8}$$

$$b = 1.33$$

A slope of 1.33 means that for every unit change in *X* (for every increase of 1 in the unemployment rate), there was a change of 1.33 units in *Y* (the number of civil disturbances increased by 1.33). The *Y* intercept (*a*) is

$$a = \overline{Y} - b\overline{X}$$

$$a = \frac{53}{5} - (1.33)\left(\frac{76}{5}\right)$$

$$a = 10.6 - (1.33)(15.20)$$

$$a = 10.6 - 20.2$$

$$a = -9.6$$

The least-squares regression equation is

$$Y = a + bX = -9.6 + (1.33)X$$

The correlation coefficient is

$$r = \frac{\Sigma(X - \overline{X})(Y - \overline{Y})}{\sqrt{[\Sigma(X - \overline{X})^2][\Sigma(Y - \overline{Y})^2]}}$$

$$r = \frac{179.40}{\sqrt{(134.80)(357.20)}}$$

$$r = \frac{179.40}{\sqrt{48150.56}}$$

$$r = \frac{179.40}{219.43}$$

$$r = 0.82$$

These variables have a strong, positive association. The number of civil disturbances increases as the unemployment rate increases. The coefficient of determination, r^2, is $(0.82)^2$, or 0.67. This indicates that 67% of the variance in civil disturbances is explained by the unemployment rate.

THE CORRELATION MATRIX

Social science research projects usually include many variables, and the data analysis phase of a project often begins with the examination of a **correlation matrix**: a table that shows the relationships between all possible pairs of variables. The correlation matrix gives a quick, easy-to-read overview of the interrelationships in the data set and may suggest strategies or "leads" for further analysis. These tables are commonly included in the professional research literature, and it will be useful to have some experience reading them.

An example of a correlation matrix, which displays some population variables for 74 nations, is presented in Table 14.5. The matrix uses variable names as rows and columns, and the cells in the table show the bivariate correlation (Pearson's *r*) for each combination of variables. Note that the row headings duplicate the column headings. To read the table, begin with birth rate, the variable in the far left-hand column (column 1) and top row (row 1). Read down column 1 or across row 1 to see the correlations of this variable with all other variables, including the correlation of birth rate with itself (1.00) in the top cell. To see the relationships between other variables, move from column to column or row to row.

Note that the diagonal from upper left to lower right of the matrix contains the correlation of each variable with itself. Values along this diagonal will always be 1.00 and, since this information is not useful, it could easily be deleted from the table. Also note that the cells below and to the left of the

TABLE 14.5   A Correlation Matrix Showing the Interrelationships of Four Variables for 74 Nations

	1	2	3	4
	Birth Rate	Infant Mortality Rate	Life Expectancy	Percent Urban
1 Birth Rate	1.00	0.88	−0.84	−0.66
2 Infant Mortality Rate	0.88	1.00	−0.89	−0.71
3 Life Expectancy	−0.84	−0.89	1.00	0.74
4 Percent Urban	−0.66	−0.71	0.74	1.00

diagonal are redundant with the cells above and to the right of the diagonal. For example, look at the second cell down (row 2) in column 1. This cell displays the correlation between birth rate and infant mortality rate, as does the cell in the top row (row 1) of column 2. In other words, the cells below and to the left of the diagonal are mirror images of the cells above and to the right of the diagonal. Commonly, research articles in the professional literature will delete the redundant cells in order to make the table more readable.

What does this matrix tell us? Starting at the upper left of the table (column 1), we can see that birth rate has a strong positive relationship with infant mortality rate. This means that infant mortality increases as birth rate increases: Nations with higher rates of birth also have higher rates of infant deaths. Birth rate has strong negative relationships with life expectancy (the higher the birth rate, the lower the life expectancy) and percent urban (nations with higher birth rates tend to be lower on urbanization).

To assess the other relationships in the data set, move from column to column and row to row, one variable at a time. For each subsequent variable, there will be one fewer cell of new information. For example, consider infant mortality rate, the variable in column 2 and row 2. We have already noted its strong positive relationship with birth rate, and, of course, we can ignore the correlation of the variable with itself. This leaves only two new relationships, which can be read by moving down column 2 or across row 2. Infant mortality rate has a strong negative relationship with life expectancy (the higher the infant mortality rate, the lower the life expectancy), and a strong negative relationship with percent urban (nations with higher infant mortality rates tend to be less urbanized).

For life expectancy, the variable in column 3, there is only one new relationship: a strong positive correlation with percent urban (the higher the life expectancy, the higher the percent urban). For column 4 (Percent Urban), there are no new relationships to consider: We have already examined all correlations between this variable and all other variables in the data set.

Before moving on, I should make two points. First, the correlations in Table 14.5 are unusually strong for social science data. This is largely because all of these variables measure a single underlying dimension: level of development or modernization. For example, low-income nations generally have higher birth rates, higher infant mortality rates, and lower life expectancy than high-income nations. The correlations are strong because they compare apples with apples and oranges with oranges.

Second, the cells in a correlation matrix will often include other information in addition to the bivariate correlations. It is common, for example, to include the number of cases on which the correlation is based and, if relevant, an indication of the statistical significance of the relationship.

Applying Statistics 14.2: Regression and Correlation

Is the desire for democracy related to a nation's level of education? Random samples of individuals from 10 nations have been asked how important it is for them to live in a country that is governed democratically. Information has also been gathered on the average years of school completed for people over 25 in each nation. The data are presented in the following table.* The X variable is average years of schooling completed. The Y variable is the average score on a scale that ranges from 1 ("democracy is not important") to 10 ("democracy is absolutely important"). Columns have been included for all necessary sums.

Nation	X	$X - \overline{X}$	Y	$Y - \overline{Y}$	$(X - \overline{X})(Y - \overline{Y})$	$(X - \overline{X})^2$	$(Y - \overline{Y})^2$
Argentina	8.8	1.37	9.1	0.44	0.60	1.88	0.19
Canada	11.3	3.87	9.0	0.34	1.32	14.98	0.12
China	6.4	−1.03	8.5	−0.16	0.17	1.06	0.03
Ghana	3.9	−3.53	9.2	0.54	−1.91	12.46	0.29
Germany	10.2	2.77	9.1	0.44	1.22	7.67	0.19
India	5.1	−2.33	7.1	−1.56	3.63	5.42	2.43
S. Africa	6.1	−1.33	8.7	0.04	−0.05	1.76	0.00
Zambia	5.5	−1.93	8.8	0.14	−0.27	3.72	0.02
U.S.	12.0	4.57	8.6	−0.06	−0.27	20.88	0.00
Indonesia	5.0	−2.43	8.5	−0.16	0.39	5.91	0.03
TOTALS	74.3	0.0	86.6	0.0	4.82	75.74	3.30

$$\overline{X} = 7.43$$
$$\overline{Y} = 8.66$$

*Data are from the World Values Survey and www.nationmaster.com.

The slope (b) is

$$b = \frac{\Sigma(X - \overline{X})(Y - \overline{Y})}{\Sigma(X - \overline{X})^2} = \frac{4.82}{75.74} = 0.06$$

A slope of 0.06 means that for every increase in years of education (a unit change in X), there is an increase of 0.06 points in the scale measuring people's desire to be ruled democratically.

The Y intercept (a) is

$$a = \overline{Y} - b\overline{X}$$
$$a = 8.66 - (.06)(7.43)$$
$$a = 8.66 - 0.45$$
$$a = 8.21$$

The least-squares regression equation is

$$Y = a + bX = 8.21 + (0.06)(X)$$

(continued)

Applying Statistics 14.2: (*continued*)

The correlation coefficient is

$$r = \frac{\Sigma(X - \overline{X})(Y - \overline{Y})}{\sqrt{[\Sigma(X - \overline{X})^2\Sigma(Y - \overline{Y})^2]}}$$

$$r = \frac{4.82}{\sqrt{(75.74)(3.3)}}$$

$$r = \frac{4.82}{\sqrt{249.94}}$$

$$r = \frac{4.82}{15.81}$$

$$r = 0.30$$

For these 10 nations, education and desire for democracy have a weak-to-moderate, positive relationship. Desire for democracy increases as education increases. The coefficient of determination, r^2, is $(0.30)^2$, or 0.09. This indicates that 9% of the variance in desire for democracy is explained by education for this sample of 10 nations.

CORRELATION, REGRESSION, LEVEL OF MEASUREMENT, AND DUMMY VARIABLES

Correlation and regression are very powerful and useful techniques, so much so that they are often used to analyze relationships between ordinal-level variables. This practice is generally not a problem, especially with ordinal variables that are "continuous" (see Chapter 13). However, this flexibility does not extend to nominal-level variables such as religious denomination or gender. The scores of nominal variables are not numbers and have no mathematical quality.

We might represent a Protestant with a score of "2" and a Catholic with a score of "1," but the former score is not "twice as much" as the latter. The scores of nominal-level variables are labels, not numbers, and it makes no sense to compute a slope or to discuss positive or negative relationships for these variables.

This is an unfortunate situation. Many of the variables that are most important in everyday social life—gender, marital status, race, and ethnicity—are nominal in level of measurement and cannot be included in a regression equation or a correlational analysis, two of the most powerful and sophisticated tools available for social science research.

Fortunately, researchers have developed a way to change the scoring on nominal-level variables by creating **dummy variables**. Dummy variables can be any level of measurement, including nominal, and have exactly two categories, one coded as 0 and the other coded as 1. Treated this way, nominal-level variables, such as gender (for example, with males coded as 0 and females coded as 1), race (with whites coded as 0 and nonwhites coded as 1), and religious denomination (Catholics coded as 1 and non-Catholics coded as 0), are commonly included in regression equations.

To illustrate, imagine that we were concerned with the relationship between racial or ethnic group and education as measured by number of years of schooling completed. If we coded whites as 0 and nonwhites as 1, we could compute a slope and Y intercept, write a regression equation using race as an independent variable, and examine the correlation between the two variables. Suppose we measured the education of a sample of nonwhite (coded as 1) and white (coded as 0) Americans and found the following regression equation:

$$Y = a + bX$$
$$Y = 12.0 + (-0.5)X$$

Education is the dependent (or Y) variable and racial or ethnic group (X) is the independent variable. The regression line crosses the vertical axis of the scattergram at the point where $Y = 12.0$. The value for the slope ($b = -0.5$) indicates a negative relationship: As racial or ethnic group "increases" (or moves toward the higher score associated with nonwhites), education tends to decrease. In other words, the nonwhite respondents in this sample averaged fewer years of schooling than the white respondents. Note that the *sign* of the slope (b) would have been positive had we reversed the coding scheme and labeled whites as 1 and nonwhites as 0, but the *value* of b would have stayed exactly the same. The coding scheme for dummy variables is arbitrary, and, as with ordinal-level variables, the researcher needs to be clear about what the values of a dummy variable indicate.

We can also use Pearson's r to assess the strength and direction of relationships with dummy variables. If we found an r of -0.23 between ethnic or racial group and education, we would conclude that there was a weak-to-moderate, "negative" relationship between these variables for this sample. Consistent with the sign of the slope, we could also say that education decreased as ethnic or racial group "increased," or moved from white to nonwhite. Also, using the coefficient of determination, we can say that racial or ethnic group explains, or accounts for, 5% ($r^2 = 0.23^2 = 0.05$) of the variance in education. *(For experience in working with dummy variables, see problem 14.9b.)*

TESTING PEARSON'S *r* FOR SIGNIFICANCE

When the relationship measured by Pearson's r is based on data from a random sample, it will usually be necessary to test r for its statistical significance. That is, we will need to know if a relationship between the variables can be assumed to exist in the population from which the sample was drawn. To illustrate this test, the r of 0.50 from the dual–wage-earner family sample will be used. As was the case when testing gamma and Spearman's rho, the null hypothesis states that there is no linear association between the two variables in the population from which the sample was drawn. The population parameter is symbolized as ρ (rho), and the appropriate sampling distribution is the t distribution.

To conduct this test, we need to make a number of assumptions in step 1. Most are familiar, but several are new. First, we must assume that both variables are normal in distribution (**bivariate normal distributions**). Second, we must assume that the relationship between the two variables is roughly linear in form.

The third assumption involves a new concept: **homoscedasticity**. Basically, a homoscedastistic relationship is one where the variance of the Y scores is uniform for all values of X. That is, if the Y scores are evenly spread above and below the regression line for the entire length of the line, the relationship is homoscedastistic.

A visual inspection of the scattergram will usually be sufficient to appraise the extent to which the relationship conforms to the assumptions of linearity and homoscedasticity. As a rule of thumb, if the data points fall in a roughly symmetrical, cigar-shaped pattern, whose shape can be approximated with a straight line, then it is appropriate to proceed with this test of significance. Any significant evidence of nonlinearity or marked departures from homoscedasticity may indicate the need for an alternative measure of association and thus a different test of significance.

Step 1. Making Assumptions and Meeting Test Requirements.

Model: Random sampling
Level of measurement is interval-ratio
Bivariate normal distributions
Linear relationship
Homoscedasticity
Sampling distribution is normal

Step 2. Stating the Null Hypothesis.

$$H_0: \rho = 0$$
$$(H_1: \rho \neq 0)$$

Step 3. Selecting the Sampling Distribution and Establishing the Critical Region. With the null hypothesis of "no relationship" in the population, the sampling distribution of all possible sample r's is approximated by the t distribution. Degrees of freedom are equal to $(N - 2)$.

Sampling distribution = t distribution
Alpha = 0.05
Degrees of freedom = $N - 2 = 10$
t(critical) = -2.28

Step 4. Computing the Test Statistic. The equation for computing the test statistic is given in Formula 14.6.

FORMULA 14.6
$$t(\text{obtained}) = r\sqrt{\frac{N-2}{1-r^2}}$$

Substituting the values into the formula, we would have:

$$t(\text{obtained}) = r\sqrt{\frac{N-2}{1-r^2}}$$
$$t(\text{obtained}) = (0.50)\sqrt{\frac{12-2}{1-(0.50)^2}}$$
$$t(\text{obtained}) = (0.50)\sqrt{\frac{10}{0.75}}$$
$$t(\text{obtained}) = (0.50)(3.65)$$
$$t(\text{obtained}) = 1.83$$

Step 5. Making a Decision and Interpreting the Results of the Test. Since the test statistic does not fall into the critical region as marked by t(critical), we fail to reject the null hypothesis. Even though the variables are substantially related in the sample, we cannot conclude that they are also related in the population. The test indicates that the sample value of $r = 0.50$ could have occurred by chance alone if the null hypothesis is true and the variables are unrelated in the population. *(For practice in conducting and interpreting tests of significance with Pearson's r, see problems 14.1, 14.2, 14.4, 14.6, 14.8, and 14.9.)*

Interpreting Statistics: The Correlates of Crime

What causes crime? Sociologists have been researching this question since the discipline was first founded, and they have conducted extensive research and have expended an enormous amount of energy on theory construction. While we cannot contribute to this voluminous body of work in a text devoted to statistical analysis, we can investigate some of the relationships and correlations that are of continuing interest to criminologists.

One prominent school of criminological thought argues that crime is related to poverty. A central proposition of this approach might be phrased as: "Crime rates will be highest among the most disadvantaged and impoverished groups, those with the highest rates of unemployment, and the lowest levels of ëeconomic viability' (job skills, levels of education) in the legitimate economy."[2] In this installment of Interpreting Statistics, we will take a look at some empirical relationships between crime and poverty using the 50 states as our sample. We'll measure crime with the homicide rate (the number of homicides per 100,000 population). The homicide rate is one of the more trustworthy measures of crime. The rates for many other types of crimes, such as rape and assault, are gross underestimates because victims of these crimes are less likely to report the incident to the police and, thus, not all incidents are included in the official counts of criminal activity. Homicides are much more likely to come to the attention of the authorities, if for no other reason than that there is a dead body that must be dealt with. For the independent variable, poverty, we will use the percent of the population of the state that is living below the poverty line.

The basic descriptive statistics for the variables are reported in Table 14.6 for 2007, the latest year for which information is available. Homicide rates ranged from a high of 14.2 murders per 100,000 population in Louisiana to a

TABLE 14.6 Homicide Rates and Poverty: Basic Statistics (2007)

Variable	Mean	Standard Deviation	Range	Number of Cases
Homicide rate	4.83	2.63	13.1	50
Percent poor	12.6	3.07	13.5	50

[2]For example, see Currie, E., and Skolnock, J. 1997. *Social Problems*. New York: Longman, p. 347.

FIGURE 14.8 Homicide Rate by Percent Poor

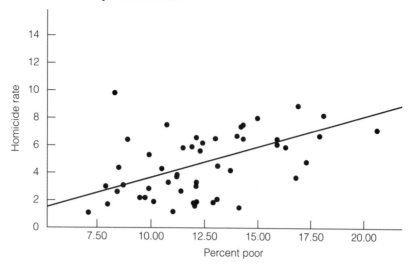

low of 1.1 murders per 100,000 population in New Hampshire. Poverty rates ranged from a high of 20.6% in Mississippi to a low of 7.1%, also in New Hampshire.

The first step in assessing an association between interval-ratio–level variables is to produce a scatterplot. Figure 14.8 plots the homicide rate on the vertical, or Y, axis and the poverty rate along the horizontal, or X, axis. What can we say about this relationship? The regression line is not horizontal, so there is a relationship between these variables. The dots (states) are fairly well scattered around the regression line, and we can see immediately that this will be, at best, a moderately strong relationship. The regression line slopes up from left to right, so the relationship is positive: States that have higher poverty rates tend to have higher homicide rates.

The next step in assessing this relationship is to specify the regression line. We'll skip the mechanics of computation here and simply report the values for the regression coefficients (a and b).

$$Y = a + bX$$
$$Y = 2.13 + (0.44)X$$

The regression line will cross the Y axis at the point where $Y = 2.13$, and every unit increase in X (percent poor) produces an increase of 0.44 in the homicide rate.

Next, we will calculate and interpret Pearson's r. Again, we'll skip the details of the computation and report that $r = 0.52$, which reinforces the impression that there is a moderately strong relationship between homicide and poverty. The coefficient of determination (r^2) is 0.27, which means that poverty rate, by itself, explains 27% of the variation in the homicide rate.

In sum, the linear regression equation, the correlation coefficient, and the coefficient of determination suggest that there is a moderately strong, positive relationship between poverty and crime. The amount of unexplained variation (73%) suggests that many other variables besides poverty have an important influence on the homicide rate.

We should note two limitations of this simple test. First, correlation is not the same thing as causation. Just because two variables are correlated does not mean that they have a causal relationship. These results could be taken as support for the proposition that higher rates of poverty lead to higher rates of violence, but the mere existence of a relationship—even a strong relationship in the direction predicted by the theory—does not prove that one variable causes the other.

Second, consider a related point: The crime of homicide is by definition an individual act of behavior, but the data we used in this test were collected from states. Just because there is an association between the variables at the macro (state) level does not necessarily mean that the variables are related in the same way at the micro (individual) level.[3] All we know from our analysis is that states with higher rates of poverty tend to have higher rates of homicide. Our theory would lead us to assume that it is the victims of inequality (poor people) who are the offenders, but this conclusion is not proven by this analysis. In fact, it may be the wealthier residents of the poorer states who are committing the murders. We would need much more information—on both the micro and macro levels—before we could come to any final conclusions with respect to the theory that poverty and crime are related.

---

[3]This problem is called the *ecological fallacy.*

SUMMARY

This summary is based on the example used throughout the chapter.

1. We began with a question: Is the number of children in dual–wage-earner families related to the number of hours per week husbands devote to housework? We presented the observations in a scattergram (Figure 14.1), and our visual impression was that the variables were associated in a positive direction. The pattern formed by the observation points in the scattergram could be approximated with a straight line; thus, the relationship was roughly linear.

2. Values of Y can be predicted with the freehand regression line, but predictions are more accurate if the least-squares regression line is used. The least-squares regression line is the line that best fits the data by minimizing the variation in Y. Using the formula that defines the least-squares regression line ($Y = a + bX$), we found a slope (b) of 0.69, which indicates that each additional child (a unit change in X) is accompanied by an increase of 0.69 hours of housework per week for the husbands. We also predicted, based on this formula, that in a dual–wage-earner family with six children ($X = 6$), husbands would contribute 5.63 hours of housework a week ($Y = 5.63$ for $X = 6$).

3. Pearson's r is a statistic that measures the overall linear association between X and Y. Our impression from the scattergram of a substantial positive relationship was confirmed by the computed r of 0.50. We also saw that this relationship yields an r^2 of 0.25, which indicates that 25% of the total variation in Y (husband's housework) is accounted for, or explained, by X (number of children).

4. Assuming that the 12 families represented a random sample, we tested the Pearson's r for its statistical significance and found that, at the 0.05 level, we could not assume that these two variables were also related in the population.

5. We acquired a great deal of information about this bivariate relationship. We know the strength and direction of the relationship and have also identified the regression line that best summarizes the effect of X on Y. We know the amount of change we can expect in Y for a unit change in X. In short, we have a greater volume of more precise information about this association between interval-ratio variables than we ever did about associations between ordinal or nominal variables. This is possible, of course, because the data generated by interval-ratio measurement are more precise and flexible than those produced by ordinal or nominal measurement techniques.

SUMMARY OF FORMULAS

Least-squares regression line:

14.1 $Y = a + bX$

Slope (b):

14.2. $b = \dfrac{\Sigma(X - \overline{X})(Y - \overline{Y})}{\Sigma(X - \overline{X})^2}$

Y intercept (a):

14.3 $a = \overline{Y} - b\overline{X}$

Pearson's r:

14.4 $r = \dfrac{\Sigma(X - \overline{X})(Y - \overline{Y})}{\sqrt{[\Sigma(X - \overline{X})^2][\Sigma(Y - \overline{Y})^2]}}$.

Coefficient of determination:

14.5 $r^2 = \dfrac{\Sigma(Y' - \overline{Y})}{\Sigma(Y - \overline{Y})}$

14.6 $t(\text{obtained}) = r\sqrt{\dfrac{N - 2}{1 - r^2}}$

GLOSSARY

Bivariate normal distributions. The model assumption in the test of significance for Pearson's r that both variables are normally distributed.

Coefficient of determination (r^2). The proportion of all variation in Y that is explained by X. Found by squaring the value of Pearson's r.

Conditional means of Y. The mean of all scores on Y for each value of X.

Correlation matrix. A table that shows the correlation coefficients between all possible pairs of variables.

Dummy variable. A nominal-level variable dichotomized so that it can be used in regression analysis. A dummy variable has two scores, one coded as 0 and the other as 1.

Explained variation. The proportion of all variation in Y that is attributed to the effect of X. Equal to $\Sigma(Y' - \overline{Y})^2$.

Homoscedasticity. The model assumption in the test of significance for Pearson's r that the variance of the Y scores is uniform across all values of X.

Linear relationship. A relationship between two variables in which the observation points (dots) in the scattergram can be approximated with a straight line.

Pearson's r. A measure of association for variables that have been measured at the interval-ratio level; ρ (Greek letter rho) is the symbol for the *population* value of Pearson's r.

Regression line. The single best-fitting straight line that summarizes the relationship between two variables. Regression lines are fitted to the data points by the least-squares criterion, whereby the line touches all conditional means of Y or comes as close to doing so as possible.

Scattergram. Graphic display device that depicts the relationship between two variables.

Slope (b). The amount of change in one variable per unit change in the other; b is the symbol for the slope of a regression line.

Total variation. The spread of the Y scores around the mean of Y. Equal to $\Sigma(Y - \overline{Y})^2$.

Unexplained variation. The proportion of the total variation in Y that is not accounted for by X. Equal to $\Sigma(Y' - \overline{Y})^2$.

Y intercept (a). The point where the regression line crosses the Y axis.

Y'. Symbol for predicted score on Y.

PROBLEMS

14.1 PS Why does voter turnout vary from election to election? For municipal elections in five different cities, information has been gathered on the percent of registered voters who actually voted, unemployment rate, average years of education for the city, and the percentage of all political ads that used "negative campaigning" (personal attacks, negative portrayals of the opponent's record, etc.). For each relationship:
   a. Draw a scattergram and a freehand regression line.
   b. Compute the slope (b) and find the Y intercept (a). (*HINT: Remember to compute b before*

computing *a*. A computing table such as Table 14.3 is highly recommended.)

c. State the least-squares regression line and predict the voter turnout for a city in which the unemployment rate was 12%, a city in which the average years of schooling was 11, and an election in which 90% of the ads were negative.

d. Compute *r* and r^2. (*HINT: A computing table such as Table 14.3 is highly recommended. If you constructed one for computing b, you already have most of the quantities you will need to solve for r.*)

e. Assume these cities are a random sample and conduct a test of significance for each relationship.

f. Describe the strength and direction of the relationships in a sentence or two. Which (if any) relationships were significant? Which factor had the strongest effect on turnout?

TURNOUT AND UNEMPLOYMENT

City	Turnout	Unemployment Rate
A	55	5
B	60	8
C	65	9
D	68	9
E	70	10

TURNOUT AND LEVEL OF EDUCATION

City	Turnout	Average Years of School
A	55	11.9
B	60	12.1
C	65	12.7
D	68	12.8
E	70	13.0

TURNOUT AND NEGATIVE CAMPAIGNING

City	Turnout	% of Negative Ads
A	55	60
B	60	63
C	65	55
D	68	53
E	70	48

14.2 SOC Occupational prestige scores for a sample of fathers and their oldest son and oldest daughter are presented below.

Family	Father's Prestige	Son's Prestige	Daughter's Prestige
A	80	85	82
B	78	80	77
C	75	70	68
D	70	75	77
E	69	72	60
F	66	60	52
G	64	48	48
H	52	55	57

Analyze the relationship between father's and son's prestige and the relationship between father's and daughter's prestige. For each relationship:

a. Draw a scattergram and a freehand regression line.

b. Compute the slope (*b*) and find the *Y* intercept (*a*).

c. State the least-squares regression line. What prestige score would you predict for a son whose father had a prestige score of 72? What prestige score would you predict for a daughter whose father had a prestige score of 72?

d. Compute *r* and r^2.

e. Assume these families are a random sample and conduct a test of significance for both relationships.

f. Describe the strength and direction of the relationships in a sentence or two. Does the occupational prestige of the father have an impact on his children? Does it have the same impact for daughters as it does for sons?

14.3 GER The residents of a housing development for senior citizens have completed a survey in which they indicated how physically active they are and how many visitors they receive each week. Are these two variables related for the 10 cases reported here? Draw a scattergram and compute *r* and r^2. Find the least-squares regression line. What would be the predicted number of visitors for a person whose level of activity was a 5? How about a person who scored 18 on level of activity?

Case	Level of Activity	Number of Visitors
A	10	14
B	11	12
C	12	10
D	10	9
E	15	8
F	9	7
G	7	10
H	3	15
I	10	12
J	9	2

14.4 PS The following variables were collected for a random sample of 10 precincts during the last national election. Draw scattergrams and compute r and r^2 for each combination of variables and test the correlations for their significance. Write a paragraph interpreting the relationship between these variables.

Precinct	Percent Democrat	Percent Minority	Voter Turnout
A	49	9	36
B	78	15	60
C	62	18	64
D	50	10	56
E	45	12	55
F	56	8	52
G	85	20	25
H	25	0	42
I	13	5	89
J	33	9	88

14.5 SOC/CJ The table below presents the scores of 10 states on each of six variables—three measures of criminal activity as dependent variables and three possible independent variables: unemployment rate, population density, and population growth rate. Crime rates are number of incidents per 100,000 population in 2007.

For each combination of crime rate and the three independent variables:
a. Draw a scattergram and a freehand regression line.
b. Compute the slope (b) and find the Y intercept (a).
c. State the least-squares regression line. What homicide rate would you predict for a state with a growth rate of –1? What robbery rate would you predict for a state with a population density of 250? What auto theft rate would you predict for a state in which 12% of the population was unemployed?
d. Compute r and r^2.
e. Assume these states are a random sample and conduct a test of significance for both relationships.
f. Describe the strength and direction of each of these relationships in a sentence or two.

Table for Problem 14.5

State	Crime Rates			Independent Variables		
	Homicide	Robbery	Car Theft	Growth*	Density†	Unemployment
Maine	1	27	96	3.3	42.7	5.4
New York	4	161	145	2.7	412.8	5.5
Ohio	4	159	295	1.2	280.5	6.5
Iowa	1	44	164	2.6	53.7	4.0
Virginia	5	143	182	9.7	196.2	4.0
Kentucky	5	96	205	5.6	107.5	6.3
Texas	6	162	393	16.7	92.9	4.8
Arizona	7	152	763	26.7	57.2	5.9
Washington	3	94	582	11.1	98.4	5.3
California	6	193	600	8.5	235.7	7.1

*Percentage change in population from 2000 to 2008.
†Population per square mile of land area, 2008.
Source: United States Bureau of the Census. *Statistical Abstracts of the United States: 2009.* Washington, DC, 2007.

14.6 Data on three variables have been collected for 15 nations. The variables are fertility rate (average number of children born to each woman), education for females (expressed as a percentage of all students at the secondary level who are female), and percent of married women ages 15–49 using all forms of contraception.

Nation	Fertility Rate*	Education of Females*	Percent of Women Using Contraception†
Niger	7.3	38.4	11
Cambodia	3.0	35.6	40
Guatemala	3.6	46.9	43
Ghana	3.8	44.7	24

(continued next page)

(*continued*)

Nation	Fertility Rate*	Education of Females*	Percent of Women Using Contraception†
Bolivia	2.7	48.2	61
Kenya	4.7	47.6	39
Dominican Republic	2.8	54.8	73
Mexico	2.4	50.5	71
Vietnam	1.9	47.1	79
Turkey	1.9	37.5	71
United States	2.1	49.0	73
China	1.8	45.3	90
United Kingdom	1.7	52.7	84
Japan	1.2	49.0	52
Italy	1.3	48.1	60

*Nationmaster.com.
†World Population Data Sheet, 2009. Population Reference Bureau (www.prb.org).

    a. Compute r and r^2 for each combination of variables.
    b. Summarize these relationships in terms of strength and direction.

14.7 The basketball coach at a small local college believes that his team plays better and scores more points in front of larger crowds. The number of points scored and attendance for all home games last season are reported here. Do these data support the coach's argument?

Game	Points Scored	Attendance
1	54	378
2	57	350
3	59	320
4	80	478
5	82	451
6	75	250
7	73	489
8	53	451
9	67	410
10	78	215
11	67	113
12	56	250
13	85	450
14	101	489
15	99	472

14.8 The following table presents the scores of 15 states on three variables. Compute r and r^2 for each combination of variables. Assume that these 15 states are a random sample of all states, and test the correlations for their significance. Write a paragraph interpreting the relationship among these three variables.

State	Per Capita Expenditures on Education, 2007	Percent High School Graduates, 2007	Rank in Per Capita Income
Arkansas	1546	80.4	14
Colorado	1792	88.9	6
Connecticut	2391	88.0	1
Florida	1672	84.9	8
Illinois	1887	85.7	5
Kansas	1913	89.1	9
Louisiana	1650	79.9	11
Maryland	1959	87.4	3
Michigan	1908	87.4	12
Mississippi	1313	78.5	15
Nebraska	1536	89.6	10
New Hampshire	1863	90.5	4
North Carolina	1456	83.0	13
Pennsylvania	1943	86.8	7
Wyoming	2712	91.2	2

14.9 Fifteen individuals were randomly selected from the respondents to a public opinion survey, and their scores are reproduced here.
    a. Calculate Pearson's r for age (X) and four dependent variables (Y's): prestige, number of children, support for abortion, and hours of TV watching.
    b. Calculate Pearson's r for gender (X) and four dependent variables (Y's): support for abortion, prestige, hours of TV watching, and number of children.
    c. Test the four relationships in either *a* or *b* for significance.

Occupational Prestige	Number of Children	Age	Support for Legal Abortion	Hours of TV per Day	Gender
32	3	34	3	1	1
50	0	41	1	3	1
17	0	52	5	2	1
69	3	67	1	5	0
17	0	40	1	5	0
52	0	22	2	3	1
32	3	31	3	4	0
50	0	23	4	4	0
19	9	64	1	6	1
37	4	55	4	2	0
14	3	66	5	5	1
51	0	22	2	0	0
45	0	19	3	7	0
44	0	21	4	1	1
46	4	58	2	0	1

KEY TO VARIABLES:

"Number of Children," "Age," and "Hours of TV" are actual values. "Occupational Prestige" is a continuous ordinal scale that indicates the amount of respect or esteem associated with occupation. The higher the score, the greater the prestige. "Support for Legal Abortion" is a collapsed ordinal scale that measures respondent's strength of agreement or disagreement with the statement "A woman should be able to get a legal abortion for any reason," where 1 = strongly agree, 2 = agree, 3 = neither agree nor disagree, 4 = disagree, and 5 = strongly disagree. "Gender" is a nominal-level variable coded as a dummy variable, with 0 = female and 1 = male.

YOU ARE THE RESEARCHER

Who Watches TV? Who Succeeds in Life?

Two projects are presented to help you apply the statistical skills developed in this chapter. In the first, you will analyze the correlates of time spent watching TV. You will select four independent variables and assess their impact on *tvhours* (number of hours per day spent watching TV). In the second project, you will choose either *income06* or *prestg80* as your dependent variable. Both variables measure social class standing and you will select independent variables you believe might be associated with an individual's level of success.

In both projects, you will have the option to conduct an additional analysis in which you analyze males and females separately to see if the pattern of correlation varies by sex. You will use an SPSS command called **Split File**, which allows us to examine subgroups within the sample independently.

There is no need for a detailed demonstration of how to generate Pearson's *r* with SPSS. We will use essentially the same command used in Chapter 13 to generate Spearman's rho. Briefly: Click **Analyze → Correlate → Bivariate**. The **Bivariate Correlations** dialog box will open and "Pearson" will already be checked. Find your variables in the list on the left and transfer them to the **Variables:** window, click **OK** and SPSS will produce a correlation matrix showing the relationship of all variables with each other. Recall that the cells in the output present three pieces of information: the value of the measure of association (Pearson's *r* in this case), its statistical significance, and the number of cases.

Optional Analysis

The **Split File** command is used when we want to see if variables are related in the same way for various subgroups in the sample. To demonstrate this command, let's look at the relationship between religiosity (measured by *attend*) and support for gay marriage (*marhomo*). The correlation between these variables for the entire sample is 0.32. This is a moderate, positive relationship: As frequency of church attendance rises (as score on *attend* goes up), opposition to gay marriage also rises (score on *marhomo* increases). Would the same relationship hold for both males and females?

To observe the effect of sex on the bivariate relationship, we split the GSS sample into two subfiles. Men and women will then be processed separately, and SPSS will produce one correlation matrix for men and another for women.

Click **Data** from the main menu and then click **Split File**. On the **Split File** window, click the button next to "organize output by groups." This will generate separate outputs for men and women. Select *sex* from the variable list and click the arrow to move the variable name into the **Groups Based On** window. Click **OK**, and all procedures requested will be done separately for men and women. To restore the full sample, call up the **Split File** window again and click the **Reset** button. For now, click **Statistics** → **Correlate** → **Bivariate** and rerun the correlation *attend* and *marhomo*.

The output shows a correlation of 0.26 for males and 0.40 for females. This result suggests that religiosity has a slightly more powerful impact for females than for males. For both groups, there is a weak-to-moderate, positive relationship between religiosity and opposition to gay marriage. Be sure to return to the **Split File** window and click **Reset** to restore the full sample before starting the projects.

PROJECT 1: Who Watches TV?

In this exercise, your dependent variable will be *tvhours*. Scores on this variable range from zero to a high of 24 hours (all day!), with most cases clustered around two to three hours of TV watching per day.

Step 1: Choosing Independent Variables

Select four variables from the 2008 GSS that you think might be important causes of *tvhours*. Your independent variables *cannot* be nominal in level of measurement unless you recode the variable into a dummy variable (see below). You may use any interval-ratio or ordinal level variable.

Dummy variables. To include a dummy variable in your analysis, recode the variable so that it has only two values: 0 and 1. Some possibilities include recoding *sex* so that males = 0 and females = 1, *racecen1* so that whites = 0 and nonwhites = 1, or *relig* so that Protestants = 0 and non-Protestants = 1. See Chapter 10 for instructions on recoding.

Once you have selected your variables, list them in the table below and describe exactly what they measure.

SPSS Variable Name	What exactly does this variable measure?

Step 2: Stating Hypotheses

State hypotheses about the relationships you expect to find between your independent variables and *tvhours*. State these hypotheses in terms of the direction of the relationship you expect to find. For example, you might hypothesize that TV time will increase as age increases.

Step 3: Running Bivariate Correlations

Click **Analyze** → **Bivariate** → **Correlate** and place all variables in the **Variables:** box. Click **OK** to get your results.

Step 4: Recording Results

Use the table below to summarize your results. Enter the *r* for each independent variable in each cell. As you read the correlation matrix, ignore correlations of variables with themselves and any redundant information.

	Independent Variables			
	1. _____	2. _____	3. _____	4. _____
tvhours				

Step 5: Analyzing and Interpreting Results

Write a short summary of results for each independent variable. Your summary needs to identify the variables being tested and the strength and direction of the relationship. It is probably best to characterize the relationship in general terms and then cite the statistical values in parentheses. Be sure to note whether or not your hypotheses were supported. *Be careful when interpreting direction* and refer back to the coding scheme to make sure you understand the relationship.

Step 6: Optional Analysis Using Split Files Command

You can analyze the correlations for different subgroups by splitting the sample. *Do not* use any of your independent variables to split the sample, and follow the instructions above to use the command. Disregard the results for very small groups. Are the variables related in essentially the same way for the subgroups as for the entire sample? If not, what experiences might account for the differences?

PROJECT 2: Who Succeeds?

For this exercise, choose either income (*income06*) or occupational prestige (*prestg80*) as your dependent variable. Most Americans would regard these variables as measures of success in life. What are the correlates and antecedents of affluence and prestige?

Step 1: Choosing Independent Variables

Select four variables that you think might be important causes of the dependent variable you selected. An obvious choice is education (use *educ*, the interval-ratio version, not *degree*, the ordinal version). Remember that independent variables *cannot* be nominal in level of measurement unless you recode the variable into a dummy variable (see below). You may use any interval-ratio or ordinal-level variable.

Dummy variables.   To include a dummy variable in your analysis, recode the variable so that it has only two values: 0 and 1. Some possibilities include recoding *sex* so that males = 0 and females = 1, *racecen1* so that whites = 0 and nonwhites = 1, or *relig* so that Protestants = 0 and non-Protestants = 1. See Chapter 10 for instructions on recoding.

Once you have selected your variables, list them in the table below and describe exactly what they measure.

SPSS Variable Name	What exactly does this variable measure?

Step 2: Stating Hypotheses

State hypotheses about the relationships you expect to find between your independent variables and the dependent variable you selected. State these hypotheses in terms of the direction of the relationship you expect to find. For example, you might hypothesize that income will increase as age increases.

Step 3: Running Bivariate Correlations

Click **Analyze** → **Bivariate** → **Correlate** and place all variables in the **Variables:** box. Click **OK** to get your results.

Step 4: Recording Results

Use the table below to summarize your results. Enter the *r* for each independent variable in each cell. As you read the correlation matrix, ignore correlations of variables with themselves and any redundant information.

	Independent Variables			
Dependent Variable:	1. _____	2. _____	3. _____	4. _____
_____				

Step 5: Analyzing and Interpreting Results

Write a short summary of results for each independent variable. Your summary needs to identify the variables being tested and the strength and direction of the relationship.

It is probably best to characterize the relationship in general terms and then cite the statistical values in parentheses. Be sure to note whether or not your hypotheses were supported. *Be careful when interpreting direction* and refer back to the coding scheme to make sure you understand the relationship.

Step 6: Optional Analysis Using Split Files Command

You can analyze the correlations for different subgroups by splitting the sample. *Do not* use any of your independent variables to split the sample, and follow the instructions above to use the command. Disregard the results for very small groups. Are the variables related in essentially the same way for the subgroups as for the entire sample? If not, what experiences might account for the differences?

1. For each situation, compute and interpret the appropriate measure of association. Also, compute and interpret percentages for bivariate tables. Describe relationships in terms of the strength and pattern or direction.

   a. For 10 cities, data have been gathered on total crime rate (major felonies per 100,000 population) and the percentage of people who are new immigrants (arrived in the United States within the past five years). Are the variables related?

City	Total Crime Rate	Percent Immigrants
A	1500	10
B	1200	18
C	2000	9
D	1700	11
E	1600	15
F	1000	20
G	1700	9
H	1300	22
I	900	10
J	700	15

   b. There is some evidence that people's involvement in their communities (membership in voluntary organizations, participation in local politics, and so forth) has been declining, and television has been cited as the cause. Do the following data support the idea that TV is responsible for the decline?

Hours of Community Service	Television Viewing			Totals
	Low	Moderate	High	
Low	5	10	18	33
Moderate	10	12	10	32
High	15	8	7	30
Totals	30	30	35	95

   c. A national magazine has rated the states in terms of "quality of life" (a scale that includes health care, availability of leisure facilities, unemployment rate, and a number of other variables) and the quality of the state system of higher education. Both scales range from 1 (low) to 20 (high). Is there a correlation between these two variables for the 10 states listed?

State	Quality of Life	Quality of Higher Education
A	10	10
B	12	13
C	15	18
D	18	20
E	10	15
F	9	11
G	11	12
H	8	6
I	13	9
J	6	8

d. Racial intermarriages are increasing in the United States, and the number of mixed-race people is growing. The following table shows the relationship between racial mixture of parents and the racial category, if any, with which a sample of people of mixed race identifies. Is there a relationship between these variables?

Do You Consider Yourself to Be:	Racial Mixture of Parents			Totals
	Black/White	White/Asian	Black/Asian	
Black	2	0	3	5
White	8	4	0	12
Asian	0	3	3	6
None of the Above	10	8	4	22
Totals	20	15	10	45

2. A number of research questions are stated here. Each can be answered by at least one of the techniques presented in Chapters 11 through 14. For each research situation, compute the most appropriate measure of association and write a sentence or two in response to the question. The questions are presented in random order. In selecting a measure of association, you need to consider the number of possible values and the level of measurement of the variables.

The research questions refer to the database shown at the end of the exercise, which is taken from the 2008 General Social Survey (GSS). The actual questions asked and the complete response codes for the GSS are presented in Appendix G. Abbreviated codes are listed here. Some variables have been recoded for this exercise.

a. Are scores on "frequency of sex during the last year" associated with income or age? Compute a measure of association, assuming that the variables are interval-ratio in level of measurement.

b. Is fear associated with sex? Is it associated with marital status?

c. Is support for spanking associated with church attendance? Is it associated with marital status?

Survey Items:

1. Marital status of respondent (*marital*):
   1. Married
   2. Not married (includes widowed, divorced, etc.)

2. How often do you attend church (*attend*)? See Appendix G for original codes:
   1. Never
   2. Rarely
   3. Often

3. Age of respondent (*age*) (Values are actual numbers.)

4. Respondent's total family income (*income98*) See Appendix G for codes.

5. Sex:
   1. Male
   2. Female

6. Support for spanking (*spanking*):
   1. Favor
   2. Oppose

7. Fear of walking alone at night (*fear*):
   1. Yes
   2. No

8. Frequency of sex during past year. See Appendix G for codes.

Case	Marital Status	Church Attendance	Age	Income	Gender	Spanking	Fear	Frequency of Sex
1	2	0	22	12	2	1	2	5
2	1	0	52	13	1	1	2	2
3	1	2	44	16	1	2	1	3
4	1	0	56	10	1	1	2	1
5	1	2	61	8	2	2	2	0
6	1	0	28	19	2	1	2	5
7	2	1	59	9	1	2	1	0
8	1	1	69	11	1	2	2	5
9	2	0	23	4	1	1	2	6
10	2	2	31	20	2	1	1	6
11	1	2	67	21	2	2	1	1
12	1	1	46	9	1	1	1	2
13	2	0	19	10	2	1	1	5
14	1	2	34	11	1	2	2	2
15	2	0	29	18	2	1	1	4
16	1	1	31	16	1	1	1	5
17	2	1	88	6	2	2	2	1
18	1	0	24	15	2	1	1	4
19	2	2	69	11	1	1	1	2
20	1	2	60	14	1	2	2	3
21	1	1	29	12	2	1	2	4
22	1	2	43	13	2	2	1	2
23	1	0	35	20	1	1	1	4
24	2	2	38	9	2	2	2	4
25	2	0	83	19	1	2	1	0

Part IV

Multivariate Techniques

The two chapters in this part introduce multivariate techniques—statistics that allow us to analyze the relationships among more than two variables at a time. These statistics are extremely useful for probing possible causal relationships between variables and are commonly reported in the literature. In particular, Chapter 16 introduces regression analysis, which is the basis for many of the most popular and powerful statistical techniques used in social science research today.

Chapter 15 covers multivariate analysis for nominal- and ordinal-level variables that have been organized into table format. The chapter discusses the logic and procedures for analyzing the effect of a third (or control) variable on the relationship between independent and dependent variables. The possible outcomes of controlling for a third variable are introduced and analyzed one at a time, and the discussion is summarized in Table 15.5.

Chapter 16 introduces partial and multiple correlation and regression for variables at the interval-ratio level of measurement. Partial correlation is analogous to controlling for a third variable with bivariate tables, as presented in Chapter 15. I make several references in Chapter 16 to the concepts presented in Chapter 15 and to Table 15.5. However, it is not necessary to read Chapter 15 in order to understand the material presented in Chapter 16.

Multiple regression and correlation are some of the most powerful and useful tools available to social science researchers. The mathematics underlying these techniques can become very complicated, and the chapter focuses on the simplest possible applications.

15

Elaborating Bivariate Tables

LEARNING OBJECTIVES   By the end of this chapter, you will be able to

1. Explain the purpose of multivariate analysis in terms of observing the effect of a control variable.
2. Construct and interpret partial tables.
3. Compute and interpret partial measures of association.
4. Recognize and interpret direct, spurious or intervening, and interactive relationships.
5. Compute and interpret partial gamma.
6. Cite and explain the limitations of elaborating bivariate tables.

Few questions can be answered by a statistical analysis of only two variables, and the typical research project will include many variables. In Chapters 12–14, we saw how various statistical techniques can be applied to bivariate relationships. In this chapter and Chapter 16, we see how some of these techniques can be extended to probe the relationships among three or more variables. This chapter presents some multivariate techniques appropriate for variables that have been measured at the nominal or ordinal level and organized into table format. Chapter 16 presents some techniques that can be used when the variables have been measured at the interval-ratio level.

Before considering the techniques themselves, we should consider why they are important and what they might be able to tell us. There are two general reasons for utilizing multivariate techniques. First, and most fundamental, is the goal of simply gathering additional information about a specific bivariate relationship by observing how that relationship is affected (if at all) by the presence of a third variable (or a fourth or a fifth). Multivariate techniques will increase our information on the bivariate relationship and (we hope) enhance our understanding of that relationship.

A second and very much related rationale for multivariate statistics involves the issue of causation. While multivariate statistical techniques cannot prove that causal connections exist between variables, they can provide valuable evidence in support of causal arguments and are very important tools for testing and revising theory.

Controlling for a Third Variable

For variables arrayed in tables, multivariate analysis proceeds by observing the effects of other variables on the bivariate relationship. That is, we observe the relationship between the independent (X) and dependent (Y) variables after a third variable (which we will call Z) has been controlled. We do this by reconstructing the relationship between X and Y for each value or score of Z. If the control variable has an effect, the relationship between X and Y will change under the various conditions of Z.[1]

THE BIVARIATE RELATIONSHIP

To illustrate, suppose that a researcher wishes to analyze the relationship between how well an individual is integrated into a group or organization and that individual's level of satisfaction with the group. The researcher has decided to focus on college students and their level of satisfaction with the college as a whole. The necessary information on satisfaction (Y) is gathered from a sample of 270 students, and all are asked to list the student clubs or organizations to which they belong. Integration (X) is measured by dividing the students into two categories: The first includes students who are not members of any organizations and the second includes students who are members of at least one organization. The researcher suspects that students who are members will report higher levels of satisfaction than nonmembers. The relationship between these two variables is displayed in Table 15.1.

The table suggests that these two variables are associated. The conditional distributions of satisfaction (Y) change across the two conditions of membership (X), and membership is associated with high satisfaction, whereas nonmembership is associated with low satisfaction. The existence and direction of the relationship are confirmed by the computation of a gamma of $+0.40$ for this table.[2] In short,

TABLE 15.1 **Satisfaction with College by Number of Memberships in Student Organizations**

Satisfaction (Y)	Memberships (X)		Totals
	None	At Least One	
Low	57 (*54.3%*)	56 (*33.9%*)	113
High	48 (*45.7%*)	109 (*66.1%*)	157
Totals	105 (*100.0%*)	165 (*100.0%*)	270
	Gamma = 0.40		

---

[1]For the sake of brevity, we focus on the simplest application of multivariate statistical analysis, the case where the relationship between an independent (X) and dependent (Y) variable is analyzed in the presence of a single control variable (Z). Once the basic techniques are grasped, they can be readily expanded to situations involving more than one control variable.

[2]We will not display the computation of gamma here. See Chapter 14 for a review of this measure of association.

FIGURE 15.1 A Direct Relationship Between Two Variables

$X \longrightarrow Y$

PARTIAL TABLES

Table 15.1 strongly suggests that integration and satisfaction are related, and these results can be taken as evidence for a causal or direct relationship between the two variables. The causal relationship is summarized symbolically in Figure 15.1, where the arrow represents the effect of X on Y.

Of course, membership is not the sole cause of satisfaction (if it were, the gamma would be +1.00). Other variables affect this relationship, and the researcher will need to consider these third variables (Z's) in a systematic and logical way. By doing so, the researcher will gather more information about the bivariate relationship and can further probe the possible causal connection between X and Y.

For example, perhaps both satisfaction (Y) and membership (X) are affected by grades (Z) and students with higher grade-point averages are more satisfied with the university and are also more likely to participate actively in the life of the campus by joining organizations. How can the effects of this third variable on the bivariate relationship be investigated?

Consider Table 15.1 again. This table provides information about the distribution of 270 cases on two variables. We can see that 165 students are members; that 48 students are both nonmembers and highly satisfied; that the majority of the students are highly satisfied; and so forth. What the table cannot show us, of course, is the distribution of these students on GPA. For all we know at this point, the 109 highly satisfied members could all have high GPAs, low GPAs, or any combination of scores on this third variable. GPA is "free to vary" in this table, since the distribution of the cases on this variable is not accounted for.

We can control the effect (if any) of a third variable by fixing its distribution so it is no longer free to vary. We do this by sorting the cases according to their score on the third variable (Z) and then observing the relationship between X and Y for each value (or score) of Z. In our example, we construct separate tables displaying the relationship between membership and satisfaction for each category of GPA. Tables that display the relationship between X and Y for each value of Z are called **partial tables**. In Table 15.2, the student sample has been subdivided into two groups based on GPA (Z): 135 students had "low" GPAs and the same number had "high" GPAs. The two partial tables in Table 15.2 show the relationship between membership (X) and satisfaction (Y) for each value of the control variable.

This type of multivariate analysis is called **elaboration** because the partial tables present the original bivariate relationship in a more detailed or elaborate form. The cell frequencies in the partial tables are subdivisions of the cell frequencies reported in Table 15.1. For example, if the frequencies of the cells in the partial tables are added together, the original frequencies of Table 15.1 will be reproduced.[3]

Also note how this method of controlling for other variables can be extended. In our example, the control variable (grades) had two categories, and, thus, there

---

[3]The total of the cell frequencies in the partial tables will always equal the corresponding cell frequencies in the bivariate table, except when, as often happens in "real-life" research situations, the researcher is missing scores on the third variable for some cases. These cases must be deleted from the analysis; as a consequence, the partial tables will have fewer cases than the bivariate table.

TABLE 15.2 Satisfaction by Membership, Controlling for GPA

A. High GPA	Memberships (X)		
Satisfaction (Y)	None	At Least One	Totals
Low	29 (54.7%)	28 (34.1%)	57
High	24 (45.3%)	54 (65.9%)	78
Totals	53 (100.0%)	82 (100.0%)	135
	Gamma = 0.40		

B. Low GPA	Memberships (X)		
Satisfaction (Y)	None	At Least One	Totals
Low	28 (53.8%)	28 (33.7%)	56
High	24 (46.2%)	55 (66.3%)	79
Totals	52 (100.0%)	83 (100.0%)	135
	Gamma = 0.39		

were two partial tables, one for each value of the control variable. If the control variable had had more than two categories, we would have had additional partial tables. Also, we can control for more than one variable at a time by sorting the cases on all scores of all control variables and producing partial tables for each combination of scores on the control variables. Thus, if we had controlled for both GPA and gender, we would have had four partial tables to consider. There would have been one partial table for males with low GPAs, a second for males with high GPAs, and two partial tables for females with high and low GPAs.

So far we have seen that for nominal and ordinal variables we begin multivariate analysis by constructing partial tables or tables that display the relationship between X and Y for each value of Z. The next step is to trace the effect of Z by comparing the partial tables with each other and with the original bivariate table. By analyzing the partial tables, we can observe the effects (if any) of the control variable on the original bivariate relationship.

Interpreting Partial Tables

The cell frequencies in the partial tables can follow a variety of forms, but we will concentrate on three basic patterns that are determined by comparing the partial tables with each other and with the original bivariate table:

1. **Direct relationships:** The relationship between X and Y is the same in all partial tables and in the bivariate table.

2. **Spurious relationships or intervening relationships:** The relationship between X and Y is the same in all partial tables but much weaker than in the bivariate table.

3. **Interaction:** The partial tables and the bivariate table all show different relationships between X and Y.

Each pattern has different implications for the relationships among the variables and for the subsequent course of the statistical analysis. We will consider each pattern in detail and then summarize the discussion in Table 15.5.

DIRECT RELATIONSHIPS

This pattern is often called **replication**, because the partial tables reproduce (or replicate) the bivariate table; the cell frequencies are the same in the partial tables, and the bivariate table and the measures of association are the same for all tables.

This outcome indicates that the control variable (Z) has no effect on the relationship between X and Y. Table 15.2 provides an example of this outcome. In this table, the relationship between membership (X) and satisfaction (Y) was investigated with GPA (Z) controlled. Table 15.2A shows the relationship for high-GPA students, and Table 15.2B shows the relationship for low-GPA students. The partial tables show the same conditional distributions of Y: For both high- and low-GPA students, about 45% of the nonmembers are highly satisfied, versus about 66% of the members. This same pattern was observed in the bivariate table (Table 15.1). Thus, the conditional distributions of Y are the same in each partial table as they were in the bivariate table.

Patterns will be easier to identify if we calculate measures of association. For Table 15.2, we see that the gamma for high-GPA students is 0.40 and that the gamma for low-GPA students is 0.39. The bivariate gamma (from Table 15.1) is 0.40. The fact that the partial and bivariate gammas are essentially the same value reinforces our conclusion that the relationship between X and Y is the same in the partial tables and the bivariate table.

This pattern of outcomes indicates that the control variable has no important impact on the bivariate relationship (if it did, the pattern in the partial tables would be different from that in the bivariate table) and may be ignored in any further analysis. In terms of the original research problem, the researcher may conclude that students who are members are more likely to express high satisfaction with the university regardless of their GPA. In other words, GPA has no effect on the relationship between membership and satisfaction. Low-GPA students who are members of at least one organization are just as likely to report high satisfaction as high-GPA students who are members of at least one organization. *(For practice in dealing with direct relationships, see problems 15.1 and 15.2.)*

SPURIOUS OR INTERVENING RELATIONSHIPS

In this pattern, the relationship between X and Y is much weaker in the partial tables than in the bivariate table but the same across all partials. Measures of association for the partial tables are much lower in value (perhaps even dropping to 0.00) than the measure computed for the bivariate table. This outcome is consistent with two different causal relationships among the three variables. The first is called a spurious relationship or **explanation**; in this situation, Z is conceptualized as being antecedent to both X and Y (that is, Z is thought to occur earlier in time than the other two variables). In this pattern, Z causes both X and Y, and the original bivariate relationship is said to be spurious. The apparent bivariate relationship between X and Y is explained by, or due to, the effect of Z. Once Z is controlled, the association between X and Y disappears and the measures of association for the partial tables are dramatically lower than the bivariate measure of association.

To illustrate, suppose that the researcher in our example had also controlled for class standing by dividing the sample into upperclass (seniors and juniors) and underclass (sophomores and freshmen) students. The reasoning of the researcher might be that dissatisfied students would have transferred or dropped out before they achieved upperclass standing and upperclass students are more satisfied because of simple self-selection. Also, students who have been on campus longer will be more likely to locate an organization of sufficient appeal or interest to join. This might especially be the case for organizations based on major field (such as the Accounting Club), which underclass students are less likely to join, or honorary organizations, for which underclass students are unlikely to qualify.

These thoughts about the possible relationships among these variables are expressed in diagram form in Figure 15.2. The absence of an arrow from membership (X) to satisfaction (Y) indicates that these variables are not truly associated. Rather, both are caused by class standing (Z). If this causal diagram is a correct description of the relationship between these three variables (if the association between X and Y is spurious), then the association between membership and satisfaction should disappear once class standing has been controlled. That is, even though the bivariate gamma was 0.40 (Table 15.1), the gammas computed on the partial tables will approach 0. Table 15.3 displays the partial tables generated by controlling for class standing.

FIGURE 15.2 A Spurious Relationship

TABLE 15.3 Satisfaction by Membership, Controlling for Class Standing

A. Upperclass Students			
	Memberships (X)		
Satisfaction (Y)	None	At Least One	Totals
Low	8 (25.0%)	32 (24.8%)	40
High	24 (75.0%)	97 (75.2%)	121
Totals	32 (100.0%)	129 (100.0%)	161
	Gamma = 0.01		

B. Underclass Students			
	Memberships (X)		
Satisfaction (Y)	None	At Least One	Totals
Low	49 (67.1%)	24 (66.7%)	73
High	24 (32.9%)	12 (33.3%)	36
Totals	73 (100.0%)	36 (100.0%)	109
	Gamma = 0.01		

The partial tables indicate that, once class standing is controlled, membership is no longer related to satisfaction. Regardless of their number of memberships, upperclass students are more likely to be high on satisfaction and underclass students are more likely to be low on satisfaction. In the partial tables, the distributions of Y no longer vary by the conditions of X, and the gammas computed on the partial tables are virtually 0. These results indicate that X and Y have no direct relationship: Their apparent relationship is spurious. Class standing (Z) is a key factor in accounting for satisfaction, and the analysis must be reoriented with class standing as an independent variable.

FIGURE 15.3 An Intervening Relationship

This outcome (partial measures much weaker than the original measure but equal to each other) is consistent with a second causal model in which Z connects X with Y. This pattern is called **interpretation** and is illustrated in Figure 15.3: X causes Z, which in turn causes Y. This pattern indicates that X and Y are associated primarily through the control variable Z.

The pattern of outcome displayed in Table 15.3 does not allow the researcher to distinguish between spurious relationships (Figure 15.2) and intervening relationships (Figure 15.3). The differentiation between these two types of causal patterns may be made on temporal or theoretical grounds but not on statistical grounds. *(For practice in dealing with spurious or intervening relationships, see problem 15.6.)*

INTERACTION

In this pattern, also called **specification**, the relationship between X and Y changes markedly for the different values of the control variable. The partial

---

STATISTICS
IN EVERYDAY LIFE

Spurious relationships are part of everyday life, not just scientific research. We sometimes leap to incorrect or premature conclusions about "what causes what" as we consider the news or chat with friends, and being aware of spurious relations can enhance your appreciation of the complexity of social life and the subtlety of casual relationships. As you think about spuriousness, it may be helpful to be able to refer to some simplified examples. Here are three statements, each of which asserts that one variable (X) causes another (Y). Each relationship is spurious. Can you identify the common cause (Z) of both X and Y for each statement? (The answers appear below.)

1. People in hospitals are more likely to die. Therefore, being in a hospital (X) causes death (Y).

2. The greater the number of firefighters at a fire, the greater the damage. Therefore, firefighters (X) cause fire damage (Y).

3. Men who listen to a lot of dance-band music from the World War II era are more likely to suffer from sexual impotence. Therefore, dance-band music (X) causes sexual impotence (Y).

Answers:

1. Z = illness. People who are very sick are more likely to be in a hospital and more likely to die.

2. Z = extent of the fire. More serious fires attract more firefighters and result in more damage.

3. Z = age. Older men are both more likely to listen to this genre of music and more likely to experience sexual impotence.

FIGURE 15.4   An Interactive Relationship

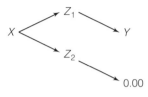

tables differ from each other and from the bivariate table. Interaction can appear in a variety of ways in the partial tables. One possible pattern, for example, is for one partial table to display a stronger relationship between X and Y than that displayed in the bivariate table, while the relationship between X and Y drops to 0 in a second partial table. Symbolically, this outcome could be represented as in Figure 15.4, which would indicate that X and the first category of Z (Z_1) have strong effects on Y, but, for the second category of Z (Z_2), there is no association between X and Y.

Interaction might be found, for example, in a situation in which *all* employees of a corporation were required to attend a program (X) designed to reduce prejudice against African Americans (Y). Such a program would be likely to have stronger effects on white employees (Z_1) than on African American employees (Z_2). That is, the program would have an effect on prejudice only for certain types of subjects. The partial tables for white employees (Z_1) might show a strong relationship between program attendance and reduction in prejudice, whereas the partial tables for African American employees might show no relationship. African American employees, being unprejudiced against themselves in the first place, would be less affected by the program.

FIGURE 15.5   An Interactive Relationship

Interaction can take other forms. For example, the relationship between X and Y can vary not only in strength but also in direction between the partial tables. This causal relationship is symbolically represented in Figure 15.5, where X and Y are positively related for the first category of Z (Z_1) and negatively related for the second category of Z (Z_2).

The researcher investigating the relationship between club membership and satisfaction with college life establishes a final control for race and divides the sample into white and black students. The partial tables are displayed in Table 15.4 and show an interactive relationship. The relationship between

TABLE 15.4   Satisfaction by Membership, Controlling for Race

A. White Students	Memberships (X)		
Satisfaction (Y)	None	At Least One	Totals
Low	40 (*50.0%*)	20 (*16.7%*)	60
High	40 (*50.0%*)	100 (*83.3%*)	140
Totals	80 (*100.0%*).	120 (*100.0%*)	200

Gamma = 0.67

B. Black Students	Memberships (X)		
Satisfaction (Y)	None	At Least One	Totals
Low	17 (*68.0%*)	36 (*80.0%*)	53
High	8 (*32.0%*)	9 (*20.0%*)	17
Totals	25 (*100.0%*).	45 (*100.0%*)	70

Gamma = −0.31

membership and satisfaction is different for white students (Z_1) and for black students (Z_2), and each partial table is different from the bivariate table. For white students, the relationship is positive and stronger than in the bivariate table. White students who are also members are much more likely to express high overall satisfaction with the university, as indicated by both the percentage distribution (83% of the white students who are members are highly satisfied) and the measure of association (gamma = 0.67 for white students).

For black students, the relationship between membership and satisfaction is very different, nearly the reverse of the pattern shown by white students.

Applying Statistics 15.1: Elaborating Bivariate Tables

Seventy-eight juvenile males in a sample have been classified as high or low on a scale that measures involvement in delinquency. Also, each subject has been classified, using school records, as either a good or a poor student. The following table displays a strong relationship between these two variables for this sample ($G = -0.69$).

Delin-quency	Academic Record		Totals
	Poor	Good	
Low	13 (27.1%)	20 (66.7%)	33 (42.3%)
High	35 (72.9%)	10 (33.3%)	45 (57.7%)
Totals	48 (100.0%)	30 (100.0%)	78 (100.0%)

Gamma = -0.69

Judging by the column percentages and the gamma, it is clear that juvenile males with poor academic records are especially prone to delinquency. Is this relationship between delinquency and academic record affected by whether the subject resides in an urban or nonurban area?

A. Urban Areas

Delin-quency	Academic Record		Totals
	Poor	Good	
Low	10 (27.8%)	3 (30.0%)	13 (28.3%)
High	26 (72.2%)	7 (70.0%)	33 (71.7%)
Totals	36 (100.0%)	10 (100.0%)	46 (100.0%)

Gamma = -0.05

B. Nonurban Areas

Delin-quency	Academic Record		Totals
	Poor	Good	
Low	3 (25.0%)	17 (85.0%)	20 (62.5%)
High	9 (75.0%)	3 (15.0%)	12 (37.5%)
Totals	12 (100.0%)	20 (100.0%)	32 (100.0%)

Gamma = -0.89

For urban juvenile males, the relationship between academic record and delinquency disappears. The gamma for this table is 0.05, and the column percentages are very similar. For this group, delinquency is not associated with experience in school.

For nonurban males, on the other hand, there is a very strong relationship between the two variables. Gamma is 0.89, and there is a dramatic difference in the column percentages. Poor students living in nonurban areas are especially prone to delinquency.

Comparing the partial tables with each other and with the bivariate table reveals an interactive relationship. Although urban juvenile males are more delinquent than nonurban males (71.7% of the urban males were highly delinquent, as compared to 37.5% of the nonurban males), their delinquency is not associated with academic record. For nonurban males, academic record is very strongly associated with delinquency. Urban males are more delinquent than nonurban males, but not because of their experience in school. Nonurban males who are also poor students are especially likely to become involved in delinquency.

Applying Statistics 15.2: Support for Cohabitation by Religiosity, Controlling for Gender

Cohabitation has become increasingly popular in U.S. society over the past several decades. What are the sources of support for this living arrangement? One obvious hypothesis would be that attitude toward cohabitation is related to religiosity and is less supported by people who attend church more frequently.

A recent public opinion survey asked respondents if they agree or disagreed with the statement "It's a good idea for a couple who intend to get married to live together first." The following table shows the relationship between church attendance and response to this item.

| | Church Attendance | | | |
Opinion on Cohabitation	Never (1)	Yearly or Monthly (2)	Daily or Weekly (3)	Totals
(1) Agree	141 *(63.8%)*	347 *(54.7%)*	60 *(20.5%)*	548 *(47.8%)*
(2) Neutral	47 *(21.3%)*	130 *(20.5%)*	49 *(16.8%)*	226 *(19.7%)*
(3) Disagree	33 *(14.9%)*	157 *(24.8%)*	183 *(62.7%)*	373 *(32.5%)*
Totals	221 *(100.0%)*	634 *(100.0%)*	292 *(100.0%)*	1147 *(100.0%)*

$$\chi^2 = 179.74 \ (\alpha < 0.001) \qquad G = 0.51$$

The chi square test indicates that the relationship is statistically significant, and the gamma of 0.51 indicates a moderate-to-strong, positive relationship. Given the way the responses are coded (higher scores indicate greater *dis*approval), the positive gamma means that, consistent with the hypothesis stated earlier, disapproval increases as church attendance increases. Disapproval is highest among frequent attendees (62.7%) and lowest among non-churchgoers (14.9%).

Is this relationship affected by gender? The gammas for the partial tables are fairly close in value to the bivariate gamma (0.51), but show a weaker relationship for males and a slightly stronger relationship for females, a pattern that might suggest an interactive relationship among the three variables. These results are somewhat ambiguous and do not precisely match the pattern we would expect in either a direct relationship or an interactive relationship (see Table 15.5). Such messy results are extremely common in real-life research situations, and they call for some caution when making conclusions.

A. Males

| | Church Attendance | | | |
Opinion on Cohabitation	Never (1)	Yearly or Monthly (2)	Daily or Weekly (3)	Totals
(1) Agree	81 *(68.6%)*	149 *(52.5%)*	21 *(25.6%)*	251 *(51.9%)*
(2) Neutral	17 *(14.43%)*	62 *(21.8%)*	13 *(15.9%)*	92 *(19.0%)*
(3) Disagree	20 *(16.9%)*	73 *(25.7%)*	48 *(58.5%)*	141 *(29.1%)*
Totals	118 *(100.0%)*	284 *(100.0%)*	82 *(100.0%)*	484 *(100.0%)*

$$G = 0.44$$

(continued)

Applying Statistics 15.2: (*continued*)

B. Females

Opinion on Cohabitation	Church Attendance			
	Never (1)	Yearly or Monthly (2)	Daily or Weekly (3)	Totals
(1) Agree	60 (*58.3%*)	198 (*56.6%*)	39 (*18.6%*)	297 (*44.8%*)
(2) Neutral	30 (*29.1%*)	68 (*19.4%*)	36 (*17.1%*)	134 (*20.2%*)
(3) Disagree	13 (*12.6%*)	84 (*24.0%*)	135 (*64.3%*)	232 (*35.0%*)
Totals	103 (*100.0%*)	350 (*100.0%*)	210 (*100.0%*)	663 (*100.0%*)

$$G = 0.54$$

What can we say? Church attendance has a weaker effect on support for cohabitation for males. Males are generally more supportive of cohabitation (compare the top rows—"agree"—of the two partial tables) and non-churchgoing males are especially approving of cohabitation (68.6% agreed).

Although females are less supportive, the higher value for gamma means that church attendance makes a bigger difference for them. Also, females who attend church frequently are especially disapproving of cohabitation (64.3% disagreed).

Looking at the overall patterns in the partial and bivariate tables, we would probably conclude that this relationship is closer to a direct than to an interactive relationship. The difference between the gammas for the partial tables is certainly worthy of note but is not dramatic enough to call this a case of interaction. We might eliminate gender from further analysis (see Table 15.5) and select another control variable (e.g., age or education) to test the relationship between church attendance and support for cohabitation further.

TABLE 15.5 A Summary of the Possible Results of Controlling for Third Variables

Partial Tables Show (compared with bivariate table)	Pattern	Implications for Further Analysis	Likely Next Step in Statistical Analysis	Theoretical Implications
Same relationship between X and Y	Direct relationship, replication	Disregard Z	Analyze another control	Theory that X causes Y is supported
Weaker relationship between X and Y	Spurious relationship	Incorporate Z	Focus on relationship between Z and Y	Theory that X causes Y is not supported
	Intervening relationship	Incorporate Z	Focus on relationships between X, Y, and Z	Theory that X causes Y is partially supported but must be revised to take Z into account
Mixed	Interaction	Incorporate Z	Analyze subgroups (categories of Z) separately	Theory that X causes Y is partially supported but must be revised to take Z into account

The great majority (80%) of the black students who are members report low satisfaction, and the gamma for this partial table is negative (-0.31). Thus, satisfaction increases with membership for white students but decreases with membership for black students. Based on these results, one might conclude that the implications of joining student organizations vary by race and has different effects for black students. It may be, for example, that black students join different kinds of organizations than do white students. If the black students belonged primarily to a black student association that had an antagonistic relationship with the university, then belonging to such an organization could increase dissatisfaction with the university as it increased awareness of racial problems. *(For practice in dealing with interaction, see problem 15.4.)*

In closing this section, let me stress that, for the sake of clarity, I have presented examples that have been unusually "clean." In an actual research project, controlling for third variables will probably have results that are considerably more ambiguous and open to interpretation than the examples presented here. In the case of spurious relationships, for example, the measures of association computed for the partial tables will probably not actually drop to zero, even

STATISTICS
IN EVERYDAY LIFE

It might be helpful to think about interaction as an unexpectedly intense reaction when variables are combined. For example, suppose a study of the causes of street crime found moderately strong associations with poverty ("areas with high poverty have higher levels of street crimes") and age ("areas with high percentages of teenagers and young adults have higher levels of street crime") but very strong relationships when the variables were combined ("areas that have *both* high poverty *and* younger populations have *much* higher levels of street crime"). This may be a case of interaction: Each independent variable has its own relationship with crime, but their combined effects have very strong impacts.

Here's an example of interaction from a public opinion poll. In Chapter 13, we looked at a poll conducted in the spring of 2010 that showed high levels of frustration and anger with the federal government. There was a moderate-to-weak, positive relationship (gamma = 0.31) between anger and age. Anger increased as age increased and about 28% of older respondents expressed anger versus only 17% of younger respondents.

What happens when we combine age and political party affiliation? Looking first at Republicans only, age and anger have a positive relationship that is about the same strength (gamma = 0.29) as the bivariate relationship. About 38% of older Republicans are angry versus 24% of younger Republicans. For Independents, we find a similar but somewhat stronger (gamma = 0.44) relationship. About 35% of older Independents are angry versus only 20% of younger independents.

Thus, for Republicans and Independents, age and feelings have about the same relationship as in the bivariate table. Democrats, however, display a very different relationship. The gamma for Democrats drops almost to zero (gamma = 0.05) and only 12% of older Democrats (vs. 7% of younger Democrats) are angry.

What does this mean? Political party affiliation and age interact in their effect on anger. Older Republicans and Independents are more likely to be angry, but neither young nor old Democrats share this emotion.

Of course, these results must be seen in the context of their times. The poll was taken shortly after Democrats, working with President Obama, passed a controversial health care reform package and in an atmosphere of intense partisan politics.

though they may be dramatically lower than the bivariate measure. (The pattern where the partial measures are roughly equivalent and much lower than the bivariate measure but not zero is sometimes called *attenuation*.) It is probably best to consider the foregoing examples as ideal types against which any given empirical result can be compared.

Table 15.5 summarizes this discussion by outlining guidelines for decision making for each of the three outcomes discussed in this section. Since your own results will probably be more ambiguous than the ideal types presented here, you should regard this table as a set of suggestions and not as a substitute for your own creativity and sensitivity to the problem under consideration.

Partial Gamma (G_p)

When the results of controlling for a third variable indicate a direct, spurious, or intervening relationship, it is often useful to compute an additional measure that indicates the overall strength of the association between X and Y after the effects of the control variable (Z) have been removed. This statistic, called **partial gamma (G_p)**, is somewhat easier to compare with the bivariate gamma than are the gammas computed on the partial tables separately. G_p is computed across all partial tables by Formula 15.1:

FORMULA 15.1
$$G_p = \frac{\Sigma N_s - \Sigma N_d}{\Sigma N_s + \Sigma N_d}$$

where N_s = the number of pairs of cases ranked the same across all partial tables
N_d = the number of pairs of cases ranked differently across all partial tables

In words, ΣN_s is the total of all N_s's from the partial tables, and ΣN_d is the total of all N_d's from all partial tables. (See Chapter 13 to review the computation of N_s and N_d.)

To illustrate the computation of G_p, let us return to Table 15.2, which showed the relationship between satisfaction and membership while controlling for GPA. The gammas computed on the partial tables were essentially equal to each other and to the bivariate gamma. Our conclusion was that the control variable had no effect on the relationship, and this conclusion can be confirmed by also computing partial gamma.

From Table 15.2A High GPA	From Table 15.2B Low GPA
$N_s = (29)(54) = 1566$	$N_s = (28)(55) = 1540$
$N_d = (28)(24) = 672$	$N_d = (28)(24) = 672$

$$\Sigma N_s = 1566 + 1540 = 3106$$

$$\Sigma N_d = 672 + 672 = 1344$$

$$G_p = \frac{\Sigma N_s - \Sigma N_d}{\Sigma N_s + \Sigma N_d}$$

$$G_p = \frac{3106 - 1344}{3106 + 1344}$$

$$G_p = \frac{1762}{4450}$$

$$G_p = 0.40$$

The partial gamma measures the strength of the association between X and Y once the effects of Z have been removed. In this instance, the partial gamma is the same value as the bivariate gamma ($G_p = G = 0.40$) and indicates that GPA has no effect on the relationship between satisfaction and membership.

When class standing was controlled (Table 15.3), clear evidence of a spurious relationship was found, since the gammas computed on the partial tables dropped almost to zero. Let us see what the value of partial gamma (G_p) would be for this second control:

From Table 15.3A Upperclassmen	From Table 15.3B Underclassmen
$N_s = (8)(97) = 776$	$N_s = (49)(12) = 588$
$N_d = (32)(24) = 768$	$N_d = (24)(24) = 576$

$$\Sigma N_s = 776 + 588 = 1364$$

$$\Sigma N_d = 768 + 576 = 1344$$

$$G_p = \frac{\Sigma N_s - \Sigma N_d}{\Sigma N_s + \Sigma N_d}$$

$$G_p = \frac{1364 - 1344}{1364 + 1344}$$

$$G_p = \frac{20}{2708}$$

$$G_p = 0.01$$

ONE STEP AT A TIME **Computing Partial Gamma (G_p)**

To Compute N_s and N_d for all Partial Tables

1: Add up all values for N_s from all partial tables to find ΣN_s.

2: Add up all values for N_d from all partial tables to find ΣN_d.

To Find Partial Gamma (G_p) by Solving Formula 15.1

1: Subtract ΣN_d from ΣN_s.

2: Add ΣN_d to ΣN_s.

3: Divide the quantity you found in step 1 by the quantity you found in step 2. The result is G_p.

Once the effects of the control variable are removed, there is no relationship between X and Y. The very low value of G_p confirms our previous conclusion that the bivariate relationship between membership and satisfaction is spurious and actually due to the effects of class standing.

In a sense, G_p tells us no more about the relationships than we can see for ourselves from a careful analysis of the percentage distributions of Y in the partial tables or by a comparison of the measures of association computed on the partial tables. The advantage of G_p is that it tells us, in a single number, the precise effects of Z on the relationship between X and Y. While G_p is no substitute for the analysis of the partial tables per se, it is a convenient way of stating our results when working with direct or spurious relationships.

Although G_p can be calculated in cases of interactive relationships (see Table 15.4), it is rather difficult to interpret in these instances. If substantial interaction is found in the partial tables, this indicates that the control variable has a profound effect on the bivariate relationship. Thus, we should not attempt to separate the effects of Z from the bivariate relationship, and, since G_p involves exactly this kind of separation, it should not be computed.

Where Do Control Variables Come From?

In one sense, this question is easy to answer: Control variables come mainly from theory. Social research proceeds in many different ways and is begun in response to a variety of problems. However, virtually all quantitative research projects are guided by a more or less explicit theory or by some question about the relationship between two or more variables. The ultimate goal of social research is to develop defensible generalizations that improve our understanding of the variables under consideration and link back to theory at some level (see Figure 1.1).

Thus, research projects are anchored in theory; and the concepts of interest, which will later be operationalized as variables, may be first probed at the theoretical level. The social world is exceedingly complex, and any attempt to explain it with a simple bivariate relationship will fail. Theories are, almost by definition, multivariate, even when they focus on a particular bivariate relationship. Thus, theory may suggest the control variables that need to be incorporated into the analysis. In the example used throughout this chapter, I tried to suggest that any researcher attempting to probe the relationship between involvement in an organization and satisfaction would, in the course of thinking over the possibilities, identify a number of additional variables that needed to be explicitly incorporated into the analysis.

Of course, textbook descriptions of the research process are simplified and suggest that research flows smoothly from conceptualization to operationalization to quantification to generalization. In reality, research is full of surprises, unexpected outcomes, and unanticipated results. Research in "real life" is more loosely structured and requires more imagination and creativity than textbooks can fully convey. My point is that the control variables that might be appropriate to incorporate in the data-analysis phase will be suggested or implied in the theoretical backdrop of the research project, along with the researcher's imagination and sensitivity to the problem being addressed.

These considerations have taken us well beyond the narrow realm of statistics and back to the planning stages of the research project. At this early time the researcher must make decisions about which variables to measure during the data-gathering phase and, thus, which variables might be incorporated as potential controls. Careful thinking and an extended consideration of possible outcomes at the planning stage will pay significant dividends during the data-analysis phase. Ideally, all relevant control variables will be incorporated and readily available for statistical analysis.

The Limitations of Elaborating Bivariate Tables

The most important limitation of elaborating bivariate tables is sample size. Elaboration is relatively inefficient because it requires that the researcher divide the sample into a series of partial tables. If the control variable has more than two or three possible values or if we attempt to control for more than one variable at a time, many partial tables will be produced. The greater the number of partial tables, the more likely we are to run out of cases to fill all of the cells. Empty or small cells, in turn, can create serious problems in terms of generalizability and confidence in our findings.

To illustrate, the example used throughout this chapter began with two dichotomized variables and a four-cell table (Table 15.1). Each of the control variables was also dichotomized, and we never confronted more than two partial tables with four cells each, or a total of eight cells for each control variable. If we had used a control variable with three values, we would have had 12 cells to fill up, and if we had attempted to control for two dichotomized variables, we would have had 16 cells in four different partial tables. Clearly, as control variables become more elaborate and/or as the process of controlling becomes more complex, empty or small cells will increasingly become a problem.

Two potential solutions to this problem immediately suggest themselves. The easy solution is to reduce the number of cells in the partial tables by collapsing categories within variables. If all variables are dichotomized, for example, the number of cells will be kept to a minimum. A better solution is to work with only very large samples so that there will be plenty of cases to fill up the cells. Unfortunately, the easy solution will often violate common sense (collapsed categories are more likely to group dissimilar elements together); and the better solution is not always feasible (mundane matters of time and money rear their ugly heads).

A third solution to the problem of empty cells requires the (sometimes risky) assumption that the variables of interest are measured at the interval-ratio level. At that level the techniques of partial and multiple correlation and regression, to be introduced in Chapter 16, are available. These multivariate techniques are more efficient than elaboration because they utilize all cases simultaneously and do not require that the sample be divided among the various partial tables.

Interpreting Statistics: Analyzing the Sources of Democracy

In Chapter 13, we analyzed the relationship between income level—which reflects level of development or industrialization—and political and civil rights for a group of nations. In this section, we analyze how that bivariate relationship behaves after controlling for a third variable.

As you recall, our theory was that high-income, more industrialized nations, would have a more educated, politically active population and stronger guarantees of political and civil rights (such as free speech and freedom of choice between rival political parties and candidates for office). We found a strong, positive relationship (gamma = +0.81) between income level and political and civil rights, consistent with these expectations.

In this chapter, we will extend this analysis by incorporating a third variable: the level of income inequality or the extent to which income is concentrated among the wealthy or distributed evenly across the population. Nations with higher levels of inequality might be less open to democracy, regardless of their level of affluence, as the elites who enjoy a greater share of resources may seek to protect their high status by stifling dissent and political protest.

Before we conduct this analysis, we need to confront one of the important limitations of elaboration. With only 89 nations in the sample, we need to minimize the number of categories for our variables so we can avoid the problem of empty cells. We can't change the three income levels but we can reduce the other variables to dichotomies by splitting them at their medians. Thus, both the dependent and control variables will have only two categories: high and low. Our bivariate table will be different from the one used in Chapter 13 and is presented here as Table 15.6.

As you can see, the new scoring doesn't have much effect on the relationship: The extent of political and civil rights is still strongly and positively related to the income level of the nations. Will this relationship hold regardless of the level of income inequality in the nations? Table 15.7 presents the partial tables for the relationship between income level and political and civil rights for nations with lower and higher levels of income inequality.

A quick glance at the partial gammas and the column percentages shows strong and positive relationships: Regardless of degree of inequality, political and civil rights increase in strength as income level increases. However, we also see that the partial gammas are different from each other and from the bivariate gamma. There seems to be some interaction between these three variables. What does this mean?

Looking first at the 39 nations with lower levels of inequality, there is a perfect relationship between level of development and political and civil rights. All five (100%) of the low-inequality, low-income nations are also low on rights, the middle-income nations in this group are split, and all 23 (100%) of the most affluent, low-inequality nations are high on rights.

TABLE 15.6 **Political and Civil Rights by Income Level (frequencies and percentages)**

Rights	Income Level		
	Low	Middle	High
Lower	20 (87.0%)	19 (48.7%)	2 (7.4%)
Higher	3 (13.0%)	20 (51.3%)	25 (92.6%)
	23 (100.0%)	39 (100.0%)	27 (100.0%)

Chi square = 31.83, df = 2, $p < 0.000$, gamma = 0.86.

TABLE 15.7 Political and Civil Rights by Income Level, Controlling for Income Inequality (frequencies and percentages)

A. Low Inequality	Income Level		
Rights	Low	Middle	High
Lower	5 (*100.0%*)	6 (*54.5%*)	0 (*0.0%*)
Higher	0 (*0.0%*)	5 (*45.5%*)	23 (*100.0%*)
	5 (*100.0%*)	11 (*100.0%*)	23 (*100.0%*)

Gamma = 1.00

B. High Inequality	Income Level		
Rights	Low	Middle	High
Lower	11 (*78.6%*)	9 (*37.5%*)	0 (*0.0%*)
Higher	3 (*21.4%*)	15 (*62.5%*)	1 (*100.0%*)
	14 (*100.0%*)	24 (*100.0%*)	1 (*100.0%*)

Gamma = 0.75

Turning to the 39 higher-inequality nations, we see the same general pattern: The great majority of low-income nations are low on rights and the middle-income nations are split but more likely to be high on rights. However, when we turn to the high-income category, we find an important problem: There's only one case in the column. All other high-income nations are in the low-inequality group. What is this nation that is affluent, highly democratic, and highly unequal in its distribution of income? It may come as no surprise to learn that the nation with this unusual combination of scores is the United States, often identified as the most unequal of all advanced, industrial nations and a society in which income inequality has been increasing for some time (see Chapter 4). At any rate, the fact that there is only one nation in this column means that we have to exercise caution in our interpretation.

Overall, and basing our conclusion more on the low- and middle-income nations, it seems reasonable to conclude that there is a direct relationship between the extent of rights and the levels of development regardless of level of income inequality

SUMMARY

1. Most research questions require the analysis of the interrelationship among many variables, even when the researcher is primarily concerned with a specific bivariate relationship. Multivariate statistical techniques provide the researcher with a set of tools by which additional information can be gathered about the variables of interest and by which causal interrelationships can be probed.

2. When variables have been organized in bivariate tables, multivariate analysis proceeds by controlling for a third

variable. Partial tables are constructed and compared with each other and with the original bivariate table. Comparisons are made easier if appropriate measures of association are computed for all tables.

3. A direct relationship exists between the independent (X) and dependent (Y) variables if, after controlling for the third variable (Z), the relationship between X and Y is the same across all partial tables and the same as in the bivariate table. This pattern suggests a causal relationship between X and Y.

4. If the relationship between X and Y is the same across all partial tables but much weaker than in the bivariate table, the relationship is either spurious (Z causes both X and Y) or intervening (X causes Z, which causes Y). Either pattern suggests that Z must be explicitly incorporated into the analysis.

5. Interaction exists if the relationship between X and Y varies across the partial tables and between each partial and the bivariate table. This pattern suggests that no simple or direct causal relationship exists between X and Y and that Z must be explicitly incorporated into the causal scheme.

6. Partial gamma (G_p) is a useful summary statistic that measures the strength of the association between X and

Y after the effects of the control variable (Z) have been removed. Partial gamma should not be computed when an analysis of the partial tables shows substantial interaction.

7. Potential control variables must be identified before the data-gathering phase of the research project. The theoretical backdrop of the research project, along with creative thinking and some imagination, will suggest the variables that should be controlled for and measured.

8. Controlling for third variables by constructing partial tables is inefficient, in that the cases must be spread out across many cells. If the variables have many categories and/or the researcher attempts to control for more than one variable simultaneously, "empty cells" may become a problem. It may be possible to deal with this problem by either collapsing categories or gathering very large samples. If interval-ratio level of measurement can be assumed, the multivariate techniques presented in the next chapter will be preferred, since they do not require the partitioning of the sample.

SUMMARY OF FORMULAS

Partial gamma: 15.1

$$G_p = \frac{\Sigma N_s - \Sigma N_d}{\Sigma N_s + \Sigma N_d}$$

GLOSSARY

Direct relationship. A multivariate relationship in which the control variable has no effect on the bivariate relationship.

Elaboration. The basic multivariate technique for analyzing variables arrayed in tables. Partial tables are constructed to observe the bivariate relationship in a more detailed or elaborated format.

Explanation. See **Spurious relationship**.

Interaction. A multivariate relationship wherein a bivariate relationship changes across the categories of the control variable.

Interpretation. See **Intervening relationship**.

Intervening relationship. A multivariate relationship wherein a bivariate relationship becomes substantially weaker after a third variable is controlled for. The

independent and dependent variables are linked primarily through the control variable.

Partial gamma (G_p). A statistic that indicates the strength of the association between two variables after the effects of a third variable have been removed.

Partial tables. Tables produced when controlling for a third variable.

Replication. See **Direct relationship**.

Specification. See **Interaction**.

Spurious relationship. A multivariate relationship in which a bivariate relationship becomes substantially weaker after a third variable is controlled for. The independent and dependent variables are not causally linked. Rather, both are caused by the control variable.

Z. Symbol for any control variable.

PROBLEMS

15.1 [SOC] Problem 13.1 concerned the assimilation of a small sample of immigrants. One table showed the relationship between length of residence in the United States and facility with the English language:

	Length of Residence		
English Facility	Less Than Five Years (Low)	More Than Five Years (High)	Totals
Low	20	10	30
High	5	15	20
Totals	25	25	50

a. If necessary, find the column percentages and compute gamma for this table. Describe the bivariate relationship in terms of strength and direction. Is the relationship between residence and language affected by the gender of the immigrant? Here are the partial tables showing the bivariate relationship for males and females:

A. Males

	Length of Residence		
English Facility	Less Than Five Years (Low)	More Than Five Years (High)	Totals
Low	10	5	15
High	2	8	10
Totals	12	13	25

B. Females

	Length of Residence		
English Facility	Less Than Five Years (Low)	More Than Five Years (High)	Totals
Low	10	5	15
High	3	7	10
Totals	13	12	25

b. Find the column percentages and compute gamma for each of the partial tables. Compare these percentages and gammas with each other and with the bivariate table. Does controlling for gender change the bivariate relationship?

Is the relationship between residence and language affected by the origin of the immigrant? Here are the partial tables showing the bivariate relationship for Asian and Hispanic immigrants:

A. Asian Immigrants

	Length of Residence		
English Facility	Less Than Five Years (Low)	More Than Five Years (High)	Totals
Low	8	4	12
High	2	5	7
Totals	10	9	19

B. Hispanic Immigrants

	Length of Residence		
English Facility	Less Than Five Years (Low)	More Than Five Years (High)	Totals
Low	12	6	18
High	3	10	13
Totals	15	16	31

c. Find the column percentages and compute gamma for each of the partial tables. Compare these percentages and gammas with each other and with the bivariate table. Does controlling for the origin of the immigrants change the bivariate relationship?

15.2 [SOC] Data on suicide rates, age structure, and unemployment rates have been gathered for 100 census tracts. Suicide and unemployment rates have been dichotomized at the median so that each tract could be rated as high or low. Age structure is measured in terms of the percentage of the population age 65 and older. This variable has also been dichotomized, and tracts have been rated as high or low. The following tables display the bivariate relationship between suicide rate and age structure and the same relationship controlling for unemployment.

Suicide Rate	Population 65 and Older (%)		
	Low	High	Totals
Low	45	20	65
High	10	25	35
Totals	55	45	100

a. Calculate percentages and gamma for the bivariate table. Describe the bivariate relationship in terms of strength and direction.

SUICIDE RATE BY AGE,
CONTROLLING FOR UNEMPLOYMENT

A. High Unemployment

| Suicide Rate | Population 65 and Older (%) | | Totals |
	Low	High	
Low	23	10	33
High	5	12	17
Totals	28	22	50

B. Low Unemployment

| Suicide Rate | Population 65 and Older (%) | | Totals |
	Low	High	
Low	22	10	32
High	5	13	18
Totals	27	23	50

b. Calculate percentages and gamma for each partial table. Compare the partial tables with each other and with the bivariate table. Compute partial gamma (G_p).

c. Summarize the results of the control. Does unemployment rate have any effect on the relationship between age and suicide rate? Describe the effect of the control variable in terms of the pattern of percentages, the value of the gammas, and the possible causal relationships among these variables.

15.3 **SW** Do long-term patients in mental health facilities become more withdrawn and reclusive over time? A sample of 608 institutionalized patients was rated by a standard "reality orientation scale." Is there a relationship with length of institutionalization? Does gender have any effect on the relationship?

REALITY ORIENTATION BY LENGTH
OF INSTITUTIONALIZATION

| Reality Orientation | Length of Institutionalization | | Totals |
	Less Than five Years	More Than five Years	
Low	200	213	413
High	117	78	195
Totals	317	291	608

REALITY ORIENTATION BY LENGTH
OF INSTITUTIONALIZATION, CONTROLLING
FOR GENDER

A. Females

| Reality Orientation | Length of Institutionalization | | Totals |
	Less Than five Years	More Than five Years	
Low	95	120	215
High	60	37	97
Totals	155	157	312

B. Males

| Reality Orientation | Length of Institutionalization | | Totals |
	Less Than five Years	More Than five Years	
Low	105	93	198
High	57	41	98
Totals	162	134	296

15.4 **SOC** Is there a relationship between attitudes on sexuality and age? Are older people more conservative with respect to questions of sexual morality? A national sample of 925 respondents has been questioned about attitudes on premarital sex. Responses have been collapsed into two categories: those who believe that premarital sex is "always wrong" and those who believe it is not wrong under certain conditions ("sometimes wrong"). These responses have been cross-tabulated by age, and the results are reported here:

ATTITUDE TOWARD PREMARITAL
SEX BY AGE

| Premarital Sex Is: | Age | | Totals |
	Younger Than 35	35 and Older	
Always wrong	90	235	325
Sometimes wrong	420	180	600
Totals	510	415	925

a. Calculate percentages and gamma for the table. Is there a relationship between these two variables? Describe the bivariate relationship in terms of its strength and direction. Next, the bivariate relationship is reproduced after controlling for the sex of the respondent.

Does gender have any effect on the relationship?

ATTITUDE TOWARD PREMARITAL SEX BY AGE, CONTROLLING FOR GENDER

A. Males

Premarital Sex Is:	Age		Totals
	Younger Than 35	35 and Older	
Always wrong	70	55	125
Sometimes wrong	190	80	270
Totals	260	135	395

B. Females

Premarital Sex Is:	Age		Totals
	Younger Than 35	35 and Older	
Always wrong	20	180	200
Sometimes wrong	230	100	330
Totals	250	280	530

b. Summarize the results of this control. Does gender have any effect on the relationship between age and attitude? If so, describe the effect of the control variable in terms of the pattern of percentages, the value of the gammas, and the possible causal interrelationships among these variables.

15.5 SOC A job-training center is trying to justify its existence to its funding agency. To this end, data on four variables have been collected for each of the 403 trainees served over the past three years: (1) whether or not the trainee completed the program, (2) whether or not the trainee got and held a job for at least a year after training, (3) the sex of the trainee, and (4) the race of the trainee. Is employment related to completion of the program? Is the relationship between completion and employment affected by race or sex?

EMPLOYMENT BY TRAINING

Held Job for at Least One Year?	Training Completed?		Totals
	Yes	No	
Yes	145	60	205
No	72	126	198
Totals	217	186	403

EMPLOYMENT BY TRAINING, CONTROLLING FOR RACE

A. Whites

Held Job for at Least One Year?	Training Completed?		Totals
	Yes	No	
Yes	85	33	118
No	38	47	85
Totals	123	80	203

B. Blacks

Held Job for at Least One Year?	Training Completed?		Totals
	Yes	No	
Yes	60	27	87
No	34	79	113
Totals	94	106	200

EMPLOYMENT BY TRAINING, CONTROLLING FOR SEX

C. Males

Held Job for at Least One Year?	Training Completed?		Totals
	Yes	No	
Yes	73	30	103
No	36	62	98
Totals	109	92	201

D. Females

Held Job for at Least One Year?	Training Completed?		Totals
	Yes	No	
Yes	72	30	102
No	36	64	100
Totals	108	94	202

15.6 SOC What are the social sources of support for the environmental movement? A recent survey gathered information on level of concern for such issues as global warming and acid rain. Is concern for the environment related to level of education? What effects do the control variables have? Write a paragraph summarizing your conclusions.

CONCERN FOR THE ENVIRONMENT BY LEVEL OF EDUCATION

| | Level of Education | | |
Concern	Low	High	Totals
Low	27	35	62
High	22	48	70
Totals	49	83	132

CONCERN FOR THE ENVIRONMENT BY LEVEL OF EDUCATION, CONTROLLING FOR GENDER

A. Males

| | Level of Education | | |
Concern	Low	High	Totals
Low	14	17	31
High	11	22	33
Totals	25	39	64

B. Females

| | Level of Education | | |
Concern	Low	High	Totals
Low	13	18	31
High	11	26	37
Totals	24	44	68

CONCERN FOR THE ENVIRONMENT BY LEVEL OF EDUCATION, CONTROLLING FOR "LEVEL OF TRUST IN THE NATION'S LEADERSHIP"

A. Low Levels of Trust

| | Level of Education | | |
Concern	Low	High	Totals
Low	6	22	28
High	10	40	50
Totals	16	62	78

B. High Levels of Trust

| | Level of Education | | |
Concern	Low	High	Totals
Low	21	13	34
High	12	8	20
Totals	33	21	54

CONCERN FOR THE ENVIRONMENT BY LEVEL OF EDUCATION, CONTROLLING FOR RACE

A. Whites

| | Level of Education | | |
Concern	Low	High	Totals
Low	19	11	30
High	18	44	62
Totals	37	55	92

B. Blacks

| | Level of Education | | |
Concern	Low	High	Totals
Low	8	24	32
High	4	4	8
Totals	12	28	40

15.7 In problem 13.14, we investigated the relationships between income and five dependent variables. Let's return to two of those relationships and see if sex has any effect on the bivariate relationships. The bivariate and partial tables are presented here along with the bivariate gammas that were computed in the previous exercise. Compute percentages and gammas for each of the partial tables and state your conclusions. Does controlling for sex have any effect?

a. Support for abortion by income (from problem 13.14a):

| Support the Right to an Abortion? | Income | | | |
	(1) Low	(2) Moderate	(3) High	Totals
Yes (1)	220	218	226	664
No (2)	366	299	250	915
Totals	586	517	476	1579

Gamma = −0.14

(HINT: Note that the higher score on the variable that measures support for abortion is associated with "No." The negative sign of gamma means that as score on income increases, score on the dependent variable decreases. In other words, people with higher incomes were more supportive of the legal right to abortion [more likely to say "Yes," or have a score of 1]. This is actually a positive relationship: As income

increases, support for the legal right to abortion increases.)

SUPPORT FOR THE LEGAL RIGHT TO ABORTION BY INCOME, CONTROLLING FOR SEX

A. Males

Right to an Abortion?	Income			
	(1) Low	(2) Moderate	(3) High	Totals
Yes (1)	89	96	100	285
No (2)	130	141	122	393
Totals	219	237	222	678

B. Females

Right to an Abortion?	Income			
	(1) Low	(2) Moderate	(3) High	Totals
Yes (1)	131	122	126	379
No (2)	236	158	128	522
Totals	367	280	254	901

b. Support for the right to commit suicide for people with an incurable disease by income (from problem 13.15c):

Right to Suicide?	Income			
	(1) Low	(2) Moderate	(3) High	Totals
Approve (1)	343	341	338	1022
Oppose (2)	227	194	147	568
Totals	570	535	485	1590

Gamma = −0.14

(HINT: Be careful in interpreting direction for this table.)

APPROVAL OF SUICIDE FOR PEOPLE WITH AN INCURABLE DISEASE BY INCOME, CONTROLLING FOR SEX:

A. Males

Right to Suicide?	Income			
	(1) Low	(2) Moderate	(3) High	Totals
Approve (1)	140	165	170	475
Oppose (2)	82	80	61	223
Totals	222	245	231	698

B. Females

Right to Suicide?	Income			
	(1) Low	(2) Moderate	(3) High	Totals
Approve (1)	203	176	168	547
Oppose (2)	145	114	86	345
Totals	348	290	254	892

15.8 SOC Are attitudes about homosexuality related to age? Does the relationship vary by nation? The bivariate relationship between age and approval of homosexuality are presented for Canada and the United States, along with the bivariate gammas. For both nations, there is a weak, negative relationship. Given the way the variable was scored, the negative direction means that approval decreases as age increases. Are the bivariate relationships affected by gender? For each partial table, find gamma and write a paragraph of analysis comparing the partial gammas to the bivariate gammas. Does the relationship between age and approval change from nation to nation? How? *(HINT: It will be easier to analyze the relationship and trace the effects of controlling for sex by getting percentages for all tables.)*

a. Approval of homosexuality by age (Canada):

Attitude Toward Homosexuality	Age			
	15–29	30–49	50 and Older	TOTALS
Disapprove	88	234	288	610
Neither approve nor disapprove	70	181	215	466
Approve	225	357	337	919
TOTALS	383	772	840	1995

Gamma = −0.14

b. Approval of homosexuality by age (United States):

Attitude Toward Homosexuality	Age			
	15–29	30–49	50 and Older	TOTALS
Disapprove	93	176	251	520
Neither approve nor disapprove	66	114	99	279
Approve	88	131	144	363
TOTALS	247	421	494	1162

Gamma = −0.13

c. Approval of homosexuality by age, controlling for sex (Canada):

A. Males

	Age			
Attitude Toward Homosexuality	15–29	30–49	50 and Older	TOTALS
Disapprove	56	126	143	325
Neither approve nor disapprove	44	103	103	250
Approve	90	151	152	393
TOTALS	190	380	398	968

B. Females

	Age			
Attitude Toward Homosexuality	15–29	30–49	50 and Older	TOTALS
Disapprove	32	109	145	286
Neither approve nor disapprove	25	78	111	214
Approve	136	205	186	527
TOTALS	193	392	442	1027

d. Approval of homosexuality by age, controlling for sex (United States):

A. Males

	Age			
Attitude Toward Homosexuality	15–29	30–49	50 and Older	TOTALS
Disapprove	42	96	119	257
Neither approve nor disapprove	32	56	50	138
Approve	46	63	62	171
TOTALS	120	215	231	566

B. Females

	Age			
Attitude Toward Homosexuality	15–29	30–49	50 and Older	TOTALS
Disapprove	50	81	132	263
Neither approve nor disapprove	34	58	49	141
Approve	43	67	82	192
TOTALS	127	206	263	596

Data are from the World Values survey. The original wording of the survey item was "Please tell me . . . whether you think homosexuality can always be justified, never be justified, or something in between."

YOU ARE THE RESEARCHER

A Multivariate Analysis of Sexual Attitudes

In Chapter 13, you investigated the possible causes of attitudes and opinions about premarital sex. Here, we will expand on that project by bringing control variables into the analysis. Attitudes toward premarital sex (*premarsx*) will once again be the dependent variable. Remember that the coding scheme for this variable: Higher scores indicate greater support for premarital sex and lower scores mean greater disapproval.

Step 1: Choosing Independent Variables

Previously, you selected four possible independent variables, all of which were measured at the ordinal level. Choose any two of those or, if you wish, choose two new (ordinal-level) independent variables. List your independent variables here:

SPSS Variable Name	What exactly does this variable measure?
_____	_____
_____	_____

Step 2: Choosing a Control Variable

Select a variable from the 2008 GSS that you think might affect the relationships between attitudes toward premarital sex and your independent variables. Your control variables can be any level of measurement but should have *only* two or three scores. If necessary, you can recode variables with more categories to fit this description. List your control variable here, along with its categories or scores.

SPSS Variable Name	What exactly does this variable measure?	Category 1 (e.g., male for *sex*)	Category 2 (e.g., female for *sex*)	Category 3 (If needed)

Step 3: Stating Hypotheses

State hypotheses about the relationships you expect to find between your independent variables and *premarsx*. You can use the same hypotheses you used in Chapter 13. State these hypotheses in terms of the direction of the relationship you expect to find. For example, you might hypothesize that approval of premarital sex will decline as age increases (a negative relationship).

Step 4: Predicting the Effects of the Control Variables

For each combination of dependent, independent, and control variables, state your expectations for the relationship. State an expectation for each category of the control variable. For example, if the control variable is "nation of birth," you might say that you expect approval of premarital sex to decline as age increases for Americans but not for Canadians.

Step 5: Running Crosstabs

a. Bivariate tables: *If necessary*, generate the bivariate tables for each of your independent variables and *premarsx*. Click **Analyze → Descriptives → Crosstabs** and place *premarsx* in the **Rows:** box and both of your independent variables in the **Columns:** box. Click the **Statistics** button to get chi square and gamma and the **Cells** button to get column percentages.

The output will consist of two bivariate tables and statistics, one for each combination of *premarsx* and the independent variables.

b. Multivariate tables: Click **Analyze → Descriptives → Crosstabs** again. If necessary, place the dependent variable (*premarsx*) in the **Rows:** box and both of your independent variables in the **Columns:** box. Next, place the control variable in box at the bottom of the **Crosstabs** window. If necessary, click the **Statistics** button to get chi square and gamma and the **Cells** button to get column percentages.

The output will consist of two sets of partial tables showing the relationship between *premarsx* and each independent variable for each value of the control variable. Partial gammas will be reported in a separate box below the partial tables.

Step 4: Recording Resuls

These commands will generate a lot of output, and it will be helpful to summarize your results in the following table.

First Independent Variable (SPSS name _____)	Control Variable (SPSS name _____)		
Value of Bivariate Gamma = _____	Name of 1st category of control variable _____	Name of 2nd category of control variable _____	Name of 3rd category of control variable _____
	Partial Gamma = _____	Partial Gamma = _____	Partial Gamma = _____

Second Independent Variable (SPSS name _____)	Control Variable (SPSS name _____)		
Value of Bivariate Gamma = _____	Name of 1st category of control variable _____	Name of 2nd category of control variable _____	Name of 3rd category of control variable _____
	Partial Gamma = _____	Partial Gamma = _____	Partial Gamma = _____

Step 5: Analyzing and Interpreting Results

For each independent variable, interpret the impact, if any, of the control variable. Look for evidence of direct, spurious, or interactive relationships using Table 15.5 as a guide. Remember that your results may not perfectly resemble any of these outcomes: Real-life research often generates ambiguous, difficult-to-understand results. Write a paragraph of interpretation for each set of partial tables. Were your expectations confirmed? How?

16

Partial Correlation and Multiple Regression and Correlation

LEARNING OBJECTIVES   By the end of this chapter, you will be able to

1. Compute and interpret partial correlation coefficients.
2. Find and interpret the least-squares multiple regression equation with partial slopes.
3. Calculate and interpret the multiple correlation coefficient (R^2).
4. Explain the limitations of partial and multiple regression analysis.

As mentioned at the beginning of Chapter 15, social science research is, by nature, multivariate, involving the simultaneous analysis of scores of variables. Some of the most powerful and widely used statistical tools for multivariate analysis are introduced in this chapter. We cover techniques that are used to analyze causal relationships and to make predictions, both crucial endeavors in any science.

These techniques are based on Pearson's r (see Chapter 14) and are most appropriately used with high-quality, precisely measured interval-ratio variables. As we have noted on many occasions, such data are relatively rare in social science research, and the techniques presented in this chapter are commonly used on variables measured at the ordinal level and with nominal-level variables in the form of dummy variables (see Chapter 14).

The techniques presented in this chapter are generally more flexible than those presented in Chapter 15. They produce more information and provide a wider variety of ways of disentangling the interrelationships among the variables.

We first consider partial correlation analysis, a technique analogous to controlling for a third variable by constructing partial tables (see Chapter 15). The second technique, multiple regression and correlation, allows researchers to assess the effects, separately and in combination, of more than one independent variable on the dependent variable.

Throughout this chapter, we focus on research situations involving three variables. This is the simplest application, but extensions to situations involving four or more variables are relatively straightforward. However, computations become very complex when dealing with more than two independent variables, and you should use computerized statistical packages (such as SPSS) for these situations.

Partial Correlation

In Chapter 14, we used Pearson's r to measure the strength and direction of bivariate relationships. To provide an example, we looked at the relationship between husband's contribution to housework (the dependent, or Y, variable) and number of children (the independent, or X, variable) for a sample of 12 families. We found a positive relationship of moderate strength ($r = 0.50$) and concluded that husband's contribution to housework tends to increase as the number of children increases.

You might wonder, as researchers commonly do, if this relationship holds true for *all* types of families? For example, might husbands in strongly religious families respond differently than those in less religious families? Would politically conservative husbands behave differently than husbands who are politically liberal? How about more educated husbands? Would they respond differently than less educated husbands? We can address these kinds of issues by means of a technique called **partial correlation**, in which we observe how the bivariate relationship changes when a third variable, such as religiosity, education, or ethnicity, is introduced. Third variables are often referred to as Z variables or **control variables**.

Partial correlation proceeds by first computing Pearson's r for the bivariate (or zero-order) relationship and then computing the partial (or first-order) correlation coefficient. If the partial correlation coefficient differs from the **zero-order correlation** coefficient, we conclude that the third variable has an effect on the bivariate relationship. If, for example, well-educated husbands respond differently to an additional child than less well-educated husbands, the partial correlation coefficient will differ in strength (and perhaps in direction) from the bivariate correlation coefficient.

Before considering matters of computation, we'll consider the possible relationships between the partial and bivariate correlation coefficients and what they might mean.

TYPES OF RELATIONSHIPS

Three types of relationships were introduced in Chapter 15 and summarized in Table 15.5. I repeat the information here for the convenience of those readers who did not cover Chapter 15 and to place these patterns in the context of partial correlation.

Direct Relationship. One possible outcome is that the partial correlation coefficient is essentially the same value as the bivariate coefficient. Imagine, for example, that after we controlled for husband's education, we found a partial correlation coefficient of $+0.49$, compared to the zero-order Pearson's r of $+0.50$. This would mean that the third variable (husband's education) has no effect on the relationship between number of children and husband's hours of housework. In other words, regardless of their education, husbands respond in a similar way to additional children. This outcome is consistent with the conclusion that there is a *direct* or *casual* relationship (see Figure 16.1) between X and Y and that the third variable (Z) is irrelevant and should be discarded from further consideration, although the researcher would probably run additional

FIGURE 16.1 A Direct Relationship Between X and Y

$X \longrightarrow Y$

FIGURE 16.2   A Spurious Relationship Between *X* and *Y*

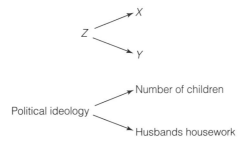

tests with other likely control variables (e.g., the researcher might control for the religion or ethnicity of the family).

Spurious and Intervening Relationships. A second possible outcome occurs when the partial correlation coefficient is much weaker than the bivariate correlation, perhaps even dropping to zero. This outcome is consistent with two different relationships. The first is called a *spurious* relationship: The control variable (*Z*) is a cause of both the independent (*X*) and the dependent (*Y*) variable (see Figure 16.2). This outcome would mean that *X* and *Y* are not actually related. They appear to be related only because both are dependent on a common cause (*Z*). Once *Z* is taken into account, the apparent relationship between *X* and *Y* disappears.

What would a spurious relationship look like? Imagine that we controlled for the political ideology of parents and found that the partial correlation coefficient was much weaker than the bivariate Pearson's *r*. This might indicate that the number of children does not actually change the husband's contribution to housework (that is, the relationship between *X* and *Y* is not direct). Rather, political ideology is the mutual cause of both of the other variables: More conservative families are more likely to follow traditional gender role patterns (in which husbands contribute less to housework) *and* have more children.

FIGURE 16.3   An Intervening Relationship Between *X* and *Y*

This pattern (partial correlation much weaker than the bivariate correlation) is also consistent with an *intervening* relationship between the variables (see Figure 16.3). In this situation, *X* and *Y* are not linked directly but are connected through the control variable. Again, once *Z* is controlled, the apparent relationship between *X* and *Y* disappears.

How can we tell the difference between spurious and intervening relationships? This distinction cannot be made on statistical grounds: Spurious and intervening relationships look exactly the same in terms of statistics. The researcher may be able to distinguish between these two relationships in terms of the time order of the variables (i.e., which came first) or on theoretical grounds, but not on statistical grounds.

Interaction. A third possible relationship between variables should be mentioned, even though it *cannot* be detected by partial correlation analysis. This relationship, called interaction, occurs when the relationship between *X* and *Y* changes markedly under the various values of *Z*. For example, if we controlled

FIGURE 16.4 An Interactive Relationship Between X, Y, and Z

for social class and found that husbands in middle-class families increased their contribution to housework as the number of children increased while husbands in working-class families did just the reverse, we would conclude that there was interaction between these three variables. In other words, there would be a positive relationship between X and Y for one category of Z and a negative relationship for the other category, as illustrated in Figure 16.4.

Interactive relationships are explored in Chapter 15. We will not examine them further here, however, since they cannot be detected by partial correlation.

COMPUTING AND INTERPRETING THE PARTIAL CORRELATION COEFFICIENT

The logic and computation of the partial correlation coefficient requires some new concepts and terminology. We'll introduce the terminology first and then move on to the formula.

Terminology and Formula. Partial correlation requires that we deal with more than one bivariate relationship and we need to differentiate between them with subscripts. Thus, the symbol r_{yx} will refer to the correlation coefficient between variable Y and variable X; r_{yz} will refer to the correlation coefficient between Y and Z; and r_{xz} will refer to the correlation coefficient between X and Z. Recall that correlation coefficients calculated for bivariate relationships are often referred to as zero-order correlations.

Partial correlation coefficients, or first-order partials, are symbolized as $r_{yx.z}$. The variable to the right of the dot is the control variable. Thus, $r_{yx.z}$ refers to the partial correlation coefficient that measures the relationship between variables X and Y while controlling for variable Z. The formula for the first-order partial is

FORMULA 16.1
$$r_{yx.z} = \frac{r_{yx} - (r_{yz})(r_{xz})}{\sqrt{1 - r_{yz}^2}\sqrt{1 - r_{xz}^2}}$$

Note that you must first calculate the zero-order coefficients between all possible pairs of variables (variables X and Y, X and Z, and Y and X) before solving this formula.

Computation. To illustrate the computation of a first-order partial, we will return to the relationship between number of children (X) and husband's contribution to housework (Y) for 12 dual-career families. The zero-order r between these two variables ($r_{yx} = 0.50$) indicated a moderate, positive relationship (as number of children increased, husbands tended to contribute more to housework). Suppose the researcher wished to investigate the possible effects of husband's education on the bivariate relationship. The original data (from Table 14.1) and the scores of the 12 families on the new variable are presented in Table 16.1.

The zero-order correlations, as presented in the correlation matrix in Table 16.2, indicate that the husband's contribution to housework is positively related to number of children ($r_{yx} = 0.50$), that better-educated husbands tend to do less housework ($r_{yz} = -0.30$), and that families with better-educated husbands have fewer children ($r_{xz} = -0.47$).

TABLE 16.1   Scores on Three Variables for 12 Dual–Wage-Earner Families

Family	Husband's Housework (Y)	Number of Children (X)	Husband's Years of Education (Z)
A	1	1	12
B	2	1	14
C	3	1	16
D	5	1	16
E	3	2	18
F	1	2	16
G	5	3	12
H	0	3	12
I	6	4	10
J	3	4	12
K	7	5	10
L	4	5	16

TABLE 16.2   Zero-Order Correlations

	Husband's Housework (Y)	Number of Children (X)	Husband's Years of Education (Z)
Husband's Housework (Y)	1.00	0.50	−0.30
Number of Children (X)		1.00	−0.47
Husband's Years of Education (Z)			1.00

Is the relationship between husband's housework (X) and number of children (Y) affected by husband's years of education? Substituting the zero-order correlations into Formula 16.1, we would have

$$r_{yx.z} = \frac{r_{yx} - (r_{yz})(r_{xz})}{\sqrt{1 - r_{yz}^2}\sqrt{1 - r_{xz}^2}}$$

$$r_{yx.z} = \frac{(0.50) - (-0.30)(-0.47)}{\sqrt{1 - (-0.30)^2}\sqrt{1 - (0.47)^2}}$$

$$r_{yx.z} = \frac{(0.50) - (0.14)}{\sqrt{1 - 0.09}\sqrt{1 - 0.22}}$$

$$r_{yx.z} = \frac{0.36}{\sqrt{0.91}\sqrt{0.78}}$$

$$r_{yx.z} = \frac{0.36}{(0.95)(0.88)}$$

$$r_{yx.z} = \frac{0.36}{0.84}$$

$$r_{yx.z} = 0.43$$

Interpretation. The first-order partial ($r_{yx.z} = 0.43$), which measures the strength of the relationship between husband's housework (Y) and number of children (X) while controlling for husband's education (Z), is lower in value than the zero-order coefficient ($r_{yx} = 0.50$), but the difference in the two values is not great. This result suggests a direct relationship between variables X and Y. That is, when controlling for husband's education, the statistical relationship between husband's housework and number of children is essentially unchanged. Regardless of education, husband's hours of housework increase with the number of children.

Our next step in statistical analysis would probably be to select another control variable. The more the bivariate relationship retains its strength across a series of controls for third variables (Z's), the stronger the evidence for a direct relationship between X and Y.

ONE STEP AT A TIME	Computing and Interpreting the Partial Correlation Coefficient

To begin, compute Pearson's r for all pairs of variables. Be clear about which variable is independent (X), which is dependent (Y), and which is the control (Z).

To Find the Partial Correlation Coefficient, Solve Formula 16.1

1: Multiply r_{yz} by r_{xz}.

2: Subtract the value you found in step 1 from r_{yx}. *This value is the numerator of Formula 16.1.*

3: Square the value of r_{yz}.

4: Subtract the quantity you found in step 3 from 1.

5: Take the square root of the quantity you found in step 4.

6: Square the value of r_{xz}.

7: Subtract the quantity you found in step 6 from 1.

8: Take the square root of the quantity you found in step 7.

9: Multiply the quantity you found in step 8 by the value you found in step 5. *This value is the denominator of Formula 16.1.*

10: Divide the quantity you found in step 2 by the quantity you found in step 9. *This is the partial correlation coefficient.*

To Interpret the Partial Correlation Coefficient, Compare Its Value to the Zero-Order Correlation

Choose the description below that comes closest to matching the relationship between the two values:

1. The partial correlation coefficient is roughly the same value as the zero-order or bivariate correlation. A good rule of thumb for "roughly the same" is a difference of less than 0.10. This outcome is evidence that that the control variable (Z) has no effect and that the relationship between X and Y is direct.

2. The partial correlation coefficient is much less (say, more than 0.10 less) than the bivariate correlation. This is evidence that the control variable (Z) changes the relationship between X and Y. The relationship between X and Y is either spurious (Z causes both X and Y) or intervening (X and Y are linked by Z).

Be aware that that X, Y, and Z may have an interactive relationship in which the relationship between X and Y changes for each category of Z. Partial correlation analysis cannot detect interactive relationships.

STATISTICS
IN EVERYDAY LIFE

An analysis of a representative sample of Americans age 25 to 65 found a correlation of 0.43 between years of schooling and income, a moderate, positive relationship. Income increases as education increases, and education explains almost 19% of the variation in income.

Is this relationship affected by age? The partial correlation between education and income while controlling for age is also 0.43, the same value as the zero-order correlation. This indicates that the relationship between education and income is direct: Regardless of age, income increases as education increases.

People with higher levels of education tend to have higher income regardless of their age, a finding that you may find reassuring as you move to the completion of your degree.

FIGURE 16.5 A Possible Causal Relationship Among Three Variables

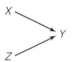

In closing, I should mention an additional possible outcome, in which the partial correlation coefficient is greater in value than the zero-order coefficient ($r_{yx.z} > r_{yx}$). This outcome would be consistent with a causal model in which the variable taken as independent and the control variable each had a separate effect on the dependent variable and were uncorrelated with each other. This relationship is depicted in Figure 16.5. The absence of an arrow between X and Z indicates that they have no mutual relationship.

This pattern means that both X and Z should be treated as independent variables, and the next step in the statistical analysis would probably involve multiple correlation and regression, techniques that will be presented in the remainder of this chapter. *(For practice in computing and interpreting partial correlation coefficients, see problems 16.1 to 16.3.)*

Multiple Regression: Predicting the Dependent Variable

In Chapter 14, the least-squares regression line was introduced as a way of describing the overall linear relationship between two interval-ratio variables and of predicting scores on Y from scores on X. This line was the best-fitting line to summarize the bivariate relationship and was defined by this formula:

FORMULA 16.2
$$Y = a + bX$$

where a = the Y intercept
b = the slope

The least-squares regression line can be modified to include (theoretically) any number of independent variables. This technique is called **multiple regression**. For ease of explication, we will confine our attention to the case involving two independent variables. The least-squares multiple regression equation is

FORMULA 16.3
$$Y = a + b_1X_1 + b_2X_2$$

where b_1 = the partial slope of the linear relationship between the first independent variable and Y
b_2 = the partial slope of the linear relationship between the second independent variable and Y

Some new notation and some new concepts are introduced in this formula. First, while the dependent variable is still symbolized as Y, the independent variables are differentiated by subscripts. Thus, X_1 identifies the first independent variable and X_2 the second. The symbol for the slope (b) is also subscripted to identify the independent variable with which it is associated.

PARTIAL SLOPES

A major difference between the multiple and bivariate regression equations concerns the slopes (b's). In the case of multiple regression, the b's are called **partial slopes**, and they show the amount of change in Y for a unit change in the independent variable while controlling for the effects of the other independent variables in the equation. The partial slopes are thus analogous to partial correlation coefficients and represent the direct effect of the associated independent variable on Y.

Computing Partial Slopes. The partial slopes for the independent variables are determined by Formulas 16.4 and 16.5:[1]

FORMULA 16.4
$$b_1 = \left(\frac{s_y}{s_1}\right)\left(\frac{r_{y1} - r_{y2}r_{12}}{1 - r_{12}^2}\right)$$

FORMULA 16.5
$$b_2 = \left(\frac{s_y}{s_2}\right)\left(\frac{r_{y2} - r_{y1}r_{12}}{1 - r_{12}^2}\right)$$

where b_1 = the partial slope of X_1 on Y
b_2 = the partial slope of X_2 on Y
s_y = the standard deviation of Y
s_1 = the standard deviation of the first independent variable (X_1)
s_2 = the standard deviation of the second independent variable (X_2)
r_{y1} = the bivariate correlation between Y and X_1
r_{y2} = the bivariate correlation between Y and X_2
r_{12} = the bivariate correlation between X_1 and X_2

To illustrate the computation of the partial slopes, we will assess the combined effects of number of children (X_1) and husband's education (X_2) on husband's contribution to housework. All the relevant information can be calculated from Table 16.1 and is noted here:

Husband's Housework	Number of Children	Husband's Education
$\bar{Y} = 3.3$	$\bar{X}_1 = 2.7$	$\bar{X}_2 = 13.7$
$s_y = 2.1$	$s_1 = 1.5$	$s_2 = 2.6$

Zero-order correlations
$r_{y1} = 0.50$
$r_{y2} = -0.30$
$r_{12} = -0.47$

[1]Partial slopes can be computed from zero-order slopes, but Formulas 16.4 and 16.5 are somewhat easier to use.

The partial slope for the first independent variable (X_1) is

$$b_1 = \left(\frac{s_y}{s_1}\right)\left(\frac{r_{y1} - r_{y2}r_{12}}{1 - r_{12}^2}\right)$$

$$b_1 = \left(\frac{2.1}{1.5}\right)\left(\frac{0.50 - (-0.30)(-0.47)}{1 - (-0.47)^2}\right)$$

$$b_1 = (1.4)\left(\frac{0.50 - 0.14}{1 - 0.22}\right)$$

$$b_1 = (1.4)\left(\frac{0.36}{0.78}\right)$$

$$b_1 = (1.4)(0.46)$$

$$b_1 = 0.64$$

For the second independent variable (X_2), the partial slope is

$$b_2 = \left(\frac{s_y}{s_2}\right)\left(\frac{r_{y2} - r_{y1}r_{12}}{1 - r_{12}^2}\right)$$

$$b_2 = \left(\frac{2.1}{2.6}\right)\left(\frac{-0.30 - (+.50)(-0.47)}{1 - (-0.47)^2}\right)$$

$$b_2 = (0.81)\left(\frac{-0.30 - (-0.24)}{1 - 0.22}\right)$$

$$b_2 = (0.81)\left(\frac{-0.30 + 0.24}{0.78}\right)$$

$$b_2 = (0.81)\left(\frac{-0.06}{0.78}\right)$$

$$b_2 = (0.81)(-0.08)$$

$$b_2 = -0.07$$

Finding the Y Intercept. Now that partial slopes have been determined for both independent variables, the Y intercept (a) can be found. Note that a is calculated from the mean of the dependent variable (symbolized as $\overline{Y}$) and the means of the two independent variables $(\overline{X}_1$ and $\overline{X}_2)$.

FORMULA 16.6

$$a = \overline{Y} - b_1\overline{X}_1 - b_2\overline{X}_2$$

Substituting the proper values for the example problem at hand, we would have

$$a = \overline{Y} - b_1\overline{X}_1 - b_2\overline{X}_2$$
$$a = 3.3 - (0.64)(2.7) - (-0.07)(13.7)$$

ONE STEP AT A TIME   **Computing and Interpreting Partial Slopes**

These procedures apply when there are two independent variables and one dependent variable. For more complex situations, use a computerized statistical package such as SPSS to do the calculations.

To Compute the Partial Slope Associated with the First Independent Variable by Using Formula 16.4

1: Divide s_y by s_1.

2: Multiply r_{y2} by r_{12}.

3: Subtract the value you computed in step 2 from r_{y1}.

4: Square the value of r_{12}.

5: Subtract the value you computed in step 4 (r_{12}^2) from 1.

6: Divide the value you computed in step 3 by the value you computed in step 5.

7: Multiply the value you computed in step 6 by the value you computed in step 1. *This value is the partial slope associated with the first independent variable.*

To Compute the Partial Slope Associated with the Second Independent Variable by Using Formula 16.5

1: Divide s_y by s_2.

2: Multiply r_{y1} by r_{12}.

3: Subtract the value you computed in step 2 from r_{y2}.

4: Square the value of r_{12}.

5: Subtract the value you computed in step 4 from 1.

6: Divide the value you computed in step 3 by the value you computed in step 5.

7: Multiply the value you computed in step 6 by the value you computed in step 1. *This value is the partial slope associated with the second independent variable.*

To Interpret Partial Slopes

The value of a partial slope is the increase in the value of Y for a unit increase in the value of the associated independent variable while controlling for the effects of the other independent variable.

$$a = 3.3 - (1.7) - (-1.0)$$
$$a = 3.3 - 1.7 + 1.0$$
$$a = 2.6$$

The Least-Squares Multiple Regression Line and Predicting Y. For our example problem, the full least-squares multiple regression equation would be

$$Y = a + b_1X_1 + b_2X_2$$
$$Y = 2.6 + (0.64)X_1 + (-0.07)X_2$$

ONE STEP AT A TIME   **Computing the Y Intercept**

To Find the Y Intercept by Using Formula 16.6

1: Multiply the mean of X_2 by b_2.

2: Multiply the mean of X_1 by b_1.

3: Subtract the quantity you found in step 1 from the quantity you found in step 2.

4: Subtract the quantity you found in step 3 from the mean of Y. *The result is the value of a, the Y intercept.*

ONE STEP AT A TIME	Using the Multiple Regression Line to Predict Scores on *Y*

1: Choose a value for X_1. Multiply this value by the value of b_1.

2: Choose a value for X_2. Multiply this value by the value of b_2.

3: Add the values you found in steps 1 and 2 to the value of *a*, the *Y* intercept. *The resulting value is the predicted score on Y.*

As was the case with the bivariate regression line, this formula can be used to predict scores on the dependent variable from scores on the independent variables. For example, what would be our best prediction of husband's housework (Y') for a family of four children ($X_1 = 4$) where the husband had completed 11 years of schooling ($X_2 = 11$)? Substituting these values into the least-squares formula, we would have

$$Y' = 2.6 + (0.64)(4) + (-0.07)(11)$$
$$Y' = 2.6 + 2.6 - 0.8$$
$$Y' = 4.4$$

Our prediction would be that this husband would contribute 4.3 hours per week to housework. This prediction is, of course, a kind of "educated guess," which is unlikely to be perfectly accurate. However, we will make fewer errors of prediction using the least-squares line (and, thus, incorporating information from the independent variables) than we would using any other method of prediction (assuming, of course, that there is a linear association between the independent and the dependent variables). *(For practice in predicting Y scores and in computing slopes and the Y intercept, see problems 16.1 to 16.6.)*

Multiple Regression: Assessing the Effects of the Independent Variables

The least-squares multiple regression equation (Formula 16.3) is used to isolate the separate effects of the independents and to predict scores on the dependent variable. However, in many situations, using this formula to determine the relative importance of the various independent variables will be awkward—especially when the independent variables differ in terms of units of measurement (e.g., number of children vs. years of education). When the units of measurement differ, a comparison of the partial slopes will not necessarily tell us which independent variable has the strongest effect and is thus the most important. Comparing the partial slopes of variables that differ in units of measurement is a little like comparing apples and oranges.

We can make it easier to compare the effects of the independent variables by converting all variables in the equation to a common scale and thereby eliminating variations in the values of the partial slopes that are solely a function of differences in units of measurement. We can, for example, standardize all distributions by changing the scores of all variables to Z scores. Each distribution of scores would then have a mean of 0 and a standard deviation of 1 (see Chapter 5), and comparisons between the independent variables would be much more meaningful.

COMPUTING THE STANDARDIZED REGRESSION COEFFICIENTS

To standardize the variables to the normal curve, we could actually convert all scores into the equivalent Z scores and then recompute the slopes and the Y intercept. This would require a good deal of work; fortunately, a shortcut is available for computing the slopes of the standardized scores directly. These **standardized partial slopes**, called **beta-weights**, are symbolized **b***.

Beta-Weights. The beta-weights show the amount of change in the standardized scores of Y for a one-unit change in the standardized scores of each independent variable while controlling for the effects of all other independent variables.

Formulas and Computation for Beta-Weights. When we have two independent variables, the beta-weight for each is found by using Formulas 16.7 and 16.8:

FORMULA 16.7
$$b_1^* = b_1\left(\frac{s_1}{s_y}\right)$$

FORMULA 16.8
$$b_2^* = b_2\left(\frac{s_2}{s_y}\right)$$

We can now compute the beta-weights for our sample problem to see which of the two independent variables has the stronger effect on the dependent. For the first independent variable, number of children (X_1):

$$b_1^* = b_1\left(\frac{s_1}{s_y}\right)$$

$$b_1^* = (0.64)\left(\frac{1.5}{2.1}\right)$$

$$b_1^* = (0.64)(0.71)$$

$$b_1^* = 0.45$$

For the second independent variable, husband's education (X_2):

$$b_2^* = b_2\left(\frac{s_2}{s_y}\right)$$

$$b_2^* = (-0.07)\left(\frac{2.6}{2.1}\right)$$

$$b_2^* = (-0.07)(1.24)$$

$$b_2^* = -0.09$$

ONE STEP AT A TIME | **Computing and Interpreting Beta-Weights (b^*)**

These procedures apply when there are two independent variables and one dependent variable. For more complex situations, use a computerized statistical package such as SPSS to do the calculations.

To Compute the Beta-Weight Associated with the First Independent Variable Using Formula 16.7

1: Divide s_1 by s_y.

2: Multiply the value you found in step 1 by the partial slope of the first independent variable (b_1). *This value is the beta-weight associated with the first independent variable.*

To Compute the Beta-Weight Associated with the Second Independent Variable Using Formula 16.8

1: Divide s_2 by s_y.

2: Multiply the value you found in step a by the partial slope of the second independent variable (b_2). *This value is the beta-weight associated with the second independent variable.*

To Interpret Beta-Weights

A beta-weight (or standardized partial slope) shows the increase in the value of Y for a unit increase in the value of the associated independent variable while controlling for the effects of the other independent variable after all variables have been standardized (or transformed to Z scores).

Comparing the value of the beta-weights, we see that number of children has a stronger effect than husband's education on husband's housework. Furthermore, the net effect (after controlling for the effect of husband's education) of the first independent variable is positive, while the net effect of the second independent variable is negative.

The Standardized Least-Squares Regression Line. Using standardized scores, the least-squares regression equation can be written as

FORMULA 16.9

$$Z_y = a_z + b_1^* Z_1 + b_2^* Z_2$$

where Z indicates that all scores have been standardized to the normal curve

---

STATISTICS

IN EVERYDAY LIFE

Much of the research literature on the crime of rape finds that the underlying motivation is about power and dominance. However, an analysis by economist Todd Kendall suggests that sexual desire may be the important force behind these attacks. He finds a correlation between the availability of pornography on the Internet and the rate of rape. In fact, his regression analysis ". . . implies that an increase in home internet access of 10 percentage points is associated with an 7.3% decline in rape" (p. 23), net of the effect of other variables in the regression equation (poverty, unemployment, alcohol consumption, and so forth). These results suggest that pornography is a substitute for rape, a finding that is in sharp contrast to most previous literature.

Kendall, Todd. 2006. "Pornography, Rape, and the Internet." Available from http://www.toddkendall.net/internetcrime.pdf.

Applying Statistics 16.1: Multiple Regression and Correlation

Five recently divorced men have been asked to rate subjectively the success of their adjustment to single life, on a scale ranging from 5 (very successful adjustment) to 1 (very poor adjustment). Is adjustment related to the length of time married? Is adjustment related to socioeconomic status as measured by yearly income?

Case	Adjustment (Y)	Years Married (X_1)	Income (dollars) (X_2)
A	5	5	30,000
B	4	7	45,000
C	4	10	25,000
D	3	2	27,000
E	1	15	17,000

$$\overline{Y} = 3.4 \quad \overline{X}_1 = 7.8 \quad \overline{X}_2 = 28,800$$
$$s = 1.4 \quad s = 4.5 \quad s = 9173.88$$

The zero-order correlations among these three variables are

	Adjustment (Y)	Years Married (X_1)	Income (X_2)
Adjustment (Y)	1.00	−0.62	0.62
Years married (X_1)		1.00	−0.49
Income (X_2)			1.00

These results suggest a strong but opposite relationship between each independent variable and adjustment. Adjustment decreases as years married increases, and it increases as income increases.

To find the multiple regression equation, we must find the partial slopes.

For years married (X_1):

$$b_1 = \left(\frac{s_y}{s_1}\right)\left(\frac{r_{y1} - r_{y2}r_{12}}{1 - r_{12}^2}\right)$$

$$b_1 = \left(\frac{1.4}{4.5}\right)\left(\frac{(-0.62) - (0.62)(-0.49)}{1 - (-0.49)^2}\right)$$

$$b_1 = (0.31)\left(\frac{(-0.62) - (-0.30)}{1 - 0.24}\right)$$

$$b_1 = (0.31)\left(\frac{-0.32}{0.76}\right)$$

$$b_1 = (0.31)(-0.42)$$

$$b_1 = -0.13$$

For income (X_2):

$$b_2 = \left(\frac{s_y}{s_2}\right)\left(\frac{r_{y2} - r_{y1}r_{12}}{1 - r_{12}^2}\right)$$

$$b_2 = \left(\frac{1.4}{9173.88}\right)\left(\frac{(0.62) - (-0.62)(-0.49)}{1 - (-0.49)^2}\right)$$

$$b_2 = (0.00015)\left(\frac{(0.62) - (0.30)}{1 - 0.24}\right)$$

$$b_2 = (0.00015)\left(\frac{0.32}{0.76}\right)$$

$$b_2 = (0.00015)(0.42)$$

$$b_2 = 0.000063$$

The Y intercept would be

$$a = \overline{Y} - b_1\overline{X}_1 - b_2\overline{X}_2$$

$$a = 3.4 - (-0.13)(7.8) - (0.000063)(28,800)$$

$$a = 3.4 - (-1.01) - (1.81)$$

$$a = 3.4 + 1.01 - 1.81$$

$$a = 2.60$$

(continued)

Applying Statistics 16.1: (*continued*)

The multiple regression equation is

$$Y = a + b_1X_1 + b_2X_2$$

$$Y = 2.60 + (-0.13)X_1 + (0.000063)X_2$$

What adjustment score could we predict for a male who had been married 30 years ($X_1 = 30$) and had an income of \$50,000 ($X_2 = 50,000$)?

$$Y' = 2.60 + (-0.13)(30) + (0.000063)(50,000)$$

$$Y' = 2.60 + (-3.90) + (3.15)$$

$$Y' = 1.85$$

To assess which of the two independent variables has the stronger effect on adjustment, the standardized partial slopes must be computed. For years married (X_1):

$$b_1^* = b_1\left(\frac{s_1}{s_y}\right)$$

$$b_1^* = (-0.13)\left(\frac{4.5}{1.4}\right)$$

$$b_1^* = -0.42$$

For income (X_2):

$$b_2^* = b_2\left(\frac{s_2}{s_y}\right)$$

$$b_2^* = (0.000063)\left(\frac{9173.88}{1.4}\right)$$

$$b_2^* = 0.41$$

The standardized regression equation is

$$Z_y = b_1^*Z_1 + b_2^*Z_2$$

$$Z_y = (-0.42)Z_1 + (0.41)Z_2$$

and the independent variables have nearly equal but opposite effects on adjustment. To assess the combined effects of the two independents on adjustment, the coefficient of multiple determination must be computed:

$$R^2 = r_{y1}^2 + r_{y2.1}^2(1 - r_{y1}^2)$$

$$R^2 = (-0.62)^2 + (0.46)^2(1 - (-0.62)^2)$$

$$R^2 = 0.38 + (0.21)(1 - 0.38)$$

$$R^2 = 0.38 + (0.21)(0.62)$$

$$R^2 = 0.38 + 0.13$$

$$R^2 = 0.51$$

The first independent variable, years married, explains 38% of the variation in adjustment by itself. To this quantity, income explains an additional 13% of the variation in adjustment. Taken together, the two independent variables explain a total of 51% of the variation in adjustment.

The standardized regression equation can be further simplified by dropping the term for the Y intercept, since this term will always be zero when scores have been standardized. This value is the point where the regression line crosses the Y axis and is equal to the mean of Y when all independent variables equal 0. This relationship can be seen by substituting 0 for all independent variables in Formula 16.6:

$$a = \overline{Y} - b_1\overline{X}_1 - b_2\overline{X}_2$$
$$a = \overline{Y} - b_1(0) - b_2(0)$$
$$a = \overline{Y}$$

Since the mean of any standardized distribution of scores is zero, the mean of the standardized Y scores will be zero and the Y intercept will also be zero ($a = \overline{Y} = 0$). Thus, Formula 16.9 simplifies to

FORMULA 16.10

$$Z_y = b_1^*Z_1 + b_2^*Z_2$$

The standardized regression equation, with beta-weights noted, would be

$$Z_y = (0.45)Z_1 + (-0.09)Z_2$$

and it is immediately obvious that the first independent variable has a much stronger direct effect on Y than does the second independent variable.

As we have seen, multiple regression analysis permits the researcher to summarize the linear relationship among two or more independents and a dependent variable. The unstandardized regression equation (Formula 16.2) permits values of Y to be predicted from the independent variables in the original units of the variables. The standardized regression equation (Formula 16.10) allows the researcher to assess easily the relative importance of the various independent variables by comparing the beta-weights. *(For practice in computing and interpreting beta-weights, see any of the problems at the end of this chapter. It is probably a good idea to start with problem 16.1 since it has the smallest data set and the least complex computations.)*

Applying Statistics 16.2: R^2 and Beta-Weights

The following table presents information on three variables for a small sample of 10 nations. The dependent variable is the percentage of respondents who said they are "very happy" on a survey administered to random samples from each nation. The independent variables measure health and physical well-being (life expectancy, or the number of years the average citizen can expect to live at birth) and income inequality (the amount of total income that goes to the richest 20% of the population). Our expectation is that happiness will have a positive correlation with life expectancy (the greater the health, the happier the population) and a negative relationship with inequality (the greater the inequality, the greater the discontent and the lower the level of happiness). In this analysis, we focus on R^2 and the beta-weights only.

Nation	Percent "Very Happy" (Y)	Life Expectancy (X_1)	Income Inequality (X_2)
Brazil	34.1	73	56.99
Canada	46.4	81	32.56
Germany	19.9	80	29.31
Ghana	50.1	59	40.79
India	29.0	64	36.80
Japan	29.2	83	24.85
Mexico	58.5	75	46.05
Russia	21.2	68	39.93
Rwanda	11.9	48	46.79
United States	34.4	78	40.81
Mean =	33.47	70.90	39.48
Standard deviation =	13.82	10.61	8.83

(continued)

Applying Statistics 16.2: (*continued*)

A correlation matrix presents the zero-order correlations:

	Happiness	Life Expectancy	Inequality
Happiness	1.00	0.29	0.13
Life expectancy		1.00	−0.46
GDP per capita			1.00

Consistent with our expectations, there is a positive, weak-to-moderate relationship between life expectancy and happiness. Unexpectedly, however, the relationship between inequality and happiness is positive, although weak in strength. The relationship between the two independent variables is moderate and negative, indicating that nations with more income inequality have lower life expectancy.

The combined effect of life expectancy and inequality on happiness is found by computing R^2:

$$R^2 = r_{y1}^2 + r_{y2.1}^2(1 - r_{y1}^2)$$

$$R^2 = (0.29)^2 + (0.31)^2(1 - 0.29^2)$$

$$R^2 = 0.08 + (0.10)(0.92)$$

$$R^2 = 0.08 + 0.09$$

$$R^2 = 0.17$$

By itself, life expectancy explains 8% of the variance in happiness. To this, income inequality adds another 9%, for a total of 17%. This leaves about 83% of the variance unexplained, a sizeable proportion but not unusually large in social science research.

We must calculate the beta-weights to assess the separate effects of the two independent variables. The unstandardized partial slopes needed for this computation are 0.58 for X_1 (life expectancy) and 0.53 for X_2 (income inequality).

For the first independent variable (life expectancy):

$$b_1^* = b_1\left(\frac{s_1}{s_y}\right)$$

$$b_1^* = (0.58)\left(\frac{10.61}{13.82}\right)$$

$$b_1^* = (0.58)(0.77)$$

$$b_1^* = 0.45$$

For the second independent variable (income inequality):

$$b_2^* = b_2\left(\frac{s_2}{s_y}\right)$$

$$b_2^* = (0.53)\left(\frac{8.83}{13.82}\right)$$

$$b_2^* = (0.53)(0.64)$$

$$b_2^* = 0.34$$

Recall that the beta-weights show the effect of each independent variable on the dependent variable while controlling for the other independent variables in the equation. In this case, life expectancy has the stronger effect, and the relationship is positive. The effect of income inequality is also positive.

In summary, for these nations, level of happiness has a moderate positive relationship with life expectancy and a weaker positive relationship with income inequality. Taken together, the independent variables explain 17% of the variation in happiness.

Multiple Correlation

We use the multiple regression equations to disentangle the separate direct effects of each independent variable on the dependent variable. Using **multiple correlation** techniques, we can also ascertain the combined effects of all independent variables on the dependent variable. We do so by computing the

multiple correlation coefficient (R) and the **coefficient of multiple determination (R)^2**. The value of the latter statistic represents the proportion of the variance in Y that is explained by all the independent variables combined.

We have seen that number of children (X_1) explains 25% of the variance in Y ($r_{y1}^2 = (.50)^2 = .25 \times 100 = 25\%$) by itself and that husband's education explains 9% of the variance in Y ($r_{y2}^2 = (-0.30)^2 = .09 \times 100 = 9\%$). The two zero-order correlations cannot simply be added together to ascertain their combined effect on Y, because the two independent variables are also correlated with each other and, therefore, they will "overlap" in their effects on Y and explain some of the same variance. This overlap is eliminated in Formula 16.11:

FORMULA 16.11

$$R^2 = r_{y1}^2 + r_{y2.1}^2(1 - r_{y1}^2)$$

where R^2 = the coefficient of multiple determination

r_{y1}^2 = the zero-order correlation between Y and X_1, the quantity squared

$r_{y2.1}^2$ = the partial correlation of Y and X_2, while controlling for X_1, the quantity squared

The first term in this formula (r_{y1}^2) is the coefficient of determination for Y and X_1, or the amount of variation in Y explained by X_1 by itself. To this quantity we add the amount of the variation remaining in Y (given by $1 - r_{y1}^2$) that can be explained by X_2 after the effect of X_1 is controlled ($r_{y2.1}^2$). Basically, Formula 16.11 allows X_1 to explain as much of Y as it can and then adds in the effect of X_2 after X_1 is controlled (thus eliminating the "overlap" in the variance of Y that X_1 and X_2 have in common).

COMPUTING AND INTERPRETING R AND R^2

To observe the combined effects of number of children (X_1) and husband's education (X_2) on husband's housework (Y), we need two quantities. The correlation between X_1 and Y ($r_{y1} = 0.50$) has already been found. Before we can solve Formula 16.11, we must first calculate the partial correlation of Y and X_2 while controlling for X_1 ($r_{y2.1}$):

$$r_{y2.1} = \frac{r_{y2} - (r_{y1})(r_{12})}{\sqrt{1 - r_{y1}^2}\sqrt{1 - r_{12}^2}}$$

$$r_{y2.1} = \frac{(-0.30) - (0.50)(-0.47)}{\sqrt{1 - (0.50)^2}\sqrt{1 - (-0.47)^2}}$$

$$r_{y2.1} = \frac{(-0.30) - (-0.24)}{\sqrt{0.75}\sqrt{0.78}}$$

$$r_{y2.1} = \frac{-0.06}{0.77}$$

$$r_{y2.1} = -0.08$$

ONE STEP AT A TIME	**Computing and Interpreting the Coefficient of Multiple Determination**

These procedures apply when there are two independent variables and one dependent variable. For more complex situations, use a computerized statistical package such as SPSS to do the calculations.

To Compute the Coefficient of Multiple Determination by Using Formula 16.11

1: Find the value of the partial correlation coefficient for $r_{y2.1}$.

2: Square the value you found in step 1.

3: Square the value of r_{y1}.

4: Subtract the value you found in step 3 from 1.00.

5: Multiply the value you found in step 4 by the value you found in step 2.

6: Add the value you found in step 5 to r_{y1}^2. The result is the coefficient of multiple determination (R^2).

To Interpret the Coefficient of Multiple Determination (R^2)

The coefficient of multiple determination is the total amount of the variation in Y explained by all independent variables combined.

Formula 16.11 can now be solved for our sample problem:

$$R^2 = r_{y1}^2 + r_{y2.1}^2(1 - r_{y1}^2)$$
$$R^2 = (0.50)^2 + (-0.08)^2(1 - 0.50^2)$$
$$R^2 = 0.25 + (0.006)(1 - 0.25)$$
$$R^2 = 0.25 + 0.005$$
$$R^2 = 0.255$$

The first independent variable (X_1), number of children, explains 25% of the variance in Y by itself. To this total, the second independent (X_2), husband's education, adds only a half a percent, for a total explained variance of 25.5%. In combination, the two independent variables explain a total of 25.5% of the variation in the dependent variable. *(For practice in computing and interpreting R and R^2, see any of the problems at the end of this chapter. It is probably a good idea to start with problem 16.1 since it has the smallest data set and the least complex computations.)*

Interpreting Statistics: Another Look at the Correlates of Crime

In Chapter 14, we assessed the association between poverty and crime. We found a moderate-to-strong, positive relationship ($r = 0.52$) between a measure of poverty and the homicide rate for the 50 states, an indication that there may be an important relationship between these two variables. In this installment of

Interpreting Statistics, we return to this relationship and add several independent variables to the analysis.[2]

The second independent variable is a measure of age: the percentage of the population less than 18. Research on street crimes and serious felonies like homicide commonly find moderate-to-strong relationships with age. Rates of violent crime are highest for people in their teens and twenties but decline dramatically as age rises. Thus, we can expect a substantial positive relationship between age and homicide rate (the higher the proportion of younger people in a population, the higher the homicide rate).

The third independent variable comes from a body of research that argues that there is a subculture of violence, a stronger tendency to use force and aggression, in the South. In other words, the norms and values of the Southern regional subculture sanction and endorse the everyday use of force more strongly than in other regions of the nation. Thus, we should expect to find higher rates of homicide in Southern states.

Although there are compelling reasons for including region in this analysis, we must first confront an important problem. Region (South, North, Midwest, etc.) is a nominal-level variable, but regression analysis requires that all variables be measured at the interval-ratio level. We can resolve this problem by treating region as a dummy variable (see Chapter 14). As you recall, dummy variables have exactly two categories, one coded as a score of 0 and the other as a score of 1. Treated this way, nominal-level variables such as gender and race are commonly included in regression equations. In this case, we will create a variable that is coded so that Southern states are scored as "1" and non-Southern states as "0." Since "South" is coded as a 1 and all other regions as zero, region should have a positive relationship with homicide. Before beginning the multivariate analysis, we should examine the correlation coefficients between all possible pairs of variables. Table 16.3 reports the zero-order correlations.

As we saw in Chapter 15, there is a moderate-to-strong, positive relationship between poverty and homicide: States with higher rates of poverty have higher rates of homicide ($r = 0.52$). Turning to the other variables, there are, as expected, weak-to-moderate, positive relationships between age and homicide rate and between region and homicide rate. This indicates that Southern states and states with a higher percentage of young people have higher homicide rates.

TABLE 16.3 Zero-Order Correlations

	Homicide	Poverty	Age	Region
Homicide	1.00	0.52	0.23	0.50
Poverty		1.00	0.17	0.63
Age			1.00	−0.04

---

[2]This analysis uses three independent variables, one more than earlier examples in this chapter. The additional independent variable will not complicate interpretation unduly but the additional mathematics for three (or more) independent variables are complex and will not be shown in this analysis. Use a computerized statistical package—like SPSS—for situations involving three or more independent variables.

There is a moderate relationship between poverty and region ($r = 0.63$), and no relationship between region and age ($r = -0.04$). Finally, there is a weak, positive relationship between age and poverty: States with a higher percentage of young people tend to be more impoverished.

We begin the multivariate analysis by asking if the original relation between homicide and poverty persists after controlling for the effects of age and region, controlling for these variables one at a time. Recall that a partial correlation coefficient indicates the strength and direction of a bivariate relation after controlling for the effects of a third variable. Table 16.4 reports the results of controlling for age and then region.

The partial correlation using age as the control variable is almost exactly equal to the zero-order correlation. Thus, we can conclude that age has no effect, an outcome that would increase our confidence that there is a direct, causal relationship between poverty and the homicide rate.

However, the partial correlation when controlling for region (0.30) is noticeably weaker than the zero-order correlation (0.52). On one hand, this outcome weakens the argument that there is a direct causal relationship between poverty and violence: Region has a substantial effect on the bivariate relationship and should be taken into account in any further analysis. On the other hand, the bivariate relationship between homicide and poverty is positive and moderate in strength even after controlling for region. Thus, we would probably conclude that there is a relationship between poverty and homicide rate, even though the relationship is affected by region.

Next, we can assess the relative importance of the three independent variables on homicide. Table 16.5 presents the essential information—beta-weights and R^2—that will help us sort out the relative importance and total impact of these variables on the homicide rate.

By itself, poverty explained 27% of the variance in homicide rates among the states ($r^2 = (0.52)^2 = 0.27$). To this, age and region add another 8%, for a total of 35% ($R^2 = 0.35$). These three variables together account for a little more than a third of the variation in homicide rates from state to state. This is quite a substantial percentage, even though about two-thirds of the variation remains unaccounted for or unexplained.

We might be able to raise the value of R^2 by adding more independent variables to the equation. However, it is common to find that additional independent variables explain smaller and smaller proportions of the remaining variance and have a diminishing effect on R^2. This can be caused by many factors, including the fact that the independent variables will usually be correlated with each other

TABLE 16.4 Partial Correlation Coefficients for the Relationship Between Homicide Rate and Poverty

Zero-Order Correlation Between Homicide and Poverty	Partial Correlation after Controlling for	
	Age	Region
0.52	0.50	0.30

TABLE 16.5 Regression Analysis with Homicide Rate as the Dependent Variable

Variable	Beta-Weight
Poverty	0.27
Age	0.20
Region	0.34
	$R^2 = 0.35$

(e.g., see the relationship between region and poverty in Table 16.3) and will overlap in their effect on the dependent variable.

The beta-weights in Table 16.5 show the effect of each variable while controlling for all other variables in the equation (as opposed to controlling for other variables one at a time as we did in Table 16.4). The direction of all three independent variables is positive, and the three independent variables have roughly equal effects on homicide rate when controlling for the other variables. In conclusion, this multivariate analysis indicates that poverty, age, and region all have important, positive effects on homicide rate. The beta-weights show that the independent variables are roughly equal in their effects on homicide rate, with region having a slightly stronger effect. Taken together, the three variables account for 35% of the variance in homicide rate from state to state.

The Limitations of Multiple Regression and Correlation

Multiple regression and correlation are very powerful tools for analyzing the interrelationships among three or more variables. The techniques presented in this chapter permit the researcher to predict scores on one variable from two or more other variables, to distinguish between independent variables in terms of the importance of their direct effects on a dependent variable, and to ascertain the total effect of a set of independent variables on a dependent variable. In terms of the flexibility of the techniques and the volume of information they can supply, multiple regression and correlation represent some of the most powerful statistical techniques available to social science researchers.

Powerful tools are not cheap. They demand high-quality data, and measurement at the interval-ratio level is difficult to accomplish in social science research. Furthermore, these techniques assume that the interrelationships among the variables follow a particular form. First, they assume that each independent variable has a linear relationship with the dependent variable. How well a given set of variables meets this assumption can be quickly checked with scattergrams.

Second, these techniques assume that there is no interaction among the variables in the equation. If there is interaction among the variables, it will not be possible to estimate or predict the dependent variable accurately by simply adding the effects of the independents. There are techniques for handling interaction among the variables in the set, but these techniques are beyond the scope of this text.

Third, multiple regression and correlation assume that the independent variables are uncorrelated with each other. Strictly speaking, this condition means that the zero-order correlation among all pairs of independents should be zero; but, practically, we act as if this assumption has been met if the intercorrelations among the independents are low.

To the extent that these assumptions are violated, the regression coefficients (especially partial and standardized slopes) and the coefficient of multiple determination (R^2) become less and less trustworthy and the techniques less and less useful. If the assumptions of the model cannot be met, the alternative might be to turn to the multivariate techniques described in Chapter 15. Unfortunately, those techniques, in general, supply a lower volume of less precise information about the interrelationships among the variables.

STATISTICS IN EVERYDAY LIFE	Baseball is a paradise for statistics junkies. Virtually every aspect of what happens on the field is recorded, and there are decades of data to analyze. Can this mountain of information be used for purposes other than settling arguments over trivia among baseball fans?

The answer is a resounding yes, according to Michael Lewis, author of *Moneyball*. Lewis relates the story of Billy Beane, who, as general manager of the Oakland A's, was faced with the challenge of fielding a competitive team with one of the smallest budgets in the sport. Beane accomplished this, in part, by hiring statisticians who conducted a series of regression analyses with won/lost records as the dependent variable and every conceivable indicator of performance as independent variables. Beane and his statisticians were able to identify a number of measures of player effectiveness that had strong relationships with success (i.e., had high beta-weights) but were not widely valued by other baseball executives. He used these criteria to identify and recruit highly effective players for relatively low wages. Under Beane's management, the A's were remarkably successful for many years, although they did not reach the World Series.

Lewis, Michael. 2004. *Moneyball*. New York: Norton.

READING STATISTICS 10: Regression and Correlation

Research projects that analyze the interrelationships among many variables are particularly likely to employ regression and correlation as central statistical techniques. The results of these projects will typically be presented in summary tables that report the multiple correlations, the slopes, and, if applicable, the significance of the results. The zero-order correlations are often presented in the form of a matrix that displays the value of Pearson's *r* for every possible bivariate relationship in the data set.

Usually, tables that summarize the multivariate analysis will report R^2 and the slope for each independent variable in the regression equation. An example of this kind of summary table would look like this:

Independent Variables	Multiple R^2	Beta-Weights
X_1	0.17	0.47
X_2	0.23	0.32
X_3	0.27	0.16
⋮	⋮	⋮

This table reports that the first independent variable, X_1, has the strongest direct relationship with the dependent variable and explains 17% of the variance in the dependent variable by itself ($R^2 = 0.17$). The second independent variable, X_2, adds 6% to the explained variance ($R^2 = 0.23$ after X_2 is entered into the equation). The third independent variable, X_3, adds 4% to the explained variance of the dependent ($R^2 = 0.27$ after X_3 is entered into the equation).

STATISTICS IN THE PROFESSIONAL LITERATURE

Researchers have documented a sharp racial difference in support for the death penalty in American society. In public opinion surveys, for example, capital punishment is supported by large majorities of white respondents but by only a minority of blacks. What accounts for this difference? One possibility is that support among whites is associated with anti-black racism and that opposition among blacks is associated with the perception that capital punishment—indeed the entire criminal justice system—is biased

(continued)

READING STATISTICS 10: (*continued*)

TABLE 1   **Multiple Regression Results with Support for Capital Punishment as the Dependent Variable ($N = 1117$)**[†]

	Model 1		Model 4	
	Unstandardized Coefficient	Beta-Weight	Unstandardized Coefficient	Beta-Weight
Race (black = 1)	−0.60	−0.18***	−0.36	−0.11***
Gender (male = 1)			−0.04	−0.01
Education			−0.01	−0.03
Religiosity			−0.05	−0.17***
Political conservatism			0.04	0.07**
Income			0.02	0.04
Racism			0.05	0.11***
R^2	0.03***		0.15***	

*$p < 0.05$; **$p < 0.01$; ***$p < 0.001$.

[†]Adapted from Table 1 in the original article.

against blacks. The former possibility was investigated by sociologists James Unnever and Francis Cullen. Using a nationally representative sample of over 1000 respondents, they attempted to ascertain the extent to which white support for the death penalty reflects a perception of blacks as threatening, criminally dangerous, and unintelligent.

The researchers constructed several different regression models with support for capital punishment as the dependent variable. They began their analysis by reporting the bivariate relationship between race and support, summarized as Model 1 in Table 1.

Note that race is a dummy variable, with blacks coded as 1 and whites as 0. Thus, the negative values of the regression coefficients indicate that, as expected, whites are more supportive. Note also that the relationship is statistically significant at the 0.001 level.

Unnever and Cullen introduced almost a score of independent variables in the equation, including many measures of social and political traits that previous research had shown are correlated with support for

capital punishment. For example, support has been shown to be greater among males, Southerners, political conservatives, and the more religious. The great power of regression analysis is that the researchers can examine the effect of race on support with all of these other factors controlled.

Model 4 summarizes some of their results and shows that support for the death penalty is significantly associated with racism. Indeed, the beta-weight shows that racism is one of the strongest predictors of support for the death penalty. Note that the beta-weight associated with race loses about a third of its strength (that is, it declines from −0.18 to −0.11) after the measure of racism is added to the equation. This indicates that, with all other variables controlled, white racism by itself accounts for about 33% of the racial divide in support for capital punishment.

Unnever, James, and Cullen, Francis. 2007. "The Racial Divide in Support for the Death Penalty: Does Racism Matter?" *Social Forces*, 85:1281–1301.

Finally, we should note that we have covered only the simplest applications of partial correlation and multiple regression and correlation. In terms of logic and interpretation, the extensions to situations involving more variables are relatively straightforward. However, the computations for these situations are extremely complex. If you are faced with a situation involving more than three variables, turn to one of the computerized statistical packages commonly available on college campuses (e.g., SPSS or SAS). These programs require minimal computer literacy and can handle complex calculations in, literally, the blink of an eye. Efficient use of these packages will enable you to avoid drudgery and will free you to do what social scientists everywhere enjoy doing most: pondering the meaning of your results and, by extension, the nature of social life.

SUMMARY

1. Partial correlation involves controlling for third variables in a manner analogous to that introduced in the previous chapter. Partial correlations permit the detection of direct and spurious or intervening relationships between X and Y.
2. Multiple regression includes statistical techniques by which predictions of the dependent variable from more than one independent variable can be made (by partial slopes and the multiple regression equation) and by which we can disentangle the relative importance of the independent variables (by standardized partial slopes).
3. The multiple correlation coefficient (R^2) summarizes the combined effects of all independent variables on

the dependent variable in terms of the proportion of the total variation in Y that is explained by all of the independent variables.
4. Partial correlation and multiple regression and correlation are some of the most powerful tools available to the researcher and demand high-quality measurement and relationships among the variables that are linear and noninteractive. Further, correlations among the independent variables must be low (preferably zero). Although the price is high, these techniques pay considerable dividends in the volume of precise and detailed information they generate about the interrelationships among the variables.

SUMMARY OF FORMULAS

Partial correlation coefficient:

16.1 $$r_{yx.z} = \frac{r_{yx} - (r_{yz})(r_{xz})}{\sqrt{1 - r_{yz}^2}\sqrt{1 - r_{xz}^2}}$$

Least-squares regression line (bivariate):

16.2 $Y = a + bX$

Least-squares multiple regression line:

16.3 $Y = a + b_1X_1 + b_2X_2$

Partial slope for X_1:

16.4 $$b_1 = \left(\frac{s_y}{s_1}\right)\left(\frac{r_{y1} - r_{y2}r_{12}}{1 - r_{12}^2}\right)$$

Partial slope for X_2:

16.5 $$b_2 = \left(\frac{s_y}{s_2}\right)\left(\frac{r_{y2} - r_{y1}r_{12}}{1 - r_{12}^2}\right)$$

Y intercept:

16.6 $a = \overline{Y} - b_1\overline{X}_1 - b_2\overline{X}_2$

Standardized partial slope (beta-weight) for X_1:

16.7 $$b_1^* = b_1\left(\frac{s_1}{s_y}\right)$$

Standardized partial slope (beta-weight) for X_2:

16.8 $$b_2^* = b_2\left(\frac{s_2}{s_y}\right)$$

Standardized least-squares regression line:

16.9 $Z_y = a_z + b_1^*Z_1 + b_2^*Z_2$

Standardized least-squares regression line (simplified):

16.10 $Z_y = b_1^*Z_1 + b_2^*Z_2$

Coefficient of multiple determination:

16.11 $R^2 = r_{y1}^2 + r_{y2.1}^2(1 - r_{y1}^2)$

GLOSSARY

Beta-weight (*b**). See **Standardized partial slope**.

Coefficient of multiple determination (*R*²). A statistic that equals the total variation explained in the dependent variable by all independent variables combined.

Control variables. In multivariate statistical analysis, a third variable (*Z*) whose effect on a bivariate relationship is held constant.

Multiple correlation. A multivariate technique for examining the combined effects of more than one independent variable on a dependent variable.

Multiple correlation coefficient (*R*). A statistic that indicates the strength of the correlation between a dependent variable and two or more independent variables.

Multiple regression. A multivariate technique that breaks down the separate effects of the independent variables on the dependent variable; used to make predictions of the dependent variable.

Partial correlation. A multivariate technique for examining a bivariate relationship while controlling for other variables.

Partial correlation coefficient. A statistic that shows the relationship between two variables while controlling for other variables; $r_{yx.z}$ is the symbol for the partial correlation coefficient when controlling for one variable.

Partial slope. In a multiple regression equation, the slope of the relationship between a particular independent variable and the dependent variable while controlling for all other independent variables in the equation.

Standardized partial slope (beta-weight). The slope of the relationship between a particular independent variable and the dependent variable when all scores have been normalized.

Zero-order correlation. Correlation coefficient for bivariate relationships.

PROBLEMS

16.1 [PS] In problem 14.1, data regarding voter turnout in five cities was presented. For the sake of convenience, the data for three of the variables are presented again here along with descriptive statistics and zero-order correlations.

City	Turnout	Unemployment Rate	% Negative Ads
A	55	5	60
B	60	8	63
C	65	9	55
D	68	9	53
E	70	10	48
Mean =	63.6	8.2	55.8
s =	5.5	1.7	5.3

	Unemployment Rate	Negative Ads
Turnout	0.95	−0.87
Unemployment Rate		−0.70

a. Compute the partial correlation coefficient for the relationship between turnout (*Y*) and unemployment (*X*) while controlling for the effect of negative advertising (*Z*). What effect does this control variable have on the bivariate relationship? Is the relationship between turnout and unemployment direct? (*HINT: Use Formula 16.1 and see Section 16.2.*)

b. Compute the partial correlation coefficient for the relationship between turnout (*Y*) and negative advertising (*X*) while controlling for the effect of unemployment (*Z*). What effect does this have on the bivariate relationship? Is the relationship between turnout and negative advertising direct? (*HINT: Use Formula 16.1 and see Section 16.2. You will need this partial correlation to compute the multiple correlation coefficient.*)

c. Find the unstandardized multiple regression equation with unemployment (X_1) and negative ads (X_2) as the independent variables. What turnout would be expected in a city in which the unemployment rate was 10% and 75% of the campaign ads were negative? (*HINT: Use Formulas 16.4 and 16.5 to compute the partial slopes and then use Formula 16.6 to find a, the Y intercept. The regression line is stated in Formula 16.3. Substitute 10 for X_1 and 75 for X_2 to compute predicted Y.*)

d. Compute beta-weights for each independent variable. Which has the stronger impact on turnout? (*HINT: Use Formulas 16.7 and 16.8 to calculate the beta-weights.*)

e. Compute the multiple correlation coefficient (*R*) and the coefficient of multiple determination (*R*²). How much of the variance in voter turnout is explained by the two independent variables?

(*HINT: Use Formula 16.11. You calculated $r^2_{y2.1}$ in part b of this problem.*)

f. Write a paragraph summarizing your conclusions about the relationships among these three variables.

16.2 | SOC | A scale measuring support for increases in the national defense budget has been administered to a sample. The respondents have also been asked to indicate how many years of school they have completed and how many years, if any, they served in the military. Take "support" as the dependent variable.

Case	Support	Years of School	Years of Service
A	20	12	2
B	15	12	4
C	20	16	20
D	10	10	10
E	10	16	20
F	5	8	0
G	8	14	2
H	20	12	20
I	10	10	4
J	20	16	0

a. Compute the partial correlation coefficient for the relationship between support (Y) and years of school (X) while controlling for the effect of years of service (Z). What effect does this have on the bivariate relationship? Is the relationship between support and years of school direct?

b. Compute the partial correlation coefficient for the relationship between support (Y) and years of service (X) while controlling for the effect of years of school (Z). What effect does this have on the bivariate relationship? Is the relationship between support and years of service direct? (*HINT: You will need this partial correlation to compute the multiple correlation coefficient.*)

c. Find the unstandardized multiple regression equation with school (X_1) and service (X_2) as the independent variables. What level of support would be expected in a person with 13 years of school and 15 years of service?

d. Compute beta-weights for each independent variable. Which has the stronger impact on turnout?

e. Compute the multiple correlation coefficient (R) and the coefficient of multiple determination (R^2). How much of the variance in support is explained

by the two independent variables? (*HINT: You calculated $r^2_{y2.1}$ in part b of this problem.*)

f. Write a paragraph summarizing your conclusions about the relationships among these three variables.

16.3 | SOC | Data on civil strife (number of incidents), unemployment, and urbanization have been gathered for 10 nations. Take civil strife as the dependent variable. Compute the zero-order correlations among all three variables.

Number of Incidents of Civil Strife	Unemployment Rate	Percentage of Population Living in Urban Areas
0	5.3	60
1	1.0	65
5	2.7	55
7	2.8	68
10	3.0	69
23	2.5	70
25	6.0	45
26	5.2	40
30	7.8	75
53	9.2	80

a. Compute the partial correlation coefficient for the relationship between strife (Y) and unemployment (X) while controlling for the effect of urbanization (Z). What effect does this have on the bivariate relationship? Is the relationship between strife and unemployment direct?

b. Compute the partial correlation coefficient for the relationship between strife (Y) and urbanization (X) while controlling for the effect of unemployment (Z). What effect does this have on the bivariate relationship? Is the relationship between strife and urbanization direct? (*HINT: You will need this partial correlation to compute the multiple correlation coefficient.*)

c. Find the unstandardized multiple regression equation with unemployment (X_1) and urbanization (X_2) as the independent variables. What level of strife would be expected in a nation in which the unemployment rate was 10% and 90% of the population lived in urban areas?

d. Compute beta-weights for each independent variable. Which has the stronger impact on strife?

e. Compute the multiple correlation coefficient (R) and the coefficient of multiple determination (R^2).

How much of the variance in strife is explained by the two independent variables?

f. Write a paragraph summarizing your conclusions about the relationships among these three variables.

16.4 SOC/CJ In problem 14.5, crime and population data were presented for each of 10 states. The data are reproduced here.

State	Crime Rates			Independent Variables		
	Homicide	Robbery	Car Theft	Growth*	Density†	Unemployment
Maine	1	27	96	3.3	42.7	5.4
New York	4	161	145	2.7	412.8	5.5
Ohio	4	159	295	1.2	280.5	6.5
Iowa	1	44	164	2.6	53.7	4.0
Virginia	5	143	182	9.7	196.2	4.0
Kentucky	5	96	205	5.6	107.5	6.3
Texas	6	162	393	16.7	92.9	4.8
Arizona	7	152	763	26.7	57.2	5.9
Washington	3	94	582	11.1	98.4	5.3
California	6	193	600	8.5	235.7	7.1

*Percentage change in population from 2000 to 2008.
†Population per square mile of land area, 2008.
Source: United States Bureau of the Census. *Statistical Abstracts of the United States: 2009.* Washington, DC, 2007.

Take the three crime variables as the dependent variables (one at a time) and do the following:

a. Find the multiple regression equations (unstandardized) with each crime variable (one at a time) and growth and density as independent variables.

b. Make a prediction for each crime variable for a state with a 5% growth rate and a density of 500.

c. Compute beta-weights for each independent variable in each equation and compare their relative effect on each dependent.

d. Compute R and R^2 for each crime variable. Write a paragraph summarizing your findings.

16.5 PS Problem 14.4 presented data on 10 precincts. The information is reproduced here.

Percent	Percent Democrat	Precinct Minority	Voter Turnout
A	50	10	56
B	45	12	55
C	56	8	52
D	78	15	60
E	13	5	89
F	85	20	25
G	62	18	64
H	33	9	88
I	25	0	42
J	49	9	36

Take voter turnout as the dependent variable and do the following:

a. Find the multiple regression equations (unstandardized).

b. What turnout would you expect for a precinct in which 0% of the voters were Democrats and 5% were minorities?

c. Compute beta-weights for each independent variable and compare their relative effect on turnout. Which was the more important factor?

d. Compute R and R^2.

e. Write a paragraph summarizing your findings.

16.6 SW Twelve families have been referred to a counselor, and she has rated each of them on a cohesiveness scale. Also, she has information on family income and number of children currently living at home. Take family cohesion as the dependent variable.

Family	Cohesion Score	Income	Number of Children
A	10	30,000	5
B	10	70,000	4
C	9	35,000	4
D	5	25,000	0
E	1	55,000	3

(*continued on next page*)

(*continued*)

Family	Cohesion Score	Income	Number of Children
F	7	40,000	0
G	2	60,000	2
H	5	30,000	3
I	8	50,000	5
J	3	25,000	4
K	2	45,000	3
L	4	50,000	0

a. Find the multiple regression equations (unstandardized).
b. What level of cohesion would be expected in a family with an income of $20,000 and 6 children?
c. Compute beta-weights for each independent variable and compare their relative effect on cohesion. Which was the more important factor?
d. Compute R and R^2.
e. Write a paragraph summarizing your findings.

16.7 Problem 14.8 presented per capita expenditures on education for 15 states, along with rank on income per capita and the percentage of the population that has graduated from high school. The data are reproduced here.

State	Per Capita Expenditures on Education, 2007	Percent High School Graduates, 2007	Rank in Per Capita Income
Arkansas	1546	80.4	14
Colorado	1792	88.9	6
Connecticut	2391	88.0	1
Florida	1672	84.9	8
Illinois	1887	85.7	5
Kansas	1913	89.1	9
Louisiana	1650	79.9	11
Maryland	1959	87.4	3
Michigan	1908	87.4	12
Mississippi	1313	78.5	15
Nebraska	1536	89.6	10
New Hampshire	1863	90.5	4
North Carolina	1456	83.0	13
Pennsylvania	1943	86.8	7
Wyoming	2712	91.2	2*

Take per capita expenditures as the dependent variable.
a. Compute beta-weights for each independent variable and compare their relative effect on expenditures. Which was the more important factor?
b. Compute R and R^2.
c. Write a paragraph summarizing your findings.

16.8 SOC The scores on four variables for 20 individuals are reported here: hours of TV (average number of hours of TV viewing each day), occupational prestige (higher scores indicate greater prestige), number of children, and age. Take TV viewing as the dependent variable and select two of the remaining variables as independent variables.

Hours of TV	Occupational Prestige	Number of Children	Age
4	50	2	43
3	36	3	58
3	36	1	34
4	50	2	42
2	45	2	27
3	50	5	60
4	50	0	28
7	40	3	55

(*continued on next page*)

(*continued*)

Hours of TV	Occupational Prestige	Number of Children	Age
1	57	2	46
3	33	2	65
1	46	3	56
3	31	1	29
1	19	2	41
0	52	0	50
2	48	1	62
4	36	1	24
3	48	0	25
1	62	1	87
5	50	0	45
1	27	3	62

a. Compute beta-weights for each of the independent variables you selected and compare their relative effect on the hours of television watching. Which was the more important factor?

b. Compute R and R^2.

c. Write a paragraph summarizing your findings.

YOU ARE THE RESEARCHER

A Multivariate Analysis of TV Viewing and Success

The two projects below continue the investigations begun in Chapter 14. You will use an SPSS procedure called **Regression** to analyze the combined effects of your independent variables (including dummy variables if you used any) on the dependent variables from Chapter 14. The **Regression** procedure produces a much greater volume of output than the **Correlate** procedure used in Chapter 14. Among other things, **Regression** displays the slope (*b*) and the *Y* intercept (*a*), so we can use this procedure to find least-squares regression lines. In these projects, we will use very little of the power of the **Regression** command and we will be extremely economical in our choice of all the options available. I urge you to explore some of the variations and capabilities of this powerful data-analysis procedure on your own.

Let's begin with a demonstration using *marhomo* (approval of gay marriage) as the dependent variable and *attend* (church attendance) and *educ* as independent variables.

Click **Analyze, Regression**, and **Linear**, and the **Linear Regression** window will appear. Move *marhomo* into the **Dependent:** box and *attend* and *educ* into the **Independent(s):** box. If you wish, you may click the **Statistics** button and then click **Descriptives** to get zero-order correlations, means, and standard deviations for the variables. Click **Continue** and **OK**, and the following output will appear (descriptive information about the variables and the zero-order correlations are omitted to conserve space).

Model Summary

Model	R	R Square	Adjusted R Square	Std. Error of the Estimate
1	.368[a]	.136	.134	1.396

[a]Predictors: (Constant), HIGHEST YEAR OF SCHOOL COMPLETED, HOW OFTEN R ATTENDS RELIGIOUS SERVICES

Coefficients^a

Model		Unstandardized Coefficients		Standardized Coefficients	t	Sig.
		B	Std. Error	Beta		
1	(Constant)	3.845	.211		18.261	.000
	HOW OFTEN R ATTENDS RELIGIOUS SERVICES	.168	.016	.316	10.539	.000
	HIGHEST YEAR OF SCHOOL COMPLETED	−.089	.015	−.183	−6.112	.000

^aDependent Variable: HOMOSEXUALS SHOULD HAVE RIGHT TO MARRY

The "Model Summary" block reports the multiple R (.368) and R square (.136). The ANOVA output block shows the significance of the relationship (Sig. = .000). In the last output block, under "Unstandardized Coefficients," we see the Y intercept, reported as a constant of 3.845, and the slopes (B) of the independent variables on *marhomo*. From this information, we can build a regression equation (see Formula 16.3) to predict scores on *marhomo*:

$$Y = 3.845 + (0.168)(attend) + (-0.089)(educ)$$

Under "Standardized Coefficients," we find the standardized partial slopes (Beta). The beta for *attend* (.316) is greater in value than the beta for *educ* (−.183). This tells us that religiosity (*attend*) is the more important of the two independent variables. It has a positive relationship with *marhomo* after the effects of education have been controlled. As religiosity increases, disapproval of gay marriage increases. Education has a negative relationship with the dependent variable after religiosity has been controlled: as education increases, disapproval of gay marriage decreases.

Your turn.

PROJECT 1: Who Watches TV?

This project follows up on Project 1 from Chapter 14. The dependent variable is *tvhours*, which, as you recall, measures the number of hours per day the respondent watches televsion.

Step 1: Choosing Independent Variables

Choose two of the four independent variables you selected in Project 1 from Chapter 14. All other things being equal, you might choose the variables that had the strongest relationship with *tvhours*. Remember that independent variables *cannot* be nominal in level of measurement unless you recode the variable into a dummy variable. You may use any interval-ratio or ordinal-level variables with more than three categories or scores. Once you have selected your variables, list them in the table below and describe exactly what they measure.

SPSS Variable Name	What exactly does this variable measure?
_____	_____
_____	_____

Step 2: Stating Hypotheses

Restate your hypotheses from Chapter 14 and add some ideas about the relative importance of your two independent variables. Which do you expect to have the stronger effect? Why?

Step 3: Running the Regression Procedure

Click **Analyze** → **Regression** → **Linear** and place *tvhours* in the **Dependent:** box and your selected independent variables in the **Independent(s):** box. Click **Statistics** and **Descriptives**, and then click **OK** to get your results.

Step 4: Recording Results

Summarize your results by filling in the blanks below with information from the **Coefficients** box. The Y intercept (a) is reported in the top row of the box as a "constant." The slopes (b) are reported in the "Unstandardized Coefficients" column under "B," and the beta-weights (b^*) are reported in the "Standardized Coefficients" column under "Beta."

1. $a =$ _____
2. **a.** Slope for your first independent variable (b_1) = _____.
   b. Slope for your second independent variable (b_2) = _____.
3. State the least squares multiple regression equation (see formula 16.3).

$$Y = \underline{} + (\underline{})X_1 + (\underline{})X_2$$

4. **a.** Beta-weight of your first independent variable (b_1^*) = _____.
   b. Beta-weight of your second independent variable (b_2^*) = _____.
5. State the standardized least-square regression line (see Formula 15.10).

$$Z_y = (\underline{})Z_1 + (\underline{})Z_2$$

6. $R^2 =$ _____

Step 5: Analyzing and Interpreting Results

Write a short summary of your results. Your summary needs to identify your variables and distinguish between independent and dependent variables. You also need to report the strength and direction of relationships as indicated by the slopes and beta-weights and the total explained variance (R^2). How much of the variance does your first independent variable explain? How much more is explained by your second independent variable?

PROJECT 2: What are the Correlates of Success?

In this second project, we will continue the investigation into the correlates of success, which we began in Chapter 14.

Step 1: Choosing Independent Variables

Use two of the four independent variables you selected in Project 2 from Chapter 14. All other things being equal, you might choose the variables that had the strongest relationship with the variable you selected to be the dependent variable, which might be either *income06* or *prestg80*. Remember that independent variables *cannot* be nominal in level of measurement unless you recode the variable into a dummy variable. You may use any interval-ratio or ordinal level variables with more than three categories or scores. Once you have selected your variables, list them in the table below and describe exactly what they measure.

SPSS Variable Name	What exactly does this variable measure?
_____	_____
_____	_____

Step 2: Stating Hypotheses

Restate your hypotheses from Chapter 14 and add some ideas about the relative importance of your two independent variables. Which do you expect to have the stronger effect? Why?

Step 3: Running the Regression Procedure

Click **Analyze → Regression → Linear** and place your dependent variable in the **Dependent:** box and your selected independent variables in the **Independent(s):** box. Click **Statistics** and **Descriptives**, and then click **OK** to get your results.

Step 4: Recording Results

Summarize your results by filling in the blanks below with information from the **Coefficients** box. The Y intercept (a) is reported in the top row of the box as a "constant." The slopes are reported in the "Unstandardized Coefficients" column under "B," and the beta-weights are reported in the "Standardized Coefficients" column under "Beta."

1. $a = $ _____
2. a. Slope for your first independent variable (b_1) = _____.
     b. Slope for your second independent variable (b_2) = _____.
3. State the least-squares multiple regression equation (see Formula 15.3).

$$Y = \_\_ + (\_\_)X_1 + (\_\_)X_2$$

4. a. Beta-weight of your first independent variable (b_1^*) = _____.
     b. Beta-weight of your second independent variable (b_2^*) = _____.
5. State the standardized least-squares regression line (see Formula 15.10).

$$Z_y = (\_\_)Z_1 + (\_\_)Z_2$$

6. $R^2 = $ ____

Step 5: Analyzing and Interpreting Results

Write a short summary of your results. Your summary needs to identify your variables and distinguish between independent and dependent variables. You also need to report the strength and direction of relationships as indicated by the slopes and beta-weights and the total explained variance (R^2). How much of the variance does your first independent variable explain? How much more is explained by your second independent variable?

1. Two research questions that can be answered by one of the techniques presented in Chapters 15 and 16 are stated here. For each research situation, choose either the elaboration technique or regression analysis. The level of measurement of the variables should have a strong influence on your decision.

    a. For a sample of college graduates, did having a job during the school year interfere with academic success? For 20 students, data have been gathered on college GPA, the average number of hours each student worked each week, and College Board scores (a measure of preparedness for college-level work). Take GPA as the dependent variable.

    b. Only about half of this sample graduated within four years (1 = yes, 2 = no). Was their progress affected by their level of social activity (1 = low, 2 = high)? Is the relationship between these variables the same for both males (1) and females (2)?

GPA	Hours Worked per Week	College Board Score	Graduated in Four Years?	Social Activity	Sex
3.14	13	550	1	1	1
2.00	20	375	1	2	2
2.11	22	450	1	1	2
3.00	10	575	1	2	1
3.75	0	600	1	1	1
3.11	21	650	1	1	2
3.22	7	605	1	2	1
2.75	25	630	1	1	2
2.50	30	680	1	1	1
2.10	32	610	1	1	2
2.45	20	580	2	2	2
2.01	40	590	2	1	1
3.90	0	675	2	2	2
3.45	0	650	2	2	1
2.30	25	550	2	2	1
2.20	18	470	2	1	2
2.60	25	600	2	2	1
3.10	15	525	2	2	1
2.60	27	480	2	1	2
2.20	20	500	2	2	1

2. A research project has been conducted on the audiences of Christian religious television programs. Two hypotheses have been developed:

    • The amount of time devoted to religious television will increase with age and decrease with education, but age will have the strongest effect.

- Protestants will be more likely to donate money. Protestant females will be especially likely to donate money.

Test the hypotheses with the data presented in the table.

The variables are:

a. How many hours per week do you watch religious programs on television? (actual hours)

b. What is your age? (years)

c. How many years of formal schooling have you completed? (scores are years)

d. Have you ever donated money to a religious television program?
   1. Yes
   2. No

e. What is your religious preference?
   1. Protestant
   2. Non-Protestant

f. What is your sex?
   1. Female
   2. Male

Hours	Age	Education	Ever Donate?	Religion	Sex
14	52	10	1	1	1
10	45	12	1	1	2
8	18	12	1	1	1
5	45	12	1	1	1
10	57	10	1	1	2
14	65	8	2	1	1
3	23	12	1	1	1
12	47	11	2	1	1
2	30	14	2	1	2
1	20	16	2	1	2
21	60	16	1	2	1
15	55	12	1	2	2
12	47	9	2	2	1
8	32	14	1	2	2
15	50	12	1	2	1
10	45	10	2	2	2
20	72	16	1	2	1
12	40	16	2	2	2
14	42	14	2	2	2
10	38	12	1	2	1

Area Under the Normal Curve

Column (a) lists Z scores from 0.00 to 4.00. Only positive scores are displayed, but, since the normal curve is symmetrical, the areas for negative scores will be exactly the same as areas for positive scores. Column (b) lists the proportion of the total area between the Z score and the mean. Figure A.1 displays areas of this type. Column (c) lists the proportion of the area beyond the Z score, and Figure A.2 displays this type of area.

FIGURE A.1 Area Between Mean and Z

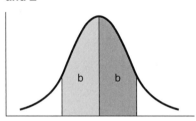

FIGURE A.2 Area Beyond Z

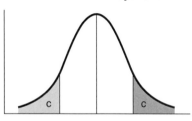

(a) Z	(b) Area Between Mean and Z	(c) Area Beyond Z	(a) Z	(b) Area Between Mean and Z	(c) Area Beyond Z
0.00	0.0000	0.5000	0.21	0.0832	0.4168
0.01	0.0040	0.4960	0.22	0.0871	0.4129
0.02	0.0080	0.4920	0.23	0.0910	0.4090
0.03	0.0120	0.4880	0.24	0.0948	0.4052
0.04	0.0160	0.4840	0.25	0.0987	0.4013
0.05	0.0199	0.4801	0.26	0.1026	0.3974
0.06	0.0239	0.4761	0.27	0.1064	0.3936
0.07	0.0279	0.4721	0.28	0.1103	0.3897
0.08	0.0319	0.4681	0.29	0.1141	0.3859
0.09	0.0359	0.4641	0.30	0.1179	0.3821
0.10	0.0398	0.4602			
			0.31	0.1217	0.3783
0.11	0.0438	0.4562	0.32	0.1255	0.3745
0.12	0.0478	0.4522	0.33	0.1293	0.3707
0.13	0.0517	0.4483	0.34	0.1331	0.3669
0.14	0.0557	0.4443	0.35	0.1368	0.3632
0.15	0.0596	0.4404	0.36	0.1406	0.3594
0.16	0.0636	0.4364	0.37	0.1443	0.3557
0.17	0.0675	0.4325	0.38	0.1480	0.3520
0.18	0.0714	0.4286	0.39	0.1517	0.3483
0.19	0.0753	0.4247	0.40	0.1554	0.3446
0.20	0.0793	0.4207			

(a) Z	(b) Area Between Mean and Z	(c) Area Beyond Z	(a) Z	(b) Area Between Mean and Z	(c) Area Beyond Z
0.41	0.1591	0.3409	0.96	0.3315	0.1685
0.42	0.1628	0.3372	0.97	0.3340	0.1660
0.43	0.1664	0.3336	0.98	0.3365	0.1635
0.44	0.1700	0.3300	0.99	0.3389	0.1611
0.45	0.1736	0.3264	1.00	0.3413	0.1587
0.46	0.1772	0.3228			
0.47	0.1808	0.3192	1.01	0.3438	0.1562
0.48	0.1844	0.3156	1.02	0.3461	0.1539
0.49	0.1879	0.3121	1.03	0.3485	0.1515
0.50	0.1915	0.3085	1.04	0.3508	0.1492
			1.05	0.3531	0.1469
0.51	0.1950	0.3050	1.06	0.3554	0.1446
0.52	0.1985	0.3015	1.07	0.3577	0.1423
0.53	0.2019	0.2981	1.08	0.3599	0.1401
0.54	0.2054	0.2946	1.09	0.3621	0.1379
0.55	0.2088	0.2912	1.10	0.3643	0.1357
0.56	0.2123	0.2877			
0.57	0.2157	0.2843	1.11	0.3665	0.1335
0.58	0.2190	0.2810	1.12	0.3686	0.1314
0.59	0.2224	0.2776	1.13	0.3708	0.1292
0.60	0.2257	0.2743	1.14	0.3729	0.1271
			1.15	0.3749	0.1251
0.61	0.2291	0.2709	1.16	0.3770	0.1230
0.62	0.2324	0.2676	1.17	0.3790	0.1210
0.63	0.2357	0.2643	1.18	0.3810	0.1190
0.64	0.2389	0.2611	1.19	0.3830	0.1170
0.65	0.2422	0.2578	1.20	0.3849	0.1151
0.66	0.2454	0.2546			
0.67	0.2486	0.2514	1.21	0.3869	0.1131
0.68	0.2517	0.2483	1.22	0.3888	0.1112
0.69	0.2549	0.2451	1.23	0.3907	0.1093
0.70	0.2580	0.2420	1.24	0.3925	0.1075
			1.25	0.3944	0.1056
0.71	0.2611	0.2389	1.26	0.3962	0.1038
0.72	0.2642	0.2358	1.27	0.3980	0.1020
0.73	0.2673	0.2327	1.28	0.3997	0.1003
0.74	0.2703	0.2297	1.29	0.4015	0.0985
0.75	0.2734	0.2266	1.30	0.4032	0.0968
0.76	0.2764	0.2236			
0.77	0.2794	0.2206	1.31	0.4049	0.0951
0.78	0.2823	0.2177	1.32	0.4066	0.0934
0.79	0.2852	0.2148	1.33	0.4082	0.0918
0.80	0.2881	0.2119	1.34	0.4099	0.0901
			1.35	0.4115	0.0885
0.81	0.2910	0.2090	1.36	0.4131	0.0869
0.82	0.2939	0.2061	1.37	0.4147	0.0853
0.83	0.2967	0.2033	1.38	0.4162	0.0838
0.84	0.2995	0.2005	1.39	0.4177	0.0823
0.85	0.3023	0.1977	1.40	0.4192	0.0808
0.86	0.3051	0.1949			
0.87	0.3078	0.1922	1.41	0.4207	0.0793
0.88	0.3106	0.1894	1.42	0.4222	0.0778
0.89	0.3133	0.1867	1.43	0.4236	0.0764
0.90	0.3159	0.1841	1.44	0.4251	0.0749
			1.45	0.4265	0.0735
0.91	0.3186	0.1814	1.46	0.4279	0.0721
0.92	0.3212	0.1788	1.47	0.4292	0.0708
0.93	0.3238	0.1762	1.48	0.4306	0.0694
0.94	0.3264	0.1736	1.49	0.4319	0.0681
0.95	0.3289	0.1711	1.50	0.4332	0.0668

(a) Z	(b) Area Between Mean and Z	(c) Area Beyond Z	(a) Z	(b) Area Between Mean and Z	(c) Area Beyond Z
1.51	0.4345	0.0655	2.06	0.4803	0.0197
1.52	0.4357	0.0643	2.07	0.4808	0.0192
1.53	0.4370	0.0630	2.08	0.4812	0.0188
1.54	0.4382	0.0618	2.09	0.4817	0.0183
1.55	0.4394	0.0606	2.10	0.4821	0.0179
1.56	0.4406	0.0594	2.11	0.4826	0.0174
1.57	0.4418	0.0582	2.12	0.4830	0.0170
1.58	0.4429	0.0571	2.13	0.4834	0.0166
1.59	0.4441	0.0559	2.14	0.4838	0.0162
1.60	0.4452	0.0548	2.15	0.4842	0.0158
1.61	0.4463	0.0537	2.16	0.4846	0.0154
1.62	0.4474	0.0526	2.17	0.4850	0.0150
1.63	0.4484	0.0516	2.18	0.4854	0.0146
1.64	0.4495	0.0505	2.19	0.4857	0.0143
1.65	0.4505	0.0495	2.20	0.4861	0.0139
1.66	0.4515	0.0485	2.21	0.4864	0.0136
1.67	0.4525	0.0475	2.22	0.4868	0.0132
1.68	0.4535	0.0465	2.23	0.4871	0.0129
1.69	0.4545	0.0455	2.24	0.4875	0.0125
1.70	0.4554	0.0446	2.25	0.4878	0.0122
1.71	0.4564	0.0436	2.26	0.4881	0.0119
1.72	0.4573	0.0427	2.27	0.4884	0.0116
1.73	0.4582	0.0418	2.28	0.4887	0.0113
1.74	0.4591	0.0409	2.29	0.4890	0.0110
1.75	0.4599	0.0401	2.30	0.4893	0.0107
1.76	0.4608	0.0392	2.31	0.4896	0.0104
1.77	0.4616	0.0384	2.32	0.4898	0.0102
1.78	0.4625	0.0375	2.33	0.4901	0.0099
1.79	0.4633	0.0367	2.34	0.4904	0.0096
1.80	0.4641	0.0359	2.35	0.4906	0.0094
1.81	0.4649	0.0351	2.36	0.4909	0.0091
1.82	0.4656	0.0344	2.37	0.4911	0.0089
1.83	0.4664	0.0336	2.38	0.4913	0.0087
1.84	0.4671	0.0329	2.39	0.4916	0.0084
1.85	0.4678	0.0322	2.40	0.4918	0.0082
1.86	0.4686	0.0314	2.41	0.4920	0.0080
1.87	0.4693	0.0307	2.42	0.4922	0.0078
1.88	0.4699	0.0301	2.43	0.4925	0.0075
1.89	0.4706	0.0294	2.44	0.4927	0.0073
1.90	0.4713	0.0287	2.45	0.4929	0.0071
1.91	0.4719	0.0281	2.46	0.4931	0.0069
1.92	0.4726	0.0274	2.47	0.4932	0.0068
1.93	0.4732	0.0268	2.48	0.4934	0.0066
1.94	0.4738	0.0262	2.49	0.4936	0.0064
1.95	0.4744	0.0256	2.50	0.4938	0.0062
1.96	0.4750	0.0250	2.51	0.4940	0.0060
1.97	0.4756	0.0244	2.52	0.4941	0.0059
1.98	0.4761	0.0239	2.53	0.4943	0.0057
1.99	0.4767	0.0233	2.54	0.4945	0.0055
2.00	0.4772	0.0228	2.55	0.4946	0.0054
2.01	0.4778	0.0222	2.56	0.4948	0.0052
2.02	0.4783	0.0217	2.57	0.4949	0.0051
2.03	0.4788	0.0212	2.58	0.4951	0.0049
2.04	0.4793	0.0207	2.59	0.4952	0.0048
2.05	0.4798	0.0202	2.60	0.4953	0.0047

(a) Z	(b) Area Between Mean and Z	(c) Area Beyond Z	(a) Z	(b) Area Between Mean and Z	(c) Area Beyond Z
2.61	0.4955	0.0045	3.11	0.4991	0.0009
2.62	0.4956	0.0044	3.12	0.4991	0.0009
2.63	0.4957	0.0043	3.13	0.4991	0.0009
2.64	0.4959	0.0041	3.14	0.4992	0.0008
2.65	0.4960	0.0040	3.15	0.4992	0.0008
2.66	0.4961	0.0039	3.16	0.4992	0.0008
2.67	0.4962	0.0038	3.17	0.4992	0.0008
2.68	0.4963	0.0037	3.18	0.4993	0.0007
2.69	0.4964	0.0036	3.19	0.4993	0.0007
2.70	0.4965	0.0035	3.20	0.4993	0.0007
2.71	0.4966	0.0034	3.21	0.4993	0.0007
2.72	0.4967	0.0033	3.22	0.4994	0.0006
2.73	0.4968	0.0032	3.23	0.4994	0.0006
2.74	0.4969	0.0031	3.24	0.4994	0.0006
2.75	0.4970	0.0030	3.25	0.4994	0.0006
2.76	0.4971	0.0029	3.26	0.4994	0.0006
2.77	0.4972	0.0028	3.27	0.4995	0.0005
2.78	0.4973	0.0027	3.28	0.4995	0.0005
2.79	0.4974	0.0026	3.29	0.4995	0.0005
2.80	0.4974	0.0026	3.30	0.4995	0.0005
2.81	0.4975	0.0025	3.31	0.4995	0.0005
2.82	0.4976	0.0024	3.32	0.4995	0.0005
2.83	0.4977	0.0023	3.33	0.4996	0.0004
2.84	0.4977	0.0023	3.34	0.4996	0.0004
2.85	0.4978	0.0022	3.35	0.4996	0.0004
2.86	0.4979	0.0021	3.36	0.4996	0.0004
2.87	0.4979	0.0021	3.37	0.4996	0.0004
2.88	0.4980	0.0020	3.38	0.4996	0.0004
2.89	0.4981	0.0019	3.39	0.4997	0.0003
2.90	0.4981	0.0019	3.40	0.4997	0.0003
2.91	0.4982	0.0018	3.41	0.4997	0.0003
2.92	0.4982	0.0018	3.42	0.4997	0.0003
2.93	0.4983	0.0017	3.43	0.4997	0.0003
2.94	0.4984	0.0016	3.44	0.4997	0.0003
2.95	0.4984	0.0016	3.45	0.4997	0.0003
2.96	0.4985	0.0015	3.46	0.4997	0.0003
2.97	0.4985	0.0015	3.47	0.4997	0.0003
2.98	0.4986	0.0014	3.48	0.4997	0.0003
2.99	0.4986	0.0014	3.49	0.4998	0.0002
3.00	0.4986	0.0014	3.50	0.4998	0.0002
3.01	0.4987	0.0013	3.60	0.4998	0.0002
3.02	0.4987	0.0013	3.70	0.4999	0.0001
3.03	0.4988	0.0012	3.70	0.4999	0.0001
3.04	0.4988	0.0012	3.80	0.4999	0.0001
3.05	0.4989	0.0011	3.90	0.4999	<0.0001
3.06	0.4989	0.0011	3.90	0.4999	<0.0001
3.07	0.4989	0.0011	4.00	0.4999	<0.0001
3.08	0.4990	0.0010			
3.09	0.4990	0.0010			
3.10	0.4990	0.0010			

Distribution of *t*

Use this table to find the critical region (step 3 of the five-step model) for tests of significance with sample means when sample size (*N*) is small. First, choose an alpha level and a one- or two-tailed test. Then find degrees of freedom to find the *t* score that marks the beginning of the critical region.

Degrees of Freedom (df)	Level of Significance for One-Tailed Test					
	.10	.05	.025	.01	.005	.0005
	Level of Significance for Two-Tailed Test					
	.20	.10	.05	.02	.01	.001
1	3.078	6.314	12.706	31.821	63.657	636.619
2	1.886	2.920	4.303	6.965	9.925	31.598
3	1.638	2.353	3.182	4.541	5.841	12.941
4	1.533	2.132	2.776	3.747	4.604	8.610
5	1.476	2.015	2.571	3.365	4.032	6.859
6	1.440	1.943	2.447	3.143	3.707	5.959
7	1.415	1.895	2.365	2.998	3.499	5.405
8	1.397	1.860	2.306	2.896	3.355	5.041
9	1.383	1.833	2.262	2.821	3.250	4.781
10	1.372	1.812	2.228	2.764	3.169	4.587
11	1.363	1.796	2.201	2.718	3.106	4.437
12	1.356	1.782	2.179	2.681	3.055	4.318
13	1.350	1.771	2.160	2.650	3.012	4.221
14	1.345	1.761	2.145	2.624	2.977	4.140
15	1.341	1.753	2.131	2.602	2.947	4.073
16	1.337	1.746	2.120	2.583	2.921	4.015
17	1.333	1.740	2.110	2.567	2.898	3.965
18	1.330	1.734	2.101	2.552	2.878	3.922
19	1.328	1.729	2.093	2.539	2.861	3.883
20	1.325	1.725	2.086	2.528	2.845	3.850
21	1.323	1.721	2.080	2.518	2.831	3.819
22	1.321	1.717	2.074	2.508	2.819	3.792
23	1.319	1.714	2.069	2.500	2.807	3.767
24	1.318	1.711	2.064	2.492	2.797	3.745
25	1.316	1.708	2.060	2.485	2.787	3.725
26	1.315	1.706	2.056	2.479	2.779	3.707
27	1.314	1.703	2.052	2.473	2.771	3.690
28	1.313	1.701	2.048	2.467	2.763	3.674
29	1.311	1.699	2.045	2.462	2.756	3.659
30	1.310	1.697	2.042	2.457	2.750	3.646
40	1.303	1.684	2.021	2.423	2.704	3.551
60	1.296	1.671	2.000	2.390	2.660	3.460
120	1.289	1.658	1.980	2.358	2.617	3.373
∞	1.282	1.645	1.960	2.326	2.576	3.291

Source: Table III of Fisher and Yates: *Statistical Tables for Biological, Agricultural and Medical Research,* published by Longman Group Ltd., London (1974), 6th edition (previously published by Oliver & Boyd Ltd., Edinburgh).

Distribution of Chi Square

Use this table to find the critical region (step 3 of the five-step model) for tests of significance with chi square. Choose an alpha level and find the degrees of freedom to find the chi square score that marks the beginning of the critical region.

df	.99	.98	.95	.90	.80	.70	.50	.30	.20	.10	.05	.02	.01	.001
1	$.0^3157$	$.0^3628$	.00393	.0158	.0642	.148	.455	1.074	1.642	2.706	3.841	5.412	6.635	10.827
2	.0201	.0404	.103	.211	.446	.713	1.386	2.408	3.219	4.605	5.991	7.824	9.210	13.815
3	.115	.185	.352	.584	1.005	1.424	2.366	3.665	4.642	6.251	7.815	9.837	11.341	16.268
4	.297	.429	.711	1.064	1.649	2.195	3.357	4.878	5.989	7.779	9.488	11.668	13.277	18.465
5	.554	.752	1.145	1.610	2.343	3.000	4.351	6.064	7.289	9.236	11.070	13.388	15.086	20.517
6	.872	1.134	1.635	2.204	3.070	3.828	5.348	7.231	8.558	10.645	12.592	15.033	16.812	22.457
7	1.239	1.564	2.167	2.833	3.822	4.671	6.346	8.383	9.803	12.017	14.067	16.622	18.475	24.322
8	1.646	2.032	2.733	3.490	4.594	5.527	7.344	9.524	11.030	13.362	15.507	18.168	20.090	26.125
9	2.088	2.532	3.325	4.168	5.380	6.393	8.343	10.656	12.242	14.684	16.919	19.679	21.666	27.877
10	2.558	3.059	3.940	4.865	6.179	7.267	9.342	11.781	13.442	15.987	18.307	21.161	23.209	29.588
11	3.053	3.609	4.575	5.578	6.989	8.148	10.341	12.899	14.631	17.275	19.675	22.618	24.725	31.264
12	3.571	4.178	5.226	6.304	7.807	9.034	11.340	14.011	15.812	18.549	21.026	24.054	26.217	32.909
13	4.107	4.765	5.892	7.042	8.634	9.926	12.340	15.119	16.985	19.812	22.362	25.472	27.688	34.528
14	4.660	5.368	6.571	7.790	9.467	10.821	13.339	16.222	18.151	21.064	23.685	26.873	29.141	36.123
15	5.229	5.985	7.261	8.547	10.307	11.721	14.339	17.322	19.311	22.307	24.996	28.259	30.578	37.697
16	5.812	6.614	7.962	9.312	11.152	12.624	15.338	18.418	20.465	23.542	26.296	29.633	32.000	39.252
17	6.408	7.255	8.672	10.085	12.002	13.531	16.338	19.511	21.615	24.769	27.587	30.995	33.409	40.790
18	7.015	7.906	9.390	10.865	12.857	14.440	17.338	20.601	22.760	25.989	28.869	32.346	34.805	42.312
19	7.633	8.567	10.117	11.651	13.716	15.352	18.338	21.689	23.900	27.204	30.144	33.687	36.191	43.820
20	8.260	9.237	10.851	12.443	14.578	16.266	19.337	22.775	25.038	28.412	31.410	35.020	37.566	45.315
21	8.897	9.915	11.591	13.240	15.445	17.182	20.337	23.858	26.171	29.615	32.671	36.343	38.932	46.797
22	9.542	10.600	12.338	14.041	16.314	18.101	21.337	24.939	27.301	30.813	33.924	37.659	40.289	48.268
23	10.196	11.293	13.091	14.848	17.187	19.021	22.337	26.018	28.429	32.007	35.172	38.968	41.638	49.728
24	10.856	11.992	13.848	15.659	18.062	19.943	23.337	27.096	29.553	33.196	36.415	40.270	42.980	51.179
25	11.524	12.697	14.611	16.473	18.940	20.867	24.337	28.172	30.675	34.382	37.652	41.566	44.314	52.620
26	12.198	13.409	15.379	17.292	19.820	21.792	25.336	29.246	31.795	35.563	38.885	42.856	45.642	54.052
27	12.879	14.125	16.151	18.114	20.703	22.719	26.336	30.319	32.912	36.741	40.113	44.140	46.963	55.476
28	13.565	14.847	16.928	18.939	21.588	23.647	27.336	31.391	34.027	37.916	41.337	45.419	48.278	56.893
29	14.256	15.574	17.708	19.768	22.475	24.577	28.336	32.461	35.139	39.087	42.557	46.693	49.588	58.302
30	14.953	16.306	18.493	20.599	23.364	25.508	29.336	33.530	36.250	40.256	43.773	47.962	50.892	59.703

Source: Table IV of Fisher and Yates: *Statistical Tables for Biological, Agricultural and Medical Research*, published by Longman Group Ltd., London (1974), 6th edition (previously published by Oliver & Boyd Ltd., Edinburgh). Reprinted by permission of Addison Wesley Longman Ltd.

Distribution of *F*

Use this table to find the critical region (step 3 of the five-step model) for analysis of variance tests. Choose an alpha level of either 0.05 or 0.01 and then find the degrees of freedom to find the F ratio that marks the beginning of the critical region.

$p = .05$

n_1 n_2	1	2	3	4	5	6	8	12	24	∞
1	161.4	199.5	215.7	224.6	230.2	234.0	238.9	243.9	249.0	254.3
2	18.51	19.00	19.16	19.25	19.30	19.33	19.37	19.41	19.45	19.50
3	10.13	9.55	9.28	9.12	9.01	8.94	8.84	8.74	8.64	8.53
4	7.71	6.94	6.59	6.39	6.26	6.16	6.04	5.91	5.77	5.63
5	6.61	5.79	5.41	5.19	5.05	4.95	4.82	4.68	4.53	4.36
6	5.99	5.14	4.76	4.53	4.39	4.28	4.15	4.00	3.84	3.67
7	5.59	4.74	4.35	4.12	3.97	3.87	3.73	3.57	3.41	3.23
8	5.32	4.46	4.07	3.84	3.69	3.58	3.44	3.28	3.12	2.93
9	5.12	4.26	3.86	3.63	3.48	3.37	3.23	3.07	2.90	2.71
10	4.96	4.10	3.71	3.48	3.33	3.22	3.07	2.91	2.74	2.54
11	4.84	3.98	3.59	3.36	3.20	3.09	2.95	2.79	2.61	2.40
12	4.75	3.88	3.49	3.26	3.11	3.00	2.85	2.69	2.50	2.30
13	4.67	3.80	3.41	3.18	3.02	2.92	2.77	2.60	2.42	2.21
14	4.60	3.74	3.34	3.11	2.96	2.85	2.70	2.53	2.35	2.13
15	4.54	3.68	3.29	3.06	2.90	2.79	2.64	2.48	2.29	2.07
16	4.49	3.63	3.24	3.01	2.85	2.74	2.59	2.42	2.24	2.01
17	4.45	3.59	3.20	2.96	2.81	2.70	2.55	2.38	2.19	1.96
18	4.41	3.55	3.16	2.93	2.77	2.66	2.51	2.34	2.15	1.92
19	4.38	3.52	3.13	2.90	2.74	2.63	2.48	2.31	2.11	1.88
20	4.35	3.49	3.10	2.87	2.71	2.60	2.45	2.28	2.08	1.84
21	4.32	3.47	3.07	2.84	2.68	2.57	2.42	2.25	2.05	1.81
22	4.30	3.44	3.05	2.82	2.66	2.55	2.40	2.23	2.03	1.78
23	4.28	3.42	3.03	2.80	2.64	2.53	2.38	2.20	2.00	1.76
24	4.26	3.40	3.01	2.78	2.62	2.51	2.36	2.18	1.98	1.73
25	4.24	3.38	2.99	2.76	2.60	2.49	2.34	2.16	1.96	1.71
26	4.22	3.37	2.98	2.74	2.59	2.47	2.32	2.15	1.95	1.69
27	4.21	3.35	2.96	2.73	2.57	2.46	2.30	2.13	1.93	1.67
28	4.20	3.34	2.95	2.71	2.56	2.44	2.29	2.12	1.91	1.65
29	4.18	3.33	2.93	2.70	2.54	2.43	2.28	2.10	1.90	1.64
30	4.17	3.32	2.92	2.69	2.53	2.42	2.27	2.09	1.89	1.62
40	4.08	3.23	2.84	2.61	2.45	2.34	2.18	2.00	1.79	1.51
60	4.00	3.15	2.76	2.52	2.37	2.25	2.10	1.92	1.70	1.39
120	3.92	3.07	2.68	2.45	2.29	2.17	2.02	1.83	1.61	1.25
∞	3.84	2.99	2.60	2.37	2.21	2.09	1.94	1.75	1.52	1.00

Values of n_1 and n_2 represent the degrees of freedom associated with the between and within estimates of variance, respectively.

Source: Table V of Fisher and Yates: *Statistical Tables for Biological, Agricultural and Medical Research*, published by Longman Group Ltd., London (1974), 6th edition (previously published by Oliver and Boyd Ltd., Edinburgh). Reprinted by permission of Addison Wesley Longman Ltd.

$$p = .01$$

n_1 n_2	1	2	3	4	5	6	8	12	24	∞
1	4052	4999	5403	5625	5764	5859	5981	6106	6234	6366
2	98.49	99.01	99.17	99.25	99.30	99.33	99.36	99.42	99.46	99.50
3	34.12	30.81	29.46	28.71	28.24	27.91	27.49	27.05	26.60	26.12
4	21.20	18.00	16.69	15.98	15.52	15.21	14.80	14.37	13.93	13.46
5	16.26	13.27	12.06	11.39	10.97	10.67	10.27	9.89	9.47	9.02
6	13.74	10.92	9.78	9.15	8.75	8.47	8.10	7.72	7.31	6.88
7	12.25	9.55	8.45	7.85	7.46	7.19	6.84	6.47	6.07	5.65
8	11.26	8.65	7.59	7.01	6.63	6.37	6.03	5.67	5.28	4.86
9	10.56	8.02	6.99	6.42	6.06	5.80	5.47	5.11	4.73	4.31
10	10.04	7.56	6.55	5.99	5.64	5.39	5.06	4.71	4.33	3.91
11	9.65	7.20	6.22	5.67	5.32	5.07	4.74	4.40	4.02	3.60
12	9.33	6.93	5.95	5.41	5.06	4.82	4.50	4.16	3.78	3.36
13	9.07	6.70	5.74	5.20	4.86	4.62	4.30	3.96	3.59	3.16
14	8.86	6.51	5.56	5.03	4.69	4.46	4.14	3.80	3.43	3.00
15	8.68	6.36	5.42	4.89	4.56	4.32	4.00	3.67	3.29	2.87
16	8.53	6.23	5.29	4.77	4.44	4.20	3.89	3.55	3.18	2.75
17	8.40	6.11	5.18	4.67	4.34	4.10	3.79	3.45	3.08	2.65
18	8.28	6.01	5.09	4.58	4.25	4.01	3.71	3.37	3.00	2.57
19	8.18	5.93	5.01	4.50	4.17	3.94	3.63	3.30	2.92	2.49
20	8.10	5.85	4.94	4.43	4.10	3.87	3.56	3.23	2.86	2.42
21	8.02	5.78	4.87	4.37	4.04	3.81	3.51	3.17	2.80	2.36
22	7.94	5.72	4.82	4.31	3.99	3.76	3.45	3.12	2.75	2.31
23	7.88	5.66	4.76	4.26	3.94	3.71	3.41	3.07	2.70	2.26
24	7.82	5.61	4.72	4.22	3.90	3.67	3.36	3.03	2.66	2.21
25	7.77	5.57	4.68	4.18	3.86	3.63	3.32	2.99	2.62	2.17
26	7.72	5.53	4.64	4.14	3.82	3.59	3.29	2.96	2.58	2.13
27	7.68	5.49	4.60	4.11	3.78	3.56	3.26	2.93	2.55	2.10
28	7.64	5.45	4.57	4.07	3.75	3.53	3.23	2.90	2.52	2.06
29	7.60	5.42	4.54	4.04	3.73	3.50	3.20	2.87	2.49	2.03
30	7.56	5.39	4.51	4.02	3.70	3.47	3.17	2.84	2.47	2.01
40	7.31	5.18	4.31	3.83	3.51	3.29	2.99	2.66	2.29	1.80
60	7.08	4.98	4.13	3.65	3.34	3.12	2.82	2.50	2.12	1.60
120	6.85	4.79	3.95	3.48	3.17	2.96	2.66	2.34	1.95	1.38
∞	6.64	4.60	3.78	3.32	3.02	2.80	2.51	2.18	1.79	1.00

Values of n_1 and n_2 represent the degrees of freedom associated with the between and within estimates of variance, respectively.

Using Statistics: Ideas for Research Projects

This appendix presents outlines for four research projects, each of which requires you to use SPSS to analyze the 2008 General Social Survey, the database that has been used throughout this text. The research projects should be completed at various intervals during the course, and each project permits a great deal of choice on the part of the student. The first project stresses description and should be done after completing Chapters 2–4. The second involves estimation and should be completed in conjunction with Chapter 7. The third project uses inferential statistics and should be done after completing Part II, and the fourth project combines inferential statistics with measures of association (with an option for multivariate analysis) and should be done after Part III (or IV).

Project 1—Descriptive Statistics

1. Select five variables from the 2008 General Social Survey *other than* those you have used for end-of-chapter exercises *(NOTE: Your instructor may specify a different number of variables)* and use the **Frequencies** command to get frequency distributions and summary statistics for each variable. Click the **Statistics** button on the **Frequencies** command window and request the mean, median, mode, standard deviation, and range. Make a note of all relevant information when it appears on screen or make a hard copy. See Appendix G for a list of variables available in the 2008 GSS.

2. For each variable, get bar or line charts to summarize the overall shape of the distribution of the variable.

3. Inspect the frequency distributions and graphs and choose appropriate measures of central tendency and, for ordinal- and interval-ratio–level variables, dispersion. Also, for interval-ratio and ordinal variables with many scores, check for skew both by using the line chart and by comparing the mean and median. Write a sentence or two of description for each variable, being careful to include a description of the overall shape of the distribution (see Chapter 2), the central tendency (Chapter 3), and the dispersion (Chapter 4). For nominal- and ordinal-level variables, be sure to explain any arbitrary numerical codes. For example, on the variable *class* in the 2008 GSS (see Appendix G), a 1 is coded as "lower class," a 2 indicates "working class," and so forth. This is an ordinal-level variable, so you might choose to report the median as a measure of central tendency. If the median score on *class* were 2.45, for example, you might place that value in context by reporting that "the median is 2.45, about halfway between 'working class' and 'middle class.'"

4. Here are examples of *minimal* summary sentences, using fictitious data: For a *nominal-level variable* (e.g., gender), report the mode and some detail about the overall distribution. For example: "Most respondents were married (57.5%), but divorced (17.4%) and single (21.3%) individuals were also common." *For an ordinal-level variable* (e.g., occupational prestige), use the median (and perhaps the mean and mode) and the range. For example: "The median prestige score was 44.3, and the range extended from 34 to 87. The most common score was 42 and the average score was 40.8."

For an *interval-ratio level variable* (e.g., age), use the mean (and perhaps the median or mode) and the standard deviation (and perhaps the range). For example: "Average age for this sample was 42.3. Respondents ranged from 18 to 94 years of age with a standard deviation of 15.37."

Project 2: Estimation

In this exercise, you will use the 2008 GSS sample to estimate characteristics of the U.S. population. You will use SPSS to generate the sample statistics and then use either Formula 7.2 or Formula 7.3 to find the confidence interval and state each interval in words.

A. ESTIMATING MEANS

1. There are relatively few interval-ratio variables in the 2008 GSS, and for this part of the project you may use ordinal variables that have *at least* three categories or scores. Choose a total of three variables that fit this description *other than* the variables you used in Chapter 7. *(NOTE: Your instructor may specify a different number of variables.)*

2. Use the **Descriptives** command to get means, standard deviations, and sample size (N), and use this information to construct 95% confidence intervals for each of your variables. Make a note of the mean, standard deviation, and sample size or keep a hard copy. Use Formula 7.2 to compute the confidence intervals. Repeat this procedure for the remaining variables.

3. For each variable, write a summary sentence reporting the variable, the interval itself, the confidence level, and the sample size. Write in plain English, as if you were reporting results in a newspaper. Most importantly, you should make it clear that you are estimating characteristics of the population of the entire United States. For example, a summary sentence might look like this: "Based on a random sample of 1231, I estimate at the 95% level that U.S. drivers average between 64.46 and 68.22 miles per hour when driving on interstate highways."

B. ESTIMATING PROPORTIONS

1. Choose three variables that are nominal or ordinal *other than* the variables you used in Chapter 7. *(NOTE: Your instructor may specify a different number of variables.)*

2. Use the **Frequencies** command to get the percentage of the sample in the various categories of each variable. Change the percentages (remember to

use the "valid percents" column) to proportions and construct confidence intervals for one category of each variable (e.g., the percent female for *sex*) using Formula 7.3.

3. For each variable, write a summary sentence reporting the variable, the interval, the confidence level, and the sample size. Write in plain English, as if you were reporting results in a newspaper. Remember to make it clear that you are estimating a characteristic of the United States population.

4. For any one of the intervals you constructed in either Part A or Part B, identify each of the following concepts and terms and briefly explain their role in estimation: sample, population, statistic, parameter, EPSEM, representative, confidence level.

Project 3: Significance Testing

In this exercise, you will use the 2008 GSS sample to conduct tests of significance. You will use the five-step model to conduct the tests.

A. TWO-SAMPLE *t* TEST (CHAPTER 9)

1. Choose two different dependent variables from the interval-ratio or ordinal variables that have three or more scores. Choose independent variables that might logically be a cause of your dependent variables. Remember that, for a *t* test, independent variables can have *only* two categories, but you can still use independent variables with more than two categories by (a) using the **Grouping Variable** box to specify the exact categories (e.g., select scores of 1 and 5 on *marital* to compare married with never-married respondents), or (b) collapsing the scores of variables with more than two categories by using the **Recode** command. Independent variables can be any level of measurement, and you may use the same independent variable for both tests.

2. Click **Analyze**, **Compare Means**, and then **Independent Samples T Test**. Name your dependent variable(s) in the **Test Variable** window and your independent variable in the **Grouping Variable** window. You will also need to specify the scores used to define the groups on the independent variable. See Chapter 9 for examples. Make a note of the test results (group means, obtained *t* score, significance, sample size) or keep a hard copy. Repeat the procedure for the second dependent variable.

3. Write up the results of the test. See Reading Statistics 5 for some ideas about how to report results. At a minimum, your report should clearly identify the independent and dependent variables, the sample statistics, the value of the test statistic (step 4), the results of the test (step 5), and the alpha level you used.

B. ANALYSIS OF VARIANCE (CHAPTER 10)

1. Choose two different dependent variables from the interval-ratio or ordinal variables that have three or more scores. Choose independent variables that might logically be a cause of your dependent variables and that have between

three and five categories. You may use the same independent variables for both tests. Do not use the same variables you used in Project A.

2. Click **Analyze**, **Compare Means**, and then **One-way Anova**. The **One-way Anova** window will appear. Find your dependent variable in the variable list on the left and click the arrow to move the variable name into the **Dependent List** box. Note that you can request more than one dependent variable at a time. Next, find the name of your independent variable and move it to the **Factor** box. Click **Options** and then click the box next to **Descriptive** in the Statistics box to request means and standard deviations. Click **Continue** and **OK**. Make a note of the test results or keep a hard copy. Repeat, if necessary, for your second dependent variable.

3. Write up the results of the test. At a minimum, your report should clearly identify the independent and dependent variables, the sample statistics (category means), the value of the test statistic (step 4), the results of the test (step 5), the degrees of freedom, and the alpha level you used.

C. CHI SQUARE (CHAPTER 11)

1. Choose two different dependent variables of any level of measurement that have five or fewer (preferably two to three) scores. For each dependent variable, choose an independent variable that might logically be a cause. Independent variables can be any level of measurement as long as they have five or fewer (preferably two to three) categories. Output will be easier to analyze if you use variables with few categories. You may use the same independent variable for both tests.

2. Click **Analyze**, **Descriptive Statistics**, and then **Crosstabs**. The **Crosstabs** dialog box will appear. Highlight your first dependent variable and move it into the **Rows** box. Next, highlight your independent variable and move it into the Columns box. Click the **Statistics** button at the bottom of the window and click the box next to chi square. Click **Continue** and **OK**. Make a note of the results or a get a hard copy. Repeat for your second dependent variable.

3. Write up the results of the test. At a minimum, your report should clearly identify the independent and dependent variables, the value of the test statistic (step 4), the results of the test (step 5), the degrees of freedom, and the alpha level you used. It is almost always desirable to report the column percentages as well.

Project 4: Analyzing the Strength and Significance of Relationships

A. USING BIVARIATE TABLES

1. From the 2008 GSS data set, select either

   a. One dependent variable and three independent variables (possible causes)

   or

b. One independent variable and three possible dependent variables (possible effects).

Variables can be from any level of measurement but must have only a few (two to five) categories or scores. Develop research questions or hypotheses about the relationships between variables. Make sure the causal links you suggest are sensible and logical.

2. Use the **Crosstabs** procedure to generate bivariate tables. See Chapters 12 and 13 for examples. Click **Analyze**, **Descriptive Statistics**, and **Crosstabs** and place your dependent variable(s) in the rows and independent variable(s) in the columns. On the **Crosstabs** dialog box, click the **Statistics** button and choose chi square, phi or Cramer's *V*, and gamma for every table you request. On the **Crosstabs** dialog box, click the **Cells** button and get column percentages for every table you request. Make a note of results as they appear on the screen or get hard copies.

3. Write a report that presents and analyzes these relationships. Be clear about which variables are dependent and which are independent. For each combination of variables, report the test of significance and measure of association. In addition, for each relationship, report and discuss column percentages, pattern or direction of the relationship, and strength of the relationship.

4. *OPTIONAL MULTIVARIATE ANALYSIS:* Pick one of the bivariate relationships you produced in step 2 and find a logical control variable for this relationship. Run **Crosstabs** for the bivariate relationship again while controlling for the third variable. Compare the partial tables with each other and with the bivariate table. Is the original bivariate relationship direct? Is there evidence of a spurious or intervening relationship? Do the variables have an interactive relationship? Write up the results of this analysis and include them in your summary paper for this project.

B. USING INTERVAL-RATIO VARIABLES

1. From the 2008 GSS, select either

   a. One dependent variable and three independent variables (possible causes)

   or

   b. One independent variable and three dependent variables (possible effects).

   Variables should be interval-ratio in level of measurement, but you may use ordinal-level variables as long as they have more than three scores. Develop research questions or hypotheses about the relationships between variables. Make sure the causal links you suggest are sensible and logical.

2. Use the **Regression** and **Scatterplot** (click Graphs and then Scatter) procedures to analyze the bivariate relationships. Make a note of results (including r, r^2, slope, beta-weights, and a) as they appear on the screen or get hard copies.

3. Write a report that presents and analyzes these relationships. Be clear about which variables are dependent and which are independent. For each

combination of variables, report the significance of the relationship (if relevant) and the strength and direction of the relationship. Include r, r^2, and the beta-weights in your report.

4. *OPTIONAL MULTIVARIATE ANALYSIS:* Pick one of the bivariate relationships you produced in step 2 and find another logical independent variable. Run **Regression** again with both independent variables and analyze the results. How much improvement is there in the explained variance after the second independent variable is included? Write up the results of this analysis and include them in your summary paper for this project.

An Introduction to *SPSS for Windows*

Computers have affected virtually every aspect of human society, and, as you would expect, their impact on the conduct of social research has been profound. Researchers routinely use computers to organize data and compute statistics—activities that humans often find dull, tedious, and difficult but that computers accomplish with accuracy and ease. This division of labor allows social scientists to spend more time on analysis and interpretation—activities that humans typically enjoy but that are beyond the power of computers (so far, at least).

These days, the skills needed to use computers successfully are quite accessible, even for people with little or no experience. This appendix will prepare you to use a statistics program called *SPSS for Windows* (SPSS stands for Statistical Package for the Social Sciences). If you have used a mouse to "point and click" and run a computer program, you are ready to learn how to use this program. Even if you are completely unfamiliar with computers, you will find this program accessible. After you finish this appendix, you will be ready to do the exercises found at the end of most chapters of this text.

A word of caution before we begin: This appendix is intended only as an *introduction* to SPSS. It will give you an overview of the program and enough information so that you can complete the assignments in the text. It is unlikely, however, that this appendix will answer all your questions or provide solutions to all the problems you might encounter. So this is a good place to tell you that SPSS has an extensive and easy-to-use "Help" facility that will provide assistance as you request it. You should familiarize yourself with this feature and use it as needed. To get help, simply click on the **Help** command on the toolbar across the top of the screen.

SPSS is a **statistical package** (or **statpak**), that is, a set of computer programs that work with data and compute statistics as requested by the user (you). Once you have entered the data for a particular group of observations, you can easily and quickly produce an abundance of statistical information without doing any computations or writing any computer programs yourself.

Why bother to learn this technology? The truth is that the laborsaving capacity of computers is sometimes exaggerated, and there are research situations in which they are unnecessary. If you are working with a small number of observations or need only a few, uncomplicated statistics, then statistical packages are probably not going to be helpful. However, as the number of cases increases and as your requirements for statistics become more sophisticated, computers and statpaks will become more and more useful.

An example should make this point clearer. Suppose you have gathered a sample of 150 respondents and the *only* thing you want to know about these people is their average age. To compute an average, as you know, you add the scores and divide by the number of cases. How long do you think it would take you to add 150 two-digit numbers (ages) with a hand calculator? If you entered the scores at the rate of one per second—60 scores a minute—it would take about 3 or 4 minutes to enter the ages and get the average. Even if you worked slowly and carefully and did the addition a second and third time to check your math, you could probably complete all calculations in less than 15 or 20 minutes. If this were all the information you needed, computers and statpaks would not save you any time.

Such a simple research project is not very realistic, however. Typically, researchers deal with not one but scores or even hundreds of variables, and samples have hundreds or thousands of cases. While you could add 150 numbers in perhaps 3 or 4 minutes, how long would it take to add the scores for 1500 cases? What are the chances of adding 1500 numbers without making significant errors of arithmetic? The more complex the research situation, the more valuable and useful statpaks become. SPSS can produce statistical information in a few keystrokes or clicks of the mouse that might take you minutes, hours, or even days to produce with a hand calculator.

Clearly, this is technology worth mastering by any social researcher. With SPSS, you can avoid the drudgery of mere computation, spend more time on analysis and interpretation, and conduct research projects with very large data sets. Mastery of this technology might be very handy indeed in your senior-level courses, in a wide variety of jobs, or in graduate school.

Getting Started—Databases and Computer Files

Before statistics can be calculated, SPSS must have some data to process. A **database** is an organized collection of related information, such as the responses to a survey. For purposes of computer analysis, a database is organized into a **file**—a collection of information that is stored under the same name in the memory of the computer, on a disk or flash drive, or on some other medium. Words as well as numbers can be saved in files. If you've ever used a word processing program to type a letter or term paper, you probably saved your work in a file so that you could update or make corrections at a later time. Data can be stored in files indefinitely. Since it can take months to conduct a thorough data analysis, the ability to save a database is another advantage of using computers.

For the SPSS exercises in this text, we will use a database that contains some of the results of the General Social Survey (GSS) for 2008. This database contains the responses of a sample of adult Americans to questions about a wide variety of social issues. The GSS has been conducted regularly since 1972 and has been the basis for hundreds of research projects by professional social researchers. It is a rich source of information about public opinion in the United States and includes data on everything from attitudes about abortion to opinions on assisted suicide.

The GSS is especially valuable because the respondents are chosen so that the sample as a whole is representative of the entire U.S. population. A representative sample reproduces, in miniature form, the characteristics of the population from which it was taken (see Chapters 6 and 7). So when you analyze the 2008 General Social Survey database, you are in effect analyzing U.S. society as of 2008. The data are real, and the relationships you will analyze reflect some of the most important and sensitive issues in American life.

The complete General Social Survey for 2008 includes hundreds of items of information (age, sex, opinion about such social issues as capital punishment, and so forth) for thousands of respondents. Some of you will be using a student version of *SPSS for Windows*, which is limited in the number of cases and variables it can process. To accommodate those limits, I have reduced the database to 50 items of information and fewer than 1500 respondents.

The GSS data file is summarized in Appendix G. Please turn to this appendix and familiarize yourself with it. Note that the variables are listed alphabetically by their variable names. In SPSS, the names of variables must be no more than eight characters long. In many cases, the resultant need for brevity is not a problem, and variable names (e.g., *age*) are easy to figure out. In other cases, the eight-character limit necessitates extreme abbreviation, and some variable names (like *abany* or *fefam*) are not so obvious. Appendix G also shows the wording of the item that generated the variable. For example, the *abany* variable consists of responses to a question about legal abortion: Should it be possible for a woman to have an abortion for "any reason"? Note that the variable name is formed from the question: Should an *ab*ortion be possible for *any* reason? The *fefam* variable consists of responses to the statement "It is much better for everyone involved if the man is the achiever outside the home and the woman takes care of the home and family." Appendix G is an example of a code book for the database since it lists all the codes (or scores) for the survey items along with their meanings.

Notice that some of the possible responses to *abany*, *fefam*, and other variables in Appendix G are labeled IAP ("Not Applicable"), NA, or DK. The first of these responses means that the item in question was not given to a respondent. The full GSS is very long, and, to keep the time frame for completing the survey reasonable, not all respondents are asked every question. NA stands for "No Answer" and means that the respondent was asked the question but refused to answer. DK stands for "Don't Know," which means that the respondent did not have the requested information. All three of these scores are *Missing Values*, and, as "non-information," they should be eliminated from statistical analysis. Missing values are common on surveys, and, as long as they are not too numerous, they are not a particular problem.

It's important that you understand the difference between a statpak (SPSS) and a database (the GSS) and what we are ultimately after here. A database consists of information. A statpak processes the information in the database and produces statistics. Our goal is to apply the statpak to the database to produce output (for example, statistics and graphs) that we can analyze and use to answer questions. The process might be diagrammed as in Figure F.1.

Statpaks like SPSS are general research tools that can be used to analyze databases of all sorts; they are not limited to the 2008 GSS. In the same way, the

FIGURE F.1 **The Data Analysis Process**

DataBase	→	**Statpak**	→	**Output**	→	**Analysis**
(raw information)		(computer programs)		(statistics and graphs)		(interpretation)

2008 GSS could be analyzed with statpaks other than those used in this text. Other widely used statpaks include Microcase, SAS, and Stata—each of which may be available on your campus.

Starting *SPSS for Windows* and Loading the 2008 GSS

If you are using the complete, professional version of *SPSS for Windows*, you will probably be working in a computer lab, and you can begin running the program immediately. If you are using the student version of the program on your personal computer, the first thing you need to do is install the software. Follow the instructions that came with the program and return to this appendix when the installation is complete.

To start *SPSS for Windows*, find the icon (or picture) on the screen of your monitor that has an "SPSS" label attached to it. Use the computer mouse to move the arrow on the monitor screen over this icon and then double-click the left button on the mouse. This will start up the SPSS program.

After a few seconds the *SPSS for Windows* screen will appear and ask, at the top of the screen, "What would you like to do?" Of the choices listed, the button next to "Open an existing file" will be checked (or preselected), and there will probably be a number of data sets listed in the window at the bottom of the screen. Find the 2008 General Social Survey data set, probably labeled *GSS2008.sav* or something similar. If the data set is on a disk or flash drive, you will need to specify the correct drive. Check with your instructor to make sure you know where to find the 2008 GSS.

Once you've located the data set, click on the name of the file with the left-hand button on the mouse, and SPSS will load the data. The next screen you will see is the SPSS Data Editor screen.

Note that there is a list of commands across the very top of the screen. These commands begin with **File** at the far left and end with **Help** at the far right. This is the main menu bar for SPSS. When you click any of these words, a **menu** of commands and choices will drop down. Basically, you tell SPSS what to do by clicking on your desired choices from these menus. Sometimes submenus will appear, and you will need to specify your choices further.

SPSS provides the user with a variety of options for displaying information about the data file and output on the screen. I recommend that you tell the program to display lists of variables by name (e.g., *age, abany*) rather than by labels (e.g., AGE OF RESPONDENT, ABORTION IF WOMAN WANTS FOR ANY REASON). Lists displayed this way will be easier to read and compare to Appendix G. To do this, click **Edit** on the main menu bar and then click **Options** from the drop-down submenu. A dialog box labeled "Options" will appear with a series of "tabs" along the top. The "General" options should be displayed; if not, click on this tab. On the "General" screen, find the box labeled "Variable Lists" and, if they are not already selected, click "Display names" and "alphabetical" and then click **OK**. If you make changes, a message may appear on the screen that tells you that changes will take effect the next time a data file is opened.

In this section, you learned how to start up *SPSS for Windows,* load a data file, and set some of the display options for this program. These procedures are summarized in Table F.1.

TABLE F.1   Summary of Commands

To start *SPSS for Windows*	Click the SPSS icon on the screen of the computer monitor.
To open a data file	Double-click on the data file name.
To set display options for lists of variables	Click **Edit** from the main menu bar, then click **Options**. On the "General" tab make sure that "Display names" and "alphabetical" are selected and then click **OK**.

Working with Databases

Note that in the SPSS Data Editor Window the data are organized into a two-dimensional grid, with columns running up and down (vertically) and rows running across (horizontally). Each column is a variable or item of information from the survey. The names of the variables are listed at the tops of the columns. Remember that you can find the meaning of these variable names in the GSS 2008 code book in Appendix G.

Another way to decipher the meaning of variable names is to click **Utilities** on the menu bar and then click **Variables**. The **Variables** window opens. This window has two parts. On the left is a list of all variables in the database, arranged in alphabetical order and with the first variable highlighted. On the right is the **Variable Information** window, with information about the highlighted variable. The first variable is listed as *abany*. The **Variable Information** window displays a fragment of the question that was actually asked during the survey ("ABORTION IF WOMAN WANTS FOR ANY REASON") and shows the possible scores on this variable (a score of 1 = yes and a score of 2 = no), along with some other information.

The same information can be displayed for any variable in the data set. For example, find the variable *marital* in the list. You can do this by using the arrow keys on your keyboard or the slider bar on the right of the variable list window. You can also move through the list by typing the first letter of the variable name you are interested in. For example, type "m" and you will be moved to the first variable name in the list that begins with that letter. Now you can see that the variable measures marital status and that a score of "1" indicates that the respondent was married, and so forth. What do *prestg80* and *marhomo* measure? Close this window by clicking the **Close** button at the bottom of the window.

Examine the window displaying the 2008 GSS a little more. Each row of the window (reading across, or from left to right) contains the scores of a particular respondent on all the variables in the database. Note that the upper-left-hand cell is highlighted (outlined in a darker border than the other cells). This cell contains the score of respondent 1 on the first variable. The second row contains the scores of respondent 2, and so forth. You can move around in this window with the arrow keys on your keyboard. The highlight moves in the direction of the arrow, one cell at a time.

In this section, you learned to read information in the data display window and to decipher the meaning of variable names and scores. These commands are summarized in Table F.2. We are now prepared to perform some statistical operations with the 2008 GSS database.

TABLE F.2 Summary of Commands

To move around in the Data Editor window	1. Click the cell you want to highlight or 2. Use the arrow keys on your keyboard or 3. Move the slider buttons or 4. Click the arrows on the right-hand and bottom margins.
To get information about a variable	1. From the menu bar, click Utilities and then click Variables. Scroll through the list of variable names until you highlight the name of the variable in which you are interested. Variable information will appear in the window on the right. 2. See Appendix G.

Putting SPSS to Work: Producing Statistics

At this point, the database on the screen is just a mass of numbers with little meaning for you. That's okay because you will not actually have to read any information from this screen. Virtually all of the statistical operations you will conduct will begin by clicking the **Analyze** command from the menu bar, selecting a procedure and statistics, and then naming the variable or variables you would like to process.

To illustrate, let's have *SPSS for Windows* produce a frequency distribution for the variable *sex*. Frequency distributions are tables that display the number of times each score of a variable occurred in the sample (see Chapter 2). So when we complete this procedure, we will know the number of males and females in the 2008 GSS sample.

With the 2008 GSS loaded, begin by clicking the **Analyze** command on the menu bar. From the menu that drops down, click **Descriptive Statistics** and then **Frequencies**. The Frequencies window appears, with the variables listed in alphabetical order in the box on the left. The first variable (*abany*) will be highlighted. Use the slider button or the arrow keys on the right-hand margin of this box to scroll through the variable list until you highlight the variable *sex*, or type "s" to move to the approximate location.

Once the variable you want to process has been highlighted, click the arrow button in the middle of the screen to move the variable name to the box on the right-hand side of the screen. SPSS will produce frequency distributions for all variables listed in this box, but for now we will confine our attention to *sex*. Click the **OK** button in the upper-right-hand corner of the Frequencies window and, in seconds, a frequency distribution will be produced.

SPSS sends all tables and statistics to the Output window, or SPSS viewer. This window is now "closest" to you, and the Data Editor window is "behind" the Output window. If you want to return to the Data Editor, click on any part of it if it is visible, and it will move to the "front" and the Output window will be "behind" it. To display the Data Editor window if it is not visible, minimize the Output window by clicking the "-" box in the upper-right-hand corner.

Frequencies. The output from SPSS, slightly modified, is reproduced as Table F.3. What can we tell from this table? The score labels (male and female) are printed at the left, with the number of cases (frequency) in each category of

TABLE F.3   An Example of SPSS Output (respondent's sex)

		Frequency	Percent	Valid Percent	Cumulative Percent
Valid	MALE	683	46.0	46.0	46.0
	FEMALE	802	54.0	54.0	100.0
	Total	1485	100.0	100.0	

the variable one column to the right. As you can see, there are 644 males and 782 females in the sample. The next two columns give information about percentages and the last column to the right displays cumulative percentages. We will defer a discussion of this last column until a later exercise.

One of the percentage columns is labeled Percent and the other is labeled Valid Percent. The difference between these two columns lies in the handling of missing values. The Percent column is based on all cases, including people who did not respond to the item (NA) and people who said they did not have the requested information (DK). The Valid Percent column excludes all missing scores. Since we will almost always want to ignore missing scores, we will pay attention only to the Valid Percent column. Note that for sex, there are no missing scores (gender was determined by the interviewer), and the two columns are identical.

Printing and Saving Output

Once you've gone to the trouble of producing statistics, a table, or a graph, you will probably want to keep a permanent record. There are two ways to do this. First, you can print a hard copy of the contents of the Output Window to take with you. To do this, click on **File** and then click **Print** from the **File** menu. Alternatively, find the icon of a printer (third from the left) in the row of icons just below the menu bar and click on it.

The other way to create a permanent record of SPSS output is to save the Output window to the computer's memory or to a diskette. To do this, click **Save** from the **File** menu. The Save dialog box opens. Give the output a name (some abbreviation such as "freqsex" might do) and, if necessary, specify the name of the drive in which your diskette is located. Click **OK**, and the table will be permanently saved.

Ending Your *SPSS for Windows* Session

Once you have saved or printed your work, you may end your SPSS session. Click on **File** from the menu bar and then click **Exit**. If you haven't already done so, you will be asked if you want to save the contents of the Output window. You may save the frequency distribution at this point if you wish. Otherwise, click **NO**. The program will close, and you will be returned to the screen from which you began.

Code Book for the General Social Survey, 2008

The General Social Survey (GSS) is a public opinion poll that has been regularly conducted by the National Opinion Research Council. A version of the 2008 GSS is available at the Web site for this text and is used for all end-of-chapter exercises. Our version of the 2008 GSS includes 50 variables for a randomly selected subsample of about 1500 of the original respondents. This code book lists each item in the data set. The variable names are those used in the data files. The questions have been reproduced exactly as they were asked (with a few exceptions to conserve space), and the numbers beside each response are the scores recorded in the data file.

The data set includes variables that measure demographic or background characteristics of the respondents, including sex, age, race, religion, and several indicators of socioeconomic status. Also included are items that measure opinion on such current and controversial topics as abortion, capital punishment, and homosexuality.

Most variables in the data set have codes for "missing data." These codes are italicized in the listings below for easy identification. The codes refer to various situations in which the respondent does not or cannot answer the question and are excluded from all statistical operations. The codes are: IAP (the respondent was not asked the question), DK ("Don't know," that is, the respondent didn't have the requested information), and NA ("No answer," that is, the respondent refused to answer).

Please tell me if you think it should be possible for a woman to get a legal abortion if . . .

abany	She wants it for any reason.
	1. Yes
	2. No
	0 IAP, 8. DK, 9. NA

| abnomore | She is married and wants no more children |
| | (Same scoring as abany) |

age	Age of respondent
	18–89 Actual years
	99 NA

attend How often do you attend religious services?

0. Never	1. Less than once per year
2. Once or twice a year	3. Several times per year
4. About once a month	5. 2–3 times a month
6. Nearly every week	7. Every week
8. Several times a week	
9. DK or NA	

cappun   Do you favor or oppose the death penalty for persons convicted of murder?
1. Favor
2. Oppose
0. IAP, 8. DK, 9. NA

class   Subjective class identification
1. Lower class
2. Working class
3. Middle class
4. Upper class
0. IAP, 8. DK, 9. NA

concong   How much confidence do you have in the U.S. Congress?
1. Complete confidence
2. A great deal of confidence
3. Some confidence
4. Very little confidence
5. No confidence at all
0. IAP, 8. DK, 9. NA

degree   Respondent's highest degree
0. Less than high school
1. High school
2. Assoc./junior college
3. Bachelor's
4. Graduate
7. IAP, 8. DK, 9. NA

discaff   What do you think the chances are these days that a white person won't get a job or promotion while an equally or less qualified black person gets one instead? Is this:
1. Very likely
2. Somewhat likely
3. Not very likely
0. IAP, 8. DK, 9. NA

educ   Highest year of school completed
0–20. Actual number of years
97. IAP, 98. DK, 99. NA

evolved   Human beings, as we know them today, developed from earlier species of animals.
1. True
2. False
0. IAP, 8. DK, 9. NA

fatalism   People can't change the course of their lives.
1. Strongly agree
2. Agree
3. Neither agree nor disagree
4. Disagree
5. Strongly disagree
0. IAP, 8. DK, 9. NA

fear | Is there any area right around here—that is, within a mile—where you would be afraid to walk alone at night?
> 1. Yes
> 2. No
> 0. *IAP*, 8. *DK*, 9. *NA*

fechld | A working mother can establish just as warm and secure a relationship with her children as a mother who does not work.
> 1. Strongly agree
> 2. Agree
> 3. Disagree
> 4. Strongly disagree
> 0. *IAP*, 8. *DK*, 9. *NA*

fefam | It is much better for everyone involved if the man is the achiever outside the home and the woman takes care of the home and family.
> 1. Strongly agree
> 2. Agree
> 3. Disagree
> 4. Strongly disagree
> 0. *IAP*, 8. *DK*, 9. *NA*

goodlife | The way things are in America, people like me and my family have a good chance of improving our standard of living—do you agree or disagree?
> 1. Strongly agree
> 2. Agree
> 3. Neither agrees or disagree
> 4. Disagree
> 5. Strongly disagree
> 0. *IAP*, 8. *Can't choose*, 9. *NA*

happy | Taken all together, how would you say things are these days—would you say that you are very happy, pretty happy, or not too happy?
> 1. Very happy
> 2. Pretty happy
> 3. Not too happy
> 0. *IAP*, 8. *DK*, 9. *NA*

helpful | Would you say that most of the time people try to be helpful, or that they are mostly just looking out for themselves?
> 1. Try to be helpful
> 2. Just look out for themselves
> 0. *IAP*, 3. *Depends (volunteered by respondent)* 8. *DK*, 9. *NA*

hlthcovr | Do you currently have health insurance coverage?
> 1. Yes
> 2. No
> 0. *IAP*, 8. *DK*, 9. *NA*

hrs1 | How many hours did you work last week?
> 1–89. Actual hours
> −1. *IAP*, 98. *DK*, 99. *NA*

income06

Respondent's total family income from all sources

1. Less than 1000	2. 1000 to 2999
3. 3000 to 3999	4. 4000 to 4999
5. 5000 to 5999	6. 6000 to 6999
7. 7000 to 7999	8. 8000 to 9999
9. 10,000 to 12,499	10. 12,500 to 14,999
11. 15,000 to 17,499	12. 17,500 to 19,999
13. 20,000 to 22,499	14. 22,500 to 24,999
15. 25,000 to 29,999	16. 30,000 to 34,999
17. 35,000 to 39,999	18. 40,000 to 49,999
19. 50,000 to 59,999	20. 60,000 to 74,999
21. 75,000 to 89,999	22. 90,000 to 109,999
23. 110,000 to 129,999	24. 130,000 to 149,999
25. 150,000 or more	
98. DK, 99. NA	

intrhome

Do you have access to the Internet in your home?
1. Yes
2. No
0. IAP, 8. DK, 9. NA

letin1

Do you think the number of immigrants to America nowadays should be:
1. Increased a lot
2. Increased a little
3. Remain the same as it is
4. Reduced a little
5. Reduced a lot
0. IAP, 8. DK, 9. NA

losejb12

Thinking now about the next 12 months, how likely do you think it is that you will lose your job or be laid off for reasons other than for your performance?
1. Very likely
2. Fairly likely
3. Not too likely
4. Not at all likely
0. IAP, 8. DK, 9. NA

marblk

What about a close relative marrying a black person? Would you be:
1. Strongly in favor
2. In favor
3. Neither favor nor oppose
4. Oppose
5. Strongly oppose
0. IAP, 8. DK, 9. NA

marhisp

What about a close relative marrying an Hispanic American person? Would you be: (Same scoring as marblk)

marhomo

Homosexual couples should have the right to marry one another.
1. Strongly agree
2. Agree

3. Neither agree or disagree

4. Disagree

5. Strongly disagree

0. IAP, 8. DK, 9. NA

marital   Are you currently married, widowed, divorced, separated, or have you never been married?

1. Married

2. Widowed

3. Divorced

4. Separated

5. Never married

9. NA

news   How often do you read the newspaper?

1. Every day

2. A few times a week

3. Once a week

4. Less than once a week

5. Never

0. IAP, 8. DK, 9. NA

obey   How important is it for a child to learn obedience to prepare him or her for life?

1. Most important

2. 2nd most important

3. 3rd most important

4. 4th most important

5. Least important

0. IAP, 8. DK, 9. NA

paeduc   Father's highest year of school completed

0–20. Actual number of years

97. IAP, 98. DK, 99. NA

papres80   Prestige of father's occupation

17–86. Actual score

0. IAP, DK, NA

partnrs5   How many sex partners have you had over the past five years?

0. No partners	1. 1 partner
2. 2 partners	3. 3 partners
4. 4 partners	5. 5–10 partners
6. 11–20 partners	7. 21–100 partners
8. More than 100 partners	

9. 1 or more, don't know the number, 95 Several, 98 DK, 99 NA, −1. NAP

partyid   Generally speaking, do you usually think of yourself as a Republican, Democrat, Independent, or what?

0. Strong Democrat

1. Not very strong Democrat

2. Independent, close to Democrat

3. Independent

4. Independent, close to Republican

5. Not very strong Republican

6. Strong Republican

7. Other party, 8. DK, 9, NA

polviews   I'm going to show you a seven-point scale on which the political views that people might hold are arranged from extremely liberal to extremely conservative. Where would you place yourself on this scale?

1. Extremely liberal

2. Liberal

3. Slightly liberal

4. Moderate

5. Slightly conservative

6. Conservative

7. Extremely conservative

0. NAP, 8. DK, 9. NA

prayfreq   About how often do you pray?

1. Never

2. Less than once a year

3. About once or twice a year

4. Several times a year

5. About once a month

6. 2–3 times a month

7. Nearly every week

8. Every week

9. Several times a week

10. Once a day

11. Several times a day

0. IAP, 98. DK, 99. NA

premarsx   There's been a lot of discussion about the way morals and attitudes about sex are changing in this country. If a man and a woman have sex relations before marriage, do you think it is always wrong, almost always wrong, wrong only sometimes, or not wrong at all?

1. Always wrong

2. Almost always wrong

3. Wrong only sometimes

4. Not wrong at all

0. IAP, 8. DK, 9. NA

pres04   In 2004, did you vote for Kerry (the Democratic candidate) or Bush (the Republican candidate)? (Includes only those who said they voted in this election.)

1. Kerry

2. Bush

3. Other, 6. No presidential vote, 0. IAP, 8. DK, 9. NA

presg80   Prestige of respondent's occupation

17–86. Actual score

0. IAP, DK, NA

racecen1   Race or ethnicity of respondent
        1. White
        2. Black
        3. American Indian or Alaska Native
        4. Asian American or Pacific Islander
        5. Hispanic
        0. IAP, 98. DK

region   Region of interview
        1. New England (ME, VT, NH, MA, CT, RI)
        2. Mid-Atlantic (NY, NJ, PA)
        3. East North Central (WI, IL, IN, MI, OH)
        4. West North Central (MN, IA, MO, ND, SD, NE, KS)
        5. South Atlantic (DE, MY, WV, VA, NC, SC, GA, FL, DC)
        6. East South Central (KY, TN, AL, MS)
        7. West South Central (AK, OK, LA, TX)
        8. Mountain (MT, ID, WY, NV, UT, CO, AR, NM)
        9. Pacific (WA, OR, CA, AL, HI)

relig   What is your religious preference? Is it Protestant, Catholic, Jewish, some other religion, or no religion?
        1. Protestant
        2. Catholic
        3. Jewish
        4. None
        5. Other
        8. DK, 9. NA

satjob   All in all, how satisfied would you say you are with your job?
        1. Very satisfied
        2. Moderately satisfied
        3. A little dissatisfied
        4. Very dissatisfied
        0. IAP, 8. DK, 9. NA

sex   Respondent's gender
        1. Male
        2. Female

sexfreq   About how many times did you have sex during the last 12 months?
        0. Not at all
        1. Once or twice
        2. About once a month
        3. 2 or 3 times a month
        4. About once a week
        5. 2 or 3 times a week
        6. More than 3 times a week
        −1. IAP, 8. DK, 9. NA

spanking	Do you strongly agree, agree, disagree, or strongly disagree that it is sometimes necessary to discipline a child with a good, hard spanking?
	    1. Strongly agree
	    2. Agree
	    3. Disagree
	    4. Strongly disagree
	    0. IAP, 8. DK, 9. NA

spkmslm	There are always some people whose ideas are considered bad or dangerous by other people. Consider a Muslim clergyman who preached hatred of the United States . . . should he be allowed to speak or not?
	    1. Yes, allowed
	    2. Not allowed
	    0. IAP, 8. DK, 9. NA

suicide1	Do you think a person has the right to end his or her own life if this person has an incurable disease?
	    1. Yes
	    2. No
	    0. IAP, 8. DK, 9. NA

thnkself	How important is it for a child to learn to think for himself or herself to prepare him or her for life?
	    1. Most important
	    2. 2nd most important
	    3. 3rd most important
	    4. 4th most important
	    5. Least important
	    0. IAP, 8. DK, 9. NA

tvhours	On the average day, about how many hours do you personally watch television?
	    00–24. Actual hours
	    −1. IAP, DK, NA

Answers to Odd-Numbered End-of-Chapter Problems and Cumulative Exercises

In addition to answers, this section suggests problem-solving strategies and provides examples of how to phrase or interpret conclusions and answers. You should try to solve and interpret the problems on your own before consulting this section.

In solving these problems, I let my calculator or computer do most of the work. I worked with whatever level of precision these devices permitted and didn't round off until the end or until I had to record an intermediate sum. I always rounded off to two places of accuracy (that is, to two places beyond the decimal point, or to 100ths). If you follow these same conventions, your answers will match mine. However, you should realize that small, generally trivial discrepancies between your answer and mine still might occur. If the difference doesn't seem trivial, you should double-check to make sure you haven't made an error or solve the problem again using a greater degree of precision.

Finally, please allow me a brief disclaimer about mathematical or typographical errors in this section. Let me assure you, first of all, that I know how important this section is for most students and that I worked hard to be certain that these answers are correct. Human fallibility being what it is, however, I know that I cannot make absolute guarantees. Should you find any errors, please let me know so that I can make corrections in the future.

Chapter 1

1.5
a. Nominal
b. Ordinal (the categories can be ranked in terms of degree of honesty with "Returned the wallet with money" the "most honest")
c. Ordinal
d. Interval-ratio ("years" has equal intervals and a true zero point)
e. Interval-ratio
f. Interval-ratio
g. Nominal (the various patterns are different from each other but cannot be ranked from high to low)
h. Interval-ratio
i. Ordinal
j. Number of accidents: interval-ratio; Severity of accident: ordinal

1.7

	Variable	Level of Measurement	Type	Application
a.	Opinion	Ordinal	Discrete	Inferential
b.	Grade	Interval-ratio	Continuous	Descriptive (two variables)
c.	Party	Nominal	Discrete	Inferential
	Sex	Nominal	Discrete	
	Opinion	Ordinal	Discrete	
d.	Homicide rate	Interval-ratio	Continuous	Descriptive (two variables)
e.	Satisfaction	Ordinal	Discrete	Descriptive (one variable)

Chapter 2

2.1 **a.** Complex A: (5/20) × 100 = 25.00%
Complex B: (10/20) × 100 = 50.00%
b. Complex A: 4:5 = 0.80
Complex B: 6:10 = 0.60
c. Complex A: (0/20) = 0.00
Complex B: (1/20) = 0.05
d. (6/(4 + 6)) = (6/10) = 60.00%
e. Complex A: 8:5 = 1.60
Complex B: 2:10 = 0.20

2.3 Bank robbery rate = (47/211,732) × 100,000 = 22.20
Homicide rate = (13/211,732) × 100,000 = 6.14
Auto theft rate = (23/211,732) × 100,000 = 10.86

2.5 For sex:

Sex	Frequency
Male	9
Female	6
Total	15

For age, set $k = 10$. $R = 77 - 23$, or 54, so we can round off interval size to 5 ($i = 5$). The first interval will be 20–24, to include the low score of 23, and the highest interval will be 75–79.

Age	Frequency
20–24	1
25–29	2
30–34	3
35–39	2
40–44	1
45–49	3
50–54	1
55–59	1
60–64	0
65–69	0
70–74	0
75–79	1
	15

2.9 Set $k = 10$. $R = 92 - 5$, or 87, so set i at 10.

Score	Frequency
0–9	3
10–19	7
20–29	6
30–39	0

(continued next column)

(continued)

Score	Frequency
40–49	2
50–59	2
60–69	3
70–79	0
80–89	0
90–99	2
	25

2.11 Answers should note and describe the rise in all crime rates except burglary up to a peak in the early 1990s, followed by a dramatic decline and then a leveling off in recent years.

Chapter 3

3.1 "Region of birth" and "religion" are nominal-level variables, "support for legalization" and "opinion of food" are ordinal, and "expenses" and "number of movies" are interval-ratio. The mode, the most common score, is the only measure of central tendency available for nominal-level variables. For the two ordinal-level variables, *don't forget to array the scores from high to low* before locating the median. There are 10 freshmen (N is even), so the median for freshmen will be the score halfway between the scores of the two middle cases. There are 11 seniors (N is odd), so the median for seniors will be the score of the middle case. To find the mean for the interval-ratio variables, add the scores and divide by the number of cases.

Variable	Freshmen	Seniors
Region of birth	Mode = North	Mode = North
Legalization	Median = 3	Median = 5
Expenses	Mean = 48.50	Mean = 63.00
Movies	Mean = 5.80	Mean = 5.18
Food	Median = 6	Median = 4
Religion	Mode = Protestant	Mode = Protestant and None (4 cases each)

3.3

Variable	Level of Measurement	Measure of Central Tendency
Sex	Nominal	Mode = male
Social class	Ordinal	Median = "medium" (the middle case is in this category)

(continued next page)

(*continued*)

Variable	Level of Measurement	Measure of Central Tendency
Number of years in the party	I-R	Mean = 26.15
Education	Ordinal	Median = high school
Marital status	Nominal	Mode = married
Number of children	I-R	Mean = 2.39

3.5

Variable	Level of Measurement	Measure of Central Tendency
Marital status	Nominal	Mode = married
Race	Nominal	Mode = white
Age	I-R	Mean = 27.53
Attitude on abortion	Ordinal	Median = 7

3.7

Variable	Level of Measurement	Measure of Central Tendency
Sex	Nominal	There are 9 males and 6 females, so the mode is male
Support for gun control	Ordinal	Median = 1
Level of education	Ordinal	Median = 1
Age	I-R	Mean = 40.93

3.9 Attitude and opinion scales almost always generate ordinal-level data, so the appropriate measure of central tendency would be the median. The median is 9 for the students and 2 for the neighbors. Incidentally, the means are 7.80 for the students and 4.00 for the neighbors.

3.11 Mean = 380, median = 448.5. The lower value for the mean indicates a negative skew, or a few very *low* scores. For this small group of nations, the skew is caused by the score of China (10), which is much lower than the scores of the other seven nations.

3.13 To find the median, you must first rank the scores from high to low. Both groups have 25 cases (*N* = odd), so the median is the score of the 13th case. For freshman the median is 35, and for seniors the median is 30. The mean score for freshmen is 31.72. For seniors, the mean is 28.60.

3.15 The owners are using the mean while the players are citing the median. This distribution has a positive skew (the mean is greater than the median), as is typical for income data. Note the difference in wording in the two reports: The owners cite the "average" (mean) and the players cite the "typical player" (the score of the middle case or the median).

Chapter 4

4.1

Complex	IQV
A	0.89
B	0.99
C	0.71
D	0.74

Complex B has the highest IQV and is the most heterogeneous. Complex C is the least heterogeneous.

4.3 The high score is 50 and the low score is 10, so the range is 50 − 10, or 40. The standard deviation is 12.28.

4.5

Statistic	2000	2006
Mean	48,161.54	63,223.08
Median	47,300.00	59,000.00
Standard deviation	6753.64	10,845.50
Range	23,400.00	38,300.00

In this time period, the mean and median increase. The distributions for both years show a positive skew (the mean is greater than the median). The standard deviation and the range increase, indicating that there is greater variability in 2006.

4.7

Variable	Statistic	Males	Females
Labor force participation	Mean	77.60	58.40
	Standard deviation	2.73	6.73
% High school graduate	Mean	69.20	70.20
	Standard deviation	5.38	4.98
Mean income	Mean	33,896.60	29,462.40
	Standard deviation	4443.16	4597.93

Males and females are very similar in terms of educational level, but females are less involved in the labor force and, on average, earn almost $4500 less than males per year. The females in these 10 states are much more variable in their labor force participation but are similar to males in dispersion on the other two variables. See Section 9.6 for more on gender disparities in income.

4.9 $s = 244.87$

4.11 $R = 38 - 5$ or 33, $s = 8.98$. The score for Los Angeles (38) is higher than the other scores, so removing this score would reduce the variation in the data set. The standard deviation of the scores without Los Angeles would be lower in value.

4.13

Division	Range	Standard Deviation
A	$R = 18$	$s = 5.32$
B	$R = 8$	$s = 2.37$
C	$R = 3$	$s = 1.03$
D	$R = 25$	$s = 7.88$

4.15 The means are virtually the same, so the experimental course did not raise the average grade. However, the standard deviation for the experimental course is much lower, indicating that student grades were less diverse in these sections.

Chapter 5

5.1

X_i	Z Score	% Area Above	% Area Below
5	−1.67	95.25	4.75
6	−1.33	90.82	9.18
7	−1.00	84.13	15.87
8	−0.67	74.86	25.14
9	−0.33	62.93	37.07
11	0.33	37.07	62.93
12	0.67	25.14	74.86
14	1.33	9.18	90.82
15	1.67	4.75	95.25
16	2.00	2.28	97.72
18	2.67	0.38	99.62

5.3

	Z Scores	Area
a.	0.10 & 1.10	32.45%
b.	0.60 & 1.10	13.86%
c.	0.60	27.43%
d.	0.90	18.41%
e.	0.60 & −0.40	38.11%
f.	0.10 & −0.40	19.52%
g.	0.10	53.98%
h.	0.30	61.79%
i.	0.60	72.57%
j.	1.10	86.43%

5.5

X_i	Z Score	Number of Students Above	Number of Students Below
60	−2.00	195	5
57	−2.50	199	1
55	−2.83	199	1
67	−0.83	159	41
70	−0.33	126	74
72	0.00	100	100
78	1.00	32	168
82	1.67	10	190
90	3.00	1	199
95	3.83	1	199

Note: Number of students (a discrete variable) has been rounded off to the nearest whole number.

5.7

	Z Score	Area
a.	−2.20	1.39%
b.	1.80	96.41%
c.	−0.20 & 1.80	54.34%
d.	0.80 & 2.80	20.93%
e.	−1.20	88.49%
f.	0.80	21.19%

5.9

	Z Score	Area
a.	−1.00 & 1.50	.7745
b.	0.25 & 1.50	.3345
c.	1.50	.0668
d.	0.25 & −2.25	.5865
e.	−1.00 & −2.25	.1465
f.	−1.00	.1587

5.11 Yes. The raw score of 110 translates into a Z score of $+2.88$. 99.80% of the area lies below this score, so this individual was in the top 1% on this test.

5.13 For the first event, the probability is .0919; for the second, the probability is .0655. The first event is more likely.

Part I Cumulative Exercises

1. The level of measurement is the most important criterion for selecting descriptive statistics. The following table presents all relevant statistics for each variable. Statistics that are not appropriate for a variable are noted with an "X."

    Religion is nominal, so the only statistics available are the mode and the IQV. For the two ordinal-level variables ("strength" and "comfort"), the median and the range are the appropriate choices. However, for ordinal-level variables like "strength," which have a wide range of scores, it is common for social science researchers to report the standard deviation and mean. For interval-ratio variables like "pray" and "age," the mean and standard deviation are the preferred summary statistics.

	Religion	Strength	Pray	Comfort	Age
			Level of Measurement		
	Nominal	Ordinal	I-R	Ordinal	I-R
Mode	Protestant				
Median	X	7		1	
Mean	X	6.07	1.40		41.53
IQV	0.86				
Range	X	9	6	4	49
Stnd. dev.	X	2.77	1.62		12.54

3. As always, the level of measurement is the primary guideline for choosing descriptive statistics; the most appropriate statistics are noted in the following table.

	Children	School	Race	Spanking	TV	Religion
			Level of Measurement			
	I-R	Ordinal	Nominal	Ordinal	I-R	Nominal
Mode			White			Prot.
Median		1		2		
Mean	2.44				3.08	
IQV			0.50			0.54
Range	9	4		3	10	
Stnd. dev.	2.06				2.04	

Chapter 7

7.1 **a.** 5.2 ± 0.11
  b. 100 ± 0.71
  c. 20 ± 0.40
  d. 1020 ± 5.41
  e. 7.3 ± 0.23
  f. 33 ± 0.80

7.3

Confidence Level	Alpha	Area Beyond Z	Z score
95%	0.05	0.0250	± 1.96
94%	0.06	0.0300	± 1.88 or ± 1.89
92%	0.08	0.0400	± 1.75 or ± 1.76
97%	0.03	0.0150	± 2.17
98%	0.02	0.0100	± 2.33

7.5   **a.** 2.30 ± 0.04 **b.** 2.10 ± 0.01, $0.78 \pm .07$
     c. 6.00 ± 0.37

7.7   **a.** 345.23 ± 1.97. The students spent between $343.26 and $347.20 on books.
    b. 1.5 ± 0.04. The students visited the clinic between 1.46 and 1.54 times on the average.
    c. 2.8 ± 0.13 **d.** 3.5 ± 0.19

7.9 0.14 ± 0.07. The estimate is that between 7% and 21% of the population consists of unmarried couples living together.

7.11   **a.** $P_s = 823/1496 = 0.55$
       Confidence interval: 0.55 ± 0.03
       Between 52% and 58% of the population agrees with the statement.
    b. $P_s = 650/1496 = 0.43$
       Confidence interval: 0.43 ± 0.03
    c. $P_s = 375/1496 = 0.25$
       Confidence interval: 0.25 ± 0.03
    d. $P_s = 1023/1496 = 0.68$
       Confidence interval: 0.68 ± 0.03
    e. $P_s = 800/1496 = 0.53$
       Confidence interval: 0.53 ± 0.03

7.13

Alpha (α)	Confidence Level	Confidence Interval
0.10	90%	100 ± 0.74
0.05	95%	100 ± 0.88
0.01	99%	100 ± 1.15
0.001	99.9%	100 ± 1.49

7.15   The confidence interval is 0.51 ± 0.05. The estimate would be that between 46% and 56% of the population prefer candidate A. The population parameter (P_u) is equally likely to be anywhere in the interval (that is, it's just as likely to be 46% as it is to be 56%), so a winner cannot be predicted.

7.17   The confidence interval is 0.23 ± 0.08. At the 95% confidence level, the estimate would be that between 240 (15%) and 496 (31%) of the 1600 freshmen would be extremely interested. The estimated numbers are found by multiplying N (1600) by the upper (0.31) and lower (0.15) limits of the interval.

7.19   **a.** 43.87 ± 0.48
     b. 2.86 ± 0.08
     c. 1.81 ± 0.06

d. 0.29 ± 0.02 (According to the Census, about 25% of the U.S. population is actually Catholic so this estimate is too high.)

e. 0.18 ± 0.02 (According to the Census, about 26% of the U.S. population have never married, so this estimate is too low.)

f. 0.52 ± 0.02 (According to official records, about 52% of the electorate did vote for Bush, so this estimate is accurate.)

g. 0.81 ± 0.02

Chapter 8

8.3   **a.** Z(obtained) $= -41.10$
    b. Z(obtained) $= 29.16$

8.5 Z(obtained) $= 6.04$

8.7   **a.** Z(obtained) $= -13.66$
    b. Z(obtained) $= 25.50$

8.9   **a.** t(obtained) $= 4.50$

8.11 Z(obtained) $= 3.06$

8.13 Z(obtained) $= -1.48$

8.15   **a.** Z(obtained) $= -1.49$
    b. Z(obtained) $= 2.19$
    c. Z(obtained) $= -8.55$
    d. Z(obtained) $= -18.07$
    e. Z(obtained) $= 2.09$
    f. Z(obtained) $= -53.33$

8.17 t(obtained) $= -1.14$

Chapter 9

9.1   **a.** $\sigma = 1.39$, Z(obtained) $= -2.52$
    b. $\sigma = 1.61$, Z(obtained) $= 2.48$

9.3   **a.** $\sigma = 10.57$, Z(obtained) $= 1.70$
    b. $\sigma = 11.28$, Z(obtained) $= -2.48$

9.5   **a.** $\sigma = 0.08$, Z(obtained) $= -11.25$
    b. $\sigma = 0.12$, Z(obtained) $= -3.33$
      $\sigma = 0.15$, Z(obtained) $= 20.00$

9.7   These are small samples (combined N's of less than 100), so be sure to use Formulas 9.5 and 9.6 in step 4.
    a. $\sigma = 0.12$, t(obtained) $= -1.33$
    b. $\sigma = 0.13$, t(obtained) $= 14.85$

9.9   **a.** (France) $\sigma = 0.0095$, Z(obtained) $= -31.58$
    b. (Nigeria) $\sigma = 0.0075$, Z(obtained) $= -146.67$
    c. (China) $\sigma = 0.0065$, Z(obtained) $= 184.62$
    d. (Mexico) $\sigma = 0.0107$, Z(obtained) $= -74.77$
    e. (Japan) $\sigma = 0.0115$, Z(obtained) $= -43.48$

The large values for the Z scores indicate that the differences are significant at very low alpha levels (i.e., they are extremely unlikely to have been caused by random chance alone). Note that women are significantly happier than men in every nation except China, where men are significantly happier.

9.11 $P_u = 0.45$, $\sigma_p = 0.06$, Z(obtained) $= 0.67$

9.13 **a.** $P_u = 0.46$, $\sigma_p = 0.06$, Z(obtained) $= 2.17$
b. $P_u = 0.80$, $\sigma_p = 0.07$, Z(obtained) $= 1.43$
c. $P_u = 0.72$, $\sigma_p = 0.08$, Z(obtained) $= 0.75$

9.15 **a.** $P_u = 0.34$, $\sigma_p = 0.04$, Z(obtained) $= 1.50$
b. $P_u = 0.53$, $\sigma_p = 0.04$, Z(obtained) $= -2.75$
c. $P_u = 0.52$, $\sigma_p = 0.03$, Z(obtained) $= 4.00$
d. $\sigma = 0.43$, Z(obtained) $= 1.86$
e. $\sigma = 0.14$, Z(obtained) $= -5.71$
f. $\sigma = 0.08$, Z(obtained) $= -5.50$

Chapter 10

10.1

Problem	Grand Mean	SST	SSB	SSW	F ratio
a.	12.17	231.67	173.17	58.5	13.32
b.	6.87	455.73	78.53	377.20	1.25
c.	31.65	8362.55	5053.35	3309.2	8.14

10.3

Problem	Grand Mean	SST	SSB	SSW	F ratio
a.	4.39	86.28	45.78	40.50	8.48
b.	16.44	332.44	65.44	267.00	1.84

For problem 10.3a, with alpha $= 0.05$ and $df = 2$, 15, the critical F ratio would be 3.68. We would reject the null hypothesis and conclude that decision making *does* vary significantly by type of relationship. By inspection of the group means, it seems that the "cohabitational" category accounts for most of the differences.

10.5

Grand Mean	SST	SSB	SSW	F Ratio
9.28	213.61	2.11	211.50	0.08

10.7

Grand Mean	SST	SSB	SSW	F Ratio
5.40	427.32	124.06	303.26	6.00

10.9

Nation	Grand Mean	SST	SSB	SSW	F ratio
Mexico	3.78	300.98	154.08	146.90	12.59
Canada	6.88	156.38	20.08	136.30	1.77
United States	5.13	286.38	135.28	151.10	10.74

At alpha $= 0.05$ and $df = 3$, 36, the critical F ratio is 2.92. There is a significant difference in support for suicide by class in Mexico and the United States but not in Canada. The category means for Mexico suggest that the upper class accounts for most of the differences. For the United States, there is more variation across the category means, and the working class seems to account for most of the differences. Going beyond the ANOVA test and comparing the grand means, we see that support is highest in Canada and lowest in Mexico.

Chapter 11

11.1 **a.** 1.11 **b.** 0.00 **c.** 1.52 **d.** 1.46

11.3 A computing table is highly recommended as a way of organizing the computations for chi square:

Computational Table for Problem 11.3

(1) f_o	(2) f_e	(3) $f_o - f_e$	(4) $(f_o - f_e)^2$	(5) $(f_o - f_e)^2/f_e$
6	5	1	1	.20
7	8	−1	1	.13
4	5	−1	1	.20
9	8	1	1	.13
$N = 26$	$N = 26$	0		χ^2(obtained) $= 0.66$

There is 1 degree of freedom in a 2×2 table. With alpha set at 0.05, the critical value for the chi square would be 3.841. The obtained chi square is 0.65, so we fail to reject the null hypothesis of independence between the variables. There is no statistically significant relationship between race and services received.

11.5 a.

Computational Table for Problem 11.5

(1)	(2)	(3)	(4)	(5)
f_o	f_e	$f_o - f_e$	$(f_o - f_e)^2$	$(f_o - f_e)^2/f_e$
21	17.5	3.5	12.25	.70
29	32.5	−3.5	12.25	.38
14	17.5	−3.5	12.25	.70
36	32.5	3.5	12.25	.38
$N = 100$	$N = 100.0$	0		χ^2(obtained) = 2.16

With 1 degree of freedom and alpha set at 0.05, the critical region will begin at 3.841. The obtained chi square of 2.15 does not fall within this area, so the null hypothesis cannot be rejected. There is no statistically significant relationship between unionization and salary.

b. Column percentages:

	Status	
Salary	Union	Nonunion
High	60.00%	44.6%
Low	40.00%	55.4%
Totals	100.00%	100.0%

Although the relationship is not significant, unionized fire departments tend to have higher salary levels.

11.7 The obtained chi square is 5.13, which is significant (df = 1, alpha = 0.05). The column percentages show that more affluent communities have higher-quality schools.

11.9 The obtained chi square is 6.67, which is significant (df = 2, alpha = 0.05). The column percentages show that shorter marriages have higher satisfaction.

11.11 The obtained chi square is 12.59, which is significant (df = 4, alpha = 0.05). The column percentages show that proportionally more of the students living "off campus with roommates" are in the high-GPA category.

11.13 The obtained chi square is 19.34, which is significant (df = 3, alpha = 0.05). Legalization was not favored by a majority of any region, but the column percentages show that the West was most in favor.

11.15

Problem	Chi Square	Significant at α = 0.05?	Column Percentages
a.	25.19	Yes	The oldest age group was most opposed.
b.	1.80	No	The great majority (72%–75%) of all three age groups were in favor of capital punishment
c.	5.23	Yes	The oldest age group was most likely to say yes. Although significant, the differences in column percentages are small.
d.	28.43	Yes	The youngest age group was most likely to support legalization.
e.	14.17	Yes	The oldest age group was least likely to support suicide.

11.17 The obtained chi square is 4.43. With df = 3 and alpha = 0.05, this is not a significant relationship.

Part II Cumulative Exercises

1. One of the challenges of empirical research is to make reasonable decisions about which statistical test to use in which situation. You can minimize the confusion and ambiguity by approaching the decision systematically. We'll use the first problem of this exercise to consider some ways in which reasonable decisions can be made. The situation calls for a test of hypotheses ("Is the difference significant?"), so our choice of procedures will be limited to Chapters 8–11. Next, determine the types of variables you are working with. Number of minutes is an interval-ratio–level variable, and the research question asks us to compare two groups or samples: males and females. Which test should we use? The techniques in Chapter 8 (one-sample tests) and Chapter 10 (tests involving more than two samples or categories) are not relevant. Chi square (Chapter 11) won't work unless we collapse the scores on Internet use into a few categories. This leaves Chapter 9. A test of sample means fits the situation, and we have a large sample (combined N's greater than 100), so it looks like we're going to conduct a test of significance using Formulas 9.2 and 9.4

a. $\sigma = 0.14$, Z(obtained) $= 35.71$

The difference in Internet minutes is significant. Men, on the average, use this technology more frequently.

b. The table format is a sure tip-off that the chi square test is appropriate. The obtained chi square is 0.84, which is not significant at the 0.05 level. There is no statistically significant relationship between involvement and social class.

c. "Number of partners" sounds like an interval-ratio–level variable, and education has three categories. Analysis of variance is an appropriate test for this situation. The F ratio is 0.13—not at all significant—so we must conclude that this dimension of sexuality does not vary by level of education.

Grand Mean	SST	SSB	SSW	F Ratio
4.04	262.96	3.24	259.71	0.13

d. The problem asks for a characteristic of a population ("How many times do adult Americans move?") but gives information only for a random sample. The estimation procedures presented in Chapter 7 fit the situation, and, since the information is presented in the form of a mean, you should use Formula 7.2 to form the estimate. At an alpha level of 0.05, the confidence interval would be 3.5 ± 0.02.

e. The research question focuses on the difference between a single sample and a population, so Chapter 8 is relevant. Since the population standard deviation is unknown and we have a large sample, Formula 8.2 will be applicable. Z(obtained) is -0.26, which is not significant at alpha $= 0.05$. The sample is not significantly different from the population—rural school districts are the same as the universe of all school districts in this state.

Chapter 12

12.1

	Authoritarianism	
Efficiency	Low	High
Low	37.04%	70.59%
High	62.96%	29.41%
Totals	100.00%	100.00%

The conditional distributions change, so there is a relationship between the variables. The change from column to column is quite large, and the maximum difference is $(70.59 - 37.04) = 33.55$. Using Table 12.5 as a guideline, we can say that this relationship is moderate to strong. Phi is 0.33 and lambda is 0.32, again suggesting a moderate to strong relationship. From inspection of the percentages, we can see that efficiency decreases as authoritarianism increases—workers with dictatorial bosses are less productive (or, maybe, bosses become more dictatorial when workers are inefficient), so this relationship is negative in direction.

12.3

	2004 Election	
2008 Election	Democrat	Republican
Democrat	87.31%	11.44%
Republican	12.69%	88.56%
Totals	100.00%	100.00%

The maximum difference for this table is $(87.31 - 11.44)$, or 75.87. Phi is 0.75 and lambda is 0.71. This is a very strong relationship. People are very consistent in their voting habits.

12.5

	Director Experienced?	
Turnover	No	Yes
Low	14.29	40.91
Moderate	32.14	36.36
High	53.57	22.73
Totals	100.00	100.00

The maximum difference is (53.57 – 22.73), or 30.84, which indicates that the relationship is moderate-to-strong. Cramer's V is 0.36 and lambda is 0.13. The value of Cramer's V is more consistent with the maximum difference—both indicate a moderate-to-strong relationship—and the extent of the changes in the column percentages, so it is the preferred measure of association.

12.7 Maximum difference = 60
Phi = 0.59
Lambda = 0.54

All measures of association indicate that this is a strong relationship.

12.9 a. Maximum difference = 1.43
Phi = 0.01
Lambda = 0.00
b. Maximum difference = 9.61
Phi = 0.11
Lambda = 0.00
c. Maximum difference = 7.61
Phi = 0.08
Lambda = 0.00
d. Maximum difference = 0
Phi = 0.00
Lambda = 0.00
e. Maximum difference = 1.16
Phi = 0.02
Lambda = 0.00

Chapter 13

13.1 a. $G = 0.71$
b. $G = 0.69$
c. $G = -0.88$

These relationships are strong. Facility in English and income increase with length of residence (+0.71). Use the percentages to help interpret the direction of a relationship. In the first table, 80% of the "newcomers" were "Low" in English facility, while 60% of the "old-timers" were "High." In this relationship, low scores on one variable are associated with low scores on the other, and scores increase together (as one increases, the other increases), so this is a positive relationship. In contrast, contact with the old country decreases with length of residence (−0.88). Most newcomers have higher levels of contact, and most old-timers have lower levels.

13.3 $G = 0.40$. This is a moderate, positive relationship. As city size increases, crime rate increases.

13.5 Canada: $G = -0.28$
U.S.: $G = 0.21$
Mexico: $G = -0.13$

These relationships are weak to moderate. For Canada and Mexico, agreement increases with age, but for the United States, the reverse is true. Note that the column percentages show that most respondents do not agree with the statement.

13.7 $G = -0.17$. This is a weak, negative relationship. The more authoritarian the parents, the fewer the symptoms of depression.

13.9 $G = 0.08$. This a weak positive relationship. The higher the income, the higher the level of happiness.

13.11 $r_s = 0.77$. This is a strong, positive relationship. The higher the original score, the higher the performance evaluation.

13.13 $r_s = 0.62$. This is a moderate-to-strong, positive relationship. The rankings are very similar. Z(obtained) = 3.35.

Chapter 14

14.1 (*HINT: When finding the slope, remember that "Turnout" is the dependent, or Y, variable.*)

	For Turnout (Y) and		
	Unemployment	Education	Negative Campaigning
Slope (b)	3.00	12.67	−0.90
Y intercept (a)	39.00	−94.73	114.01
Reg. Eq.	$Y = 39 + 3X$	$Y = -94.73 + 12.67X$	$Y = 114.01 + (-0.90)X$
r	0.94	0.98	−0.87
r^2	0.89	0.96	0.76
t(obtained)	5.00	9.50	−3.08

14.3 (*HINT: When finding the slope, remember that "Number of Visitors" is the dependent, or Y, variable.*)

Slope (*b*)	-0.37
Y intercept (*a*)	13.42
r	-0.31
r^2	0.10

14.5

Dependent Variable		Independent Variable		
		Growth	Density	Unemployment
Homicide	*a*	2.65	3.69	-0.08
	b	0.18	0.00	0.78
	r	0.68	0.19	0.39
	r^2	0.47	0.04	0.15
Robbery	*a*	101.87	78.63	-2.07
	b	2.41	0.28	22.84
	r	0.35	0.62	0.42
	r^2	0.12	0.38	0.18
Car theft	*a*	143.37	400.10	-166.42
	b	22.60	-0.37	92.87
	r	0.77	0.19	0.41
	r^2	0.60	0.04	0.17

  c. The homicide rate for a state with a growth rate of -1 would be

  Homicide rate $= 2.65 + (0.18)(-1) = 2.47$
  The robbery rate for a state with a population density of 250 would be

  Robbery rate $= 78.63 + (0.28)(250) = 148.63$
  The auto theft rate for a state with an unemployment rate of 12% would be

  Auto theft rate $= -166.42 + (92.87)(12) = 948.02$

14.7 $b = 0.05$, $a = 53.18$, $r = 0.40$, $r^2 = 0.16$

14.9

		Prestige	Number of Children	Support for Abortion	Hours of TV per Day
Age	*r*	-0.30	0.67	0.08	0.16
	r^2	0.09	0.45	0.01	0.03
	t	-1.13	3.26	-0.29	0.59

(*continued next column*)

(*continued*)

		Prestige	Number of Children	Support for Abortion	Hours of TV per Day
Sex	*r*	-0.28	0.19	0.11	-0.29
	r^2	0.08	0.04	0.01	0.08
	t	-1.05	0.70	0.40	-1.09

Part III Cumulative Exercises

1. a. Assuming that crime rates and "Percent Iimmigrants" are both measured at the interval-ratio level, Pearson's *r* would be the appropriate measure of association. This is a moderate negative relationship ($r = -0.47$), and percentage of immigrants explains 22% of the variation in crime rate for these 10 cities. As the percentage of immigrants increases, the crime rate decreases.

  b. These variables are ordinal in level of measurement, so gamma would be the appropriate measure of association. Gamma is -0.43, indicating a moderate negative relationship. As TV viewing increases, involvement decreases.

  c. The appropriate measure for these variables is Spearman's rho, which is 0.74. This is a strong positive relationship, and states with higher quality of life have superior systems of higher education.

  d. Race is a nominal-level variable, and the table is larger than 2 × 2. Cramer's *V* is 0.40, indicating a strong relationship between the variables. The column percentages show that this sample tends to reject all of the singular racial categories in favor of "None of the Above." A second, weaker trend is for people with one white parent to identify with that group. Because of the uneven row totals, lambda is zero even though there is an association between these variables.

Chapter 15

15.1

Bivariate gamma: 0.71

	Partial Gammas	
Controlling for gender	Male	0.78
	Female	0.65
Controlling for origin	Asian	0.67
	Hispanic	0.74

The bivariate relationship is strong and positive. The longer the residence, the greater the facility in English. The bivariate relationship is not affected by the sex or origin of the immigrants. These results would be taken as strong evidence of a direct (causal) relationship between length of residence and facility in English.

15.3

Bivariate gamma: -0.23

Partial Gammas, Controlling for Gender	
Female	-0.34
Male	-0.10

This is an interactive relationship. Length of institutionalization has a greater effect on the reality orientation of females than of males.

15.5

Bivariate gamma: 0.62

Partial Gammas		
Controlling for race	Whites	0.52
	Blacks	0.68
	$G_p =$	0.60
Controlling for gender	Males	0.61
	Females	0.62
	$G_p =$	0.62

The bivariate relationship is strong and positive. Completion of the training program is closely associated with holding a job for at least one year. There is some interaction with race. The training has less impact for whites than for blacks. Blacks who did not complete the training were less likely than whites to have held a job for at least one year. Gender has no impact at all on the relationship. Overall, there is a direct relationship between completion of the training and holding a job for at least a year, though there is some interaction with race.

15.7 Support for the legal right to an abortion:

Bivariate gamma: -0.14

Partial Gammas, Controlling for Gender	
Males	-0.06
Females	-0.19
$G_p =$	-0.14

Support for the right to suicide:

Bivariate gamma: -0.14

Partial Gammas, Controlling for Gender	
Males	-0.16
Females	-0.11
$G_p =$	-0.13

Chapter 16

16.1 a. For turnout (Y) and unemployment (X) while controlling for negative advertising (Z), $r_{yx.z} = 0.95$. The relationship between X and Y is not affected by the control variable Z.

   b. For turnout (Y) and negative advertising (X) while controlling for unemployment (Z), $r_{yx.z} = -0.89$. The bivariate relationship is not affected by the control variable.

   c. Turnout (Y) $= 70.25 + (2.09)$ unemployment (X_1) $+ (-0.43)$ negative advertising (X_2). For unemployment (X_1) $= 10$ and negative advertising (X_2) $= 75$, turnout (Y) $= 58.09$.

   d. For unemployment (X_1): $b_1^* = 0.66$. For negative advertising (X_2): $b_2^* = -0.41$. Unemployment has a stronger effect on turnout than negative advertising. Note that the independent variables' effects on turnouts are in opposite directions.

   e. $R^2 = 0.98$

16.3 a. For strife (Y) and unemployment (X), controlling for urbanization (Z), $r_{yx.z} = 0.79$.

   b. For strife (Y) and urbanization (X), controlling for unemployment (Z), $r_{yx.z} = 0.20$.

   c. Strife (Y) $= -14.60 + (4.94)$ unemployment (X_1) $+ (0.16)$ urbanization (X_2). With unemployment $= 10$ and urbanization $= 90$, strife (Y') would be 49.19.

   d. For unemployment (X_1): $b_1^* = 0.78$. For urbanization (X_2): $b_2^* = 0.13$.

   e. $R^2 = 0.65$

16.5 a. Turnout (Y) $= 83.80 + (-1.16)$ Democrat (X_1) $+ (2.89)$ minority (X_2).

   b. For $X_1 = 0$ and $X_2 = 5$, $Y' = 98.25$

   c. $Z_y = -1.27Z_1 + 0.84Z_2$

   d. $R^2 = 0.51$

16.7 a. $Z_y = (0.16)$ HS grads (Z_1) $+ (-0.70)$ Rank (Z_2)

   b. $R^2 = 0.68$

Part IV Cumulative Exercises

1. a. The choice of multivariate procedures will depend on the level of measurement and the number of possible scores for each variable. Regression analysis is appropriate for interval-ratio, continuous variables. In this situation, we have three variables: GPA, hours of work, and College Board scores. College Board scores are arguably only ordinal in level of measurement, but the scores have a wide range and seem continuous. GPA is the dependent variable, and the zero-order correlation with "hours worked" is -0.83, a strong and negative relationship, which indicates that having a job did interfere with academic success for this sample. The zero-order correlation between GPA and College Board scores is 0.55, a relationship that is consistent with the idea that College Board scores predict success in college.

The results of the regression analysis with "hours worked" (X_1) and College Board scores (X_2) as independents:

$$Y = 1.71 + (-0.04)X_1 + (0.003)X_2$$
$$Z_y = (-0.75)Z_1 + (0.43)Z_2$$
$$R^2 = 0.86$$

The beta-weights indicate that "hours worked" has a stronger direct effect on GPA than College Board scores. Even for students who were well prepared for college and had high College Board scores, having a job had a negative impact on GPA. The high value for R^2 means that only a small percentage of the variance in GPA (14%) is left unexplained by these two independent variables.

b. In this situation, we have three dichotomous variables, two ordinal level and one nominal. With the limited variation possible, the elaboration technique is appropriate to analyze these relationships. The bivariate table shows that there is a relationship between graduating and level of social activity for this sample: 70% of the students with low levels of social activity graduated in four years versus only 30% of the students with high levels. Bivariate gamma is 0.69, indicating a strong relationship.

Controlling for sex reveals some interaction. The gamma for males (0.77) is stronger than the bivariate gamma, and the gamma for females (0.60) is weaker. In other words, while level of social activity has an effect on graduation rates, the effect is stronger for males than for females. This suggests that sex should be incorporated into the analysis, with a view to developing an understanding of why it would have different effects for males and females.

Glossary

Each entry includes a brief definition and notes the chapter that introduces the term.

Alpha (α). The probability of error, or the probability that a confidence interval does not contain the population value. Alpha levels are usually set at 0.10, 0.05, 0.01, or 0.001. Chapter 7

Alpha level (α). The proportion of area under the sampling distribution that contains unlikely sample outcomes, given that the null hypothesis is true. Also, the probability of Type I error. Chapter 8

Analysis of variance. A test of significance appropriate for situations in which we are concerned with the differences among more than two sample means. Chapter 10

ANOVA. See **Analysis of variance**. Chapter 10

Association. The relationship between two (or more) variables. Two variables are said to be associated if the distribution of one variable changes for the various categories or scores of the other variable. Chapter 12

Bar chart. A graph used for discrete variables. Categories are represented by bars of equal width, the height of each corresponding to the number (or percentage) of cases in the category. Chapter 2

Beta-weights (b^*). See **Standardized partial slope**. Chapter 16

Bias. A criterion used to select sample statistics as estimators. A statistic is unbiased if the mean of its sampling distribution is equal to the population value of interest. Chapter 7

Bivariate normal distributions. The model assumption in the test of significance for Pearson's r that both variables are normally distributed. Chapter 14

Bivariate table. A table that displays the joint frequency distributions of two variables. Chapter 11

Cells. The cross-classification categories of the variables in a bivariate table. Chapter 11

Central Limit Theorem. A theorem that specifies the mean, standard deviation, and shape of the sampling distribution, given that the sample is large. Chapter 6

χ^2(critical). The score on the sampling distribution of all possible sample chi squares that marks the beginning of the critical region. Chapter 11

χ^2(obtained). The test statistic computed from sample results. Chapter 11

Chi square (χ^2) test. A nonparametric test of hypothesis for variables that have been organized into a bivariate table. Chapter 11

Class intervals. The categories used in the frequency distributions for interval-ratio variables. Chapter 2

Cluster sampling. A method of sampling by which geographical units are randomly selected and all cases within each selected unit are tested. Chapter 6

Coefficient of determination (r^2). The proportion of all variation in Y that is explained by X. Found by squaring the value of Pearson's r. Chapter 14

Coefficient of multiple determination (R^2). A statistic that equals the total variation explained in the dependent variable by all independent variables combined. Chapter 16

Column. The vertical dimension of a bivariate table. By convention, each column represents a score on the independent variable. Chapter 11

Column percentages. Percentages computed with each column of a bivariate table. Chapter 12

Conditional distribution of Y. The distribution of scores on the dependent variable for a specific score or category of the independent variable when the variables have been organized into table format. Chapter 12

Conditional means of Y. The mean of all scores on Y for each value of X. Chapter 14

Confidence interval. An estimate of a population value which specifies a range of values. Chapter 7

Confidence level. A frequently used alternate way of expressing alpha, the probability that an interval estimate will not contain the population value. Confidence levels of 90%, 95%, 99%, and 99.9% correspond to alphas of 0.10, 0.05, 0.01, and 0.001, respectively. Chapter 7

Continuous variable. A variable with a unit of measurement that can be subdivided infinitely. Chapter 1

Control variable. In multivariate statistical analysis, a third variable (Z) whose effect on a bivariate relationship is held constant. Chapter 16

Correlation matrix. A table that shows the correlation coefficients between all possible pairs of variables.

Cramer's V. A chi square–based measure of association for nominally measured variables that have been organized into a bivariate table with any number of rows and columns. Chapter 12

Critical region (region of rejection). The area under the sampling distribution that, in advance of the test itself, is defined as including unlikely sample outcomes, given that the null hypothesis is true. Chapter 8

Cumulative frequency. An optional column in a frequency distribution that displays the number of cases within an interval and all preceding intervals. Chapter 2

Cumulative percentage. An optional column in a frequency distribution that displays the percentage of cases within an interval and all preceding intervals. Chapter 2

Data. Information expressed as numbers. Chapter 1

Data reduction. Summarizing many scores with a few statistics. Chapter 1

Deciles. The points that divide a distribution of scores into 10ths. Chapter 3

Dependent variable. A variable that is identified as an effect or outcome. The dependent variable is thought to be caused by the independent variable. Chapters 1 and 12

Descriptive statistics. The branch of statistics concerned with (1) summarizing the distribution of a single variable or (2) measuring the relationship between two or more variables. Chapter 1

Deviations. The distances between the scores and the mean. Chapter 4

Direct relationship. A multivariate relationship in which the control variable has no effect on the bivariate relationship. Chapter 15

Discrete variable. A variable with a basic unit of measurement that cannot be subdivided. Chapter 1

Dispersion. The amount of variety, or heterogeneity, in a distribution of scores. Chapter 4

Dummy variable. A nominal-level variable dichotomized so that it can be used in regression analysis. A dummy variable has two scores, one coded as 0 and the other as 1. Chapter 14

E_1. For lambda, the number of errors made when predicting which category of the dependent variable cases will fall into when ignoring the independent variable. Chapter 12

E_2. For lambda, the number of errors made when predicting which category of the dependent variable cases will fall into when taking the independent variable into account. Chapter 12

Efficiency. The extent to which the sample outcomes are clustered around the mean of the sampling distribution. Chapter 7

Elaboration. The multivariate technique used to analyze variables arrayed in tables. Partial tables are constructed to observe the bivariate relationship in a more detailed or elaborated format. Chapter 15

EPSEM. The Equal Probability of SElection Method for selecting samples. Every element or case in the population must have an equal probability of selection for the sample. Chapter 6

Expected frequency (f_e). The cell frequencies that would be expected in a bivariate table if the variables were independent. Chapter 11

Explained variation. The proportion of all variation in Y that is attributed to the effect of X. Equal to $\sum (Y' - \overline{Y})^2$ Chapter 14

Explanation. See **Spurious relationship**. Chapter 15

***F* ratio.** The test statistic computed in step 4 of the ANOVA test. Chapter 10

Five-step model. A step-by-step guideline for conducting tests of hypotheses. A framework that organizes decisions and computations for all tests of significance. Chapter 8

Frequency distribution. A table that displays the number of cases in each category of a variable. Chapter 2

Frequency polygon. A graphic display device for interval-ratio variables. Class intervals are represented by dots placed over the midpoints, the height of each corresponding to the number (or percentage) of cases in the interval. All dots are connected by straight lines. Same as a **line chart**. Chapter 2

Gamma (*G*). A measure of association appropriate for ordinal-level variables with only a few categories; G is the symbol for any sample gamma, γ is the symbol for any population gamma. Chapter 13

Goodness-of-fit test. An additional use for chi square that tests the significance of the distribution of a single variable. Chapter 11

Histogram. A graph used for interval-ratio variables. Class intervals are represented by contiguous bars of equal width (equal to the class limits), the height of each corresponding to the number (or percentage) of cases in the interval. Chapter 2

Homoscedasticity. The model assumption in the test of significance for Pearson's *r* that the variance of the Y scores is uniform across all values of X. Chapter 14

Hypothesis. A specific statement, derived from a theory, about the relationship between variables. Chapter 1

Hypothesis testing. Statistical tests that estimate the probability of sample outcomes if assumptions about the population (the null hypothesis) are true. Chapter 8

Independence. The null hypothesis in the chi square test. Two variables are independent if, for all cases, the classification of a case on one variable has no effect on the probability that the case will be classified in any particular category of the second variable. Chapter 11

Independent random samples. Random samples gathered so that the selection of a particular case for one sample has no effect on the probability that any other particular case will be selected for the other samples. Chapter 9

Independent variable. A variable that is identified as a cause. The independent variable is thought to cause the dependent variable. Chapters 1 and 12

Index of qualitative variation (IQV). A measure of dispersion for variables that have been organized into frequency distributions. Chapter 4

Inferential statistics. The branch of statistics concerned with making generalizations from samples to populations. Chapter 1

Interaction. A multivariate relationship wherein a bivariate relationship changes across the categories of the control variable. Chapter 15

Interpretation. See **Intervening relationship**. Chapter 15

Interquartile range (Q). The distance from the third quartile to the first quartile. Chapter 4

Intervening relationship. A multivariate relationship wherein a bivariate relationship becomes substantially weaker after a third variable is controlled for. The independent and dependent variables are linked primarily through the control variable. Chapter 15

Lambda (λ). A proportional reduction in error (PRE) measure for nominally measured variables that have been organized into a bivariate table. Chapter 12

Level of measurement. The mathematical characteristic of a variable and the major criterion for selecting statistical techniques. Variables can be measured at any of three levels, each permitting certain mathematical operations and statistical techniques. The characteristics of the three levels are summarized in Table 1.2. Chapter 1

Line chart. See **Frequency polygon**. Chapter 2

Linear relationship. A relationship between two variables in which the observation points (dots) in the scattergram can be approximated with a straight line. Chapter 14

Marginals. The row and column subtotals in a bivariate table. Chapter 11

Maximum difference. A way to assess the strength of an association between variables that have been organized into a bivariate table. The maximum difference is the largest difference between column percentages for any row of the table. Chapter 12

Mean. The arithmetic average of the scores. $\overline{X}$ represents the mean of a sample, and μ, the mean of a population. Chapter 3

Mean square estimate. An estimate of the variance calculated by dividing the sum of squares within (SSW) or the sum of squares between (SSB) by the proper degrees of freedom. Chapter 10

Measures of association. Statistics that summarize the strength and direction of the relationship between variables. Chapters 1 and 12

Measures of central tendency. Statistics that summarize a distribution of scores by reporting the most typical or representative value of the distribution. Chapter 3

Measures of dispersion. Statistics that indicate the amount of variety, or heterogeneity, in a distribution of scores. Chapter 4

Median (Md). The point in a distribution of scores above and below which exactly half of the cases fall. Chapter 3

Midpoint. The point exactly halfway between the upper and lower limits of a class interval. Chapter 2

Mode. The most common value in a distribution or the largest category of a variable. Chapter 3

Mu (μ). The mean of a population. Chapter 6

μ_p (mu-sub-p). The mean of a sampling distribution of sample proportions. Chapter 6

$\mu_{\overline{X}}$ (mu-sub-X-bar). The mean of a sampling distribution of sample means. Chapter 6

Multiple correlation. A multivariate technique for examining the combined effects of more than one independent variable on a dependent variable. Chapter 16

Multiple correlation coefficient (R). A statistic that indicates the strength of the correlation between a dependent variable and two or more independent variables. Chapter 16

Multiple regression. A multivariate technique that breaks down the separate effects of the independent variables on the dependent variable; used to make predictions of the dependent variable. Chapter 16

N_d. The number of pairs of cases ranked in different order on two variables. Chapter 13

N_s. The number of pairs of cases ranked in the same order on two variables. Chapter 13

Negative association. A bivariate relationship where the variables vary in opposite directions. As one variable increases, the other decreases, and high scores on one variable are associated with low scores on the other. Chapter 12

Nonparametric. A "distribution-free" test. These tests make no assumption about the shape of the sampling distribution. Chapter 11

Normal curve. A theoretical distribution of scores that is symmetrical, unimodal, and bell-shaped. The standard normal curve always has a mean of 0 and a standard deviation of 1. Chapter 5

Normal curve table. Appendix A; a detailed description of the area between a Z score and the mean of any standardized normal distribution. Chapter 5

Null hypothesis (H_0). A statement of "no difference." In the context of single-sample tests of significance, the population from which the sample was drawn is assumed to have a certain characteristic or value. Chapter 8

Observed frequency (f_0). The cell frequencies actually observed in a bivariate table. Chapter 11

One-tailed test. A type of hypothesis test used when (1) the direction of the difference can be predicted or (2) concern focuses on outcomes in only one tail of the sampling distribution. Chapter 8

One-way analysis of variance. Applications of ANOVA in which the effect of a single independent variable on a dependent variable is observed. Chapter 10

P_s (P-sub-s). A sample proportion. Chapter 6

P_u (P-sub-u). A population proportion. Chapter 6

Parameter. A characteristic of a population. Chapter 6

Partial correlation. A multivariate technique for examining a bivariate relationship while controlling for other variables. Chapter 16

Partial correlation coefficient. A statistic that shows the relationship between two variables while controlling for other variables; $r_{yx.z}$ is the symbol for the partial correlation coefficient when controlling for one variable. Chapter 16

Partial gamma (G_p). A statistic that indicates the strength of the association between two variables after the effects of a third variable have been removed. Chapter 15

Partial slope. In a multiple regression equation, the slope of the relationship between a particular independent variable and the dependent variable while controlling for all other independent variables in the equation. Chapter 16

Partial tables. Tables produced when controlling for a third variable. Chapter 15

Pearson's r. A measure of association for variables that have been measured at the interval-ratio level; ρ (Greek letter rho) is the symbol for the population value of Pearson's r. Chapter 14

Percentage (%). The number of cases in a category of a variable divided by the number of cases in all categories of the variable, the entire quantity multiplied by 100. Chapter 2

Percent change. A statistic that expresses the magnitude of change in a variable from time 1 to time 2. Chapter 2

Percentile. A point below which a specific percentage of the cases fall. Chapter 3

Phi (ϕ). A chi square–based measure of association for nominally measured variables that have been organized into a bivariate table with two rows and two columns (a 2 × 2 table). Chapter 12

Pie chart. A graph used especially for discrete variables with only a few categories. A circle (the pie) is divided into segments proportional in size to the percentage of cases in each category of the variable. Chapter 2

Pooled estimate. An estimate of the standard deviation of the sampling distribution of the difference in sample means based on the standard deviations of both samples. Chapter 9

Population. The total collection of all cases in which the researcher is interested. Chapter 1

Positive association. A bivariate relationship where the variables vary in the same direction. As one variable increases, the other also increases, and high scores on one variable are associated with high scores on the other. Chapter 12

Post hoc test. A technique for determining which pairs of means are significantly different. Chapter 10

Proportion (p). The number of cases in one category of a variable divided by the number of cases in all categories of the variable. Chapter 2

Proportional reduction in error (PRE). The logic that underlies the definition and computation of lambda. The statistic compares the number of errors made when predicting the dependent variable while ignoring the independent variable ($E1$) with the number of errors made while taking the independent variable into account ($E2$). Chapter 12

Quantitative research. Research projects that collect data or information in the form of numbers. Chapter 1

Quartiles. The points that divide a distribution into quarters. Chapter 3

Range (R). The highest score minus the lowest score. Chapter 4

Rate. The number of actual occurrences of some phenomenon or trait divided by the number of possible occurrences per some unit of time. Chapter 2

Ratio. The number of cases in one category divided by the number of cases in some other category. Chapter 2

Real class limits. The class intervals of a frequency distribution when stated as continuous categories. Chapter 2

Regression line. The single best-fitting straight line that summarizes the relationship between two variables. Regression lines are fitted to the data points by the least-squares criterion, whereby the line touches all conditional means of Y or comes as close to doing so as possible. Chapter 14

Replication. See **Direct relationship**. Chapter 15

Representative. The quality a sample is said to have if it reproduces the major characteristics of the population from which it was drawn. Chapter 6

Research. Any process of gathering information systematically and carefully to answer questions or test theories. Statistics are useful for research projects that collect numerical information or data. Chapter 1

Research hypothesis (H_1). A statement that contradicts the null hypothesis. In the context of single-sample tests of significance, the research hypothesis says that the population from which the sample was drawn does not have a certain characteristic or value. Chapter 8

Row. The horizontal dimension of a bivariate table, conventionally representing a score on the dependent variable. Chapter 11

Sample. A carefully chosen subset of a population. In inferential statistics, information is gathered from a sample and then generalized to a population. Chapter 1

Sampling distribution. The distribution of a statistic for all possible sample outcomes of a certain size. Under conditions specified in two theorems, the sampling distribution will be normal in shape, with a mean equal to the population value and a standard deviation equal to the population standard deviation divided by the square root of N. Chapter 6

Scattergram. A graph that depicts the relationship between two variables. Chapter 14

Σ (uppercase Greek letter sigma). "The summation of." Chapter 3

Significance testing. See **Hypothesis testing**. Chapter 8

Simple random sample. A method for choosing cases from a population by which every case and every combination of cases has an equal chance of being included. Chapter 6

Skew. The extent to which a distribution of scores has a few scores that are extremely high (positive skew) or extremely low (negative skew). Chapter 3

Slope (b). The amount of change in one variable per unit change in the other; b is the symbol for the slope of a regression line. Chapter 14

Spearman's rho (r_s). A measure of association appropriate for ordinally measured variables that are "continuous" in form; r_s is the symbol for a sample Spearman's rho; ρ_s is the symbol for a population Spearman's rho. Chapter 13

Specification. See **Interaction**. Chapter 15

Spurious relationship. A multivariate relationship in which a bivariate relationship becomes substantially weaker after a third variable is controlled for. The independent and dependent variables are not causally linked. Rather, both are caused by the control variable. Chapter 15

Standard deviation. The square root of the squared deviations of the scores around the mean, divided by N.

The most important and useful descriptive measure of dispersion; s represents the deviation of a sample; σ represents the standard deviation of a population. Chapter 4

σ_{p-p} (sigma-sub-p-minus-p). Symbol for the standard deviation of the sampling distribution of the differences in sample proportions. Chapter 9

$\sigma_{\overline{X}-\overline{X}}$ (sigma-sub-mean-minus-mean). Symbol for the standard deviation of the sampling distribution of the differences in sample means. Chapter 9

Standard error of the mean. The standard deviation of a sampling distribution of sample means. Chapter 6

Standardized partial slope (beta-weight). The slope of the relationship between a particular independent variable and the dependent variable when all scores have been normalized. Chapter 16

Stated class limits. The class intervals of a frequency distribution when stated as discrete categories. Chapter 2

Statistics. A set of mathematical techniques for organizing and analyzing data. Chapter 1

Stratified sample. A method of sampling by which cases are selected from sublists of the population. Chapter 6

Student's t distribution. A distribution used to find the critical region for tests of sample means when s is unknown and sample size is small. Chapter 8

Sum of squares between (SSB). The sum of the squared deviations of the sample means from the overall mean, weighted by sample size. Chapter 10

Sum of squares within (SSW). The sum of the squared deviations of scores from the category means. Chapter 10

Systematic sampling. A method of sampling by which the first case from a list of the population is randomly selected. Thereafter, every kth case is selected. Chapter 6

t(critical). The t score that marks the beginning of the critical region of a t distribution. Chapter 8

t(obtained). The test statistic computed in step 4 of the five-step model. The sample outcome expressed as a t score. Chapter 8

Test statistic. The value computed in step 4 of the five-step model that converts the sample outcome into a score that can be placed on the appropriate sampling distribution (Z, t, F, or χ^2). Chapter 8

Theory. A generalized explanation of the relationship between two or more variables. Chapter 1

Total sum of squares (SST). The sum of the squared deviations of the scores from the overall mean. Chapter 10

Total variation. The spread of the Y scores around the mean of Y. Equal to $\sum(Y' - \overline{Y})^2$ Chapter 14

Two-tailed test. A type of hypothesis test used when (1) the direction of the difference cannot be predicted or (2) concern focuses on outcomes in both tails of the sampling distribution. Chapter 8

Type I error (alpha error). The probability of rejecting a null hypothesis that is, in fact, true. Chapter 8

Type II error (beta error). The probability of failing to reject a null hypothesis that is, in fact, false. Chapter 8

Unexplained variation. The proportion of the total variation in Y that is not accounted for by X. Equal to $\sum(Y - Y')^2$ Chapter 14

Variable. Any trait that can change values from case to case. Chapter 1

Variance. The squared deviations of the scores around the mean divided by N. A measure of dispersion used primarily in inferential statistics and also in correlation and regression techniques; s^2 represents the variance of a sample; $\sum^2$ represents the variance of a population. Chapter 4

X. Symbol used for any independent variable. Chapter 12

X_i (*X*-sub-*i*). Any score in a distribution. Chapter 3

Y. Symbol used for any dependent variable. Chapter 12

Y **intercept** (*a*). The point where the regression line crosses the Y axis. Chapter 14

Y′. Symbol for predicted score on Y. Chapter 14

Z. Symbol for any control variable. Chapter 15

Z **scores.** Standard scores; the way scores are expressed after they have been standardized to the theoretical normal curve. Chapter 5

Z(**critical**). The Z score that marks the beginnings of the critical region on a Z distribution. Chapter 8

Z(**obtained**). The test statistic computed in step 4 of the five-step model. The sample outcomes expressed as a Z score. Chapter 8

Zero-order correlations. Correlation coefficient for bivariate relationships. Chapter 16

Credits

Chapter 1, Figure 1.1, p. 2: Source: Adapted from Walter Wallace, *The Logic of Science in Sociology* (Chicago: Aldine-Atherton, 1971.)

Chapter 1, p. 15: This system for determining level of measurement was suggested by Michael G. Bisciglia, Louisiana State University.

Chapter 2, p. 29: Source: U.S. Bureau of the Census: http://www.census.gov/ipc/www/idb/information Gateway.php.

Chapter 2, Table 2.18, p. 41: Source: U.S. Census Bureau, http://factfinder.census.gov/servlet/ADPTable?_bm=y&-geo_id=D&-qr_name=ACS_2008_1YR_G00_DP3&-ds_name=D&-_lang=en.

Chapter 2, p. 52: Source: Kreider, Rose and Elliot, Diana. 2009. "America's Families and Living Arrangements: 2007." U.S. Bureau of the Census: Current Population Reports. Available http://www.census.gov/prod/2009pubs/p20-561.pdf.

Chapter 3, p. 73: Source: http://www.nationmaster.com.

Chapter 4, p. 105: Source: Adapted from Kay, Fiona and Wallace, Jean, 2009. "Mentors as Social Capital: Gender, Mentors, and Career Rewards in Law Practice," Table 1, *Sociological Inquiry*, 79:418–452. Used by permission.

Chapter 10, pp. 261–262: Source: Choi, Heeseung, Meininger, Janet, and Roberts, Robert. 2006. "Ethnic Differences in Adolescents' Mental Distress, Social Stress, and Resources." *Adolescence*, 41:263–283. Based on Table 2, p. 273.

Chapter 11, p. 293: Source: Adapted with kind permission from Springer Science + Business Media: Walsh, Jeffrey and Krienert, Jessie, "Child-Parent Violence: An Empirical Analysis of Offender, Victim, and Event Characteristics in a National Sample of Reported Incidents," Table 1, *Journal of Family Violence*, 22, (2007):563–574.

Chapter 12, p. 328: Source: Augustson, E., Wanke, K., Rogers, S. "Predictors of Sustained Smoking Cessation: A Prospective Analysis of Chronic Smokers From the Alpha-Tocopherol Beta-Carotene Cancer Prevention Study," Table 1, p. 551, *American Journal of Public Health*, March 2008; 98:549–555. Used by permission.

Chapter 14, p. 386: Source: Data from www.nationmaster.com and World Values Survey.

Chapter 16, pp. 455–456: Source: Adapted from Unnever, James, and Francis Cullen, 2007. "The Racial Divide in Support for the Death Penalty: Does Racism Matter?" Table 1, *Social Forces*, 85: 1281–1301. Copyright © 2007 by the University of North Carolina Press. Used by permission of the publisher. www.uncpress.unc.edu.

Index

Note: Page numbers followed by "*n*" refer to footnotes.